DATE DUE

MAY 13 1987			
[illegible] 23 1987			
OCT 22 1987			
OCT 12 1987			
DEC 18 87			
DEC 17 87			
MAY 23			
NOV 2 - 88			
AUG 29 '88			
8/31/88			
Due			
8-25-89			
[illegible]			
MAY 23 '90			
MAY 9 1990			
JAN 26 1997			
JAN - 8 1997			
GAYLORD			PRINTED IN U.S.A.

The Marine Environment of the U.S. Atlantic Continental Slope and Rise

The Marine Environment of the U.S. Atlantic Continental Slope and Rise

Edited by
John D. Milliman
and
W. Redwood Wright

Jones and Bartlett Publishers, Inc.
Boston/Woods Hole

Editorial offices: Jones and Bartlett Publishers, Inc., 3 Water Street, Woods Hole, MA 02543

Sales and customer service offices: Jones and Bartlett Publishers, Inc., 20 Park Plaza, Boston, MA 02116

Library of Congress Cataloging-in-Publication Data

The Marine environment in the U.S. Atlantic continental slope and rise.

"Condensed version of a report prepared for the Minerals Management Service (MMS) of the U.S. Department of Interior in 1984 by Marine Geoscience Applications, Inc." — Pref.
Bibliography: p.
Includes index.
1. Oceanography — Atlantic Coast (U.S.) 2. Marine resources — Atlantic Coast (U.S.) 3. Geology — Atlantic Coast (U.S.) 4. Marine biology — Atlantic Coast (U.S.) I. Milliman John D. II. Wright, W. Redwood. III. Title: Marine environment of the U.S. Atlantic continental slope and rise.
GC511.M28 1987 551.46′144 86-56
ISBN 0–86720–066–9

Printed in the United States of America

10 9 8 7 6 5 4 3 2 1

Design/Production: Unicorn Production Services, Inc.
Composition: AccuComp Typographers
Printing and Binding: Alpine Press

Table of Contents

7

8

The Marine Environment of the U.S. Atlantic Continental Slope and Rise

Preface

The continental slope and rise off the eastern United States are perhaps better studied and documented than any other slope or rise in the world. Yet with the exception of the monograph by Emery and Uchupi (1972), no attempt has been made to summarize this extensive and diverse environment.

The following chapters are a revised and condensed version of a report prepared for the Minerals Management Service (MMS) of the U.S. Department of Interior in 1984 by Marine Geoscience Applications, Inc. (whose ongoing projects have been assumed by Battelle Memorial Institute). Both the original report and this book were reviewed by the Minerals Management Service and approved for publication. Approval does not signify that the contents necessarily reflect the views and policies of the Minerals Management Service, nor does mention of trade names or commercial products constitute endorsement or recommendation for use.

We thank MMS for funding the preparation of both the initial summary and this book. Eiji Imamura, William Lang and Jeffrey Petrino, all of MMS, were particularly supportive in facilitating the project wherever necessary. Jeffrey Zwinakis and Betsey Pratt prepared all the new illustrations. Susan Race provided much-needed and valuable editorial advice. We particularly thank Pamela Barrows for her typing and editorial assistance; without her help, this project would have been much more difficult.

John D. Milliman and W. Redwood Wright
Woods Hole, Massachusetts
May 1986

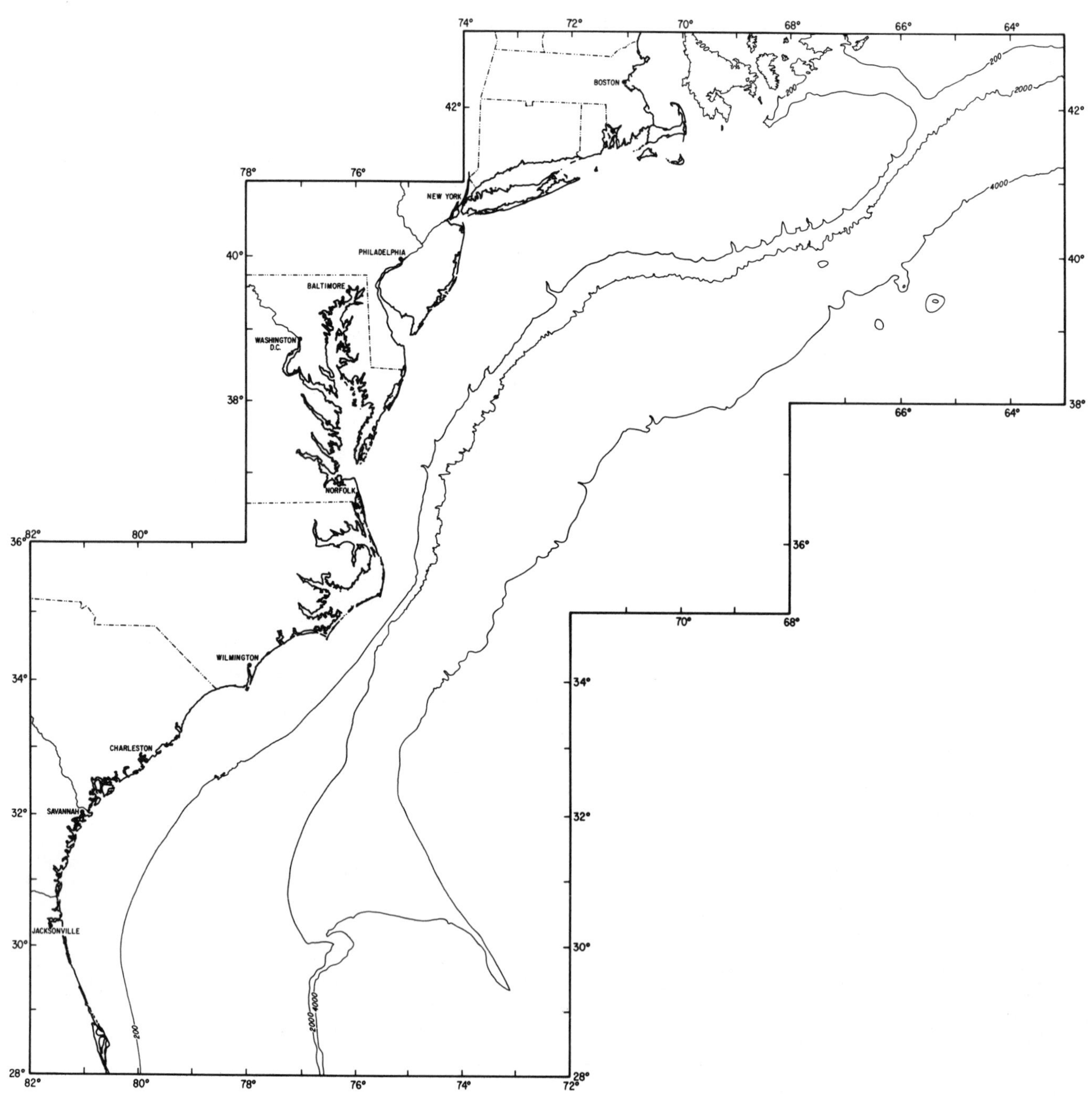

Figure 1.1 The Atlantic Continental slope and rise (ACSAR), as defined by the 200- and 4000-m isobaths and 28° and 42°N latitudes.

Introduction 1

The Atlantic continental slope and rise (in this book often referred to by its acronym, ACSAR) off the eastern United States lies primarily between latitudes 28° and 42°N and in water depths between 200 and 4000 m (Figure 1.1). To the south are the Straits of Florida and (to the southeast) the Bahamas; to the north is Canada. Offshore are the Sohm, Nares and Hatteras Aybssal Plains.

The ACSAR area is perhaps the best studied outer continental margin in the world. More than 4500 scientific articles dealing with the slope and rise have been written, and an even greater number of articles deal with the adjacent continental shelf and coastal areas. Weather has been monitored by passing ships for more than 150 years, and more recently by long-term weather buoys. Physical oceanography is well-defined in terms of both water masses and circulation patterns; there are perhaps more long-term deep-water current meter measurements in this area and the adjacent northwest Atlantic than in all other deep-sea areas in the world combined. Extensive echo-sounding, coring and geophysical surveys (e.g., Figure 1.2) give us a good picture of the marine geology in the ACSAR area. Likewise, both the chemistry and biology of the slope and rise have been studied by many investigators for more than 100 years.

In part, this knowledge reflects the close proximity of the ACSAR area to the large population base of the eastern U.S. as well as to a large number of universities and research institutes, many with long histories in marine studies. Harvard, Massachusetts Institute of Technology, Yale, Princeton, Duke, North Carolina, and Georgia are only some of the universities engaged in marine-related research, while oceanographic laboratories include the Woods Hole Oceanographic Institution, University of Rhode Island, Lamont-Doherty Geological Observatory (Columbia University), and the Rosensteil School of Marine and Atmospheric Science (University of Miami). Finally, federal and state laboratories have been involved in research off the eastern United States. The U.S. Geological Survey (USGS), National Marine Fisheries Service (NMFS), and National Ocean Survey (NOS) are the most prominent of these agencies.

The oceanographic environment on the Atlantic continental slope and rise can be discussed within six general topics: meteorology and air-sea interactions, physical oceanography, geology, chemistry, biology, and human activities and impacts. Yet a complete discussion of these individual topics also requires integration with the other topics. For instance, biological populations and productivity in the ACSAR area depend upon the physical oceanographic environment (and therefore upon air-sea interactions) and nutrient levels, as well as bathymetry and bottom sediment type.

Of necessity and design, the discussion of these six topical areas is a synthesis and evaluation of previous data (where possible) rather than presentation of new data or reinterpretation of old data. In certain sections, however, synthesis of old data has led to new interpretations, and in one chapter (Biology) new data have been included since they are necessary for the understanding of the ACSAR area.

In reading this book, many readers doubtlessly will be struck by the disparity between the content and complexity of discussion in various sections. This disparity is real, and represents marked differences in the data base. In geology and physical oceanography, for example, we have a good idea of mega-scale features, and some idea of meso- and micro-scales. In biology and chemistry our knowledge is far more rudimentary. For example, many zooplankton and phytoplankton data collected in the slope prior to the 1970s were collected (often unknowingly) from a mixture of water masses; many of these older reports are compromised (at best) or erroneous (at worst). Similarly, there are few available or reliable data concerning such diverse topics as cetaceans, microbial activity, neuston and water column and sediment chemistry.

In addition to the disparity in the data base, the reader will see three general areas in which new data must be acquired if we are to improve our understanding of the slope and rise. While these areas of future study are discussed further in Chapter 8, a brief presentation here can provide a basis for considering these points as one reads Chapters 2-7. First, we need to define distributions and processes on finer scales. For example, we know with relative accuracy the general bathymetry in the ACSAR area, but little on horizontal scales finer than 100s to 1000s of meters. Second, we need to measure fluxes of water, particles, chemical species, and organisms: how fast are these various commodities transported, transferred or produced? Third, existing data can be utilized and interpreted in better ways. In some instances, notably geology and meteorology, large bodies of data exist which, if interpreted correctly, could aid immensely in our understanding of the ACSAR environment.

Finally, we should mention that the bibliography at the end of the book represents only a small portion of the total number of papers published in the ACSAR area. Because they are all quoted in the text, we feel that they represent the key papers, but for those readers who want a more complete reference list (about 5000 references), we recommend the 1984 MGA report to MMS (Environmental Summary of the Atlantic Continental Slope and Rise, 28 to 42 °N, two volumes prepared under MMS contract no. 14-12-0001-29200).

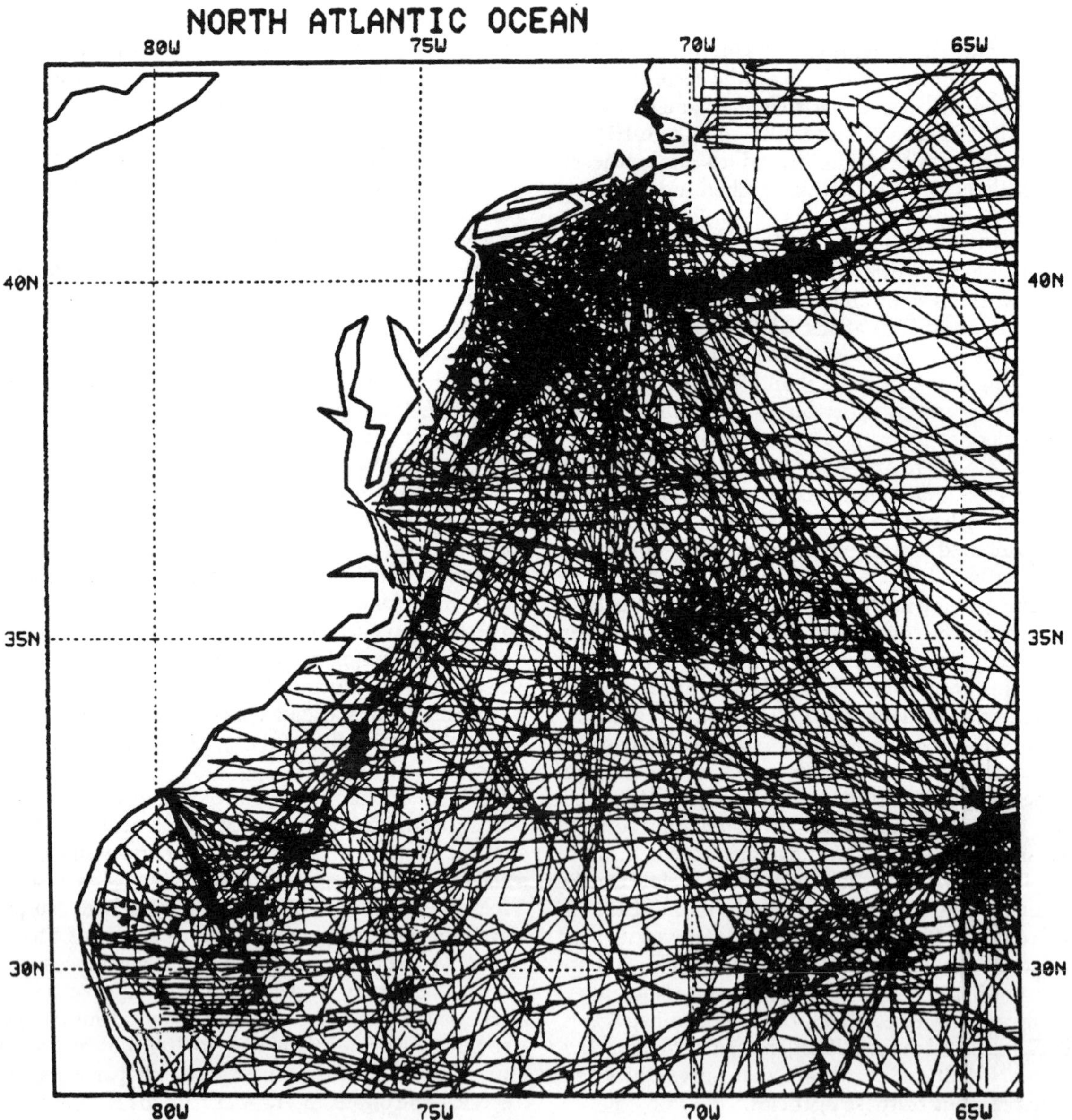

Figure 1.2 NGSDC data base of echosounding profiles in the western North Atlantic.

Meteorology and Air-Sea Interactions

2

T.M. Joyce

Over the past century, a large body of data has been accumulated by merchant and military vessels plying the waters off the U.S. east coast. Averaging these data can give mean values of air temperature, water surface temperature, winds and many other parameters, a number of which are discussed in the following paragraphs. A high pressure system located in the subtropical gyre of the North Atlantic (Fig. 2.1a) produces a northeasterly flow of air (southwesterly winds) east of the U.S. coastline in summer. In the winter this high pressure system weakens and moves southward and a low pressure system south of Greenland alters the wind direction to west and northwest (Fig. 2.1b). Superimposed on this seasonal progression of the "Bermuda High" are rapid changes with time scales of 2-5 days which represent the passage of weather systems. Fluctuations of pressure, winds, and air temperatures associated with these weather systems can exceed changes on a seasonal time scale. To state that the study area is characterized by westerlies or that climate ranges from temperate in the north to tropical/subtropical over the Gulf Stream south of Cape Hat-

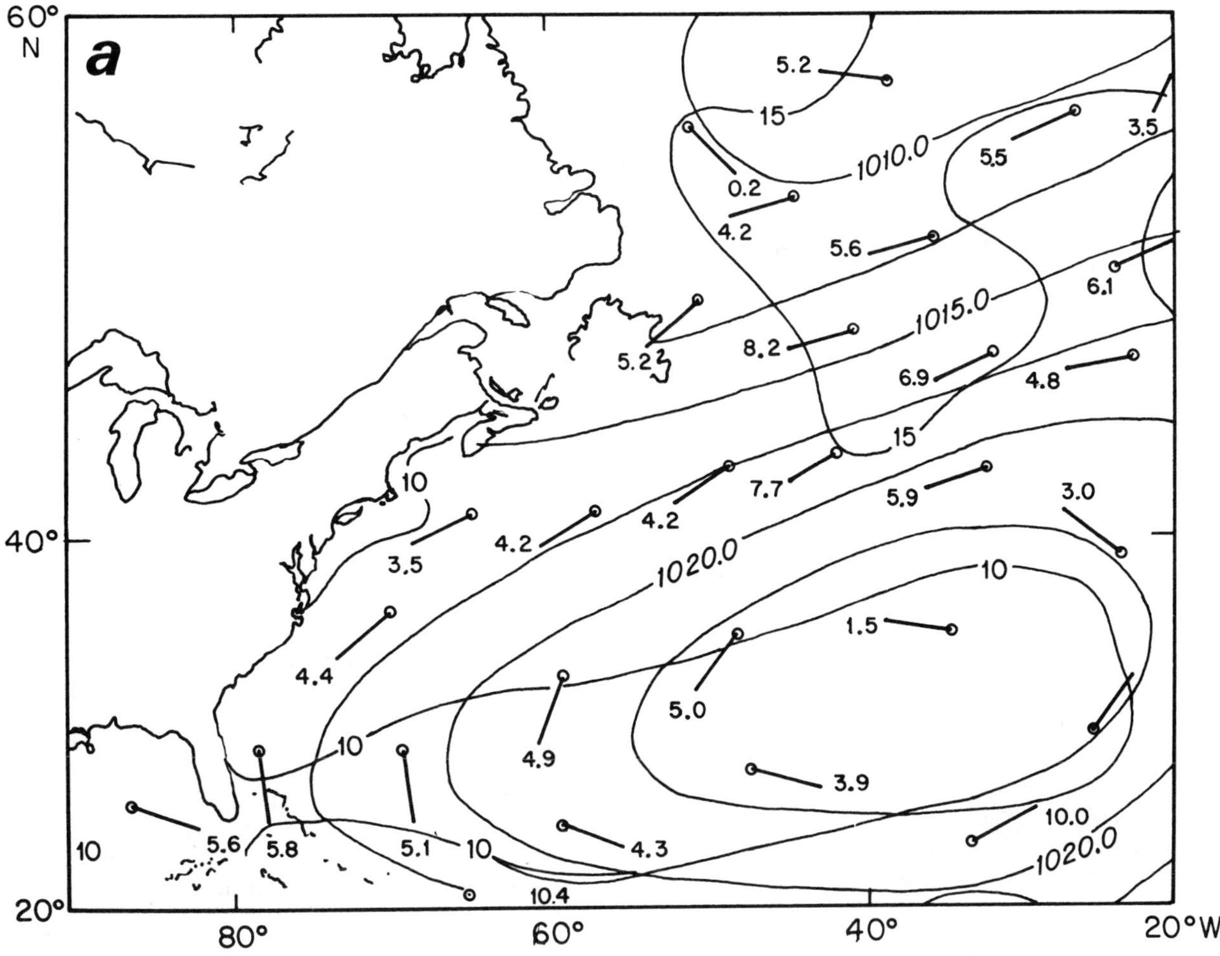

Figure 2.1a. Average atmospheric pressure (mb) and winds (kts) for June (from U.S. Navy Marine Climatic Atlas, 1974).

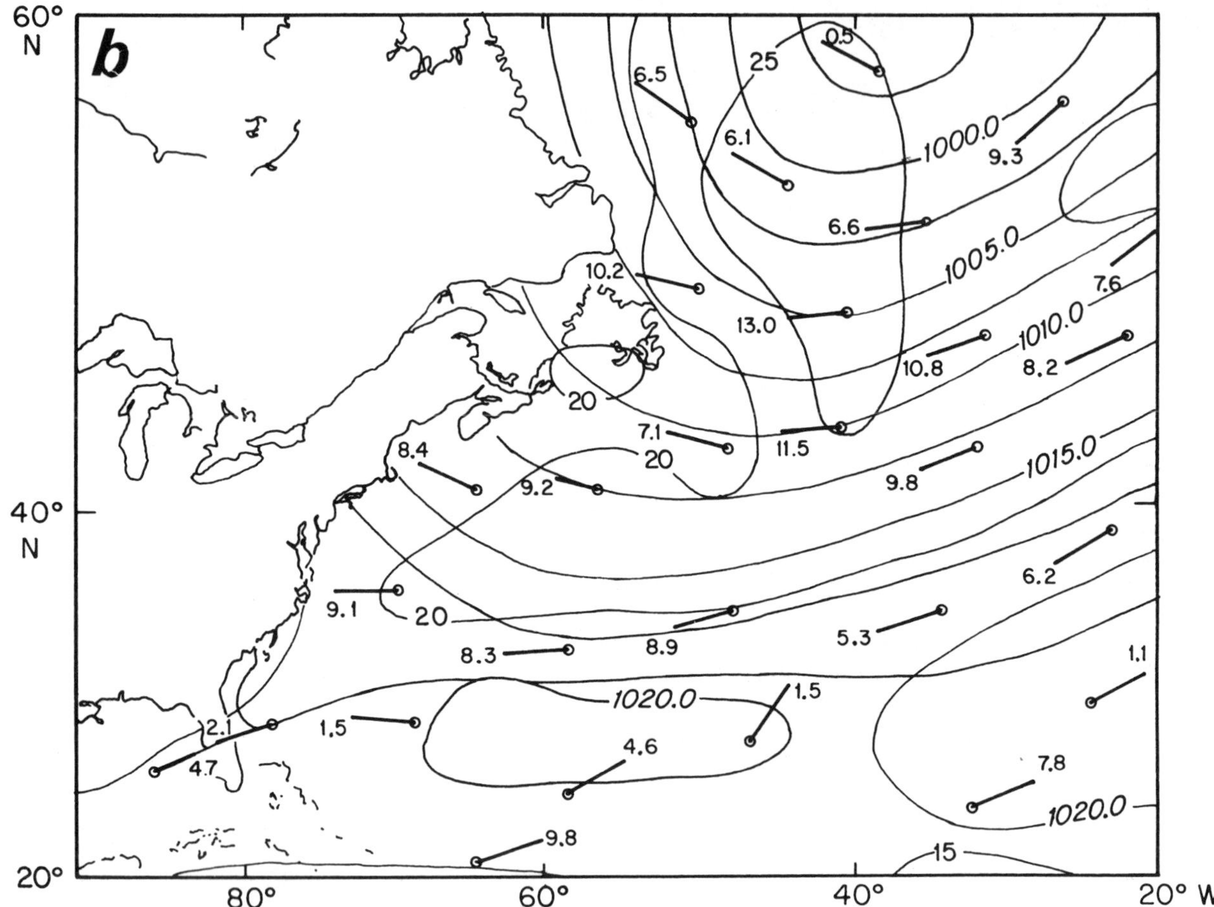

Figure 2.1b. Average atmospheric pressure and winds for the month of January (from U.S. Navy Marine Climatic Atlas).

teras belies the extreme significance of common, intense weather systems that move rapidly east/northeast across the region. Indeed, a worthwhile description of the meteorological conditions requires that one not 'average out" weather systems. Since much of the work in the last decade has been statistical in nature, one can lose sight of the structure and advection rates of these coastal storms. Thus, our review of annual and long-term inter-annual marine deck conditions stresses weather rather than general atmospheric circulation.

Annual and Secular Variations

Synoptic Scale Storm Classification

Eight types of storms can be classified by their origin, structure, and path of movement. While boundaries between these classes of storms are often vague, this classification encompasses nearly all tropical and extra-tropical storms encountered off the eastern U.S. coast.

Hurricanes and Severe Tropical Cyclones (Class 1)

A tropical storm is defined to be a warm core tropical cyclone with maximum sustained winds between 34 and 63 kts (39 and 73 mph). A hurricane has sustained winds in excess of 64 kts (74 mph). Though hurricanes have higher wind speeds than the other storms, their effects upon the coastal waters and off-shore environment are mitigated by their small size and rapid movement. Synoptic charts on successive days of Hurricane Donna in September 1960 (Fig. 2.2) typify storms of this class.

Wave Developments Forming in the Atlantic East of the U.S. or Near Cuba (Class 2)

Storms of this class are slow-moving with a pronounced east/west elongation. An example from January 1956 of this type of storm (Fig. 2.3) illustrates the blocking of the low pressure system by a high to the north. Though infrequent (9 of these were found in a 42-year period by Mather et al., 1964) these storms, with their long fetch and large scale, can be particularly harsh and damaging.

Figure 2.2 Development and movement of typical class 1 storm along east coast of the United States, 10-12 September 1960 (from Mather et al., 1964).

Figure 2.3 Development and movement of typical class 2 storm along east coast of the United States, 7-9 January 1956 (from Mather et al., 1964).

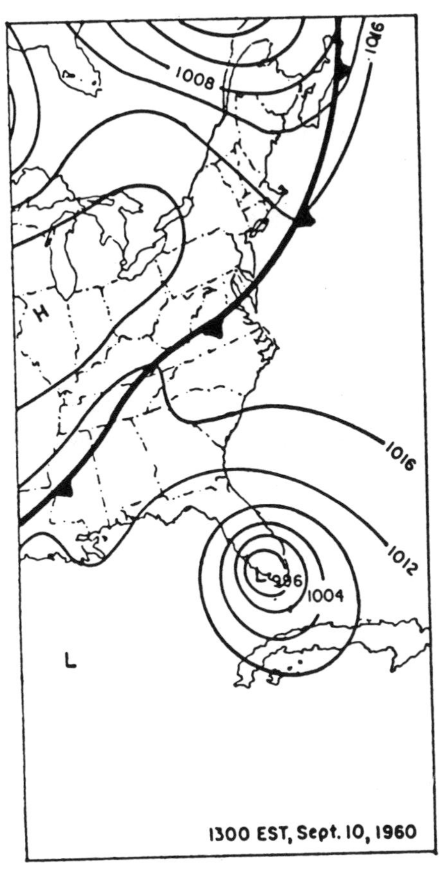
1016
1008
H
1016
1012
L
996
1004
L
1300 EST, Sept. 10, 1960

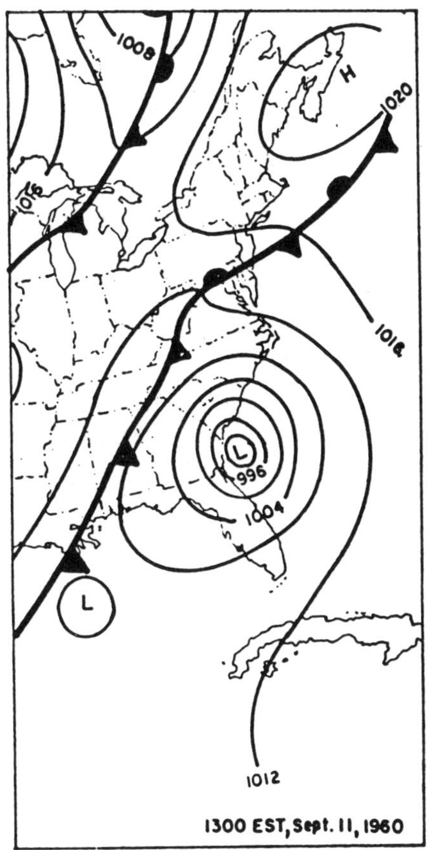
1008
H
1020
1016
1016
L
996
1004
L
1012
1300 EST, Sept. 11, 1960

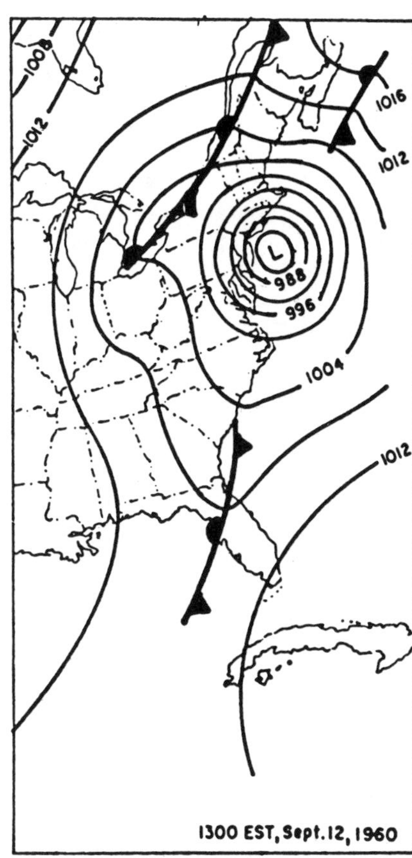
1008
1012
1016
1012
L
988
996
1004
1012
1300 EST, Sept. 12, 1960

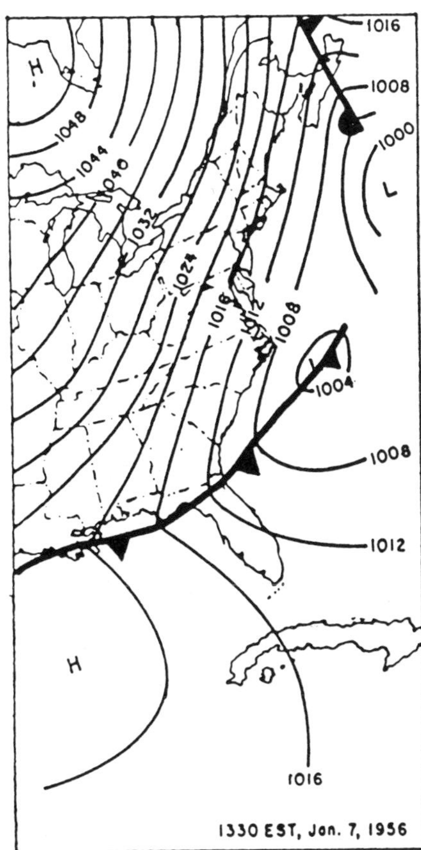
1016
H
1008
1000
1048
1044
1040
L
1032
1024
1016
1012
1008
L
1004
1008
1012
H
1016
1330 EST, Jan. 7, 1956

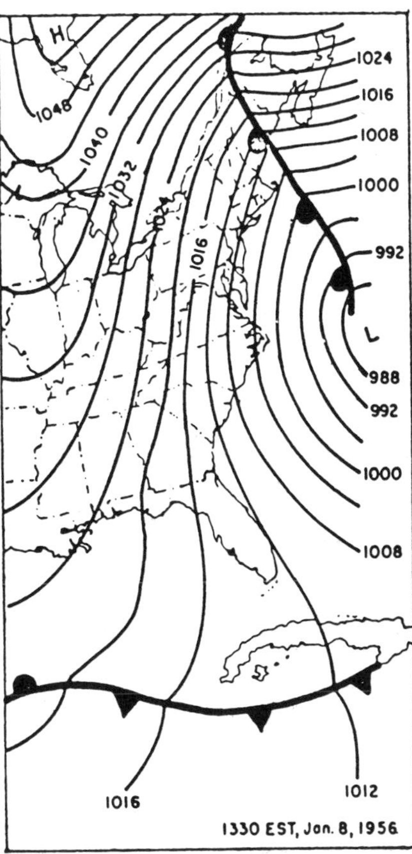
H
1024
1048
1016
1008
1040
1000
1032
1024
992
1016
L
988
992
1000
1008
1016
1012
1330 EST, Jan. 8, 1956

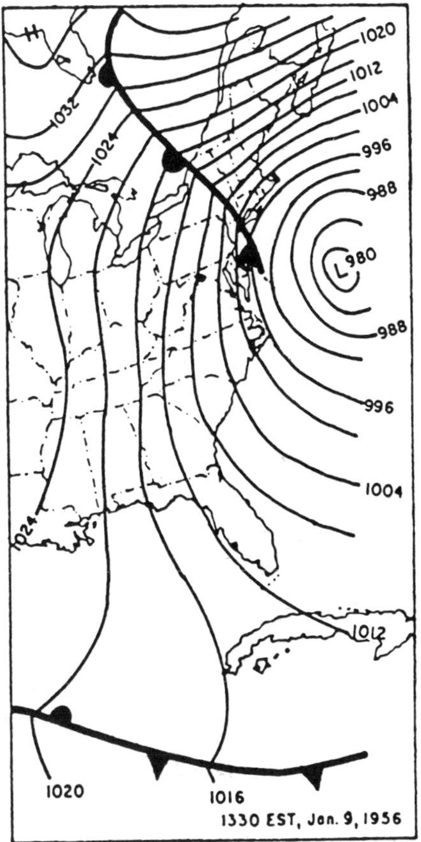
H
1020
1032
1012
1004
1024
996
988
L
980
988
996
1004
1024
1012
1020
1016
1330 EST, Jan. 9, 1956

Wave Developments Along Cold or Stationary Fronts Over or Just Offshore from the Southeast Coastal States (Class 3)

These storms are frequently blocked to the north by a high pressure system. When this occurs, a large pressure gradient is set up with strong onshore winds to the north of the storm. An example of this class of storm is shown from November 1953 in Figure 2.4. A storm of this class discussed by Bosart (1981) was not forecast and explosively deepened over a 6-hour period depositing over 60 cm of snow in Virginia, Maryland, and Delaware.

Wave Developments Along Cold or Stationary Fronts in the Gulf of Mexico Forming West of 85°W (Class 4)

These storms (Fig. 2.5) are seldom blocked and therefore move rapidly across Florida and intensify offshore. The offshore winds behind these disturbances can bring cold, dry, continental air over the warmer ocean waters, which induces a large heat loss and evaporation from the ocean to the atmosphere.

Depressions Moving Across the Southern U.S. Which Intensify Offshore: No Secondary Develops Ahead of Storm (Class 5)

As these storms pass offshore — usually near Cape Hatteras — the low pressure deepens. Again a strong offshore wind occurs behind these storms. Figure 2.6 shows an example of this storm class from January 1961.

Storms in Which Strong Secondary Cyclonic Circulation Develops: Usually off Hatteras (Class 6)

Cyclogenesis occurs near the coast on a warm or stationary front emanating from a cyclonic disturbance to the west. Resulting storms move rapidly to the northeast. An example from March 1962 is given in Figure 2.7.

Intense Cyclones Whose Entire Track is over Land Areas West of the Coastal Margin (Class 7)

These disturbances often deepen over the Great Lakes and move rapidly to the northeast over land areas. An example

Figure 2.5 Development and movement of typical class 4 storm along Gulf and east coast of the United States, 12-14 February 1960 (from Mather et al., 1964). →

Figure 2.4 Development and movement of typical class 3 storm along east coast of the United States, 4-6 November 1953 (from Mather et al., 1964).

Figure 2.6 Development and movement of typical class 5 storm along east coast of the United States, 18-20 January 1961 (from Mather et al., 1964).

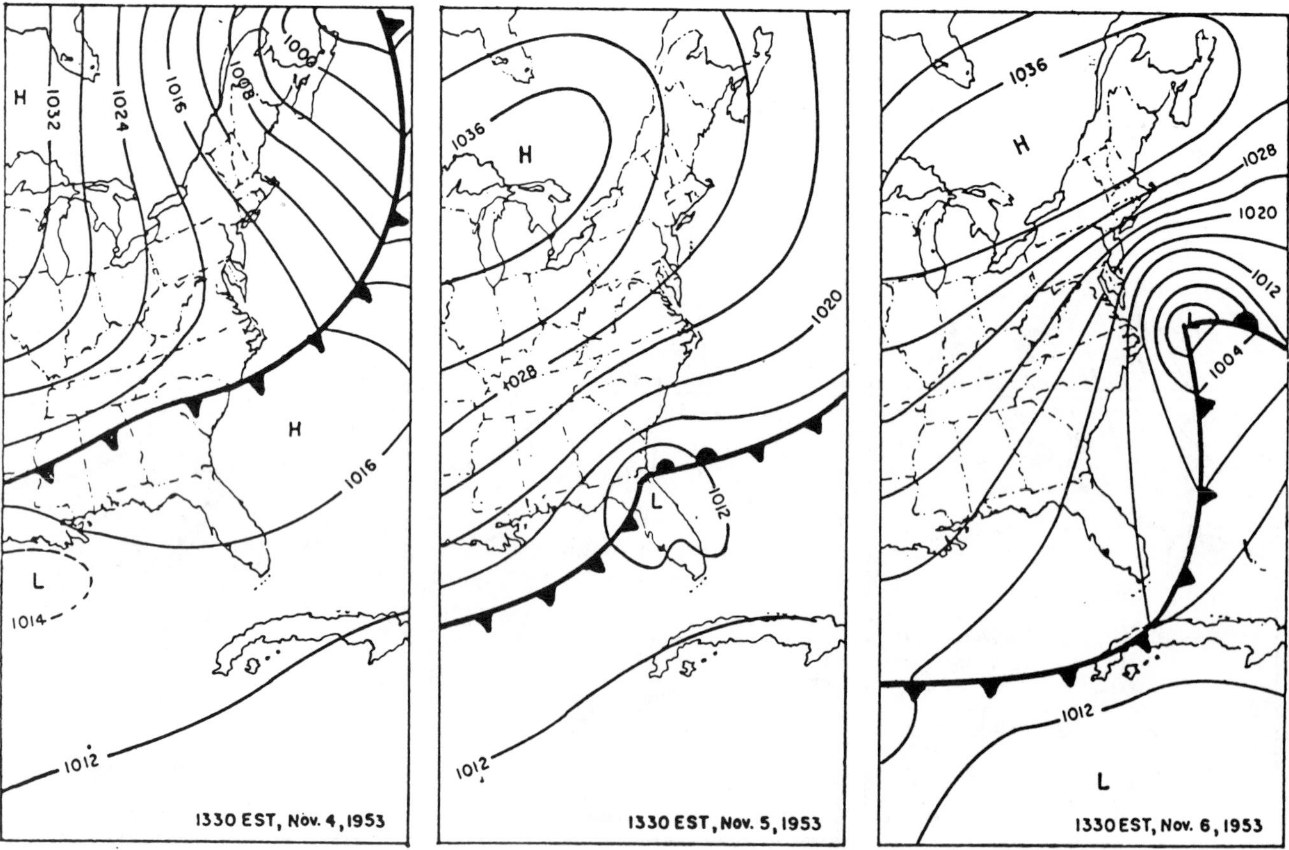

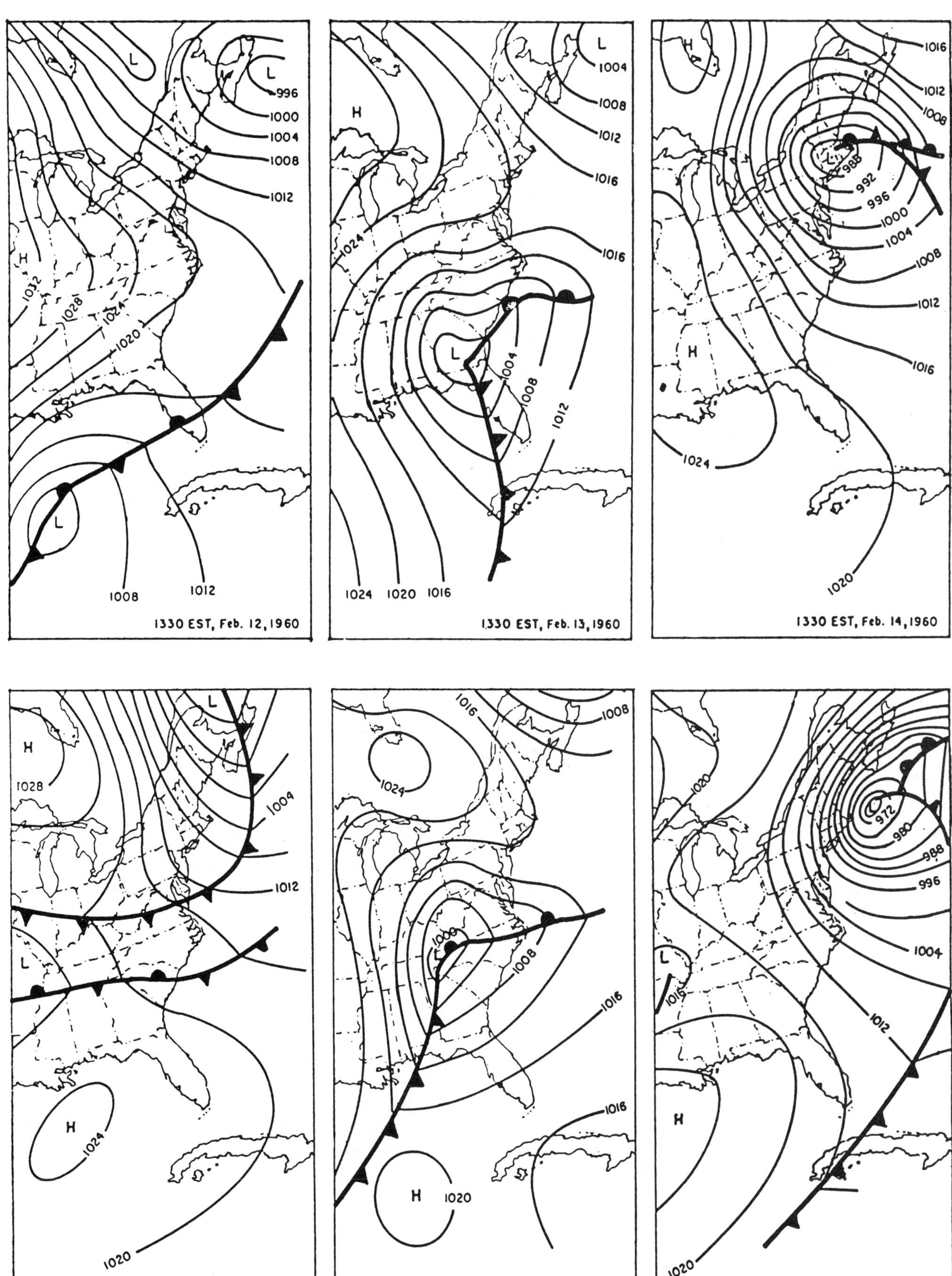

L
996
1000
1004
1008
1012
H
1032
1028
1024
1020
L
1008
1012
1330 EST, Feb. 12, 1960
L
1004
1008
1012
1016
H
1024
1016
L
1004
1008
1012
1024
1020
1016
1330 EST, Feb. 13, 1960
H
1016
1012
1008
988
992
996
1000
1004
1008
1012
H
1016
1024
1020
1330 EST, Feb. 14, 1960
H
1028
L
1004
1012
L
H
1024
1020
1300 EST, Jan. 18, 1961
1016
1008
1024
1008
1008
1016
1016
H
1020
1300 EST, Jan. 19, 1961
1020
972
980
988
996
1004
L
1016
1012
H
1020
1300 EST, Jan. 20, 1961

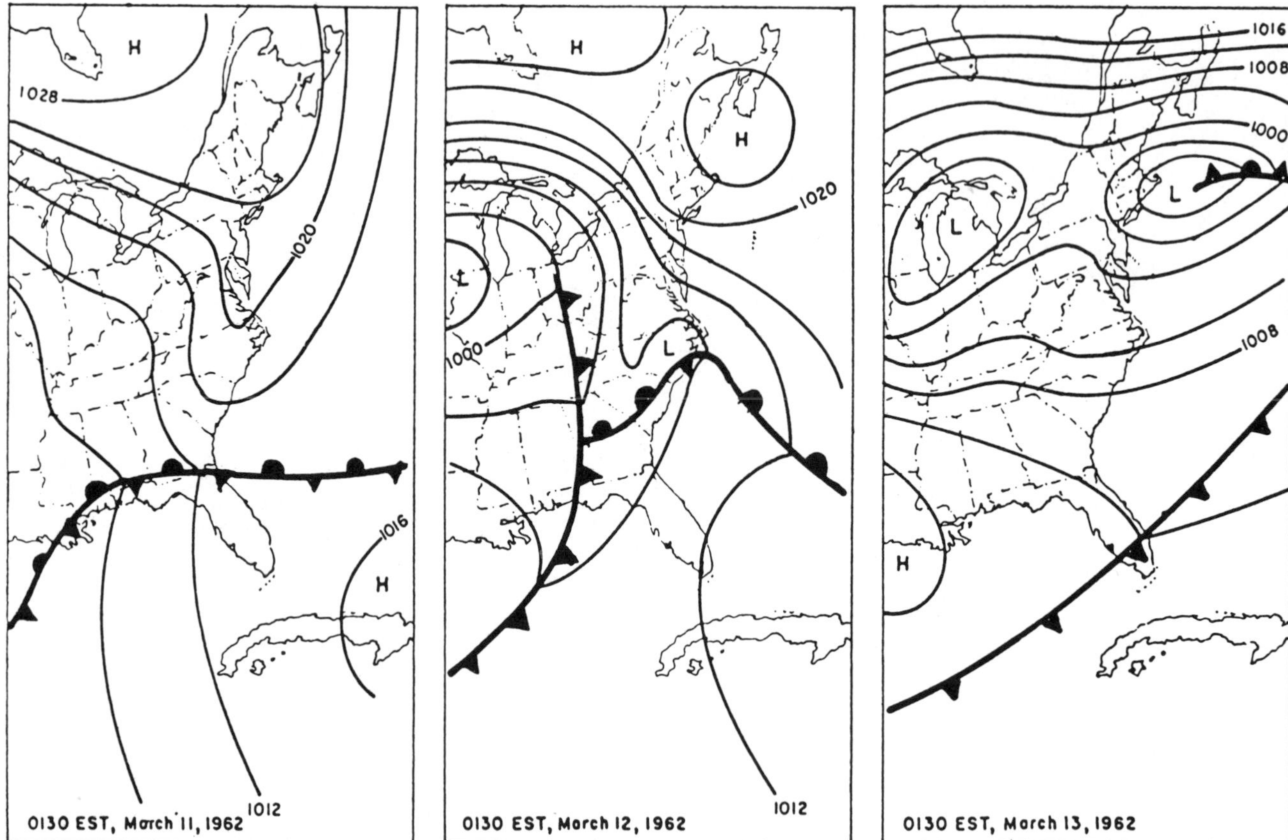

Figure 2.7 Development and movement of typical class 6 storm along east coast of the United States, 11-13 March 1962 (from Mather et al., 1964).

from February 1960 is given in Figure 2.8. While of particular concern to those inhabiting the eastern U.S., these storms are of less consequence to the offshore environment than those in classes 1-6.

Strong Cold Fronts Accompanied by Squall Lines and Severe Local Weather (Class 8)

This class of storm is associated with cold fronts but not well-developed cyclones. Winds can be high and reach hurricane force but are localized.

Statistical summaries of tropical and extra-tropical cyclones are discussed separately below. The seasonal occurrence of these severe coastal storms and their effects on the ocean will be covered in a later section.

Tropical Cyclones

Neumann et al. (1981) summarized tropical cyclone occurrence in the North Atlantic ocean for the years 1871-1980. The months of August, September, and October contain about 85 percent of these storms. In any given year the number of hurricanes (tropical cyclones with sustained winds in excess of 64 knots) in the Atlantic basin can range from zero to twelve. Many of these enter the Gulf of Mexico and do not affect the waters off the eastern U.S. coast. Neumann and Cry (1978) note that 850 tropical cyclones were recorded in a 107-year period between 1871 and 1977. Of these slightly over one half reached hurricane strength (Fig. 2.9). Shapiro (1981) analysed these hurricane records from 1899 through 1978 and found long-term trends in both the *number* of hurricanes and their *mode of occurrence:* either east coast or Gulf of Mexico and Caribbean. He found that indices measuring both of the above quantities are correlated and contain periodicities of 2.5 and 4.5 years. The maximum in east coast hurricane incidence occurs at a phase of the quasibiennial oscillation (Angill et al, 1969) when the Atlantic subtropical high pressure system is at its farthest northeasterly displacement.

Other studies have been published recently by Gray (1979) and for the Gulf of Mexico by Hebert and Taylor (1979). Scientists at the National Hurricane Center in Coral Gables, Florida, regularly summarized recent hurricane seasons (see, for example, those by Hebert [1974, 1976, 1980] and Lawrence [1977, 1979]). Although Neumann et al. (1978) show individual tracks of all tropical storms in their 107-year summary, these data have not been specifically analysed for storm frequencies in the waters of primary interest in this study. The authors, however, do consider tropical storms which affect the eastern U.S. and Gulf of Mexico coasts, citing a study made by Ho et al. (1975). Nearly 40 percent of the tropical cyclones fall into this catetory.

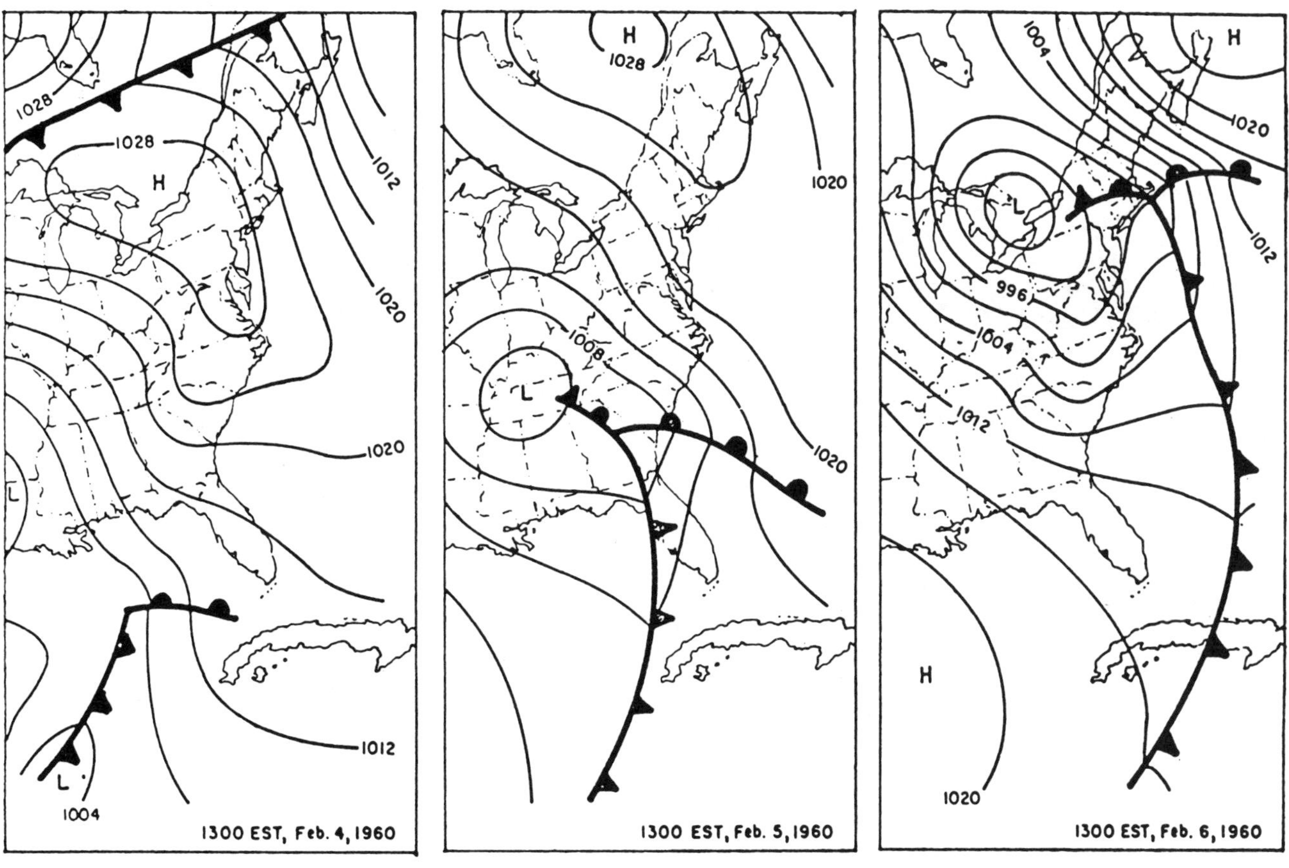

Figure 2.8 Development and movement of typical class 7 storm along east coast of the United States, 4-6 February 1960 (from Mather et al., 1964).

Extra-Tropical Cyclones

Resio and Hayden (1975), Hayden (1981), Colucci (1976) and Dickson and Namias (1976) studied the areal distribution and secular interannual climatology of extra-tropical cyclones. These storms comprise seven of the eight earlier classifications and are the most common storms off the eastern U.S. coast. Colucci (1976) analyzed winter months for a 10-year period January 1964-December 1973, and found distinct maxima in cyclone incidence (Fig. 2.10a): 1) along a band from Cape Hatteras to Cape Cod to the Gulf of Maine and 2) over the Gulf Stream after it leaves the coast at Cape

Figure 2.9 Annual distribution of the 793 recorded Atlantic tropical cyclones reaching at least tropical storm strength (open bar) and the 468 reaching hurricane strength (solid bar), 1886 through 1980. The average number of such storms is 8.4 and 4.9 per year, respectively (from Neumann et al., 1978).

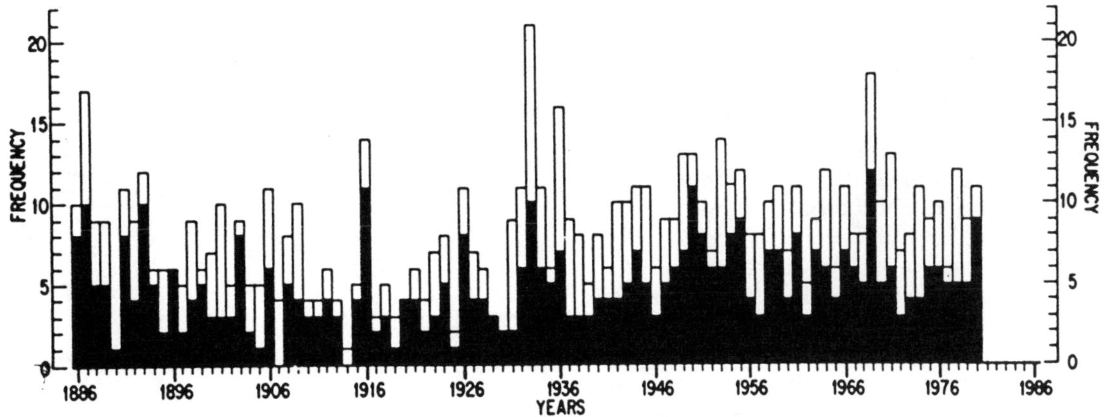

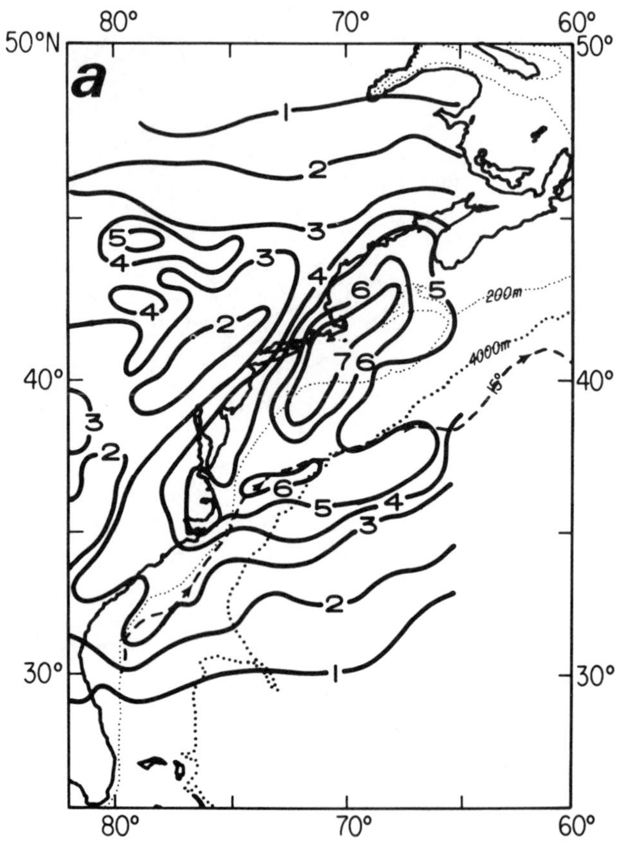

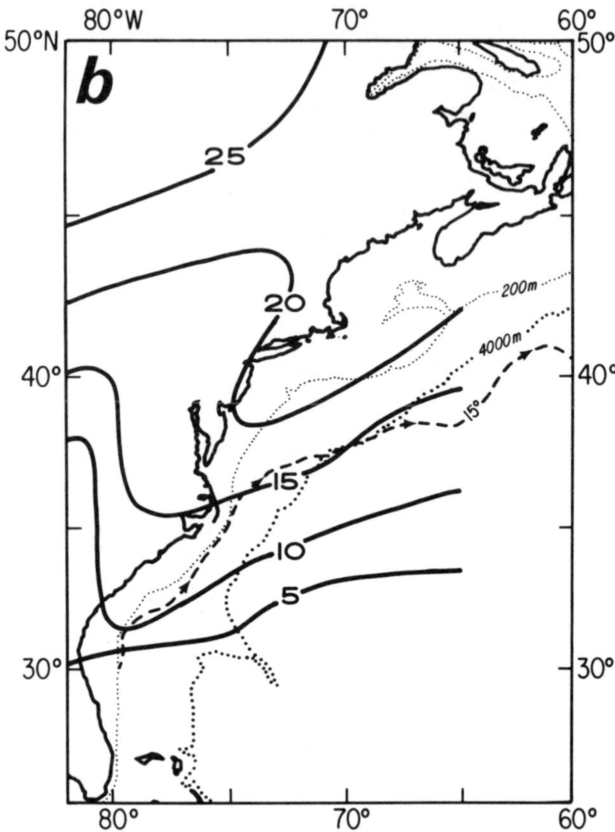

Figure 2.10a. Winter cyclone frequencies by 1° latitude-longitude quadrangles: January 1964-December 1973. Contour values should be multiplied by 10. Mean position of the northern and western boundary of the Gulf Stream. 1966-1973, is shown by dashed line. Redrafted from Collucci (1976).

Figure 2.10b. Mean cyclone frequency for the years 1885-1978, redrafted from Hayden (1981).

Figure 2.10c. Standard deviation in annual cyclone frequency for the years 1885-1978, redrafted from Hayden (19811).

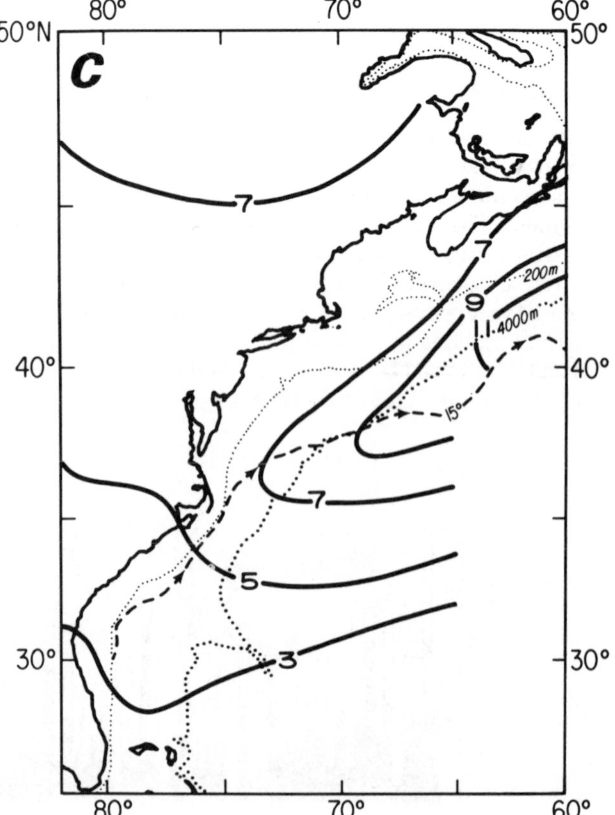

Hatteras. The first maximum encloses a broad region southwest and northeast of Cape Cod in which over 70 winter cyclones were found in the 10-year period. Over the Gulf Stream a secondary maximum of over 50-60 storms is found, with a minimum region (less than 50 cyclones) in the Slope Water to the north. In this region, however, Colucci finds that cyclones tend to deepen and become more intense. Another region of strong "Cyclogenesis" is found just off Hatteras. Hayden's (1981) study, though lacking the same spatial resolution, covers an extended time period of 87 years, from 1885-1978. Patterns of both the long-term means and standard deviation (after Hayden, 1981) are illustrated in Figure 2.10b and c.

The dominant features of the mean annual cyclone frequency distribution are the northward increase and the local maximum near the coast for a fixed latitude. The standard deviation map shows the large variability in a zone near the north wall of the Gulf Stream. Statistical analysis of the variance revealed that over 45% of the variability could be explained by 1) the tendency for cyclones to track either

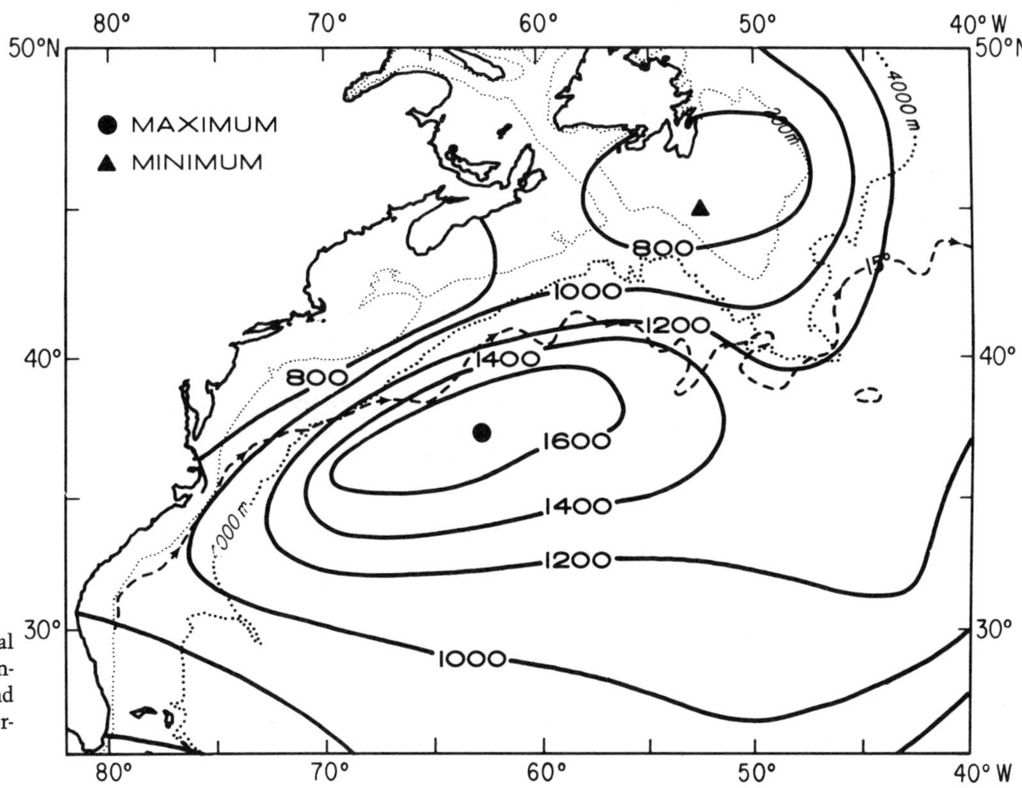

Figure 2.11 Mean annual rainfall (mm) over the Atlantic Ocean between 30°S and 70°N. Redrafted from Dorman and Bourke (1981).

offshore or over land, and 2) cyclogenesis maximum seaward of Chesapeake Bay. Both of these indices exhibit century-long secular variation undoubtedly due to changes in the intensity of the east coast baroclinic zone. This temporal structure shows that an increase in east coast cyclogenesis since the middle 1940's is part of a longer time-scale trend. Since the early 1960's, there has been a decrease in cyclogenesis and a shoreward shift of the storm tracks.

Several mechanisms have been suggested for this east coast cyclogenesis. Danard and Ellenton (1980) include a topographic orographic effect due to the Appalachian Mountains, latent heat release, and differential surface roughness between land and water. They find that this differential roughness is more important than other effects and note that, for the case studies considered, latent heat release was not large in the center of the cyclones but rather on the trailing edge where the outbreak of cold, dry air caused large evaporation behind the storm center and therefore was of little importance to storm cyclogenesis. Although warm sea surface temperature anomalies, like that of the Gulf Stream, affect the lower troposphere (Sweet, 1981), the influence of the Gulf Stream upon development and tracking of cyclones remains to be demonstrated.

Because of the large secular variation in the east-west storm climate, annually averaged data must be used with caution when anticipating "typical" off-shore meteorological conditions. We will try to include a variability index (such as the standard deviation from the mean) when quoting yearly or monthly averaged "means". Unfortunately this is not possible in many instances.

Annually Averaged Marine Precipitation

Direct marine observations of rainfall have been plagued by a number of technical problems and remain a largely unattainable goal — although, with care, more accurate measurements can be made with special instrumentation (Reed and Elliot, 1977). Recently Dorman and Bourke (1981) have employed an indirect method which uses standard marine observations made between 1950-1974 in the Atlantic Ocean from 30°S to 70°N. A detailed map for the region (Fig. 2.11) shows a rainfall maximum (1.6 meters) centered near 38°N, 62°W and annual precipitation values in excess of 1.6 meters, with values as low as 0.8 meters along the U.S. coast. The local maximum is located eastward of the regions of large cyclogenesis and cyclone variability noted earlier; it is over the mean axis of the Gulf Stream and is near the region of strong evaporation noted by Bunker (1976). This local maximum is absent in the analysis of Reed and Elliot (1979) and in earlier papers by Korzum (1974) and Baumgartner and Reichel (1975), which were based upon extrapolations of coastal and island data.

Annual Air-Sea Heat and Momentum Exchange

Indirect methods must be employed to study the heat and momentum exchange at the surface of the ocean. These methods utilize many of the standard variables readily obtained in marine weather observations. Empirical relationships have been derived which relate measurements of air temperature, dewpoint temperature, and wind velocity at some standard height above the sea surface (usually 10 meters), together with ocean surface temperature, with wind stress and latent and evaporative heat exchange. It is beyond the scope of this report to discuss these techniques. In a recent analysis of 32 years of marine weather observations between 1941 and 1972, Bunker (1975, 1976) and Bunker and Worthington (1976) summarized the seasonal and annual air-sea exchange for the North Atlantic Ocean.

Because oceanic wind stress and latent and sensible heat exchange depend in a non-linear manner upon wind speed and, for the latter two quantities, on air-sea temperature differences, it is important in the analysis not to "average out" the effect of storms *before* calculating these quantities. Bunker's spatial resolution insured that sufficient numbers of ship observations were available to take into account the different oceanic current regimes. For each Marsden Square (10° latitude by 10° longitude quadrangles) ten subregions were usually selected. For the region of interest off the eastern coast of the U.S. fourteen subareas were chosen each with typically 20,000-60,000 marine observations. Monthly data will be discussed in a later section.

The dominant feature in the annual heat budget (Fig. 2.12) is a broad region of heat loss by the ocean to the atmosphere centered over the Gulf Stream as it leaves the coast near Hatteras. The region of 200 Watt m^{-2} heat loss extends from near 32°N, 78°W in an east/northeastly direction to 40°N, 54°W with a region in excess of 264 W m^{-2} centered near 38°N, 63°W. The dominant factor in the heat balance for this region is evaporation: over 3.7 meters of water are lost each year in the maximum heat loss region near 38°N, 63°W. Note that this exceeds the maximum in precipitation of 1.6 meters found by Dorman and Bourke (1981) (see Fig. 2.11), which is located in the same area. The width of the zone of high evaporation and oceanic heat loss is greater than the Gulf Stream. However, the large meanders and eddies shed by the Gulf Stream (discussed in Chapter 2) could be a major factor in determining the size of this zone. As noted by Bunker, this region is characterized by higher than average cloudiness and by a change in wind direction: north wind in the north and south wind in the south. The mean winds for the entire region between the U.S. coast and the Gulf Stream are directed in an east/southeasterly direction with a magnitude of approximately 8 m sec^{-1} (Leetmaa and Bunker, 1978).

Seasonal Cycles

Seasonal Variation In Storm Frequency

Mather et al. (1964) grouped the eight classes (see above) of storms affecting the coastal zone by month of occurrence for the 42 years of their study (Table 2.1). All but eight hurricanes and tropical storms occurred during August, September, and October. This observation is in basic agreement with the more extensive study by Neuman et al. (1981) who cataloged Atlantic hurricanes and tropical storms between 1886 and 1980 (Fig. 2.13).

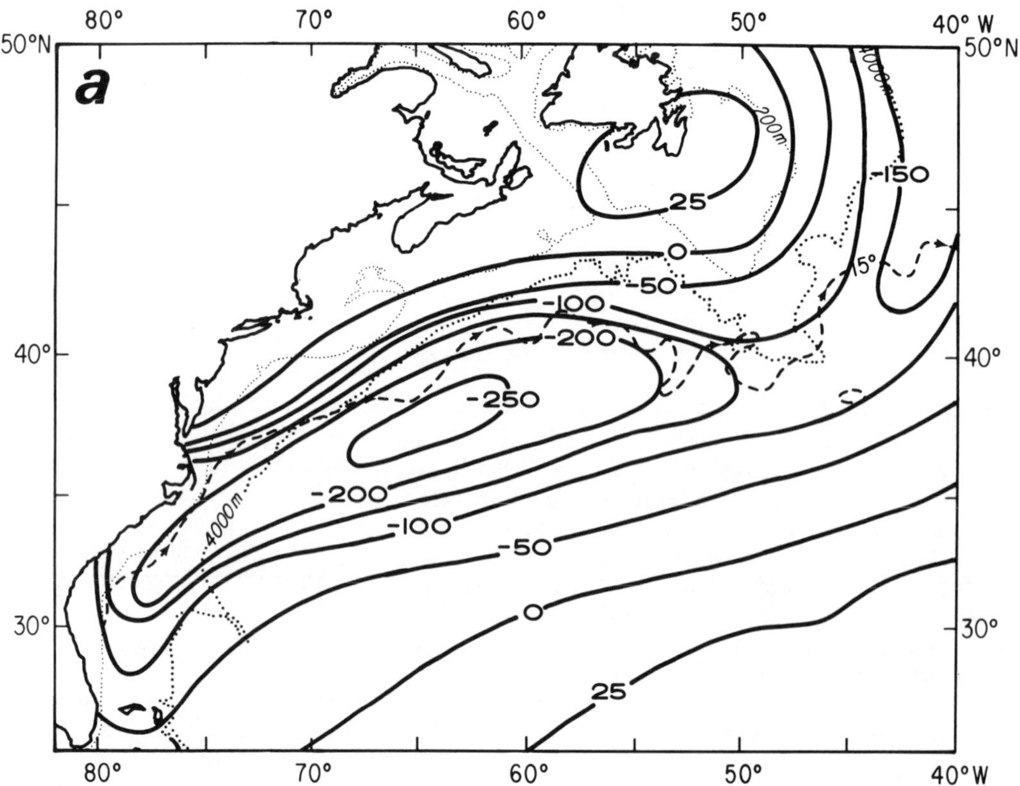

Figure 2.12a. Annual average heat gain A (W m^{-2}) for the North Atlantic. Redrafted from Bunker, 1976.

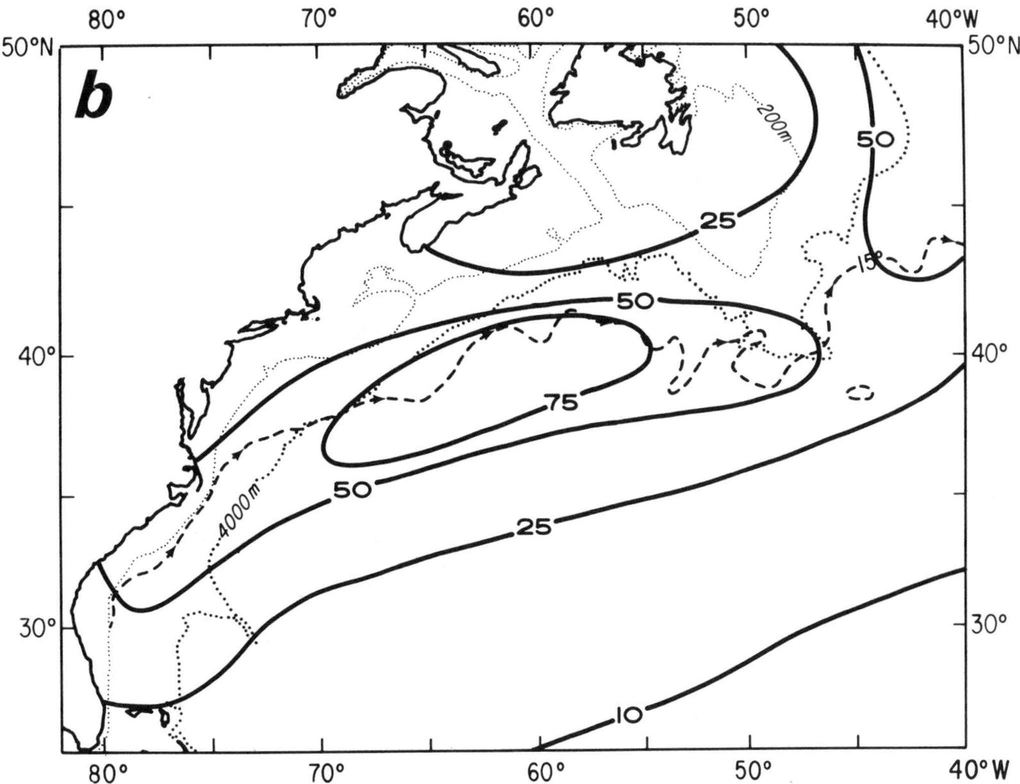

Figure 2.12b. Annual average heat flux S (W m^{-2}). Redrafted from Bunker, 1976.

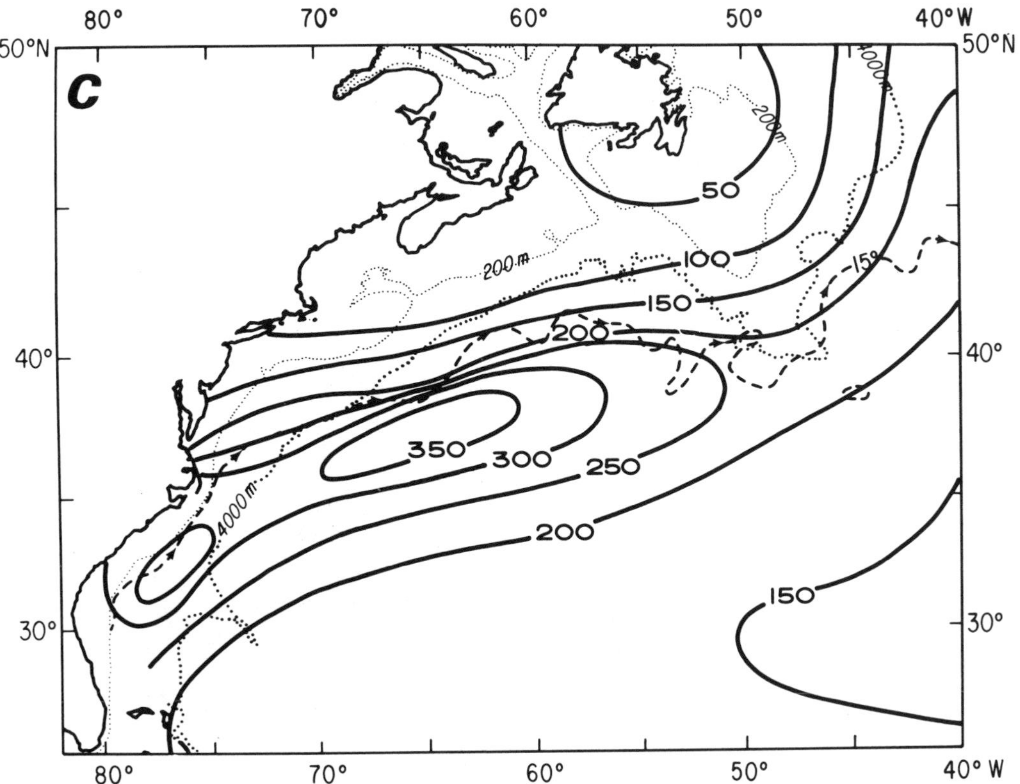

Figure 2.12c. Annual average evaporation (cm year^{-1}). Evaporation rate should be multiplied by 1.29 to get approximate heat flux (W m^{-2}). Redrafted from Bunker, 1976.

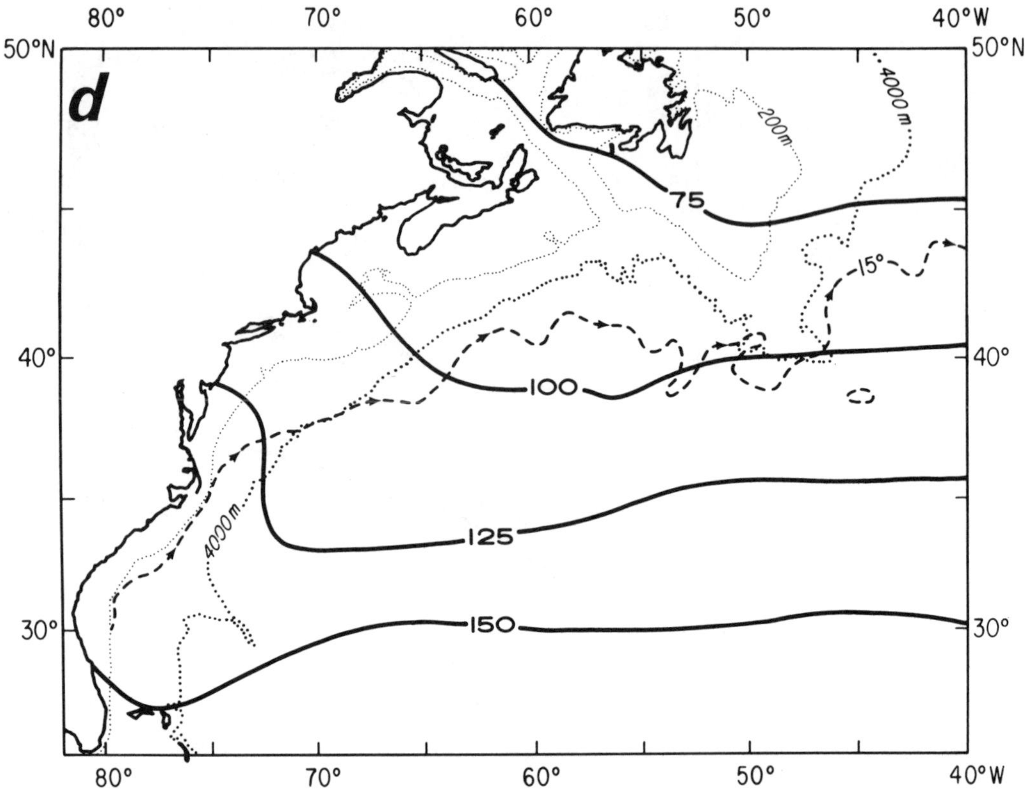

Figure 2.12d. Annual net radiation exchange R (W m^{-2}). Redrafted from Bunker, 1976.

In contrast to hurricanes, extratropical storms occur mainly in fall and winter (Table 2.1) though it is not uncommon for class 4 and 6 storms to persist into April. Winter and spring are the most favorable times for the development and intensification of cyclones according to Reitan (1974). It was this time period (November-April) that was selected for study by Colucci (1976) and discussed earlier. The variability in the marine deck weather observations for each monthly time period (see below) clearly illustrates the annual progression in the storm climate off the eastern U.S. coast.

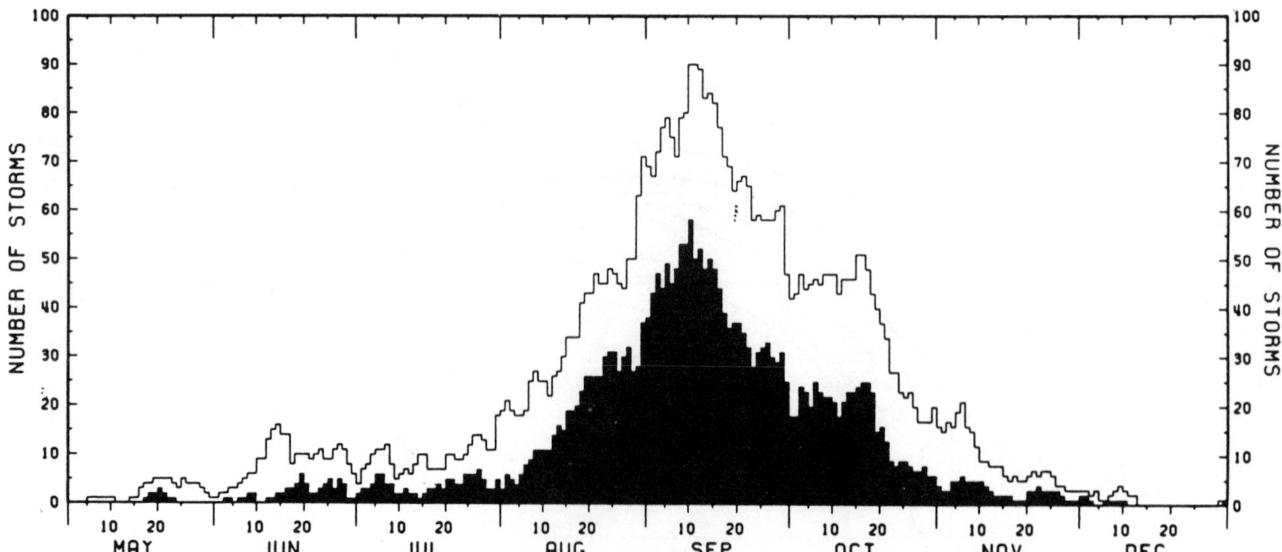

Figure 2.13 Number of tropical storms and hurricanes (open bar) and hurricanes (solid bar) observed on each day, May 1-December 31, 1886 through 1980. From Neumann (1981).

Table 2.1 Seasonal occurrence of storms by class (From Mather et al., 1964). Class 1 is hurricanes and tropical storms while the other seven classes are extratropical in nature.

Type	J	F	M	A	M	J	J	A	S	O	N	D	Y
1		1				1	3	14	22	12	3		56
2	2							1		2	3	1	9
3	2		2	1	1	1		3	3	5	5		23
4	2	3	5	3	1				2	2	2	2	22
5	2	2	2	1		1		1			3	2	14
6	2	4	3	5	1	1				1	4	4	25
7		3	3	1						3	2	2	14
8	1				1		1		3	1			7
Total	11	13	15	11	4	4	4	19	30	26	22	11	170

Seasonal Changes In Precipitation

Dorman and Bourke (1981) estimated seasonal changes in precipitation in addition to the annual means presented earlier. Summaries were made on a quarterly basis: Winter (Dec.-Feb.), Spring (Mar.-May), Summer (June-Aug.) and Fall (Sept.-Nov.). The region of maximum annual rainfall over the Gulf Stream (Fig. 2.11) also displays a relative maximum for each of the four quarters. About one-third of the annual rainfall over the Slope Water and Gulf Stream occurs during winter. The rainfall maximum south of Labrador disappears in summer, but otherwise the seasonal pattern over the northwestern Atlantic remains relatively constant compared with the tropics.

Table 2.2 Summary of Marine Weather Observations for 1941-1972 on a Monthly Basis (annual cycle)*.

Variable No.	Description	Units
1	# observations	
2	air temperature	°C
3	std. dev. of #2	°C
4	dew point temperature	°C
5	std. dev. of #4	°C
6	ocean surface temperature	°C
7	std. dev. of #6	°C
8	air-sea temp. difference	°C
9	std. dev. of #8	°C
10	total cloud cover	oktas
11	surface air pressure	mb.
12	std. dev. of #11	mb.
13	wind speed	m/sec
14	std. dev. of #13	m/sec
15	wind direction	(° T)
16	short incident wave radiation	$W\ m^{-2}$
17	net radiation into ocean	$W\ m^{-2}$
18	latent heat flux out of ocean	$W\ m^{-2}$
19	sensible heat flux out of ocean	$W\ m^{-2}$
20	total heat flux into ocean	$W\ m^{-2}$
21	wind stress	Pascals

The above are tabulated for 12 regions which cover the ocean surface over the domain of interest. They were calculated using marine weather observations taken between 1941-1972 by Bunker (1975). Solar radiation data for both short and long-wave fluxes have been derived from empirical formulae used by Bunker following Budyko (1963). These use observations of cloud cover which have been catalogued in oktas (1 okta is 1/8, 7 oktas is 7/8 cloud cover).

Seasonal Air-Sea Heat and Momentum Exchange

As noted earlier, Bunker (1976) calculated monthly averages for exchange of heat, momentum, and water vapor using 32 years of marine weather observations between 1941-1972, mainly from merchant ships. Many of these data, though available on computer printout at the Woods Hole Oceanographic Institution, have never been published. We have collected and tabulated a total of 21 variables (Table 2.2) for each of the 12 subareas in Figure 2.14 on a monthly basis, but averaged over all 32 years of data. Tables 2.3-1 to 2.3-12 contain the monthly variations of the selected variables. Those interested in the details and algorithms employed should refer to the papers by Bunker in the bibliography.

We have selected variables in which the drag coefficients for latent and sensible heat exchange and wind stress were allowed to vary with air-sea temperature difference. Use of the tables can be illustrated as follows. For the region south of Cape Cod in the Slope Water north of the Gulf Stream (Area #4) a total of 2146 observations were analyzed for the month of February. The average air temperature was 6.9 °C with standard deviation of 4.8 °C in the data. The mean wind speed was 9.94 m sec^{-1} from a direction of 303 degrees true. The wind stress magnitude of 0.33 Pascals is *directed* 303-180 = 123 degrees true.

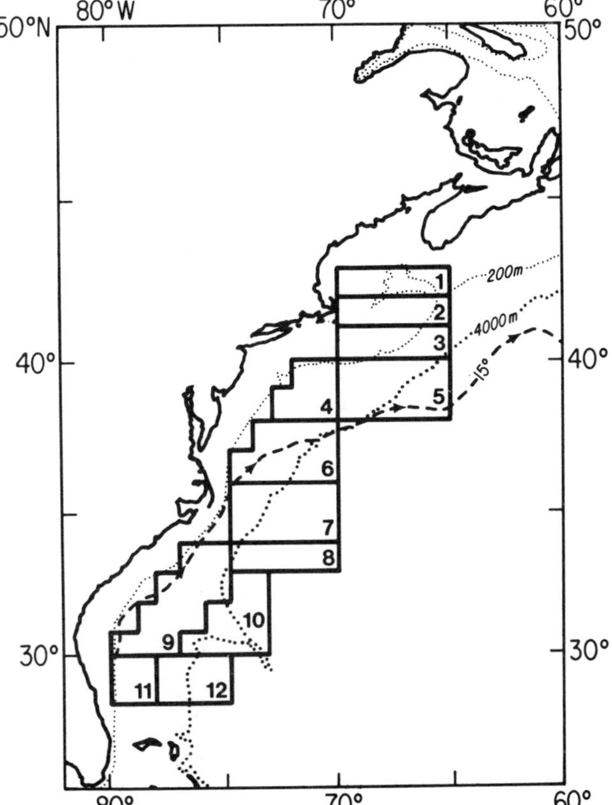

Figure 2.14 Coded areas for tabulated meteorological data. Dashed line denotes 15°C isotherm at 200 meters depth: traditionally the position of the Gulf Stream north wall.

Table 2.3 Area #: 1

Month	Jan	Feb	March	April	May	June	July	Aug	Sept	Oct	Nov	Dec
# Obs	1017	849	1076	1015	1121	979	1196	1385	1584	1409	1145	960
°C	2.4	1.2	2.8	5.7	8.7	12.9	17.0	17.9	15.8	12.7	8.9	4.2
°C	4.0	4.0	3.2	2.6	2.6	2.7	2.6	2.5	2.5	2.9	3.4	4.2
°C	−1.8	−1.8	−0.5	2.8	6.5	11.0	15.1	15.8	13.2	9.6	5.6	0.9
°C	4.9	4.8	4.3	3.7	2.9	2.8	2.5	2.8	3.6	4.2	4.9	5.3
°C	5.3	4.0	3.7	4.5	7.1	11.0	15.3	16.8	15.4	13.2	10.4	7.2
°C	2.1	2.3	2.2	2.1	2.6	3.0	3.0	2.8	2.5	2.2	2.0	2.1
°C	−2.9	−2.8	−0.9	1.2	1.6	2.0	1.7	1.0	0.4	−0.4	−1.5	−3.4
°C	3.9	4.0	3.4	2.3	2.3	2.4	2.3	2.4	2.3	2.5	3.0	4.1
oktas	6.1	6.0	5.1	4.7	4.8	4.9	5.0	4.8	4.5	4.4	5.8	6.4
mb.	1014	1013	1013	1015	1016	1015	1015	1015	1018	1017	1016	1014
mb.	11.3	11.3	10.6	10.1	7.3	6.4	5.6	5.3	7.0	8.1	9.4	10.9
m/sec	9.34	9.73	9.33	7.53	6.62	6.06	5.39	5.70	6.28	7.53	8.60	10.10
m/sec	4.82	4.68	4.87	4.26	3.62	3.24	2.80	3.25	3.75	4.17	4.70	4.97
°T	298	300	305	258	220	213	220	232	280	294	304	294
$W\ m^{-2}$	108.2	158.3	242.3	308.5	356.9	371.5	352.1	310.1	250.3	187.3	132.4	106.6
$W\ m^{-2}$	12.9	38.8	95.3	147.0	172.8	179.3	161.5	142.1	106.6	67.8	25.8	9.7
$W\ m^{-2}$	90.4	84.0	59.8	16.2	6.5	1.6	4.8	21.0	37.2	69.5	88.8	127.6
$W\ m^{-2}$	71.1	72.7	32.3	−11.3	−12.9	−12.9	−8.1	−4.9	0	14.5	37.2	90.4
$W\ m^{-2}$	−143.7	−113.1	4.9	140.5	184.1	195.4	182.5	138.9	79.1	−9.7	−100.1	−708.3
Pascals	.29	.30	.28	.16	.11	.09	.06	.08	.11	.17	.24	.34

Table 2.3 Area #: 2

Month	Jan	Feb	March	April	May	June	July	Aug	Sept	Oct	Nov	Dec
# Obs	1390	1117	1426	1484	1559	1744	2129	2438	2313	2371	1784	1438
°C	3.8	2.8	4.1	6.2	9.8	13.4	17.2	18.5	17.1	13.9	10.2	6.1
°C	3.7	3.9	3.1	2.7	3.1	3.0	3.0	2.6	2.9	3.1	3.6	4.1
°C	0.8	−0.2	1.0	3.5	7.4	11.6	15.5	16.4	14.3	10.5	6.7	2.8
°C	4.6	4.7	4.1	3.6	3.8	3.2	3.0	2.9	3.6	4.4	4.9	5.1
°C	6.4	5.3	5.0	5.5	8.1	11.2	15.4	17.5	17.0	14.7	12.1	9.2
°C	2.6	2.8	2.9	2.6	3.3	3.5	3.7	3.2	3.2	2.9	2.9	2.9
°C	−2.5	−2.4	−0.9	0.7	1.6	2.2	1.8	1.0	0.0	−0.8	−1.9	−3.1
°C	4.0	4.0	3.4	2.8	2.8	2.6	2.6	2.6	2.7	2.9	3.4	4.1
oktas	6.1	6.1	5.3	4.9	4.8	5.1	5.2	4.8	4.4	4.3	5.5	6.2
mb.	1014	1013	1013	1014	1016	1016	1016	1016	1018	1017	1016	1015
mb.	11.1	11.4	10.4	9.5	7.1	6.2	5.9	5.7	6.3	8.3	9.4	10.4
m/sec	9.89	9.82	9.26	7.64	6.45	5.70	5.04	4.98	6.17	7.53	8.58	9.53
m/sec	4.92	4.85	4.98	4.30	4.03	3.31	3.09	3.26	3.66	4.32	4.56	5.32
°T	300	294	314	296	240	224	223	238	318	311	305	293
$W\ m^{-2}$	111.4	143.7	245.5	311.7	358.5	371.5	352.1	311.7	253.6	192.2	137.3	111.4
$W\ m^{-2}$	14.5	37.2	88.8	137.3	169.6	164.7	151.8	140.5	111.4	71.1	29.1	12.9
$W\ m^{-2}$	101.8	93.7	67.8	30.7	12.9	1.6	4.9	24.2	58.1	95.3	113.1	132.4
$W\ m^{-2}$	69.5	66.2	32.3	1.6	−9.7	−12.9	−8.1	−1.6	+6.5	22.6	43.6	80.8
$W\ m^{-2}$	−151.8	−114.7	−4.9	113.1	172.8	192.2	174.4	134.1	56.5	−40.4	−127.6	−201.9
Pascals	.32	.31	.28	.17	.12	.08	.06	.06	.11	.17	.23	.32

Table 2.3 Area #: 3

Month	Jan	Feb	March	April	May	June	July	Aug	Sept	Oct	Nov	Dec
# Obs	3412	2999	3800	4092	4633	4627	4556	4705	4439	4309	3986	3724
°C	5.2	4.2	5.3	7.0	11.6	16.3	19.9	21.1	19.1	15.6	11.7	7.4
°C	4.4	4.3	3.5	3.6	3.7	3.8	3.1	2.9	3.3	3.7	4.1	4.5
°C	2.0	1.2	2.2	5.0	8.9	14.0	17.8	18.6	16.0	12.2	8.3	4.3
°C	5.1	5.2	4.4	4.3	4.2	3.9	3.2	3.4	4.2	4.8	5.0	5.4
°C	8.4	6.9	6.5	7.5	10.6	14.9	18.7	20.6	19.7	17.1	14.0	11.0
°C	3.8	3.9	4.0	4.1	4.7	4.5	3.9	3.5	3.7	3.7	3.8	3.6
°C	-3.3	-2.7	-1.2	0.4	1.0	1.4	1.2	0.5	-0.7	-1.5	-2.4	-3.6
°C	4.6	4.6	4.2	3.6	3.6	3.4	3.0	2.9	3.3	3.6	3.9	4.5
oktas	6.1	6.0	5.2	4.7	4.8	4.8	4.9	4.6	4.6	4.5	5.6	6.3
mb.	1015	1014	1013	1015	1017	1016	1017	1016	1018	1017	1016	1015
mb.	10.4	11.5	9.9	8.9	7.2	6.3	5.6	5.3	6.3	8.4	9.3	10.2
m/sec	9.51	9.71	9.17	7.69	6.24	5.57	5.04	5.34	6.45	7.54	8.68	9.77
m/sec	4.92	5.27	4.92	4.51	3.98	3.35	3.16	3.23	4.02	4.36	4.63	5.08
°T	301	299	310	305	254	229	223	246	347	330	306	302
$W\ m^{-2}$	114.7	163.1	248.7	313.3	356.9	366.6	347.2	308.5	255.2	193.8	138.9	114.7
$W\ m^{-2}$	16.2	38.8	92.1	137.3	163.1	176.0	164.7	150.2	114.7	71.1	30.7	14.5
$W\ m^{-2}$	137.3	117.9	87.2	48.5	38.8	29.1	24.2	51.7	103.4	134.1	148.6	163.1
$W\ m^{-2}$	87.2	77.5	42.0	8.1	3.2	-1.6	-3.2	4.9	19.4	37.2	56.5	93.7
$W\ m^{-2}$	-201.9	-143.7	-29.1	92.1	134.1	163.1	161.5	111.4	-1.6	-92.1	-171.2	-243.9
Pascals	.30	.32	.27	.18	.12	.08	.06	.07	.13	.18	.24	.32

Table 2.3 Area #: 4

Month	Jan	Feb	March	April	May	June	July	Aug	Sept	Oct	Nov	Dec
# Obs	2365	2146	2407	2727	2444	2525	2803	3176	3078	2762	2814	2603
°C	7.6	6.9	8.2	11.2	14.6	20.0	23.7	24.3	22.1	18.1	14.0	9.8
°C	4.6	4.8	4.3	4.0	3.7	3.3	2.2	2.4	2.9	3.7	4.2	4.8
°C	3.9	3.2	4.2	7.8	11.3	17.2	20.5	20.6	18.0	13.5	9.8	5.5
°C	5.4	5.7	5.4	5.0	4.7	3.8	3.1	3.6	4.4	5.1	5.4	5.6
°C	12.2	10.9	10.4	11.4	13.9	19.3	23.2	24.7	23.3	20.4	17.4	14.9
°C	3.7	4.1	4.4	4.5	4.4	3.6	2.4	2.2	2.6	2.8	3.2	3.5
°C	-4.6	-4.0	-2.2	-0.1	0.7	0.7	0.5	-0.4	-1.2	-2.3	-3.4	-5.1
°C	5.0	4.9	4.5	4.0	3.5	2.8	2.3	2.2	2.8	3.4	4.1	4.7
oktas	5.9	6.0	5.0	4.4	4.5	4.3	4.4	4.4	4.4	4.5	5.2	6.0
mb.	1017	1016	1015	1015	1016	1016	1017	1016	1018	1018	1017	1017
mb.	9.5	10.2	9.6	8.6	6.9	5.5	5.0	5.1	5.8	7.4	8.7	9.2
m/sec	9.35	9.94	8.73	7.86	6.42	5.85	5.55	5.97	6.69	7.86	8.67	9.36
m/sec	4.67	5.17	4.64	4.11	3.56	3.28	3.07	3.24	3.80	4.21	4.21	5.03
°T	302	303	305	278	273	225	219	239	004	355	311	300
$W\ m^{-2}$	124.4	171.2	253.6	313.3	353.7	360.2	340.8	305.2	256.8	200.3	148.6	124.4
$W\ m^{-2}$	21.0	46.8	29.5	155.0	176.0	195.4	190.6	169.6	127.6	79.1	40.4	19.4
$W\ m^{-2}$	198.7	182.5	135.7	77.5	54.9	51.7	56.5	100.1	151.8	208.3	216.4	255.2
$W\ m^{-2}$	114.7	108.2	63.0	17.8	3.2	1.6	0	8.1	24.2	51.7	75.9	126.0
$W\ m^{-2}$	-282.6	-234.2	-88.8	64.6	127.6	151.8	143.7	67.8	-43.6	-174.4	-250.3	-360.2
Pascals	.28	.33	.24	.18	.11	.09	.07	.09	.13	.19	.24	.30

Table 2.3 Area #: 5

Month	Jan	Feb	March	April	May	June	July	Aug	Sept	Oct	Nov	Dec
# Obs	2793	2831	3314	3181	3228	3462	3616	3687	3603	3365	3521	3056
°C	9.5	8.9	9.7	12.3	16.4	20.9	23.9	24.6	22.4	19.2	15.7	11.2
°C	5.0	5.0	4.6	4.4	3.9	3.3	2.4	2.5	3.1	3.7	4.3	5.0
°C	5.7	5.1	5.6	8.4	12.6	17.6	20.5	20.6	18.2	14.6	11.3	7.0
°C	5.7	5.8	5.4	5.4	5.1	4.0	3.3	3.6	4.4	4.9	5.3	5.7
°C	14.8	14.1	13.7	14.9	17.6	21.4	24.2	25.6	24.4	22.5	20.0	17.1
°C	4.5	4.7	4.9	5.0	4.7	3.8	2.6	2.3	2.6	3.0	3.5	4.0
°C	−5.3	−5.2	−4.0	−2.6	−1.1	−0.4	−0.3	−1.0	−2.0	−3.2	−4.3	−5.9
°C	5.2	5.2	5.0	4.6	3.9	3.1	2.4	2.5	2.9	3.5	4.1	4.9
oktas	6.4	6.5	5.7	5.2	4.9	4.8	4.7	4.7	4.8	5.0	5.7	6.4
mb.	1015	1013	1014	1015	1017	1016	1017	1017	1018	1017	1016	1016
mb.	10.3	11.4	9.6	9.2	6.6	6.1	5.4	4.8	6.3	7.8	9.0	9.8
m/sec	10.33	10.47	9.85	8.96	7.36	6.57	6.03	6.35	7.33	8.04	9.70	10.34
m/sec	5.66	5.53	5.21	4.89	4.06	4.13	3.53	3.51	4.65	4.46	4.99	5.32
°T	295	296	303	304	283	234	232	244	013	332	298	297
W m^{-2}	123.12	170.1	252.7	311.04	351.5	359.6	341.8	306.2	255.96	199.3	147.42	123.1
W m^{-2}	16.2	37.3	87.5	132.8	166.86	186.3	179.8	158.8	118.3	71.3	35.6	16.2
W m^{-2}	283.5	265.7	236.5	187.5	139.7	105.3	95.6	147.4	204.1	264.1	312.7	340.2
W m^{-2}	144.2	136.1	108.5	66.4	30.2	14.6	8.1	17.82	35.6	64.8	100.4	153.9
W m^{-2}	−405	−358.02	−251.1	−115.02	6.5	72.9	89.1	1.6	−116.6	−251.1	−377.5	−479.5
Pascals	.38	.38	.33	.26	.16	.13	.10	.11	.18	.70	.30	.36

Table 2.3 Area #: 6

Month	Jan	Feb	March	April	May	June	July	Aug	Sept	Oct	Nov	Dec
# Obs	6943	6227	7000	7705	7393	7315	8254	7646	7109	7512	7159	6614
°C	10.7	10.6	11.1	14.2	18.3	22.6	25.4	25.7	23.7	19.9	15.9	12.3
°C	5.3	5.1	4.8	4.4	3.9	3.1	2.1	2.1	2.7	3.5	4.2	4.9
°C	6.3	6.3	6.6	10.0	14.4	19.1	22.0	21.7	19.5	15.2	10.9	7.6
°C	6.3	6.1	6.0	5.3	5.1	3.6	2.8	3.2	3.8	5.0	5.5	5.9
°C	16.4	15.5	14.6	15.4	18.6	22.5	25.5	26.3	24.9	22.5	19.8	17.8
°C	5.0	5.4	5.8	5.9	5.3	3.9	2.3	2.0	2.6	3.4	4.0	4.4
°C	−5.7	−5.0	−3.5	−1.2	−0.3	0.1	−0.1	−0.6	−1.2	−2.5	−3.9	−5.5
°C	5.5	5.5	5.7	5.0	4.1	3.1	2.3	2.2	2.6	3.5	4.3	5.2
oktas	5.5	5.5	5.0	4.5	4.4	4.4	4.5	4.5	4.3	4.5	4.9	5.5
mb.	1018	1017	1016	1016	1017	1017	1017	1017	1018	1018	1017	1018
mb.	8.8	9.5	8.7	7.9	6.2	5.3	4.5	4.3	5.4	6.9	8.1	8.3
m/sec	9.30	9.48	8.76	8.02	6.81	5.96	5.90	5.95	6.60	7.77	8.50	8.95
m/sec	4.85	4.90	4.64	4.28	3.78	3.30	3.13	3.33	3.99	4.16	4.65	4.70
°T	307	307	307	287	243	198	214	208	043	012	310	301
W m^{-2}	137.3	184.1	260.02	314.9	347.2	353.7	337.5	306.9	261.6	206.7	158.3	134.1
W m^{-2}	30.7	58.1	106.6	156.7	184.1	197.03	192.2	171.2	134.1	85.6	48.5	27.5
W m^{-2}	298.8	276.2	226.1	158.3	122.7	92.1	93.7	129.2	169.6	237.4	279.4	311.7
W m^{-2}	135.7	122.7	88.8	38.8	19.4	8.1	4.9	11.3	22.6	50.1	83.98	127.6
W m^{-2}	−397.3	−331.1	−203.5	−37.2	46.8	104.97	101.8	35.5	−51.7	−193.8	−313.3	−411.8
Pascals	.29	.29	.24	.19	.13	.09	.09	.09	.13	.18	.23	.26

Table 2.3 Area #: 7

Month	Jan	Feb	March	April	May	June	July	Aug	Sept	Oct	Nov	Dec
# Obs	5714	5466	6235	6271	6638	6577	7070	6686	6075	6380	6130	5591
°C	14.6	14.4	15.3	18.1	21.5	24.4	26.4	26.7	25.3	22.2	18.8	15.9
°C	4.6	4.6	4.4	3.8	2.8	2.4	1.8	1.8	2.2	2.9	3.6	4.5
°C	9.8	9.6	10.3	13.2	17.1	20.7	22.9	22.9	20.7	17.1	13.4	10.8
°C	5.6	5.7	5.7	5.3	4.8	3.3	2.4	2.7	3.3	4.5	4.9	5.6
°C	20.7	20.2	20.0	20.9	22.8	25.2	27.1	27.7	27.0	25.3	23.4	21.9
°C	3.0	3.1	3.5	3.5	2.9	2.4	1.7	1.5	1.8	2.2	2.5	2.7
°C	-6.1	-5.8	-4.7	-2.8	-1.4	-0.8	-0.7	-1.0	-1.7	-3.1	-4.6	-6.0
°C	4.7	4.8	4.7	4.0	3.0	2.5	2.0	2.0	2.2	3.0	3.8	4.6
oktas	5.8	5.8	5.5	4.7	4.8	4.7	4.9	4.8	4.5	4.9	5.1	5.7
mb.	1018	1016	1016	1017	1017	1017	1018	1017	1017	1017	1017	1018
mb.	8.5	9.2	8.2	7.8	5.6	4.7	4.2	4.1	4.8	6.5	7.2	7.9
m/sec	9.59	10.37	9.38	8.57	7.24	6.52	6.47	6.24	6.66	8.00	8.57	9.15
m/sec	4.79	4.97	4.70	4.23	3.56	3.37	3.26	3.39	4.03	4.10	4.68	4.68
°T	292	290	288	274	206	200	216	208	062	035	305	291
W m^{-2}	148.6	193.8	261.6	313.3	344.0	350.5	334.3	306.9	264.9	213.2	164.7	142.1
W m^{-2}	33.9	59.8	101.8	148.6	174.4	187.3	179.3	163.1	134.1	85.6	50.1	30.7
W m^{-2}	382.8	392.5	326.2	253.6	172.8	132.4	129.2	150.2	210.0	298.8	360.2	389.2
W m^{-2}	142.1	143.7	109.8	61.4	27.5	14.5	11.3	14.5	27.5	58.1	93.7	130.8
W m^{-2}	-484.5	-470.0	-331.1	-156.7	-19.4	48.5	48.5	8.1	-93.7	-264.9	-397.3	-486.1
Pascals	.30	.35	.28	.22	.14	.11	.11	.10	.14	.19	.24	.27

Table 2.3 Area #: 8

Month	Jan	Feb	March	April	May	June	July	Aug	Sept	Oct	Nov	Dec
# Obs	2475	2302	2660	2642	2718	2516	2661	2679	2560	2556	2595	2504
°C	16.3	16.0	16.7	19.1	21.9	24.6	26.4	27.0	25.8	23.1	20.1	17.6
°C	3.7	3.7	3.5	2.8	2.4	2.0	1.6	1.6	1.8	2.4	2.9	3.6
°C	11.4	11.3	11.8	14.4	17.7	21.4	23.3	23.4	21.6	18.3	14.9	12.4
°C	5.0	5.0	5.1	4.6	4.7	2.9	1.9	2.2	2.7	4.1	4.3	5.1
°C	20.8	20.2	20.0	20.7	22.5	25.0	27.0	27.7	27.1	25.5	23.4	22.0
°C	1.9	1.9	1.9	2.1	2.1	1.7	1.3	1.3	1.5	1.6	1.6	1.8
°C	-4.5	-4.2	-3.3	-1.7	-0.6	-0.4	-0.5	-0.8	-1.4	-2.4	-3.3	-4.5
°C	3.8	3.8	3.6	2.8	2.2	1.9	1.6	1.7	1.9	2.4	2.9	3.5
oktas	5.8	5.7	5.5	4.6	4.4	4.6	4.7	4.6	4.5	4.8	5.1	5.5
mb.	1018	1017	1017	1017	1018	1017	1019	1017	1017	1017	1017	1019
mb.	8.4	8.4	8.0	7.0	5.2	4.7	4.0	3.8	4.6	6.2	7.1	7.0
m/sec	9.30	9.64	8.90	8.13	6.74	6.17	6.29	6.06	6.69	7.48	8.19	8.90
m/sec	4.80	4.77	4.41	3.98	3.31	3.44	3.14	3.26	4.11	3.91	4.25	4.51
°T	284	283	274	257	208	198	215	203	078	056	295	284
W m^{-2}	158.3	201.9	264.9	314.9	342.4	347.2	334.3	306.9	268.1	216.4	171.2	150.2
W m^{-2}	40.4	64.6	105.0	155.0	184.1	192.2	185.7	166.4	135.7	90.4	54.9	37.2
W m^{-2}	319.8	306.9	255.2	182.5	119.5	92.1	106.6	177.6	185.7	242.3	290.7	334.3
W m^{-2}	101.8	98.5	71.1	32.3	11.3	8.1	8.1	11.3	21.0	38.8	63.0	93.7
W m^{-2}	-376.3	-331.1	-218.0	-51.7	63.0	101.8	84.0	40.4	-59.8	-180.9	-292.3	-389.2
Pascals	.28	.29	.24	.19	.12	.10	.10	.10	.14	.16	.20	.25

Table 2.3 Area #: 9

Month	Jan	Feb	March	April	May	June	July	Aug	Sept	Oct	Nov	Dec
# Obs	12769	11548	13215	12423	12882	13866	14228	14000	13013	13558	13734	12736
°C	17.2	17.5	18.5	21.2	23.9	26.0	27.6	27.8	26.6	24.0	21.0	18.6
°C	4.4	4.3	4.0	3.0	2.3	2.0	1.8	1.7	1.9	2.6	3.4	4.1
°C	12.1	12.3	13.2	15.8	19.1	22.1	23.8	23.8	22.2	18.8	15.3	13.1
°C	5.4	5.6	5.5	4.7	4.0	2.8	1.9	2.0	2.6	4.0	4.7	5.2
°C	23.3	22.9	23.0	23.9	25.5	27.1	28.5	28.9	28.4	27.2	25.6	24.3
°C	1.9	2.1	2.1	2.1	1.8	1.5	1.2	1.2	1.3	1.5	1.8	1.9
°C	−6.1	−5.4	−4.4	−2.8	−1.6	−1.1	−0.9	−1.1	−1.8	−3.2	−4.6	−5.7
°C	4.3	4.2	3.9	3.0	2.4	2.0	1.8	1.8	1.9	2.57	3.4	4.1
oktas	5.3	5.2	4.9	4.2	4.2	4.5	4.5	4.6	4.6	4.7	4.6	5.1
mb.	1019	1018	1017	1018	1017	1017	1018	1017	1017	1017	1018	1020
mb.	7.0	7.3	6.5	6.3	4.9	4.0	3.5	3.5	3.9	5.3	5.8	6.2
m/sec	8.72	9.15	8.77	7.97	6.82	6.45	6.28	6.01	6.84	7.80	7.99	8.22
m/sec	4.06	4.29	4.11	3.70	3.26	3.22	3.11	3.20	3.73	4.02	4.04	4.04
°T	302	291	273	232	172	179	205	187	073	041	354	306
W m^{-2}	168.0	211.6	269.7	314.9	340.8	345.6	332.7	308.5	269.7	222.9	179.3	156.7
W m^{-2}	48.5	75.9	116.3	163.1	187.3	190.6	187.3	169.6	137.3	96.9	64.6	45.2
W m^{-2}	400.5	386.0	344.0	272.9	197.0	156.7	151.8	164.7	231.0	329.5	381.1	392.5
W m^{-2}	126.0	117.9	92.1	51.7	25.8	17.8	12.9	16.2	27.5	56.5	85.6	111.4
W m^{-2}	−471.6	−416.7	−313.3	−148.6	−24.2	29.07	37.2	1.6	−109.8	−299.4	−395.7	−455.4
Pascals	.23	.26	.23	.18	.12	.11	.10	.10	.13	.18	.19	.21

Table 2.3 Area #: 10

Month	Jan	Feb	March	April	May	June	July	Aug	Sept	Oct	Nov	Dec
# Obs	3558	3234	3761	3425	3415	3415	3440	3497	3532	3451	3424	3309
°C	18.1	18.0	18.6	20.4	23.0	25.4	27.2	27.5	26.6	24.2	21.4	19.2
°C	3.3	3.5	3.1	2.5	2.2	2.0	1.7	1.6	1.9	2.2	2.8	3.1
°C	13.2	13.3	13.9	15.8	19.0	22.1	23.8	23.9	22.5	19.6	16.2	13.9
°C	4.8	4.9	4.8	4.2	3.7	2.5	1.7	1.9	2.5	3.6	4.3	4.5
°C	21.6	21.0	21.0	21.7	23.6	25.7	27.6	28.2	27.7	26.1	24.1	22.8
°C	1.7	1.8	1.8	1.8	1.8	1.6	1.3	1.3	1.5	1.6	1.6	1.6
°C	−3.5	−3.1	−2.4	−1.3	−9.7	−0.4	−0.3	−0.7	−1.1	−1.9	−2.7	−3.5
°C	3.2	3.3	3.1	2.4	2.0	1.8	1.7	1.7	1.7	2.1	2.6	3.1
oktas	5.4	5.4	5.2	4.3	4.2	4.5	4.4	4.5	4.6	4.8	4.6	5.1
mb.	1019	1018	1017	1018	1018	1017	1019	1017	1016	1016	1018	1019
mb.	7.0	7.1	7.0	6.1	4.8	4.1	3.4	3.6	3.9	5.5	5.6	6.4
m/sec	8.57	9.05	8.48	7.41	6.11	5.96	5.76	5.72	6.47	7.54	7.47	8.00
m/sec	4.35	4.65	4.39	3.63	3.17	3.42	2.80	3.34	3.93	4.06	3.99	4.18
°T	285	270	260	228	176	189	201	191	095	058	357	293
W m^{-2}	172.8	214.8	27.13	316.5	340.8	345.6	332.7	310.1	271.3	224.5	180.9	159.9
W m^{-2}	83.3	79.1	117.9	166.4	190.6	197.0	193.8	174.4	140.5	98.5	67.8	48.5
W m^{-2}	281.0	268.1	229.3	164.7	113.1	92.1	101.8	121.1	171.2	229.3	256.8	287.5
W m^{-2}	74.3	71.1	51.7	24.2	11.3	6.5	6.5	9.7	16.2	30.7	46.8	67.8
W m^{-2}	−297.2	−251.9	−159.9	−14.5	175.9	106.6	100.1	54.9	−37.2	−153.4	−227.7	−303.6
Pascals	.23	.26	.22	.15	.10	.10	.08	.09	.12	.17	.17	.19

Table 2.3 Area #: 11

Month	Jan	Feb	March	April	May	June	July	Aug	Sept	Oct	Nov	Dec
# Obs	6993	6568	7590	6777	7274	7960	8103	8241	7844	7797	7502	7214
°C	20.2	20.3	21.4	23.2	25.1	27.0	28.3	28.4	27.6	25.8	23.2	21.2
°C	3.5	3.6	3.1	2.3	1.9	1.7	1.6	1.6	1.7	2.1	2.7	3.1
°C	15.2	15.2	16.3	18.0	20.5	23.0	24.1	24.2	23.4	20.8	17.8	15.8
°C	4.7	4.9	4.7	3.8	3.3	2.2	1.6	1.7	2.0	3.3	3.9	4.4
°C	24.3	24.0	24.2	25.0	26.3	27.7	28.9	29.3	28.9	27.9	26.4	25.2
°C	1.6	1.7	1.8	1.7	1.6	1.4	1.1	1.1	1.2	1.3	1.5	1.5
°C	-4.1	-3.6	-2.8	-1.8	-1.1	-0.7	-0.7	-0.9	-1.3	-2.1	-3.2	-4.0
°C	3.5	3.4	3.2	2.4	2.1	1.8	1.7	1.7	1.7	2.1	2.7	3.2
oktas	4.7	4.6	4.4	4.0	3.9	4.4	4.0	4.2	4.6	4.4	4.2	4.4
mb.	1020	1019	1018	1018	1017	1017	1019	1017	1016	1016	1018	1020
mb.	5.4	5.8	5.1	4.7	3.9	3.3	2.7	3.2	3.1	4.1	4.4	4.6
m/sec	7.87	7.97	7.61	7.00	6.09	5.52	5.06	5.10	6.26	7.17	7.44	7.50
m/sec	3.77	3.85	3.63	3.19	3.07	2.92	2.65	2.79	3.50	3.63	3.58	3.70
°T	330	290	241	111	109	150	171	152	076	046	027	013
W m^{-2}	187.3	227.7	276.2	316.5	337.5	340.8	331.1	311.7	277.8	234.2	193.8	174.4
W m^{-2}	67.8	96.9	135.7	174.4	197.0	195.4	200.3	182.5	147.0	111.4	80.8	63.0
W m^{-2}	374.6	302.0	264.9	216.4	164.7	127.6	126.0	140.5	185.7	264.9	314.9	329.5
W m^{-2}	80.8	71.1	53.3	30.7	17.8	11.3	9.7	11.3	16.2	33.9	54.9	74.3
W m^{-2}	-329.5	-264.9	-172.8	-59.8	29.1	69.5	82.4	45.2	-45.5	-174.4	-279.4	-334.3
Pascals	.18	.19	.16	.13	.10	.08	.06	.07	.11	.14	.16	.17

Table 2.3 Area #: 12

Month	Jan	Feb	March	April	May	June	July	Aug	Sept	Oct	Nov	Dec
# Obs	2537	2348	2802	2226	2375	2522	2540	2466	2437	2490	2348	2467
°C	20.1	20.0	20.7	22.2	24.3	26.5	27.8	28.1	27.4	25.5	22.9	20.9
°C	2.9	3.0	2.8	2.2	1.9	1.7	1.5	1.5	1.6	1.9	2.5	2.7
°C	15.1	15.0	15.7	17.4	20.2	22.9	23.9	24.2	23.5	20.8	17.6	15.5
°C	4.5	4.6	4.6	3.8	3.0	2.2	1.5	1.6	2.1	3.3	3.9	4.2
°C	23.0	22.5	22.7	23.4	24.9	26.8	28.4	28.9	28.4	27.2	25.4	24.0
°C	1.5	1.6	1.6	1.5	1.5	1.4	1.1	1.1	1.1	1.5	1.4	1.4
°C	-2.9	-2.5	-2.0	-1.2	-0.5	-0.3	-0.6	-0.7	-1.0	-1.7	-2.5	-3.1
°C	2.8	2.8	2.7	2.2	1.9	1.8	1.6	1.7	1.7	1.9	2.4	2.6
oktas	5.0	4.9	4.6	3.9	4.0	4.6	4.0	4.3	4.5	4.5	4.3	4.5
mb.	1010	1018	1018	1018	1017	1017	1019	1017	1016	1015	1018	1020
mb.	5.5	6.1	5.4	5.1	4.1	3.5	2.9	3.3	3.8	4.2	4.8	4.9
m/sec	7.70	7.80	7.68	6.78	5.94	5.76	5.40	5.59	6.07	7.00	7.17	7.17
m/sec	3.97	4.05	3.88	3.24	3.02	2.94	2.67	3.27	3.56	3.81	3.77	3.79
°T	286	246	226	137	134	167	175	172	109	062	031	343
W m^{-2}	187.3	227.7	277.8	318.2	339.2	240.8	332.7	311.7	276.2	232.6	192.2	172.8
W m^{-2}	67.8	93.7	134.1	174.4	195.4	192.2	201.9	180.9	145.4	109.8	79.1	61.4
W m^{-2}	255.2	232.6	218.0	161.5	116.3	100.1	122.7	13.7	153.4	227.7	256.8	263.3
W m^{-2}	54.9	46.8	38.8	19.4	9.7	6.5	8.1	9.7	12.9	25.8	40.4	51.7
W m^{-2}	-237.4	-176.0	-116.3	6.5	80.8	96.9	88.8	48.5	-8.1	-130.8	-206.7	-245.5
Pascals	.18	.18	.17	.12	.09	.08	.07	.09	.10	.14	.15	.15

Rather than dwell on these summaries, we will note that the areas through which the Gulf Stream passes have greater heat loss and warmer water temperatures than the others. For example, the December heat loss in Area #4 is 363 W m^{-2}, with is 5 times the annual average of 73 W m^{-2} for that area. To the east of Area #4, the Gulf Stream passes through Area #5, in which the heat loss to the atmosphere in January exceeds 400 W m^{-2}. We can see that the pattern of annual means discussed earlier is reflected on the seasonal time scale with the oceanic regime of the Gulf Stream to the northeast of Cape Hatteras playing a central role in the water vapor and thermal energy transfer to the atmosphere.

The standard deviation of various tabulated quantities is not to be interpreted as an "error" in the mean. Rather it is an index to both the intensity of the storms within any particular month and secular variations in the storm climate. In the example above for Area #4, the mean wind speed in February was 9.94 m sec^{-1} with a standard deviation of 5.17 m sec^{-1}. Since wind speed is always a positive quantity, it does not have a normal probability distribution. The standard deviation is derived from the second moment of the variable. Willebrand (1978) examined wind stress data and observed an "energy containing" frequency band at the synoptic time scale in the period range 3-4 days. We might anticipate the dominant source for wind variability (standard deviation) to be due to the cyclones discussed in earlier sections.

The U.S. Navy has broken the region into nine zones. Marine weather summaries on a monthly basis have been tabulated in three separate volumes (U.S. Naval Weather Service Command, volumes 2, 3, 4, 1970). A more recent summary (Naval Weather Service, 1976) includes monthly tabulations by 1 degree latitide, longitude squares as well as charts. These do not include calculated air/sea exchanges and wind stress, but may be of considerable value for planning purposes. Data summarized include winds, air and sea temperatures, surface currents, and wind waves. On a large scale with coarser spatial resolution, the Navy has prepared a Marine Climatic Atlas of the World (Volume 1, 1974) with monthly charts of basic meteorological variables. Charts of surface currents, tides, and sea ice are included in the form of seasonal summaries and tidal ranges and phases.

The National Data Buoy Center in Bay St. Louis, Mississippi has been collecting environmental buoy data from large, moored meteorlogical buoys for nearly a decade (see Fig. 2.15). These data have been published as seasonal summaries in the Mariners Weather Log. A recent comprehensive climatic summary of NOAA's buoy data was published in January 1983 (NOAA, 1983), sponsored jointly by the National Climatic Data Center (NCDC) and the National Data Buoy Center (NDBC). This summary includes tabulated monthly and annual averages of the buoy data as well as histograms.

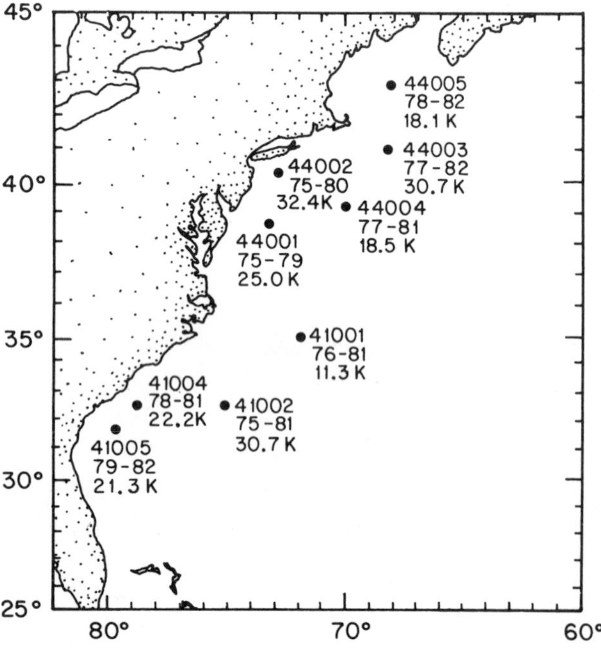

Figure 2.15 North Atlantic and Gulf of Mexico Buoys. (From Climatic Summary for NOAA Data Buoys, 1983).

Upper Ocean Response To Storms

The ocean responds to atmospheric forcing on a variety of time scales. Annually averaged winds produce a mass transport in the upper 20-40 meters which is directed ninety degrees to the right of the wind direction. This Ekman transport is given by $(\tau/\varrho f)$, where τ is the wind stress, ϱ is the density of seawater and f is the Coriolis parameter (equal to $2\ \omega \sin \theta$, ω = rotation rate of earth, θ is latitude). On monthly time scales, the ocean responds to winds in the same manner; monthly Ekman transports can be calculated directly from the wind stress and direction data tabulated at the end of this chapter.

Seasonal cycles of heat and mass exchanged with the atmosphere alter the hydrographic characteristics of the water masses in the upper 200 meters over the outer continental shelf and slope. These physical changes are discussed in the next chapter (see also Robinson et al., 1979).

Superposed on the seasonal atmospheric forcing are storm-related events of short duration and large amplitude. These storms affect mainly the surface mixed layer and are capable of generating high frequency currents with periods of one day as well as large amplitude surface gravity waves with periods of approximately ten seconds. Because this mode of oceanic response is so closely coupled with synoptic-scale winds, we include here a discussion of the transitory upper ocean gravity waves, currents and changes in the mixed layer depth which are of significance to a variety of offshore operations. We do not treat shoal waters and a nearby coastline (Beardsley and Boicourt, 1981). Provided the water column is stratified (not vertically mixed to the bottom) and the observation site is 30 km or more from shore, the effects of bottom topography are not of great importance for time scales of one day or less; nor is the response strongly site-specific.

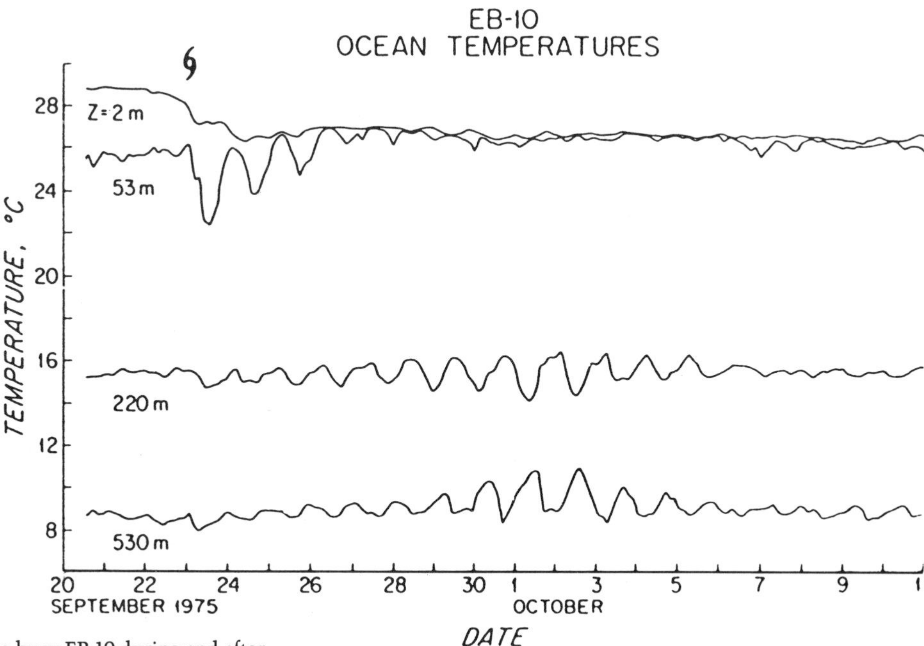

Figure 2.16 Ocean temperatures from buoy EB-10 during and after the passage of Hurricane Eloise (symbol at upper left). Nominal instrument depth is at left. Data from Withee and Johnson (1976).

Mixed Layer Response To Storms

Summer and Fall: Hurricanes

Hurricanes have relatively small radii (300 km), but are very intense cyclonic storms which can dramatically alter the local environment. Some of the best observations from continental shelf regions come from data buoys operated by the NOAA Data Buoy Office (Withee and Johnson, 1976 and Speer, 1978). At buoy EB-20, which was moored on the outer continental shelf of the Northern Gulf of Mexico and directly under the track of Hurricane Eloise (Price, 1981), the mixed-layer depth increased from 30 m to more than 70 m at locations to the right of the storm track (Fig. 2.16). This vertical mixing brought colder water into the mixed-layer and caused a drop in near surface temperature of about 2 °C. These changes in mixed layer depth and sea surface temperature persisted for at least 3 weeks, and probably until the seasonal thermocline was erased by winter cooling. The mixed-layer currents generated by Eloisde were roughly 1 m s^{-1}, and were dominated by near-inertial frequency motion (note the wave-like oscillations of temperature in Figure 2.16) associated with vertical and horizontal advection set up by these currents (Martin, 1982). Within the mixed-layer, the inertial currents decayed to one-folding value relative to the current maximum in about 5 days, due primarily to wave radiation into the thermocline. There, the current amplitude rose to a maximum, typically 0.4 m s^{-1}, in about 3 days, and then decayed slowly. Thermocline currents generated by a hurricane may remain well above normal levels for a period of 2 weeks after the hurricane passage (Brooks, 1983).

If the vertical mixing is strong enough to deepen the mixed-layer to the bottom, the details of bottom topography and bottom friction may become important (Cooper and Pearce, 1982; Forristall et al., 1978; Mayer et al., 1981). Currents are then constrained at least approximately along isobaths, and do not have the inertial wave characteristics described above.

Winter and Spring: Extratropical Cyclones

Mid-latitude storms are much larger weather systems which occur at the rate of about five per month during winter and early spring (Mooers et al., 1976). The center of the low will typically cross the East Coast around Chesapeake Bay and then move northeasterly. The center has a cyclonic circulation which may in exceptional cases approach hurrican intensity. The storm system includes a trailing cold front which may extend over the eastern coast from New England to Florida. Behind the front, winds are northerly, cold and relatively dry.

The upper ocean response to a low pressure system depends a great deal on the time of year. The first few storms of early winter occur when sea surface temperature is still high, and the mixed-layer fairly shallow. In that event, the response will include substantial mixed-layer deepening as in the hurrican case. Cold winds behind the front cause rapid heat loss from the sea surface which can reduce sea surface temperature by several degrees C. (Price et al., 1978). Mixed-layer currents are typically 0.3 m s^{-1} (Klinck et al., 1981), but may be more in certain cases. Thermocline-depth currents are not large compared to the normal background of tidal and inertial motions.

By the end of winter, the repeated cold front passages and normal seasonal heat loss together cause the mixed-layer to reach its yearly maximum value (75 m) north of Cape Hatteras. The response to late winter storms is reduced considerably by this relatively deep mixed-layer.

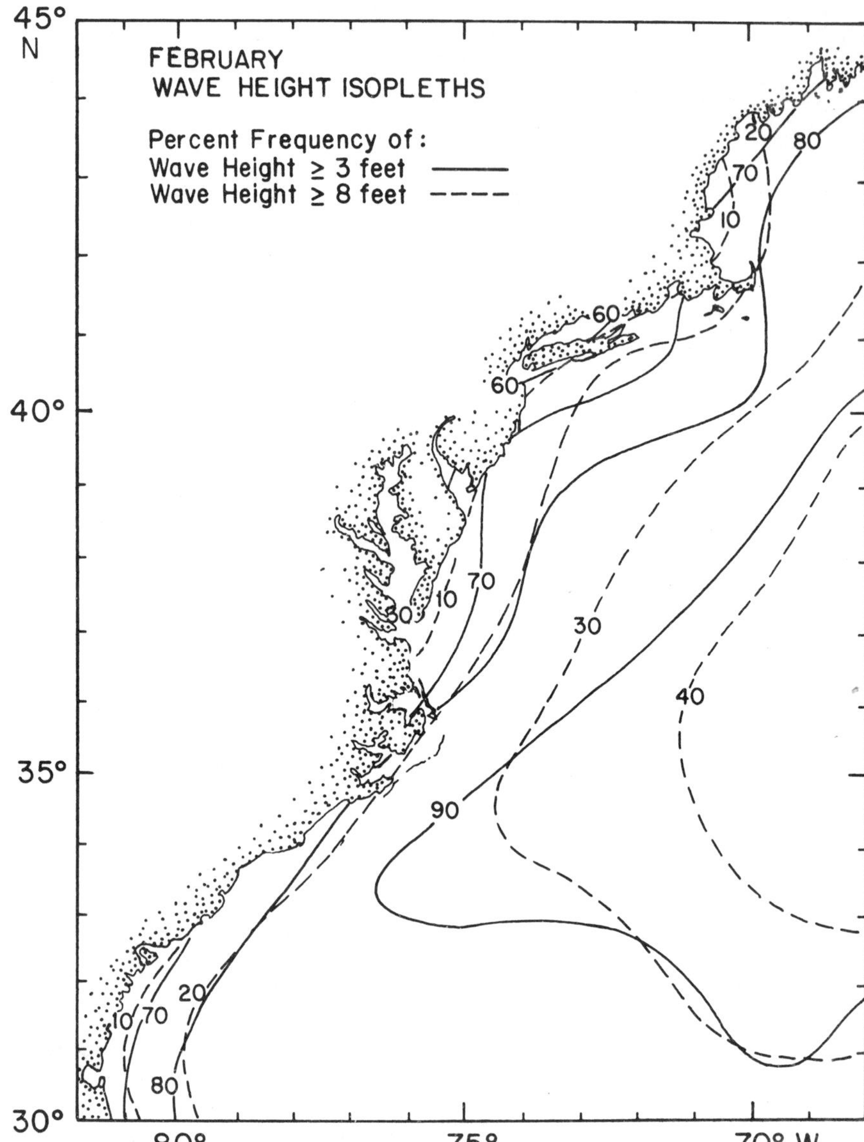

Figure 2.17 Frequencies (percent) for waves greater than or equal to 3 ft. (solid line) or 8 ft. (dashed line) February (after U.S. Naval Weather Service, 1976).

Future Developments

Two entire issues of the Journal of Geophysical Research (v. 87, 30 April 1982 and v. 88, 28 February 1983) were devoted to a geophysical evaluation of data collected from SEASAT. In these issues as well as in other papers (e.g., Parsons, 1979; Fedor et al., 1979) the use of altimetry for the estimation of significant wave height shows great promise for the wind wave prediction problem. With the possibility that surface winds can be quantitatively obtained remotely from scatterometer aboard SEASAT, it is clear that wave prediction models can be evaluated in a variety of different conditions (Queffenlou, 1983). Furthermore, it is now possible to obtain a surface wind and wave climatology as a function of time over much of the ocean, independently of wave prediciton models (see Mognard et al., 1983, for an analysis in the Southern Ocean). In the next few years we look for a growing body of surface wave literature which will draw heavily upon surface wind and wave measurements from earth orbiting satellites. Eventually, surface wave data will be routinely available from satellites much as it is now from a few NOAA environmental buoys located off the eastern coast of the United States (Fig. 2.15). Wave data from the NOAA meteorological buoy network have been recently published in Climatic Summaries for NOAA Data Buoys (1983).

While ship observations of surface waves are subjective and biased low due to avoidance, they nevertheless provide useful data in the absence of direct measurements via buoys or satellites. Monthly summaries of shipboard observations can be found in the form of tables and charts prepared by the U.S. Naval Weather Service. We show in Figure 2.17 contours of percent frequency for waves with amplitudes greater than 3 and 8 feet. Charts like these contain no information about direction, frequency content, or variability in the data, but can be useful for giving a qualitative picture of the surface wave climate.

Physical Oceanography

3

W.J. Schmitz, T.M. Joyce, W.R. Wright, and N.G. Hogg

The character of the water overlying the Atlantic continental slope and rise (ACSAR) of the United States changes dramatically at Cape Hatteras, where the Gulf Stream diverges from the coast (Fig. 3.1). Southwest of Hatteras the Gulf Stream impinges directly upon the slope while the Sargasso Sea overlies the broad Blake-Bahama Outer Ridge and associated mid-depth features. Northeast of Hatteras the Gulf Stream meanders well beyond the shelf edge, and the space between the two is occupied by Slope Water, a distinctive but variable mixture of several source waters. There is continuity between the two regions only below about 2000 m, where the Western Boundary Undercurrent (WBUC) flows to the southwest along the continental rise.

This chapter begins with a discussion of the water masses of the western North Atlantic and their importance in the ACSAR region. Following sections deal with the

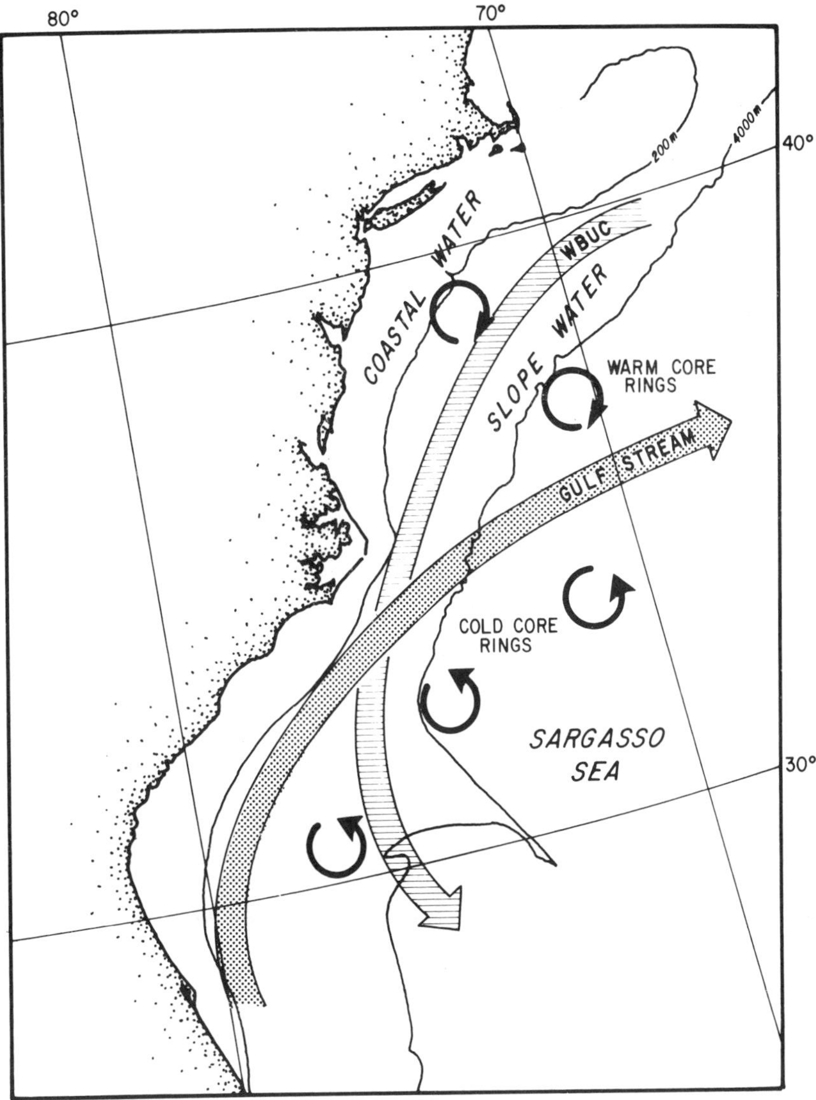

Figure 3.1 Principal hydrographic regimes of the ACSAR region.

major currents of the region and their fluctuations in space and time, and finally with two special features of the physical oceanography of the region: the warm core rings (WCRs) which are spawned from Gulf Stream meanders in the Slope Water, and the submarine canyons which affect and are in turn affected by the water moving in them.

Water Masses

Introduction

The water masses of ACSAR are essentially those of the North American Basin as defined by Wright and Worthington (1970) and elaborated by Worthington (1976) in connection with their volumetric temperature-salinity census for the entire North Atlantic Ocean. Another comprehensive report on the water masses of the western North Atlantic is provided by Emery and Uchupi (1972).

Worthington (1976) divided the water column vertically into three major layers: the deep water (colder than 4 degrees potential temperature); the thermocline (4°θ to 17°C); and the warm water (warmer than 17°C). In the North American Basin the deep water accounts for two-thirds of all the water, the thermocline for about one quarter and the warm water for about 8 per cent. These same proportions hold for the continental slope region south of Cape Hatteras; in the Slope Water region north of Cape Hatteras there is no water warmer than 17°C except in the summer.

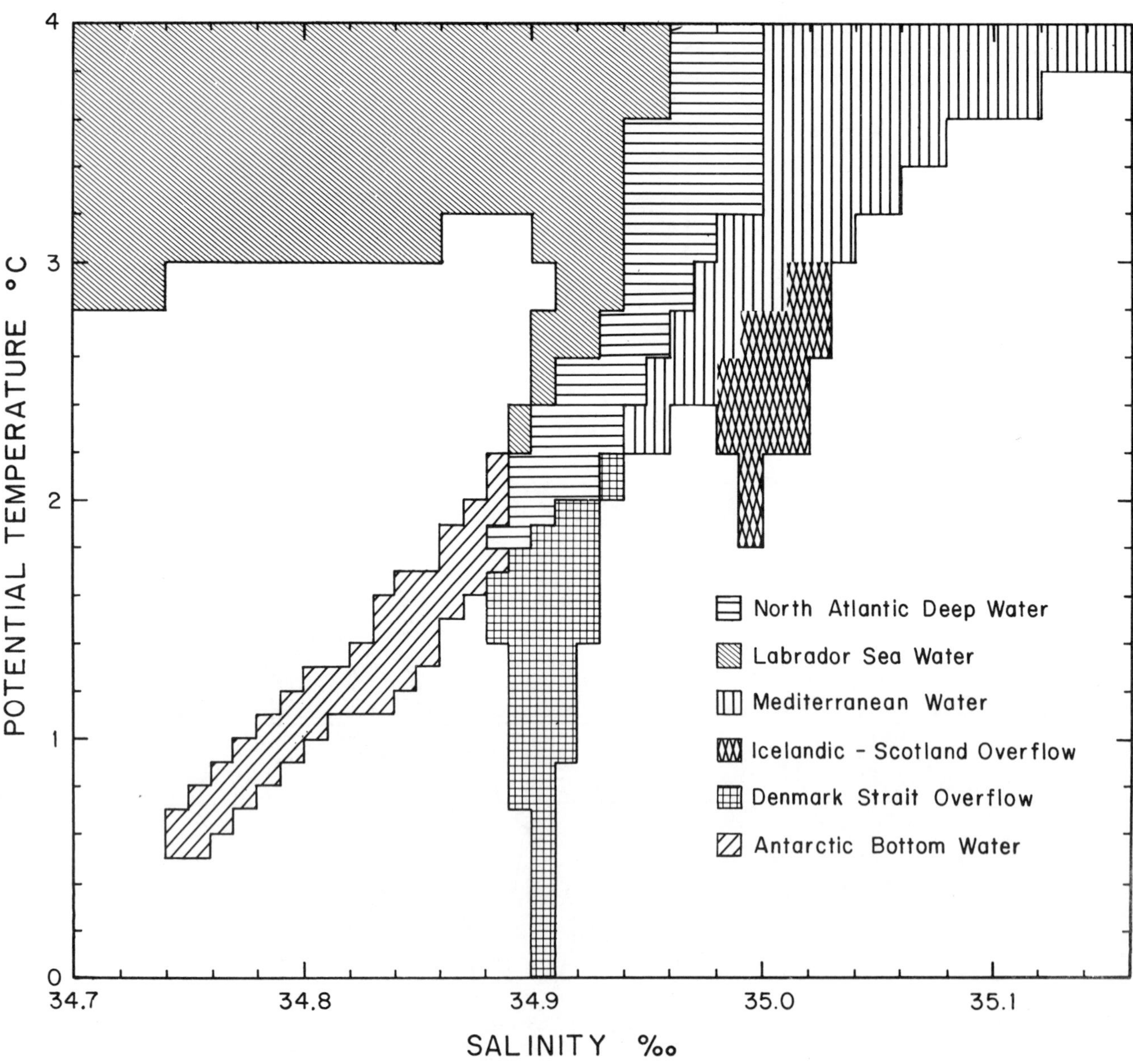

Figure 3.2 Volume (x 10^6 km^3) of the deep water masses of the North Atlantic Ocean. After Worthington, 1976.

In this discussion, shelf water masses are included only to the extent that they are found seaward of the 200-m bottom contour. Good summary discussions of the coastal water masses may be found in Bumpus (1973), Godshall et al. (1980) and Hopkins and Garfield (1981).

Deep Water

There are five sources of the deep water in the North American Basin (Fig. 3.2):

1. Antarctic Bottom Water (AABW).
2. Labrador Sea Water (LSW).
3. Mediterranean Water (MedW).
4. Denmark Strait Overflow Water
5. Iceland-Scotland Overflow Water

All of them are present to some degree in the region of interest, and all contribute to the character of the principal water mass of the region, the North Atlantic Deep Water (NADW).

1. AABW is fresher than 34.89‰, and is the only water mass colder than 1.8 °θ. It consists of water formed by wintertime convection in the Southern Ocean, which during its northward flow was gradually mixed with the warmer and more saline overlying NADW. By the time it reaches the U.S. slope the AABW is distinguished from NADW by its characteristically high-silicate signature (Metcalf, 1969) rather than its T/S characteristics. Most of the AABW in the North American Basin is deeper than 4000 m; at those depths it has been observed flowing northward along the west flank of the Mid-Atlantic Ridge, westward north of Bermuda and then southward along the continental rise (Tucholke et al., 1973). The very deepest water overlying the rise south of Cape Hatteras contains AABW.
2. LSW forms in the southern Labrador Sea where winter cooling causes relatively fresh water (less than 34.94‰, 3.5 to 4.0 °θ) to sink to depths of 1400 to 1600 m. This water spreads mostly to the east, but some of it flows around the Grand Banks and moves to the southwest along the continental shelf where it becomes part of the Slope Water, faintly discernible by a slight salinity minimum. Using potential vorticity minimum as a tracer, Talley and McCartney (1982) have found a continuous band of LSW as far south as 20 °N latitude.
3. The Mediterranean Sea outflow, although small in volume, influences most of the water in the North Atlantic between 2.4 °θ and 13 °C as it spreads westward in a broad high salinity tongue centered at about 30 °N latitude. This tongue extends south of Bermuda and intrudes toward the continental rise at depths to 2000 m. Any water colder than 4.0 °θ and more saline than 35‰ can be said to have some Mediterranean influence.

4. and 5. The Norwegian Sea Overflow through the Iceland-Scotland passage crosses the Mid-Atlantic Ridge into the western basin of the North Atlantic through the Gibbs Fracture Zone at about 53 °N latitude, picking up approximately equal parts of NADW along the way. It then joins the Denmark Strait Overflow, which has also entrained NADW. The combined flow moves northward along the west coast of Greenland and then southward along the Labrador Coast past the Grand Banks and finally along the continental slope as the WBUC. Temperatures are generally less than 3.4 °θ in the WBUC and it can generally be distinguished by a dissolved oxygen maximum (around 6.25 ml/l) at around 2.4 °θ (Fig. 3.3). By the time the WBUC reaches the U.S. continental slope its T/S characteristics are identical with NADW; it is in fact the principal source of that water mass.

The NADW, a mixture of water from all these sources, is the most abundant water mass in the North American Basin and in the entire North Atlantic Ocean, in each case accounting for about 70 per cent of all the deep water. Virtually all of the water colder than 4 °θ overlying the U.S. continental slope is NADW. On a T/S diagram, NADW lies along a line from 1.8 °θ, 34.89‰, to 4.0 °θ, 35.0‰; the potential temperature-salinity correlation is very high: if you know the potential temperature you can determine salinity to ± 0.02‰ or better (Worthington and Metcalf, 1961).

The Gulf Stream carries almost entirely NADW below 4 °θ. The Stream is identified on temperature or salinity sections by sharply sloping isolines (Fig. 3.3). A given temperature or salinity value lies about 400 to 600 m shallower at the inshore edge of the Stream than at the offshore edge and in the Sargasso Sea. As the Stream is approximately 100 km wide, the isothermal and isohaline surfaces slope at about 5 x 10^{-3}. (The sea surface gradient across the Gulf Stream is 1 m in 100 km, or 10^{-5}, sloping upward offshore.) The strong horizontal gradients in the Stream extend throughout the water column. For example, off Cape Fear where the Gulf Stream runs along the continental slope, the 3 °θ isotherm intersects the bottom at 2500 m (Fig. 3.3). About 150 km seaward, south of the Stream, the same isotherm is found at 2900 m. For the 34.94‰ isohaline, lying in roughly the same depth, the corresponding figures are 2600 and 3200 m.

Slope Water colder than 4 °θ lies very nearly on the same T/S curve as NADW but is very slightly fresher (Fig. 3.4). The T/S differences for Slope and Gulf Stream waters are minimal except warmer than 3 °θ (Fig. 3.5). However, in the Slope Water the same combination of temperature and salinity are found up to several hundred meters shallower than in the Stream.

The Thermocline (4 °θ to 17 °C)

The water mass distribution in the thermocline is straightforward compared with either the deep water or the surface waters. Salinity distribution at the 6°, 10°, and 15° C surfaces lie at about 1100, 800, and 600 m in the Sargasso Sea and at 800, 300, and 200 m, respectively, at the inshore edge of the Gulf Stream (Fig. 3.6a,b,c). Each isotherm shows a uniform salinity southwest of Cape Hatteras and some evidence of fresher water northeast of the Cape.

Worthington (1976) subdivided the thermocline into three layers: lower (4 °θ to 7 °C), mid (7 °C to 12 °C), and upper (12 °C to 17 °C) but the only important water mass

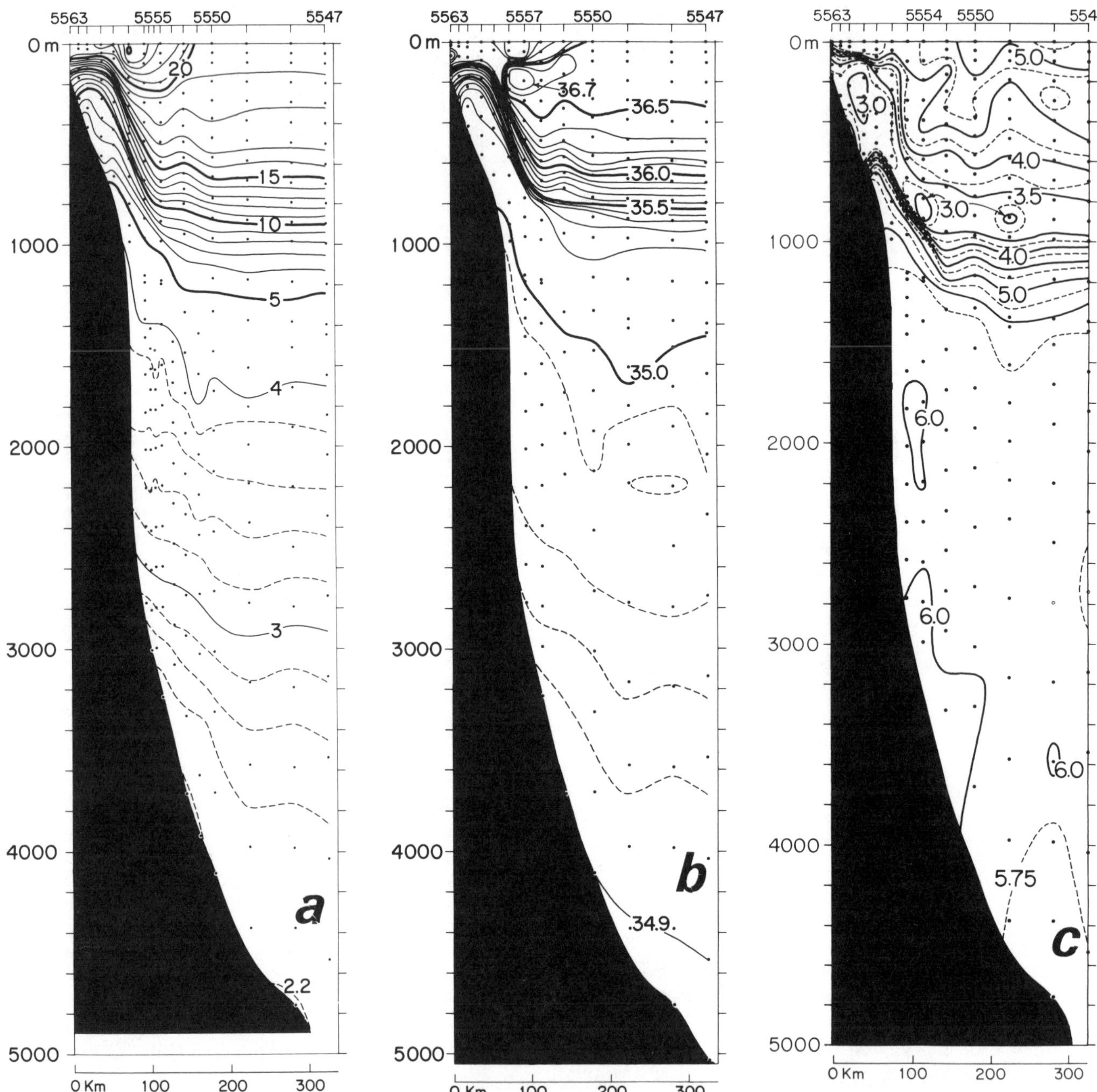

Figure 3.3 Temperature (°C), salinity (‰) and dissolved oxygen (ml/l) sections across the Gulf Stream off Cape Fear. After Worthington and Kawai, 1972.

in the region southwest of Cape Hatteras is the Western North Atlantic Water (WNAW, Wright and Worthington, 1970). This water mass, also called North Atlantic Central Water (Iselin, 1936), includes nearly 90 per cent of all the thermocline water in the North American BAsin. It is the water which circulates in the Gulf Stream in the 4 °θ to 17 ° range and forms the thermocline in the Sargasso Sea. Like the NADW it has a tight T/S relationship. An analytic T/S curve for the western North Atlantic has been developed by Armi and Bray (1982).

Northeast of Cape Hatteras in the Slope Water, the T/S curve runs parallel to that of WNAW but is about 0.1‰ fresher up to about 12 °C, presumably because it is diluted slightly by southward flowing Labrador Coastal Water (McClellan, 1957; Worthington, 1976). The influx of Labrador Coastal Water varies from year to year so the T/S relationship in the Slope Water is not as tight as in the main WNAW mass. Otherwise the difference between the Slope Water and Gulf Stream T/S curves can be discerned only by very careful analysis. However, because the Gulf Stream

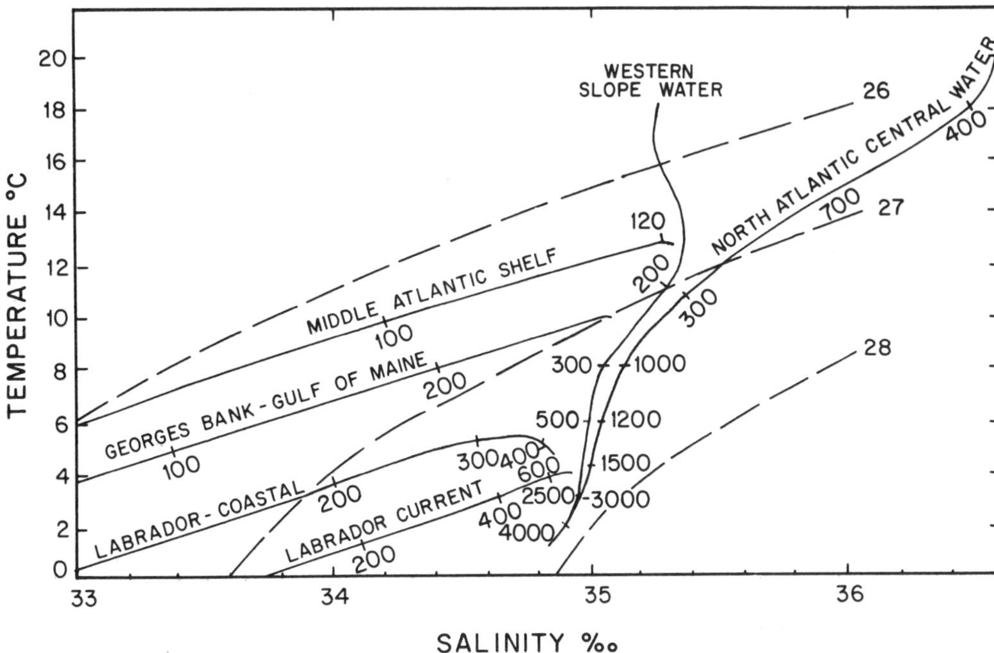

Figure 3.4 T/S relationships in the Western Slope Water and associated water masses in winter. Numbers indicate approximate depth in meters. From Wright, 1976a.

is a region of strongly sloping gradients, Slope Water of any T/S value occurs several hundred meters shallower than the corresponding value in the Gulf Stream.

Another difference in the Slope Water is that upper thermocline water and the warm water layer appear only during summer months. This is because the depth of seasonal warming coincides roughly with the top of the mid-thermocline, 200 m. Slope water temperatures at 200 m range from 10-11 °C at the shelf edge to 15 °C at the inshore edge of the Gulf Stream, with a mean around 12°C. In the summer, the T/S curves for the two regions diverge above 15°C, with the Slope Water rarely exceeding 23°C and 36.0‰, while the Gulf Stream reaches 28 °C and 36.9‰ (see Fig. 3.4).

Upper Layers

Sea surface temperatures show much less seasonal variability southwest of Cape Hatteras than in the Slope Water (Fig. 3.7). These data, however, are long-term averages, an oversimplification which tends to obliterate the convoluted fronts and sharp gradients which may exist at any time. The complexity of the actual sea surface temperature structure in the Slope Water is now being revealed by high resolution satellite imagery (Fig. 3.8). Satellite images are received at least daily, and resulting sea surface temperature charts are issued by NOAA three times weekly for the Slope Water region and twice weekly for the region south of Cape Hatteras. For several years the Atlantic Environmental Group (AEG) of Northeast Fisheries Center (NMFS/NOAA) has been describing the changes systematically. Analysis can allow a measure of the distance from the 200-m curve to the front (e.g., Fig. 3.9). Data also can be presented as "biographies" of anticyclonic warm-core Gulf Stream rings, including plots of ring position as a function of time (Fig. 3.10). Finally, monthly averages and anomalies for long-term sea surface temperatures can show average trends (Fig. 3.11).

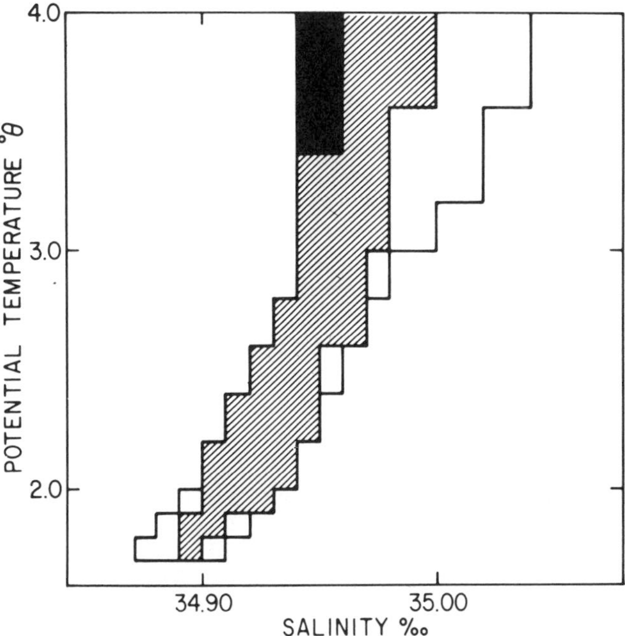

Figure 3.5 Potential temperature/salinity envelopes for the Western Slope Water and the Gulf Stream below 4°θ
Hatched area indicates overlap; black is Slope Water only and white is Gulf Stream only. After Worthington (1976).

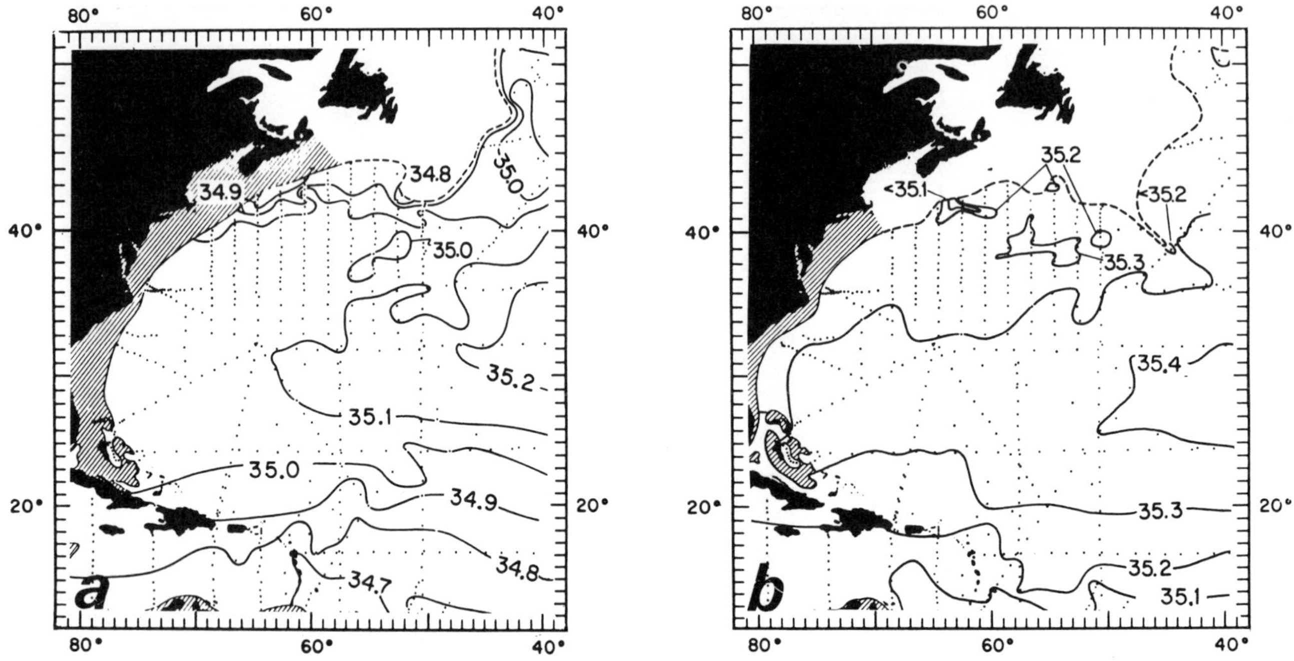

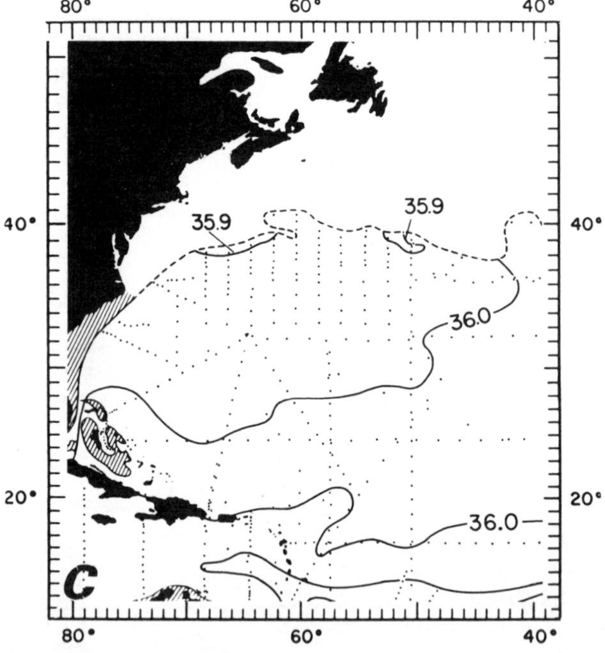

Figure 3.6 Salinities (‰) at the (a) 6°C, (b) 10°C and (c) 15°C surfaces in the northwest Atlantic. After Worthington (1976).

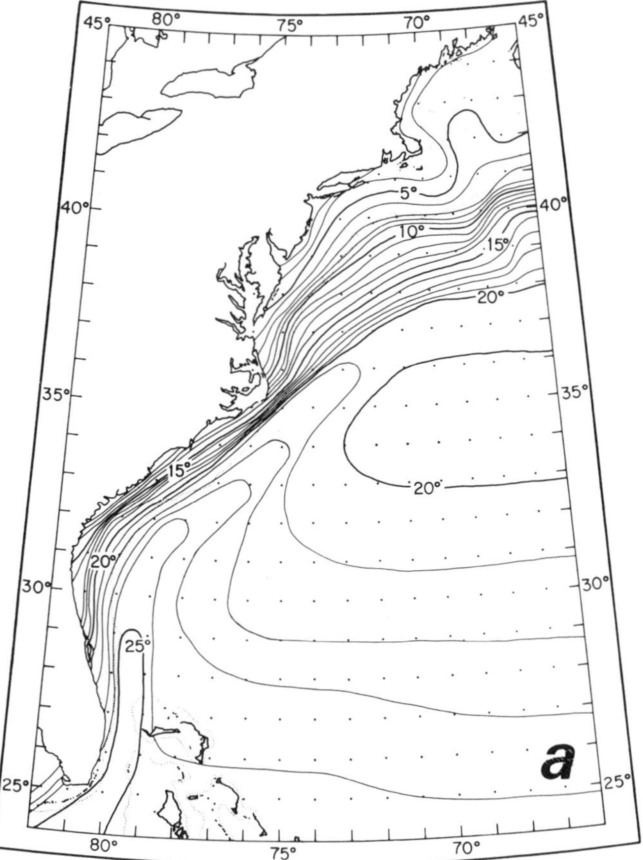

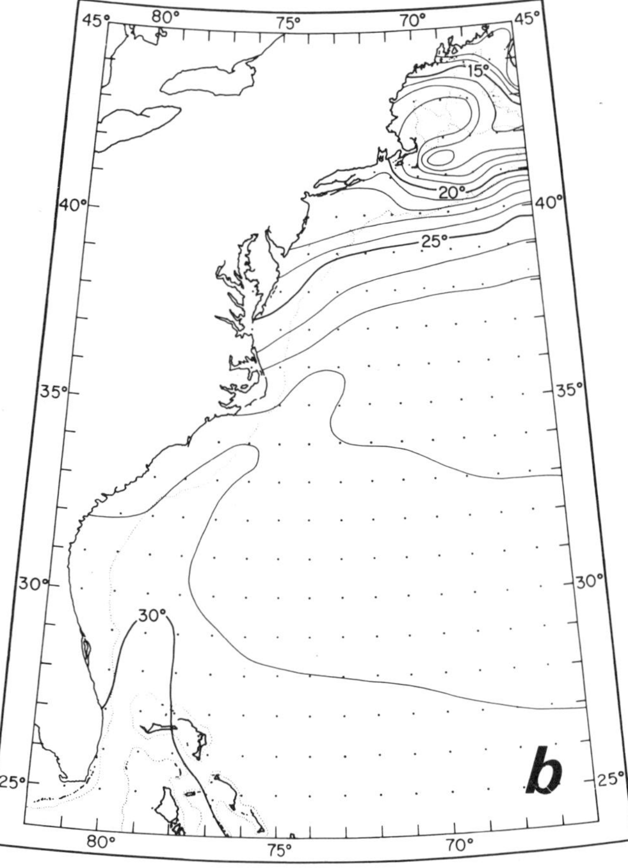

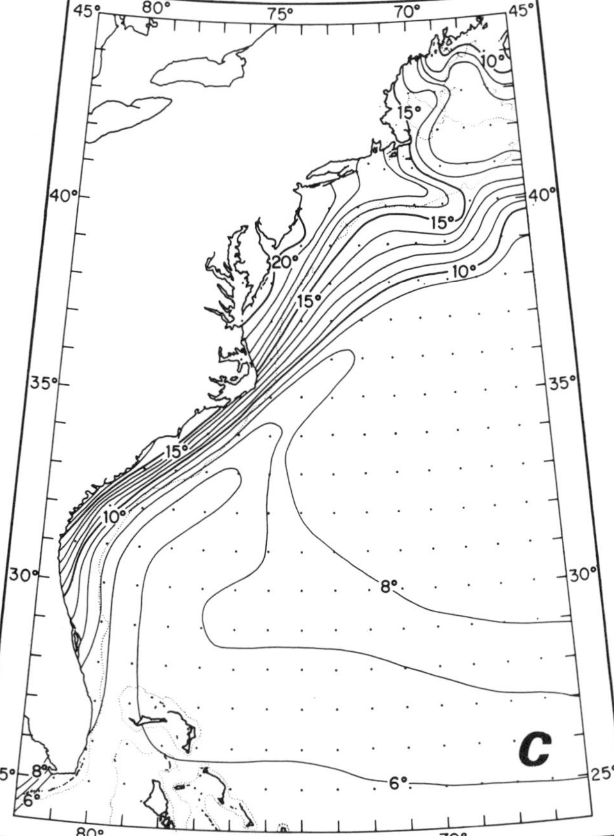

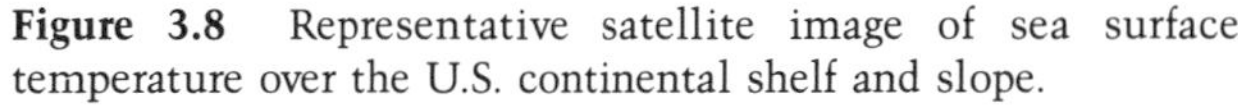
Figure 3.8 Representative satellite image of sea surface temperature over the U.S. continental shelf and slope.

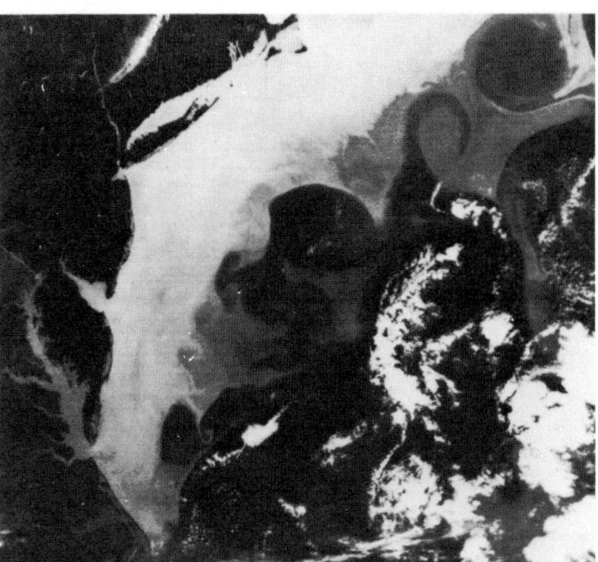

Figure 3.7 Average sea surface temperature (°C) in the western North Atlantic in (a) March, (b) August, and (c) mean range. After Schroeder (1966).

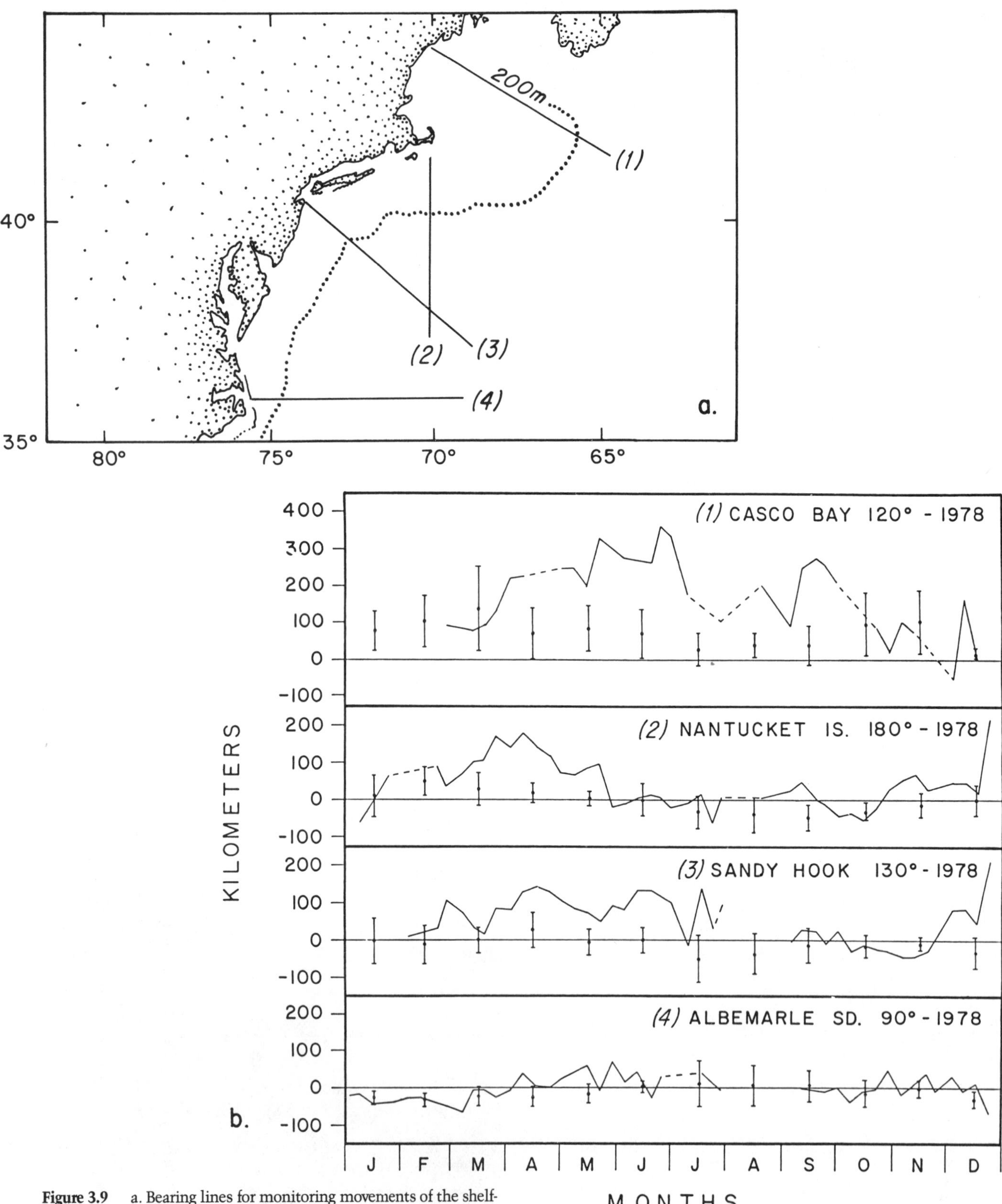

Figure 3.9 a. Bearing lines for monitoring movements of the shelf-slope front. After Hilland, 1983. b. Shelf-slope front positions in 1980 relative to the 200-m isobath along selected bearing lines. Seaward is positive. Dots and vertical lines represent monthly means and two standard deviations from a four-year base period. From Hilland, 1983.

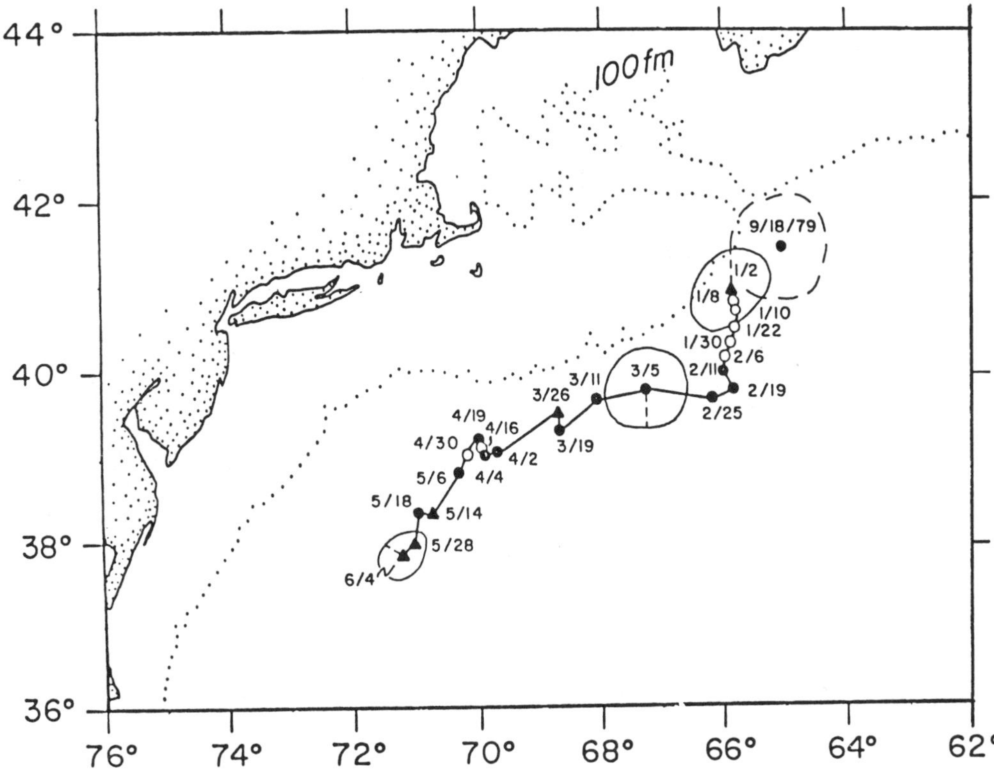

Figure 3.10 Trackline for Warm Core Ring 79-G. After Fitzgerald and Chamberlin, 1983.

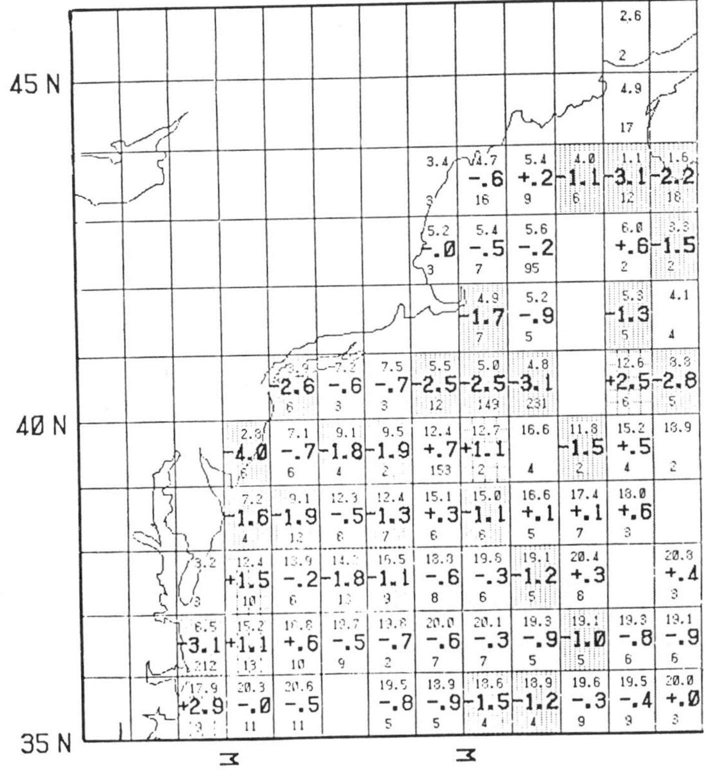

Figure 3.11 Sea surface temperature anomaly (°C) for January 1981. After Ingham and McLain, 1983.

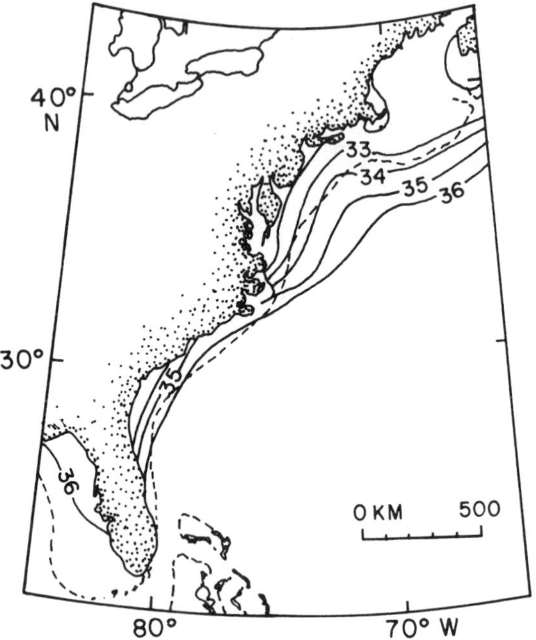

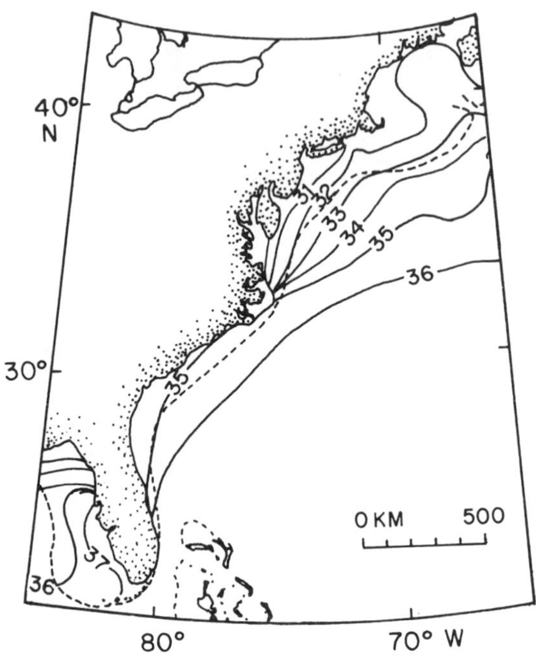

Figure 3.12 Sea surface salinity (‰) in (a) February, and (b) August. After Emery and Uchupi, 1972.

Sea surface salinity in the region is probably as variable as sea surface temperature although, since salinity is measured only by research vessels, detail cannot be plotted. Sea surface salinity charts for February and August (Fig. 3.12) show little seasonal variation except for somewhat sharper gradients in winter. Again, the averaging process wipes out the sharp gradients which occur in individual sections across the shelf and slope.

Southwest of Cape Hatteras

In surface waters the principal water mass is WNAW, and most of it is the 18° Water of Schroeder et al. (1959), Worthington (1959) and Istoshin (1961): a water mass or water type close to 18°C, 36.5‰. It occupies a layer 300 to 400 m thick centered at about 300 m depth. This water forms in the northern Sargasso Sea (west of 45°W longitude and from 33°N latitude to the Gulf Stream, according to Worthington, 1976) in the winter, when cold northwesterlies off the North American continent chill and stir the surface waters. It is carried east and then south and west in the Gulf Stream recirculation (Talley and Raymer, 1982; Worthington, 1976) to form the distribution seen in the figure. The Gulf Stream also carries water of 18°C, of course, but it does not form an anomalously thick layer as it does in the Sargasso Sea.

Below the 18° Water the T/S curve closely follows that of WNAW to about 12°, roughly at 500 m over the Blake Plateau. Between 12° and 6° most of the water in the region is also WNAW, but Atkinson (1983) reports significant amounts of Antarctic Intermediate Water (AAIW) inshore of the 1000-m curve. AAIW forms in the southern South Atlantic Ocean and flows northward in the North Atlantic into the Caribbean Sea and Gulf of Mexico, and thence through the Florida Straits north toward Cape Hatteras. It can be identified by fresh salinity anomalies as great as −0.25‰ from the WNAW curve (Fig. 3.13). Positive nutrient anomalies (up to 7 μm in nitrate) and negative dissolved oxygen anomalies (up to −1.0 ml l^{-1}) are also characteristic of AAIW in this region.

Atkinson surveyed the South Atlantic Bight from 27°N to 33°N out to 1000 m depth in September, 1980. He found salinity anomalies from −0.05 to −0.15‰ throughout the area, with a clearly defined tongue running north from the Florida Straits to at least 31°N. The depth of the maximum anomaly (Fig. 3.13) sloped sharply upward inshore indicating the presence of the Gulf Stream. Apparently the AAIW core was diluted by mixing as it moved north so that the signature had disappeared off Onslow Bay.

Slope Water

The existence of two Slope Water masses was suggested by Fuglister (1963), who noted that the fresh salinity anomaly in the northern part of the region increased to the east. Wright (1976b) identified the warmer, more saline mass as Western Slope Water because it occurs principally west of 66°W; the cooler and fresher mass is called Eastern or Canadian Slope Water. Gatien (1976) noted that the boundary between the two lies more nearly east-west than north-south; she renamed them Warm Slope Water and Labrador Slope Water. Wright's terminology is used here, in a summary of his 1976 description.

Eastern Slope Water occurs mostly east of 66°W longitude, but small amounts can occur in the vicinity of eastern

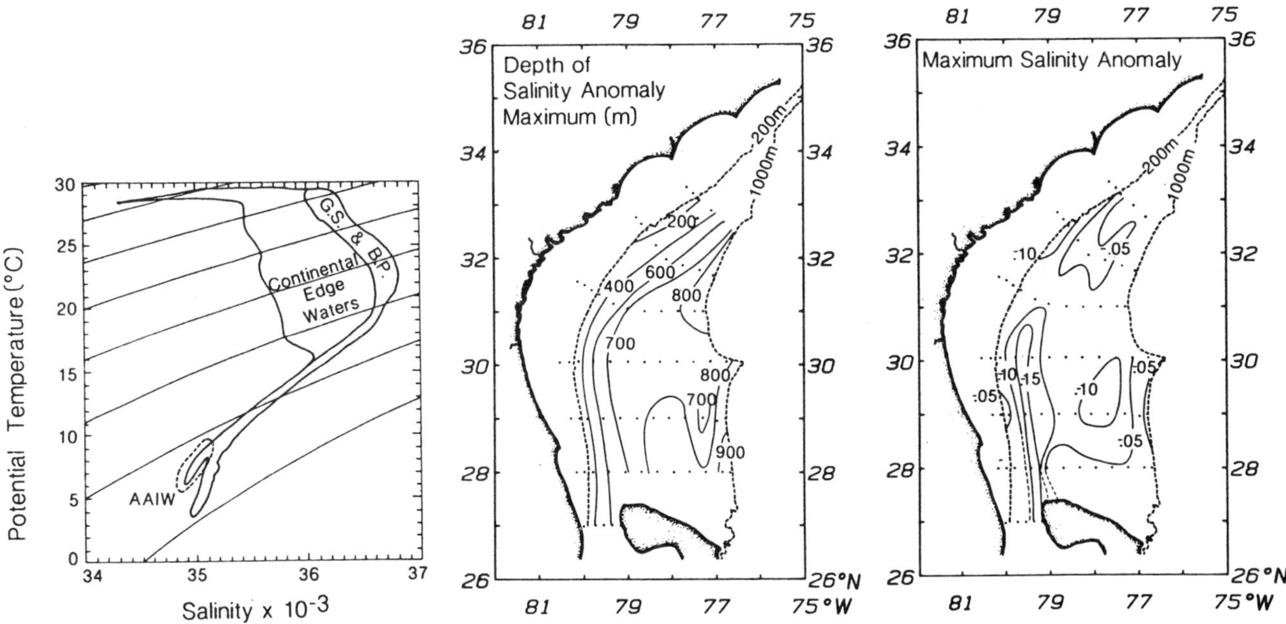

Figure 3.13 Antarctic Intermediate Water (AAIW) in the South Atlantic Bight. Top: potential temperature/salinity envelope of all data. Center: depth of salinity anomaly maximum (m). Lower right: maximum salinity anomaly (‰). From Atkinson, 1983.

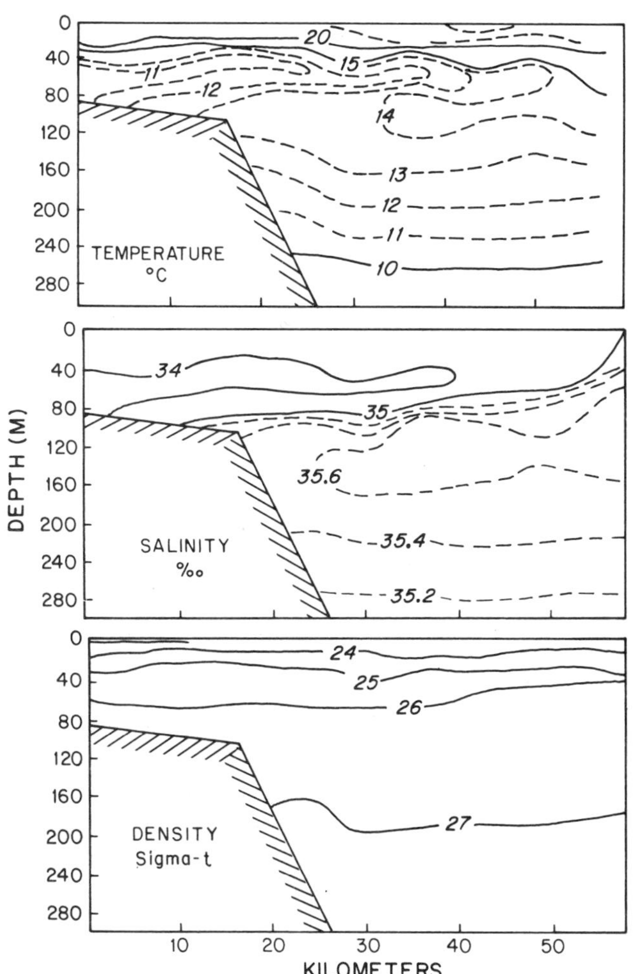

Figure 3.14 Temperature (°C), salinity (‰) and density (sigma-t) sections across the shelf edge south of New England in summer.

Georges Bank. Documentation exists for at least three instances of large-scale, mid-depth influxes into the Western Slope Water area. Such incursions have a marked effect on the organisms in the region but do not appear to cause any lasting changes in the hydrography. The disappearance of the tilefish industry in 1882 was attributed to a massive influx of water from 0° to 4°θ of Labrador Current origin (Hachey, 1955). Worthington (1964) described a similar incursion in 1959, when the Slope Water was colder at all depths, by 3° to 10°C, than in either 1950 or 1960. By a fortunate coincidence deep current measurements were made in the Slope Water in 1959 and 1960 (Volkmann, 1962); a westward flow of 60 sv was measured in 1959 while a year later the flow was only 17 sv. Worthington estimated it would take about eight months to replace half the Slope Water at the higher rate. The third (possible) incursion occurred in 1965, when both temperature and salinity were considerably lower than normal in the Slope Water (Colton, 1968). As air temperatures were not unusually cold, the water must have been advected from the east.

The first detailed survey of the Western Slope Water was made in the fall of 1981 by the Soviet R/V *Stvor* to provide background information for the multilaboratory Warm Core Ring Investigation of 1981 to 1983. On two cruises more than 100 hydrographic stations were made to 1000 m, along with a variety of plankton tows. These data have not been worked up but are available at the Northeast Fisheries Center in Woods Hole, Mass.

In the winter, storms and cold temperatures create a well-mixed layer down to about 100 or 150 m. Summer warming creates a seasonal thermocline overlain by a surface layer of low-density water. In combination with reduced storm activity in summer, this structure inhibits vertical mixing and thus reduces the upward transfer of nutrients into the photic zone. Below the thermocline is a layer of nearly uniform temperature; this represents the remnant of the winter-cooled water and has been named the Upper Slope Water Thermostad (Wright and Parker, 1976).

Along the northern edge of the Slope Water, at the shelf break, there are several notable features, which can be seen in Figure 3.14.

Shelf Water Bulge. The boundary that separates the warmer, saltier Slope Water from the colder, less saline Shelf Water is not vertical; most of the time there is a wedge of Shelf Water extending from about the 100 m curve to the sea surface some 30 to 80 km seaward — well beyond the 200-m curve (Wright, 1976). The position of the front is highly variable, especially at the sea surface. Although the front is convoluted, two basic configurations have been identified: a simple shelf water lens and the multi-layered structure in late spring and early summer. Posmentier and Houghton (1981) show that such layering results from a reversal of the offshore density gradient associated with seasonal warming; the convolutions result in shoreward transfer of salt to help balance the offshore movement of fresher coastal water (Gordon and Aikman, 1981). An intensive study of processes across the shelf break and down the continental slope, the Shelf Edge Exchange Project (SEEP), has been started by G. Csanady at Woods Hole Oceanographic Institution and J. Walsh of University of South Florida.

Cold Water Pool. Part of the Shelf Water bulge is water that has been cooled by winter convection and storms so that it is colder than the underlying Slope Water. There is no density inversion because the Shelf Water salinity is much lower than that of the Slope Water. In summer this cold water is overlain by the seasonal thermocline so that it appears as a tongue at depths of 50 to 80 m, extending well seaward of the shelf break (Houghton et al., 1983). The temperature in the tongue can be as low as 6 °C and the salinity as low as 33‰. The water warms slowly during the summer but usually retains its identity until renewed by winter cooling (Ketchum and Corwin, 1964; Wright, 1976).

Warm Band. Because of the Shelf Water bulge there is a temperature maximum zone some 40 to 80 m thick centered at about 120 m depth and usually associated with a salinity maximum of about 35.5‰. This feature, part of the Upper Slope Water Thermostad (Wright and Parker, 1976), creates a band of warmer water on the sea floor at the shelf break, usually in the range of 9° to 12 °C. Below the warm band the temperature decreases steadily and seasonal influences are absent.

Detached Parcels. From time to time, parcels of Shelf Water become detached from the parent water mass and move off into the Slope Water. Ford et al. (1952) and Ford and Miller (1952) reported such parcels along the inshore edge of the Gulf Stream; they have since been observed all along the length of the shelf edge and have been identified throughout the western Slope Water. The parcels seem to occur as thin lenses of anomalously fresh and cold water, ranging from 20 to 80 m thick and from 10 to 20 km in horizontal extent. They have not been studied in detail and little is known of their abundance, persistence, and importance to the hydrography of the Slope Water.

The Shelf Water parcels on the inshore edge of the Gulf Stream seem to occur as long strips, some with a surface expression and some without. The actual entrainment process off Cape Hatteras was observed in May, 1969 (Fisher, 1972), when a parcel of Shelf Water, originally about 14 km wide, elongated and narrowed to about 2 km as it was carried northeast with the Gulf Stream. The surface water in the entrainment region dropped below 13 °C as compared with 15 °C in the Slope Water and 24 °C in the Gulf Stream; at 100 m the entrained water was colder than 8 °C.

Shelf Water Filaments. Associated with warm core rings (see below) and often clearly visible in satellite imagery (Fig. 3.8) are parcels of Shelf Water which are entrained by the swift currents of the rings and are carried a considerable distance offshore. Depending upon the frequency of ring occurrence along the shelf edge, such filaments could account for removal of up to one-third of the Shelf Water volume east of Cape Hatteras (R. Schlitz, 1983, personal communication). One 1977 ring which was studied in detail (EG&G, 1978) drew off about 12 per cent of the water on Georges Bank. During the Warm Core Ring (WCR) project of 1981-1983, described later, physical oceanographers at the Northeast Fisheries Center (Woods Hole) concentrated on the interactions of WCRs with Shelf Water; the results of their work should add considerably to our understanding of these phenomena.

Currents

Strong Currents of the Wester North Atlantic

The ACSAR region is located in that part of the world's oceans where one finds the strongest low frequency horizontal flows: namely the Gulf Stream System including the Western Boundary Undercurrent — WBUC. These strong currents and their associated eddy fields are a prominent source of drag on structures moored on the slope and rise.

Regional Setting

The western margins of most ocean basins are the site of intense boundary currents (Stommel, 1965). Worthington's (1976) version of the North Atlantic circulation (Fig. 3.15) shows the Florida Current directly over the Atlantic margin south of Cape Hatteras (see Fofonoff, 1980, for a recent review of the nomenclature for the Gulf Stream System). North Atlantic Deep Water flows equatorward in the WBUC near the 4000 m depth contour, impinging on the seaward side of the ACSAR throughout its length. These

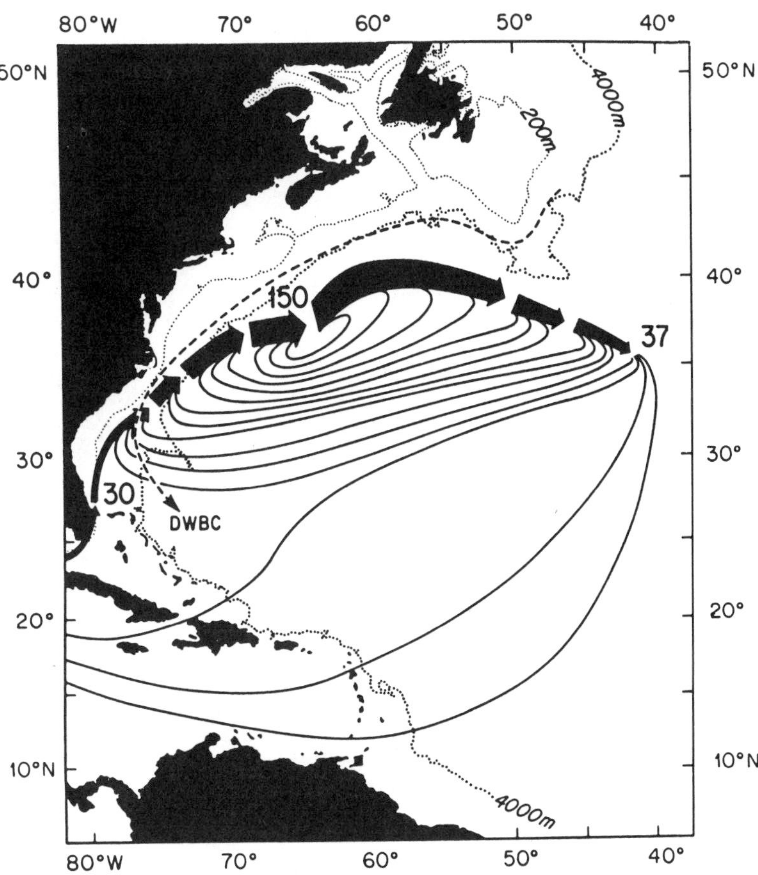

Figure 3.15 Transport streamlines for the North Atlantic circulation, as adapted from Worthington (1976). The dashed line is the WBUC. The contour interval is 10 (x 10^6 m^3 s^{-1}) and the numbers next to the contours are transports at key sections, namely Florida Straits, Nova Scotia, and Grand Banks.

two currents "cross" near Cape Hatteras. North of Cape Hatteras, the ACSAR is under the "indirect influence" of the Gulf Stream System, as noted below.

The current structure and water mass properties of the Gulf Stream vary geographically and temporally, with a variety of time-dependent changes (such as mesoscale, seasonal, and interannual) found whenever an appropriate investigation has sought such variability. The path of the Gulf Stream meanders with a broad spectrum of variability, particularly downstream of Cape Hatteras. It is now clear that the most intense eddies in the ocean are typically found in regions of strong currents (Schmitz et al., 1983). The strong current regimes also may be the source of the eddies. The strongest eddies occur near intense currents along continental east coasts and along the equator, coincident with maximum kinetic energy (K_E; Fig. 3.16a). Richardson (1983a) has found much larger K_E for the Gulf Stream System (Fig. 3.16b); the difference between the results in these figures reflects the state of the art in research on time-dependent surface currents.

The nature of the fluctuations giving rise to the high K_E values is not completely known. Meanders (lateral motion) of the Gulf Stream are important but not necessarily the dominant contributor; rings are the most clear-cut eddies of all. Generally speaking, the existence of high energy events for any portion of the ACSAR south of Cape Hatteras can be determined by the question "Is the Gulf Stream there or not?" The presence of eddies in combination with the Stream would create the most energetic situation of all. If the WBUC is present, strong bottom-intensified currents may be expected.

Meanders and eddies are associated with the Gulf Stream both along the coast and in the open ocean (Fig. 3.17). In the Florida Straits, lateral displacements of the surface front associated with the Gulf Stream are 10-50 km. As the shelf begins to widen on the Blake Plateau north of the straits, eddy dimensions are perhaps 100-200 km, and meanders of 30 km amplitude are observed (Lee et al., 1981; Bane and Brooks, 1979). North of the "Charleston Bump" (a topographic high extending seaward near 32°N), eddy dimensions can reach 300 km and meander amplitude 200 km (Legeckis, 1979; Bane and Brooks, 1979; Brooks and Bane, 1981; Bane et al., 1981). Cyclonic eddies have been observed embedded in the Gulf Stream front in the Florida Straits region (Lee, 1975; Lee and Mayer, 1977) and along the Florida/Georgia outer shelf (Lee et al., 1981). These features, first described as "shingles" by von Arx et al. (1955), occur on the shoreward side of meanders.

Temperature and current time series in early 1980 (Lee and Atkinson, 1983) indicate that such cyclonic eddies propagate northward at speeds of 50 to 70 cm sec^{-1}. They produce coherent fluctuations of cross- and along-shelf velocity components in the 5 to 9 day band with along-shelf coherence scale around 100 km. Lee and Atkinson note that frontal eddies are short-lived phenomena, the entire cycle of formation and dissipation taking place in 1 to 3 weeks. They

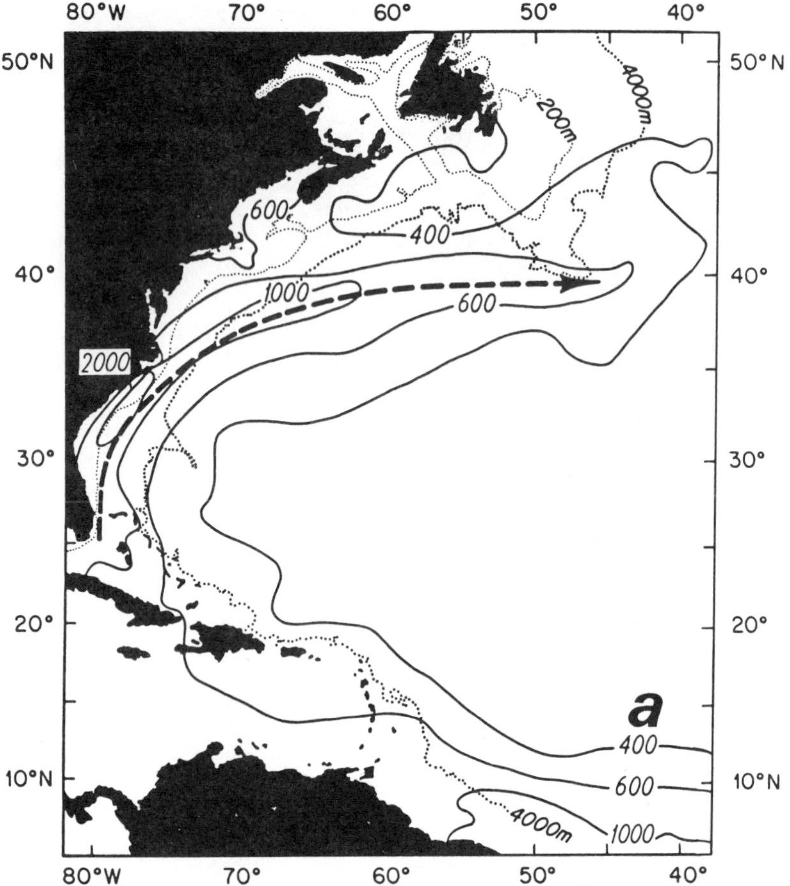

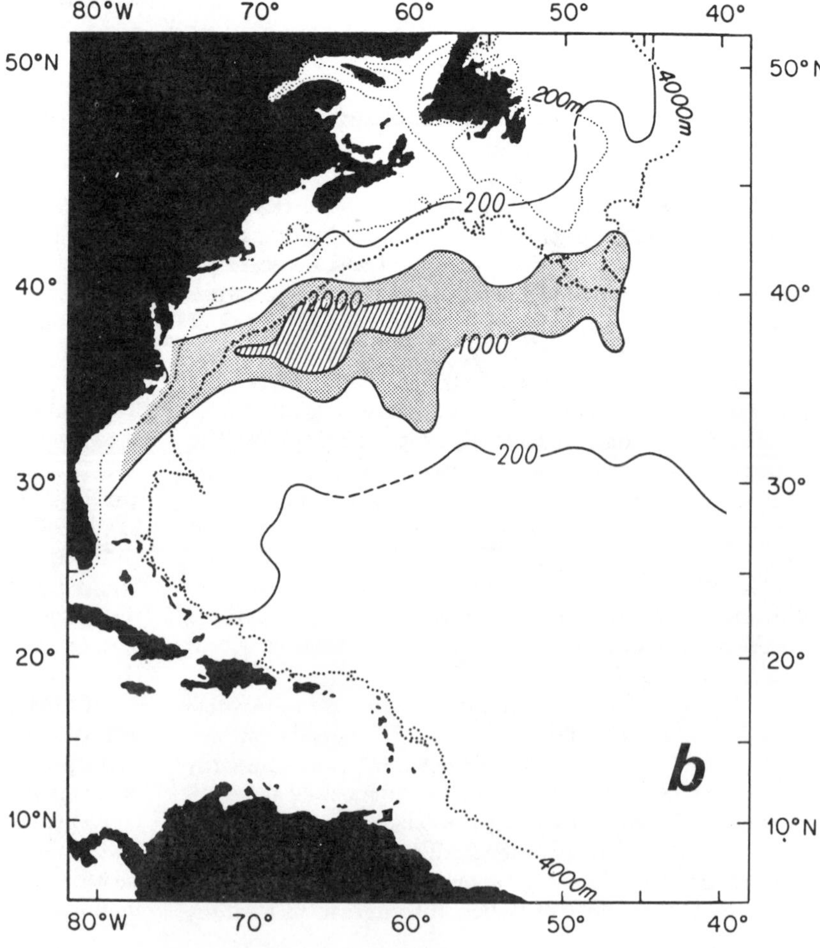

Figure 3.16 (a) Eddy kinetic energy (K_E, cm^2 s^{-2}) at the sea surface for the North Atlantic Ocean on a 1° grid. The dashed line is the locus of maximum kinetic energy of the mean flow, roughly the "axis" of the Gulf Stream. After Wyrtki *et al.* (1976). (b) Surface K_E for the North Atlantic Ocean on a 2° grid based on drogued-drifter data. After Richardson (1982).

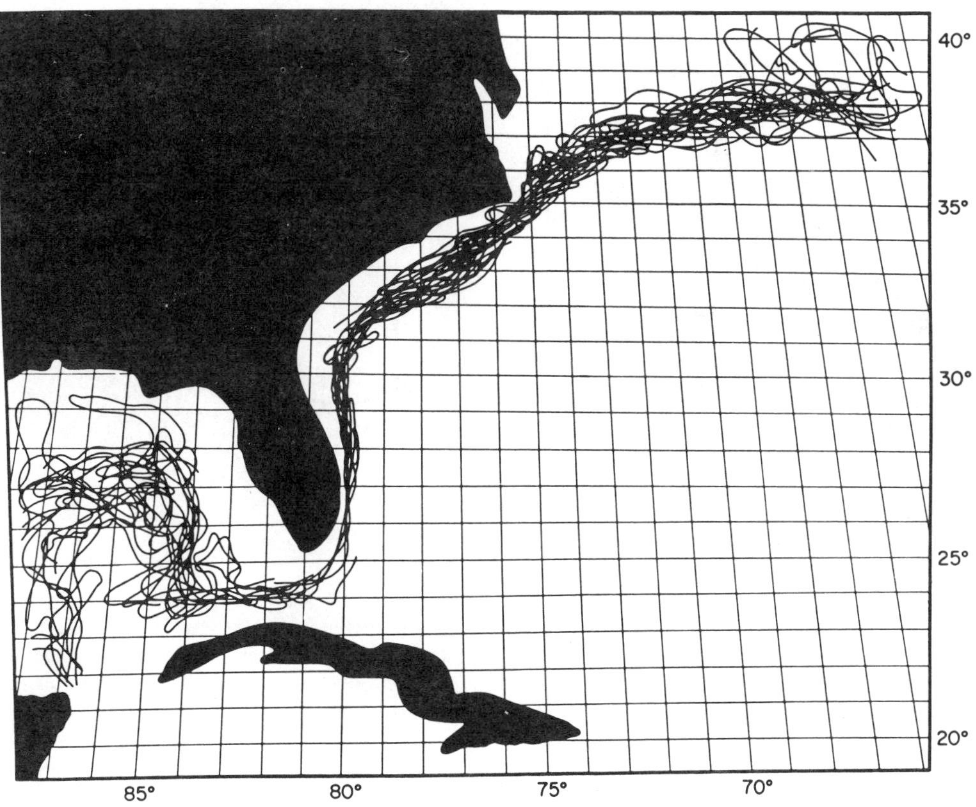

Figure 3.17 A composite of Gulf Stream surface frontal positions as determined from satellite data by Legeckis (1979). Adapted from Fofonoff (1980).

conclude that such eddies are the principal cause of low-frequency temperature and current variability at the shelf break in the South Atlantic Bight.

As the Stream flows into deeper water downstream of Cape Hatteras, the envelope and amplitude of the meanders broaden greatly. In the region 100 to 200 km beyond Cape Hatteras, where the water depth is approximately 3000 m, the rms meander amplitude increases from 15 to 30 km (Watts and Johns, 1982). Further downstream, Halliwell and Mooers (1979) have analyzed satellite imagery to find that the rms amplitude continues to grow to around 80 km near 65° West and the New England Seamount Chain. Pinched-off Gulf Stream meanders, typically occurring east of 70° West, are called rings (see below). Rings which form north of the Gulf Stream usually contain a core of warm Sargasso Sea water and are known as warm core rings (WCRs); they rotate in the clockwise or anti-cyclonic direction. They are found only in the Slope Water region northeast of Cape Hatteras and provide the strongest currents found there. Rings which form south of the Gulf Stream usually contain cool Slope Water in the core and rotate in a counter-clockwise or cyclonic direction; they are known as cold core or cyclonic rings. They are found in the ACSAR only south of Cape Hatteras (see Fig. 3.1). A cold core ring which coalesces with the Gulf Stream (Richardson, 1983b) may provide some of the strongest currents found on the Stream's offshore edge (near the 4000-m depth contour).

The "Mean" and/or Permanent Gulf Stream System

As the Gulf Stream emerges from the Florida Straits, it is roughly 100 km wide, with its axis located over water depths of 500-800 m. The width of the synoptic stream remains about 100 km (although about 300 km is possible) until the vicinity of the Grand Banks where a complex branching/decaying regime becomes dominant.

The "mean" Gulf Stream path south of Cape Hatteras has been described by Olson et al. (1983) on the basis of five years (1976-80) of satellite and XBT data. As shown in Figure 3.18 (upper) the variability in position is minimal south of 30°N but fluctuations increase considerably toward Cape Fear and then decrease somewhat toward Cape Hatteras. Almost all of the variability (at all latitudes) is accounted for by frequencies of 3 weeks and lower, with about half at frequencies greater than three months (Fig. 3.18 lower).

Passing over the Blake Plateau, the Gulf Stream runs close to shore in water depths less than 1000 m; the coldest water it carries is about 7°C. Off Onslow Bay (near 34°N) the Blake Plateau ends and the continental slope descends steeply to depths of 4000 m. Downstream of Cape Hatteras, the mean water depth below the Gulf Stream is more than 4000 m and bottom temperatures are about 2°C. In this region, the Stream is an eastward-flowing current flanked on either side by regions of westward flow. To the north is the

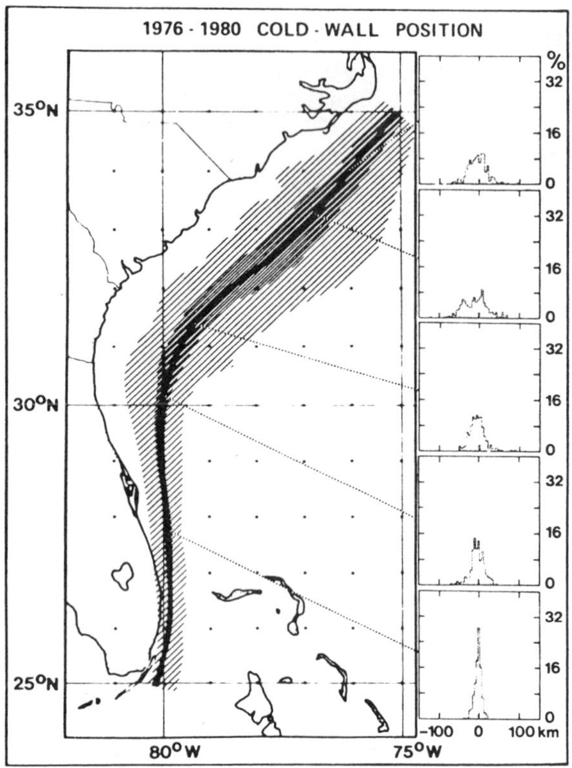

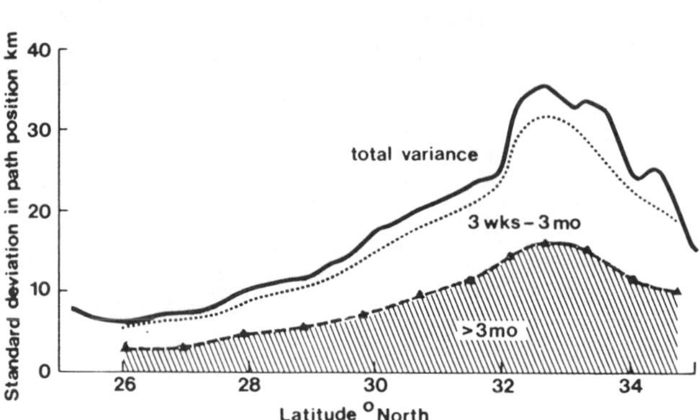

Figure 3.18 Left: mean, standard deviation and total range of Gulf Stream paths between 25° and 35° N. On the right are histograms of cross-stream frontal position for selected locations. Right Top: the distribution of the standard deviation in cross-stream frontal displacement as a function of latitude. A breakdown of the variations as a function of frequency is included. From Olson et al., 1983.

Slope Water, where westward flow has been measured by Webster (1969), Schmitz (1974, 1977) and Luyten (1977). To the south is the westward recirculation described by Worthington (1976) and Schmitz (1977, 1978, 1980, 1984).

The volume transport of the Gulf Stream was estimated to be in the range of 100-150 x 10^6 m^3 sec^{-1} south of Cape Cod (Fuglister, 1963; Warren and Volkmann, 1968; Knauss, 1969). Barrett and Schmitz (1971) found 130-200 x 10^6 m^3 sec^{-1} by direct measurement with transport floats or dropsondes. Schmitz (1977) and Stommel et al. (1978) conclude that most of the transport in the Gulf Stream is recirculated in the western North Atlantic, excepting the 30 x 10^6 m^3 sec^{-1} carried through the Straits of Florida. The volume transport of the Gulf Stream increases significantly downstream. Knauss (1969) found an increase of 7% per 100 km, roughly doubling the transport from 30 x 10^6 m^3 sec^{-1} off Miami to 60 x 10^6 m^3 sec^{-1} off Cape Hatteras and again to about 120 x 10^6 m^3 sec^{-1} near 65°W. The Gulf Stream System must somehow entrain colder heavier water after it moves off the Blake Plateau at Cape Hatteras and down the continental slope into the deep Western North Atlantic.

A number of observers have presented evidence indicating that beyond Cape Hatteras the influence of the Gulf Stream extends to the bottom (Fuglister, 1963; Warren and Volkmann, 1968; Knauss, 1969; Schmitz et al., 1970). A vertically coherent Gulf Stream, however, was not substantiated by subsequent direct current measurements. These showed a vigorous velocity field in the deep water (Schmitz et al., 1970) that is not a simple downward extension of the near-surface flow.

The mechanism of separation of the Gulf Stream from the continental slope at Cape Hatteras remains ambiguous. The early theories of mean ocean circulation (see Stommel, 1965) required an intensifying current along the western boundary only to the latitude of maximum wind-stress curl. Poleward of the maximum, the current weakened and returned into the ocean interior as a broad slow flow specified by the meridional scale of the wind-stress curl field. The inertial models that were developed subsequently indicated that an intensifying current with westward flow from the interior could be extended well past the latitude of maximum curl of the wind stress by inertial recirculation. In two-layer inertial models, the northward extent is limited by surfacing of the inshore isopycnal.

As it moves eastward the Stream encounters the New England Seamount chain, near 60°-65°W, downstream of which the upper-level current system appears to weaken and become more variable (Richardson, 1981). West of the Grand Banks, a strongly baroclinic frontal structure generally extends to the ocean bottom, no matter if depth is 500 or 5000 m, having a width scale of the internal Rossby radius of deformation characterizing the main thermocline. The structure of the currents associated with the Gulf Stream is thus best determined in the Straits of Florida and on the Blake Plateau. Richardson et al. (1969) have published the velocity structure across several sections along the continental margin from the Florida Straits to just south of Cape Hatteras, using transport floats.

Observations Near Cape Hatteras

Cape Hatteras is the key location in the Gulf Stream System: the point of separation of the western boundary current from the coast and a region where a strong relatively shallow current (depths of 1000 m and less) initially encounters the abyssal ocean. Here the Gulf Stream flows from the Blake Plateau over the continental slope and rise into the deep North Atlantic, passing over the southward-flowing WBUC. The simplest picture is that the WBUC flows obliquely under the Stream, as suggested by Barrett (1965). There is evidence, however, that at times the Gulf Stream extends to the sea floor and splits the WBUC (Richardson and Knauss, 1971). Key measurements near Cape Hatteras have been made by Richardson and Knauss (1971), Richardson (1977), and Watts and Johns (1982).

During July 1967 three measurements of the Gulf Stream volume transport were made off Cape Hatteras, using "transport floats", free falling instruments which measure the vertically averaged horizontal velocity and depth (Richardson and Knauss, 1971). The transport values were 58, 67 and 64 x 10^6 m^3 sec^{-1}, for an average of 63 x 10^6 m^3 sec^{-1}. These measurements indicated that the Gulf Stream extends to the bottom underneath and offshore of the high speed surface layer, with deep southward flow on both sides. A hydrographic section across the Stream indicated that the southward flowing water is the Western Boundary Undercurrent.

The crossover between the Gulf Stream and WBUC off Cape Hatteras was examined again for May through July of 1971 (Richardson, 1977). Current meter, hydrographic station, XBT, GEK and ship drift observations were made along a baseline normal to the Gulf Stream axis. Strong southwestward velocities were recorded under the Stream. One current meter 100 m above the bottom at a depth of 2575 m recorded instantaneous velocities as high as 47 cm sec^{-1} and a mean velocity of 10.8 cm sec^{-1} to the southwest over 54 days. The southwestward flow extended from the continental slope out under the Stream, to a depth of at least 4100 m, with an estimated transport of 24 x 10^6 m^3 sec^{-1}.

The temperature, salinity, and silicate profiles from the first section showed large horizontal gradients through the Gulf Stream and a Gulf Stream ring. At this and another section, southwestward flow was observed, with maximum speeds up to 24 cm sec^{-1}. This flow, coincident with water traceable to the Labrador and Norwegian seas, indicates that the Gulf Stream did not extend to the sea floor at this location and time.

The first long-term current meter array moored near the Gulf Stream in the vicinity of Cape Hatteras was described by Watts and Johns (1982). Their values (Watts, personal communication) of abyssal K_E ($\sim$15 cm^2 s^{-2}) are probably typical of abyssal depths near the mid-latitude jet near the western edge of the North Atlantic subtropical gyre.

Blake Plateau Observations

Fofonoff (1980) considered the relationship between the relatively well-defined Florida Current over the Blake Plateau and larger Gulf Stream beyond. The first detailed measurements of meanders over the Blake Plateau were described and discussed by Von Arx et al. (1955) and Webster (1961a, 1961b). Richardson et al. (1969) described a series of dropsonde transects made across the Florida Current in the Straits and over the Blake Plateau in the 1960s. Legeckis (1979) and Bane and Brooks (1979) have examined meanders with satellite sea surface temperature imagery (see also Olson et al., 1983). Recent moored instrument measurements have been made by Brooks and Bane (1981) and Lee and Waddell (1983).

Brooks and Bane (1981, 1983) and Bane et al. (1981) used current meter data from the continental slope along with satellite infrared images and XBT temperature profiles to describe Gulf Stream meanders off Onslow Bay. A significant part of the variability observed occurs within a period range of 2 to 14 days. Meanders were observed downstream of the area off Charleston where a seaward deflection of the Stream is often found (see also Bane, 1983). The degree of deflection was greatest near the surface; almost no deflection existed within the deeper portion of the water column. Filaments of warm Gulf Stream water extended southwestward "behind" the crests of the meanders. The filaments had relatively shallow thermal structure, extending from the surface to a depth of less than 100 m. They were oriented approximately parallel to the bottom contours over the outer shelf and upper slope, and were separated from the main body of the Gulf Stream by cool water. The presence of the cool surface water between the Stream and the filaments was due to upwelling of water from deep within or below the main Stream. The data provide simultaneous space and time descriptions of the large-amplitude meanders which propagate along the continental margin in the Gulf Stream between Charleston and Cape Hatteras. The meanders appear to be generated upstream of ACSAR and then "amplified" by the deflection process which occurs off Charleston. Energy flux calculations for the region off Onslow Bay indicate that meander kinetic energy is converted to mean kinetic energy in that area (Webster, 1961b; Brooks and Bane, 1981). It seems likely that the deflection produces meander growth within the 100 km or so immediately downstream of Charleston, with the flow subsequently becoming stabilized farther downstream. These results are consistent with Webster's (1961a, b), who found the Gulf Stream surface front and current axis to be meandering in a skewed fashion off Onslow Bay on a time scale of 4 to 7 days with estimated wave lengths of about 100 km.

Singer et al. (1983) examined all available hydrographic data for the Charleston Bump region and identified a persistent area of doming isotherms, indicating upwelling, associated with the offshore deflection of the Gulf Stream there. The greatest evidence of doming was off Long Bay, suggesting cyclonic circulation inshore of the Stream in that area most of the year. They also found seasonal fluctuations in the depth of the main thermocline, with colder water (and higher nutrient levels) at shallower depths in winter and late summer.

Current and temperature variability over the Blake Plateau along 30°N (off Jacksonville) was measured by Lee and Waddell (1983) with a cross-stream array (Fig. 3.19) of five subsurface moorings extending from the shelf edge to the eastern boundary of the Plateau. The results showed energetic current and temperature fluctuations within a period band of 2 days to 2 weeks. Flow pertur-

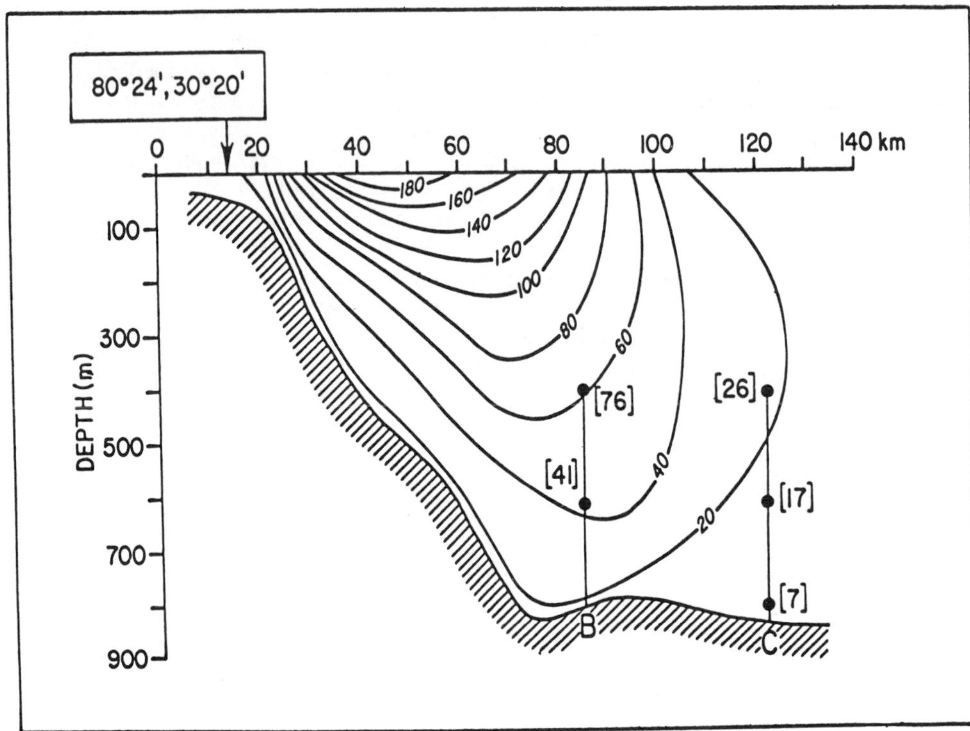

Figure 3.19 Mean velocities along a transect across the Blake Plateau near 30°N. Data were taken between August 2, 1980 and October 6, 1981. From Lee and Waddell (1983).

bations on the cyclonic side of the Stream generally had a cyclonic sense of rotation, whereas on the anticyclonic side the fluctuations were anti-cyclonic and 180° out of phase with those on the cyclonic side. These events appear to account for 45 to 70% of the total observed variability throughout the year. Fluctuations of this type can be visualized as east-west meanders, propagating to the north at speeds of 30 to 70 cm sec^{-1}, with wave lengths of 100 to 200 km and periods of several days. Fluctuations on the anticyclonic side of the Stream appear to derive perturbation kinetic and potential energy from the mean flow, whereas on the cyclonic side fluctuations tend to supply energy to the mean flow. No significant coherence was observed between the low-frequency current fluctuations and local winds. Low-frequency flow variability near the bottom at the shelf edge appears to be related to Gulf Stream meanders. Indications are that cold, cyclonic perturbations occur at times of offshore meanders due to the formation of cyclonic eddies in the frontal region. During the onshore meander stage, the western edge of the Stream is closer to the shelf and northward near-bottom flows occur with increased temperatures. Near-bottom flows on the Blake Plateau east of the Gulf Stream show prolonged southward flow events that last up to 42 days with speeds in excess of 20 cm sec^{-1}. These events were not correlated with flows at any other site and the nature of the generating mechanism is not clear. A possible explanation may be the southward undercurrent that has been observed beneath the Gulf Stream off Cape Hatteras.

Olson et al. (1983) found that the distributions of both surface frontal position and thermocline variance exhibit a similar downstream dependence. In the region just upstream and over the Charleston Bump there is a trend towards larger offshore excursions. The fact that the mean surface front is correlated with a constant water depth over long distances indicates topographic influence on the Stream. The regions of frontal shifts are also places with substantial changes in the bottom topography.

A data set along the Blake Escarpment near 28°N (Perkins and Wimbush, 1976) will soon be available in detail, including interpretation (Lai, 1983). This was the first field program located on the offshore edge of the Florida Current on the Blake Plateau. Perkins and Wimbush (1976) feel that they observed a vestigial Gulf Stream Ring passing through their array.

The Western Boundary Undercurrent (WBUC)

This current, also called the Deep Western Boundary Current (Hogg, 1983), was predicted by Stommel (1965). Using the newly developed neutrally-buoyant float (Swallow, 1955), Swallow and Worthington (1961) first measured the WBUC, off Cape Romain, South Carolina. Southward flows ranged 9 to 18 cm sec^{-1} over a period of a month, and transport was estimated to be 6.7 x 10^6 m^3 sec^{-1}. Subsequent measurements by a number of investigators (Volkmann, 1962; Barrett, 1965; Worthington and Kawai, 1972; Richardson and Knauss, 1971; Amos et al., 1971; Richardson, 1977) indicated transports ranging from 2 to 50 x 10^6 m^3 sec^{-1}, with an average of 16 x 10^6 m^3 sec^{-1}. Equatorward deep currents along the continental rise and slope north of the Gulf Stream have been reported by Webster (1969), Schmitz (1974), Luyten (1977), Schmitz (1977, 1978, 1980), and Richardson et al. (1981).

Observations suggest that deep and abyssal flows proceed equatorward away from poleward sources of sinking water, concentrated along paths near western continental or topographic boundaries. The compensating circulation polewards and upwards presumably occurs as a broad, weak flow in the interior of the ocean basin. The WBUC is thus the boundary current associated with North Atlantic Deep Water. This thin jet, 50 km in width, carries water into subtropical regions which, until recently, were free of anthropogenic tritium. A distinct core of anthropogenic tritium (from nuclear weapons testing in the atmosphere) marks the WBUC at 30°N latitude (Jenkins and Rhines, 1980). The concentrations indicate a roughly 10-fold dilution of source waters in the far north, where extensive distributions of tritium are found.

The observational basis for the WBUC was confirmed by a hydrographic and long-term moored array experiment carried out in 1977-1978 (Mills and Rhines, 1979; Jenkins and Rhines, 1980). Mean flow appears to follow the contours around the Blake-Bahama Outer Ridge, with average velocities ranging from 0.6 to 3.8 cm/sec (Fig. 3.20).

Models

Several early attempts to model the Gulf Stream (see Stommel, 1965, for a summary) were actually most relevant to that part of the western boundary current close to the shelf. These simple but informative models recently have been exploited again by Luyten and Stommel (1983, 1985).

Most modern numerical models are eddy-resolving — that is, they explicitly consider time-dependent fluctuations (at low frequencies). Gyre-scale models that resolve eddies (first developed by Holland and Lin, 1975a, b; Holland, 1978; Holland et al., 1983) have been under development for several years. Although these models are still idealized, they have been used to identify important mechanisms that contribute to the dynamics of the Gulf Stream system.

A regional numerical model of the Florida Current apparently does not exist. However, Blumberg and Mellor (1983) have developed a three-dimensional time-dependent numerical model of circulation in the South Atlantic Bight. Their 25 x 25 km grid is too coarse to resolve the important small scale events but the model produces some mesoscale activity along with recognizable features of the general regional circulation and oceanography including detailed representation of vertical structure.

Eddies generate mean flow; they help drive the deep Gulf Stream, enhancing the Stream's transport significantly. Recent eddy-resolving general circulation models contain eddies that look and act very much like rings. The large

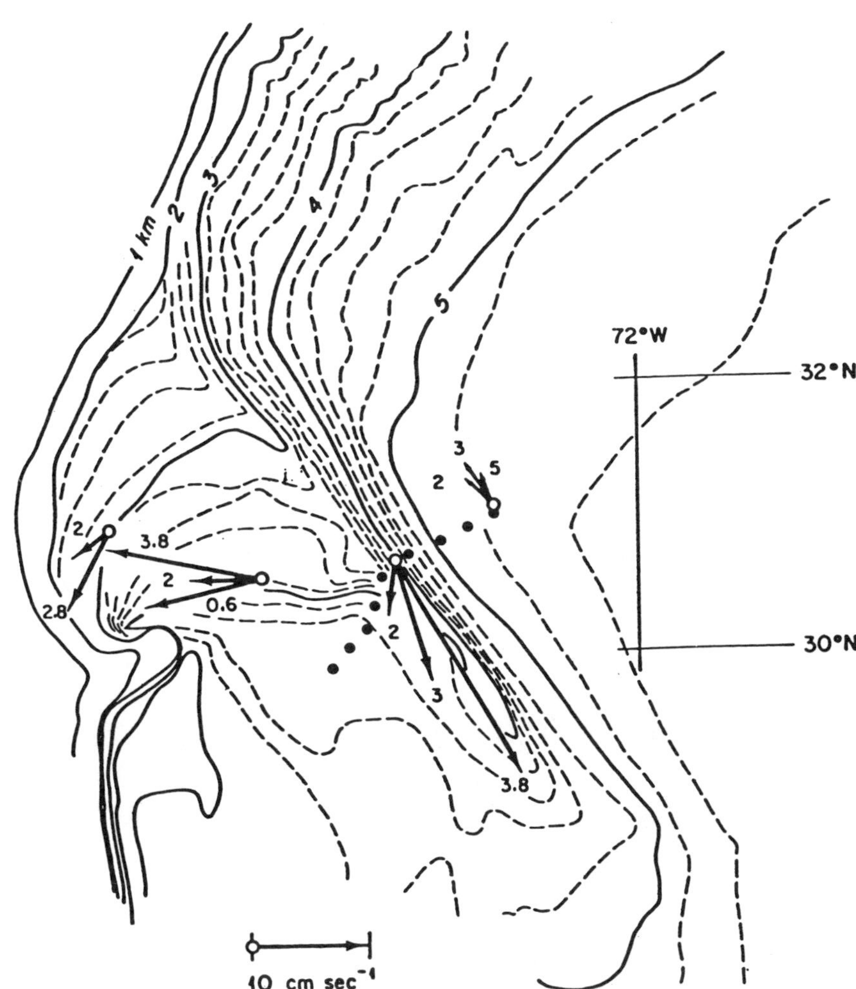

Figure 3.20 Data summary from the Blake-Bahama Outer Ridge Array: 12-month mean flows.

number of highly energetic rings that coexist near the Gulf Stream are considered to be a key parameter which the models should reproduce. Two basic recent references on large-scale models are Haidvogel (1983) and Holland et al. (1983).

Local models for individual eddies are also being pursued, particularly in the case of Gulf Stream Rings (Richardson, 1983). The movement of rings is apparently controlled by a combination of advection by large-scale flows and the tendency for propagation to the west as a packet of Rossby modes. The overall translation of rings is in the same direction as the mean flow which is to the west on either side of the Gulf Stream and which has a mean speed of ~5 cm sec^{-1} (Worthington, 1976; Luyten, 1977; Schmitz, 1980). There are periods during which rings appear to propagate normal to the inferred large-scale flow. This motion and possibly part of the regular translation is probably due to the tendency for longer scale disturbances to propagate westward on β-plane. The nonlinearity of the ring circulation can also induce a northward (southward) component of translation to a cyclonic (anticyclonic) ring.

Currents on the Continental Slope

The Slope Water region, lying east of Cape Hatteras between the shelf break and the inshore edge of the Gulf Stream, is strongly influenced by wind, tides, and instabilities of the Gulf Stream which take the form of propagating eddies (i.e., rings) and more wavelike disturbances. In turn, these are greatly modified by the underlying topography. The characteristics of current fluctuations will be discussed below from the frequency domain point of view: that is, how the nature of current fluctuation changes as a function of time scale.

Time Averaged Flows:

Wintertime cooling and evaporation in the Labrador and Norwegian Seas cause water at the surface to become dense and eventually sink and flow equatorward along the margins of the western North Atlantic. This NADW occurs between 3000 and 4000 m, with a maximum westward flow of about 5 cm sec^{-1} along longitude 70°W. More prominent is a surface intensified flow toward the west greater than 10 cm

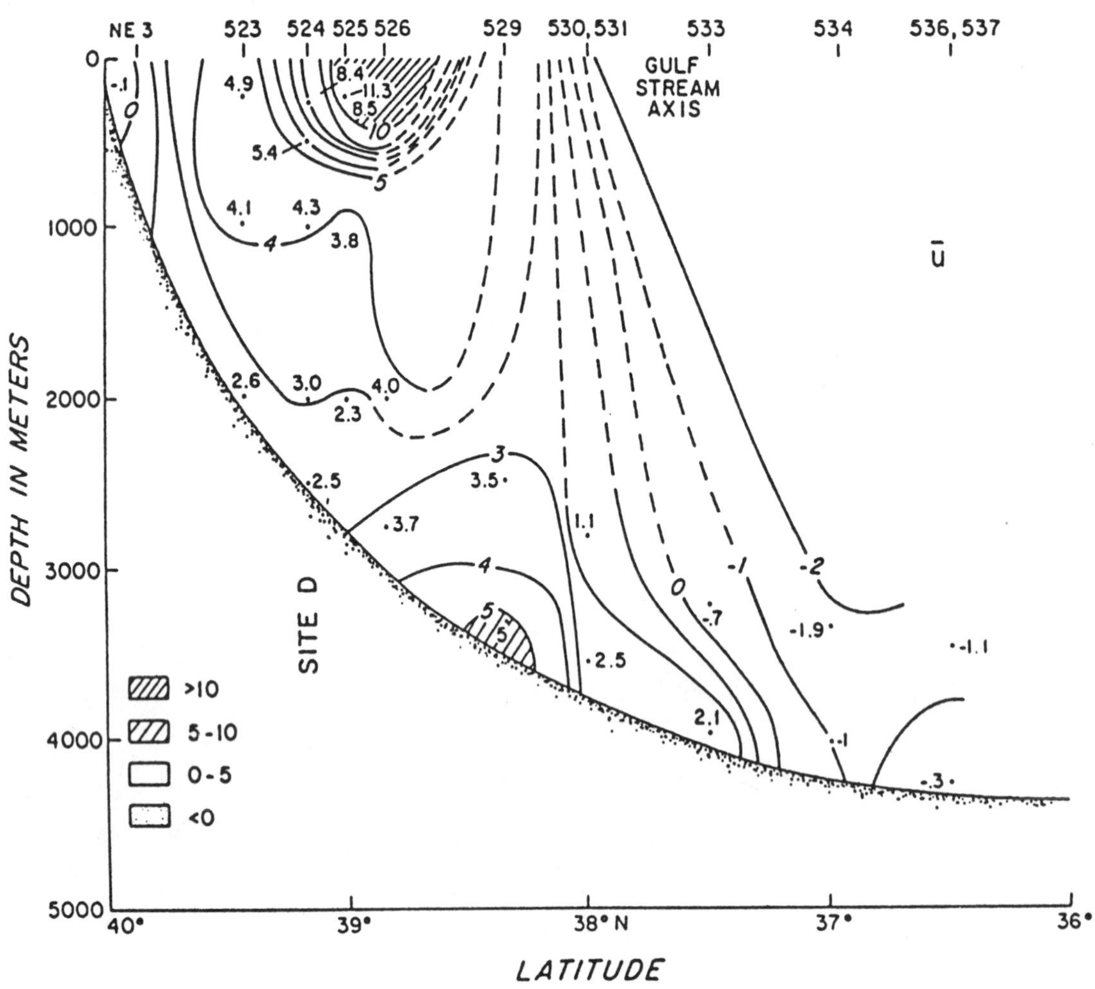

Figure 3.21 Average east-west component of current at 70°W. Observations are mainly from the Rise Array (Luyten, 1977; Spencer 1979) (from Hogg, 1983).

sec^{-1} (Fig. 3.21), a flow first discovered by Webster (1969). The source of this current is presently unknown, although it could be some extension and/or modification of the Labrador Current known to exist further to the east. The southwestward extent of the flow is also unknown although the mean current must quickly switch to a strong easterly flow as one moves into the Gulf Stream.

As revealed by Figure 3.21 practically the whole region inshore of the Gulf Stream moves, on average, toward the west. Total transport of water below 1000 m can be estimated to be about 10 x 10^6 m^3 sec^{-1}. An equal volume probably flows westward above 1000 m. For comparison, the Gulf Stream is believed to transport 30 to 150 x 10^6 m^3 sec^{-1} (Worthington, 1976).

Although other longitudes are not so well sampled, it appears that the picture of the mean flow described above is typical of this region. For example, recent measurements (Peter Smith, personal communication; Smith and Petrie, 1982) by the Bedford Institute of Oceanography near 62°W also reveal a westward mean flow less than 10 cm sec^{-1} near the bottom although there is a shallow (200 m) region of easterly mean flow (10 cm sec^{-1}, wind-driven?) between the Gulf Stream and the shelf break (there are few shallow measurements at 70°W).

Low Frequency Fluctuations

Time series observations of currents in this region also show energetic time variable flows. Careful inspection of stick vectors from the Rise Array experiment (Fig. 3.22) reveals the average bias toward westerly flows (north of instruments 5331 and 5322 on the 4000 m isobath). However these means are overwhelmed by strong fluctuations as great as 25 to 30 cm sec^{-1}. Also apparent is the horizontal inhomogeneity of this field: eddy kinetic energies drop from a maximum of 87.7 cm^2 sec^{-2} on 5342 (~37°00'N, 4138 in water depth) to around 10 cm^2 sec^{-2} at the north end of the array (39°26'N, 1991 m). Kinetic energy at lower frequencies (22-108 day periods) is at a maximum further south and at much higher values (2.8-12 days) than higher frequencies (Hogg, 1981).

The inhomogeneity of the low frequency motion field has been conclusively tied to topographic Rossby wave dynamics (Thompson, 1977). For example, events move offshore with time (Hogg, 1981). This is characteristic of Rossby waves whose energy radiates onshore away from their source. Figure 3.23 shows the horizontal phase variation for 12 to 22 day motions and the proportion of variance at each current meter which is accounted for by a single topographic Rossby wave. Hogg (1981) showed that about 50% of the low frequency variance (periods of 8.3 to 108 days) was caused by these wavelike motions. By retracing their propagation vectors, he suggested that they were generated near 68°W 38°N through spontaneous instability processes possibly linked to meandering of the Gulf Stream.

Louis and Smith (1982) and Louis et al., (1982) have shown more conclusively that topographic wave energy over the middle and deep slope can result from meandering and ring formation processes in the Gulf Stream. South of Nova Scotia the low frequency fluctuations have similar offshore propagation characteristics to those at 70°W. However motions here appear somewhat more "bursty" than at 70°W, and Louis and Smith (1982) have predicted the time evolution of a burst assuming that it arose from Rossby waves radiated by a Gulf Stream warm core ring further offshore. Figure 3.24 shows the comparison between measured and theoretical currents at a site 200 km to the north near the slope-rise junction in 1000 m of water.

Topographic Rossby waves, if small in amplitude, are linearly polarized and give rise to rectified momentum fluxes which are believed to be important in the forcing of the mean flows in the Slope Water region (Thompson, 1971a,b, 1977; Rhines, 1971). Luyten (1977) has shown that these rectified fluxes are capable of accelerating the mean flow at the rate of 1 cm sec^{-1} d^{-1}.

The earliest hard evidence that topographic Rossby waves were the dominant source of low frequency variability in the area came from analysis of system records at Site D (39°10'N, 70°00'W) by Thompson and Luyten (1976). They found that currents are anisotropic, preferring one direction as in a linearly polarized wave; the polarization direction rotates with frequency predicted by theory. Theory also predicts that these waves should be "bottom-trapped", with the vertical trapping scales proportional to wavelength. Low frequency motions beneath the thermocline do appear to be bottom-trapped as a rule. Figure 3.25 shows the ratio of energy at 1000 m to that at 2500 at Site D: between about 6- and 30-day periods there is bottom-trapping. Because different frequency components of the motion respond to the underlying topography in different ways the degree of bottom intensification of total kinetic energy is also variable. The more northern sites (up to Site D) are more dominated by high frequency, shorter waves and consequently exhibit more trapping (Hogg, unpublished data).

There are presently no simultaneous measurements of flow conditions along one longitude throughout the slope and rise region. A moored experiment was carried out in 1976 covering the region from just north of Site D to the shelf (the New England Shelf/Slope Experiment; Ou et al., 1980). Motions in this area also exhibit many of the properties of freely progagating topographic Rossby Waves (Ou and Beardsley, 1980) although the interaction with and reflection from the shelf break are also important processes. Kinetic energy apparently reaches a minimum at about Site D for it increases again to the north possibly because the shoaling topography causes energy to increase faster than it can be dissipated.

Topographic waves appear to dominate deep motions but become less energetic away from the bottom and relatively weak in the thermocline where the most energetic motions result from the passage of Gulf Stream (warm core) rings. Figure 3.26 shows time series at various depths of low-passed current and temperature at Site D. The short regular (~10-day) oscillations in the deep water result from the Rossby waves discussed above. Above 1000 m oscillations are overwhelmed by less regular, more energetic events probably resulting from the passage of warm core rings. As broken-off pieces of the Gulf Stream, the rings can have maximum velocities of several knots. What is interesting about Figure 3.26 is the rapid decrease in ring speeds with depth such that they are difficult to recognize at 1000 m.

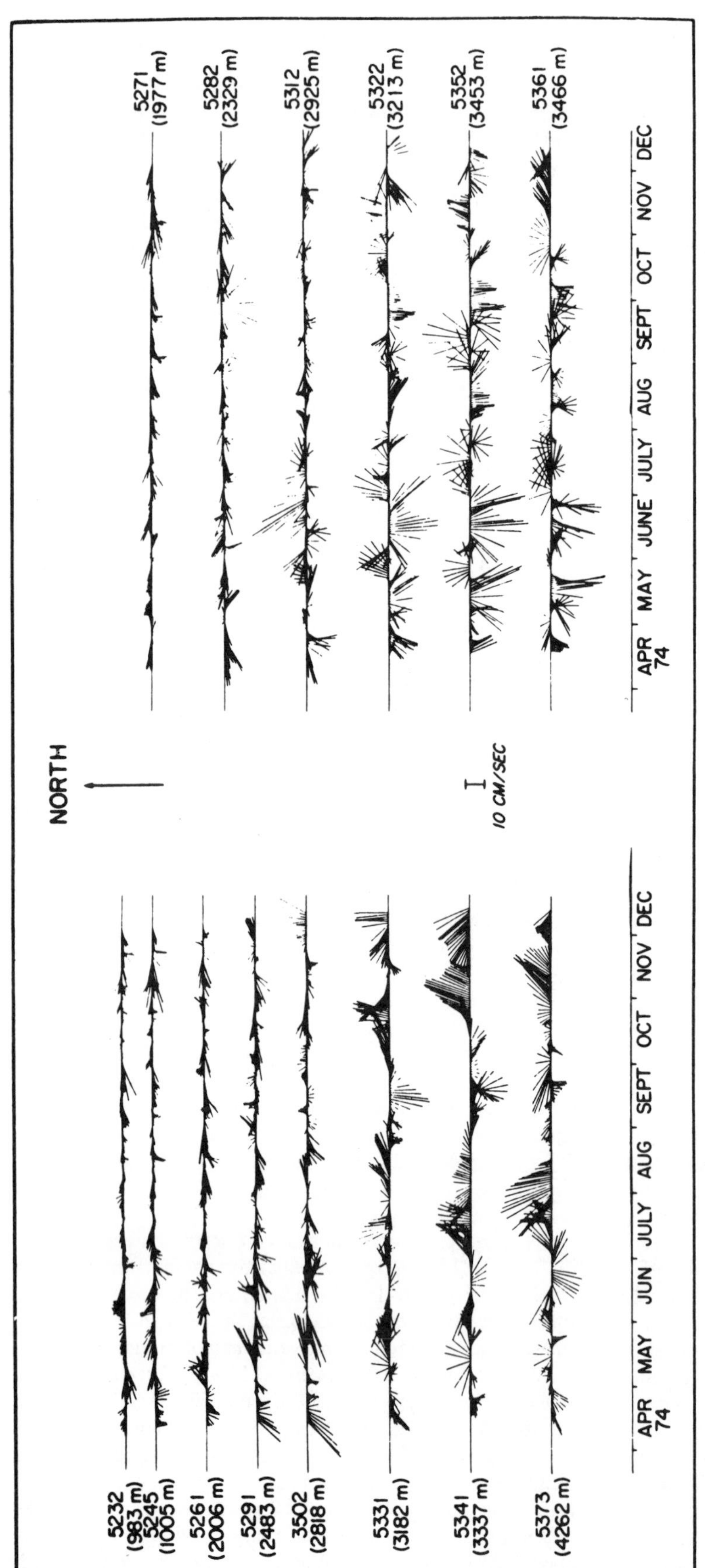

Figure 3.22 Stick plots (velocity vectors, north up) of currents measured approximately 1000 m above the bottom in the Rise Array along longitude 70°W (from Spencer, 1979, after Luyten, 1977).

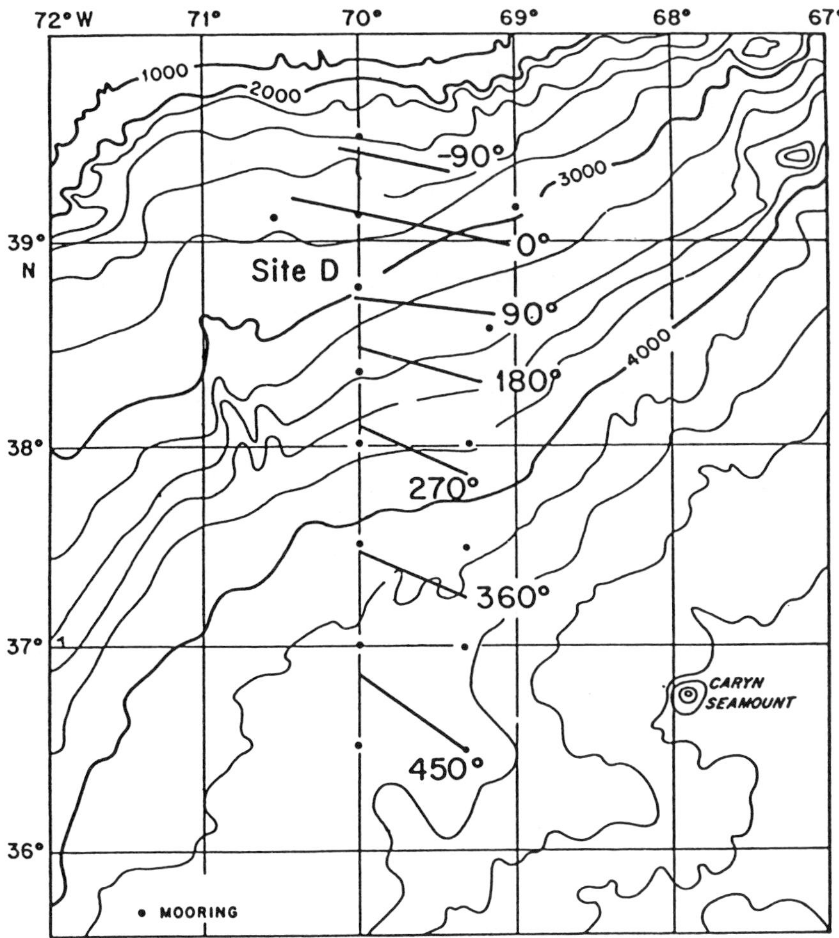

Figure 3.23 Variation in phase for motions with periods from 12-22 days in the Rise Array. Wave crests are propagating onshore; wavelength varies but is around 300 km (from Hogg, 1981).

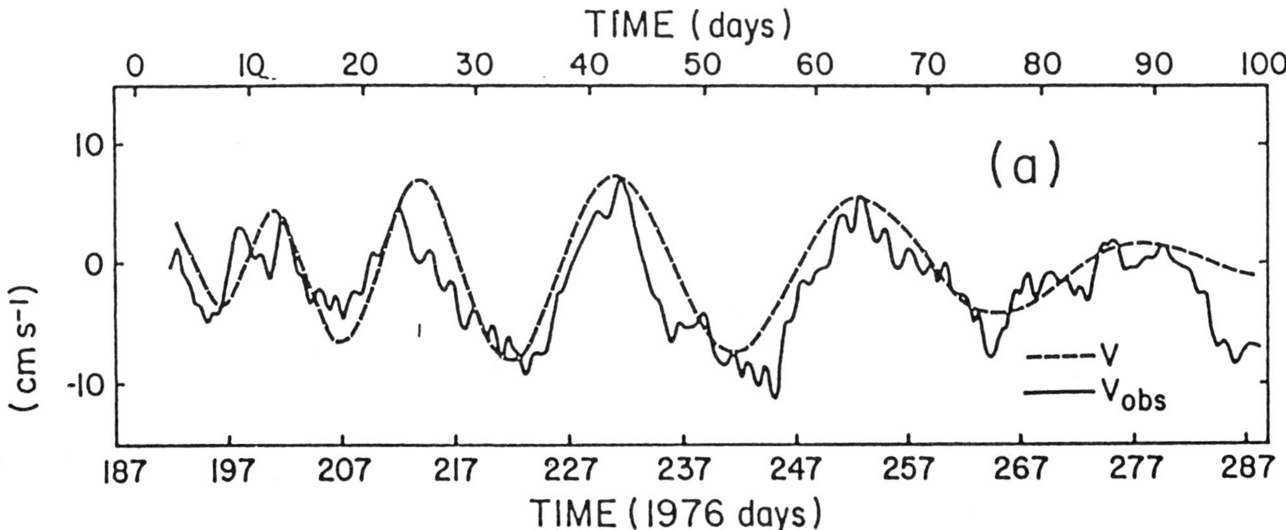

Figure 3.24 Comparison of predicted velocity from a Ring source with that observed (from Louis and Smith, 1982).

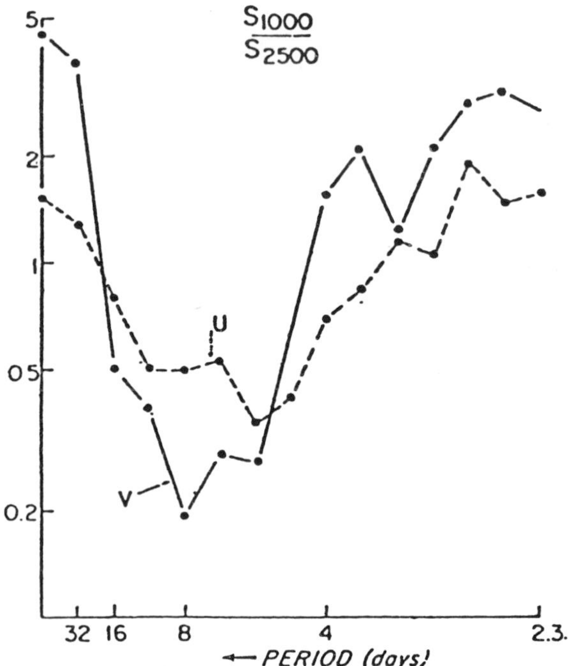

Figure 3.25 Ratio of energy density at 1000 m depth to that at 2500 m depth from a site near Site D (from Thompson and Luyten, 1976).

High Frequency Currents:

By definition "high-frequency" currents are those associated with time scales shorter than about two days. The ocean is stratified so that particles displaced vertically experience a buoyancy force returning them to their initial level. This internal spring constant causes motions to be inhibited that oscillate at a frequency greater than the natural internal frequency — the period of which varies from tens of minutes in the main and seasonal thermoclines to several hours deeper. Typical spectra have sharp peaks at the semidiurnal, diurnal and inertial (half-pendulum day) periods and steady decreases in energy density with frequency above these periods (Fig. 3.27).

Garrett and Munk (1972, 1975) proposed that the internal wave section of the spectrum (excluding tides and the inertial period) can be made into a "universal" spectrum if the kinetic energies and frequency are scaled by factors related to the natural buoyancy frequency described above (the Brunt-Vaisala frequency). Wunsch (1976) made a systematic search of available deep records in the North Atlantic from below 2000 m and found the Garrett-Munk proposal to be satisfactory except near an isolated seamount. Deep records from the Slope Water region departed by no more than a factor of two from the universal form, a conclusion reached earlier by Webster (1971).

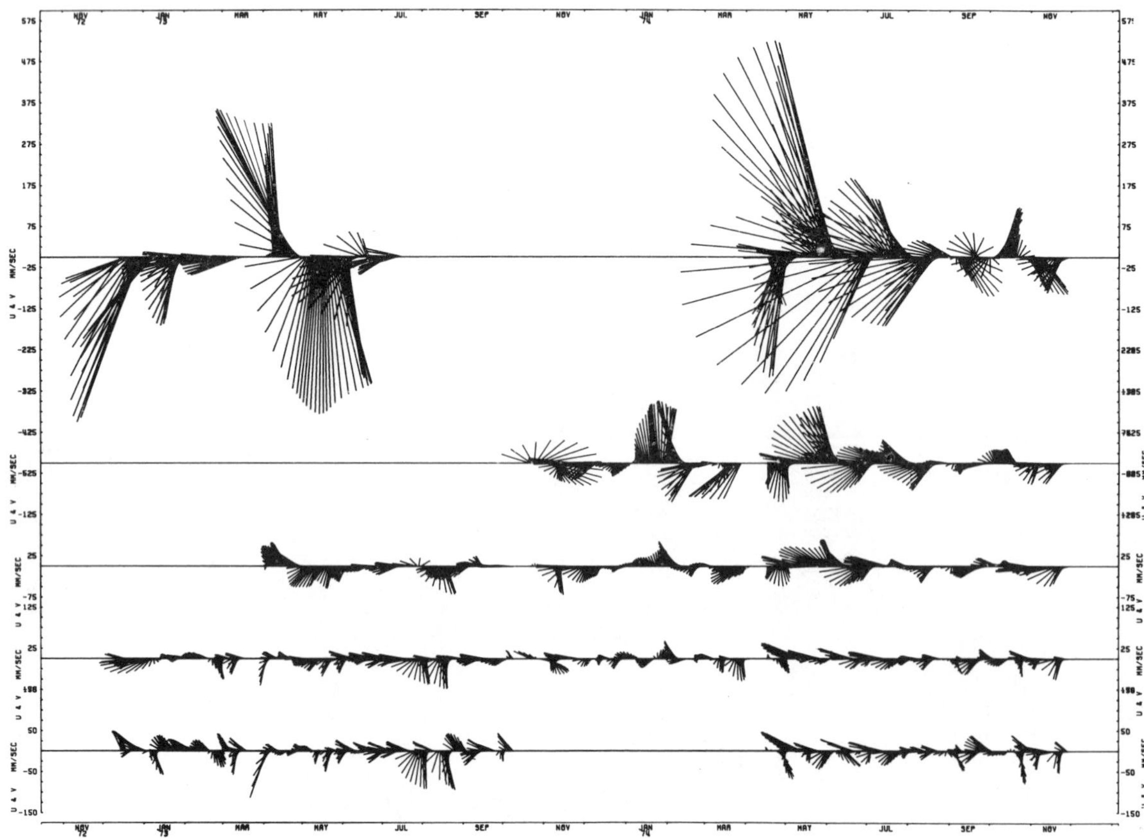

Figure 3.26 Time series of currents at various depths at Site D (Hogg, unpublished document).

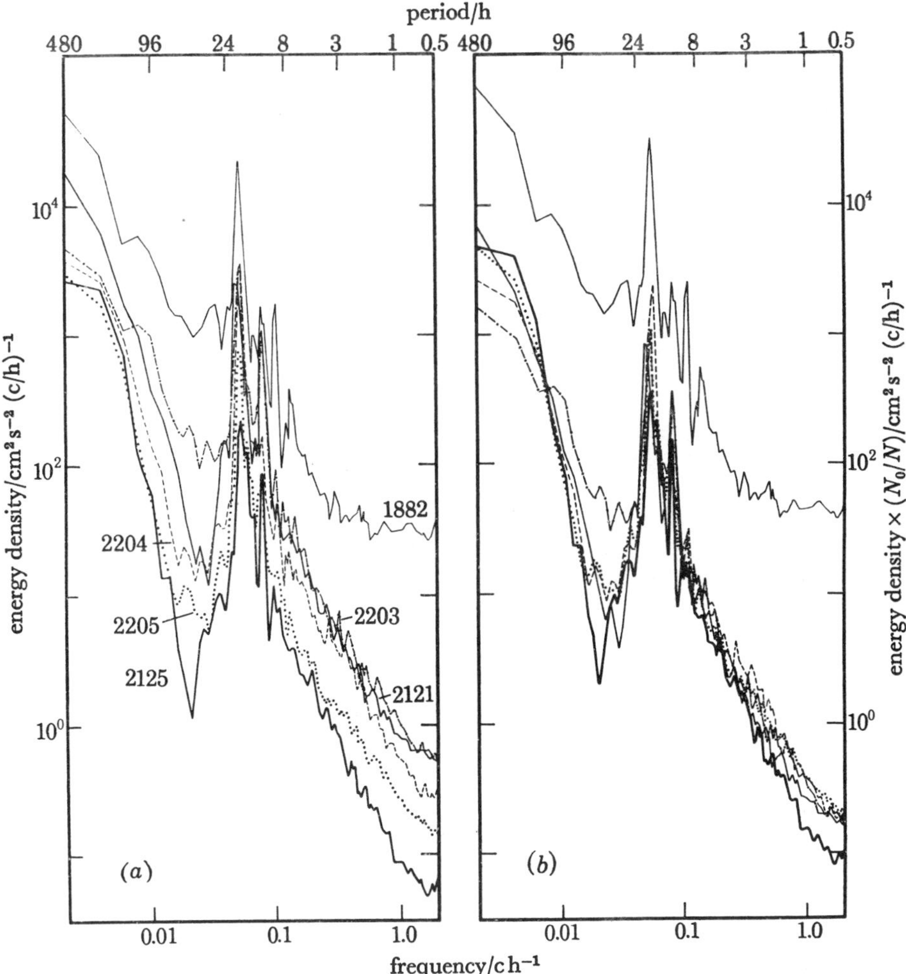

Figure 3.27 Power density spectra at Site D (a) before and (b) after scaling by the Brunt-Vasalla frequency (from Fofonoff and Webster, 1971).

Wunsch and Hendry (1972) analyzed deep records from a small scale "coherent" array set somewhat to the northwest of Site D in 1970 (two months). They found energy levels away from the semidiurnal and inertial bands to be roughly independent of bottom depth over ranges from 900 m to 2500 m in line with the Garrett-Munk universal spectrum hypothesis. Significant variations existed in the dominant energetic bands in the high frequency spectrum — the inertial and tidal peaks. Inertial energy dropped by over an order of magnitude at the shallower depths while the semidiurnal band increased by a factor of five or so.

When the bottom slope equals the characteristic slope of an internal gravity wave, $(\sigma^2 - f^2)^{1/2}/(N^2 - \sigma^2)^{1/2}$, where σ, f, and N are wave, inertial, and buoyancy frequencies, the inviscid reflection of internal gravity waves from a sloping bottom becomes singular (Wunsch, 1969). At this point one can expect amplified motion damped only by dissipation. Consistent with this idea Wunsch and Hendry (1972) found the bottom slope to satisfy this condition for the semidiurnal period at the location of most energetic tidal energy — roughly 1000 m depth, the most northerly measurement location. Energy was most amplified (factor of 3) in the bottom 10 m.

A summary of tidal constants in this area, useful for computing the barotropic tide, is contained in the pamphlet by Cartwright et al. (1979). In a study of tides on the Scotian shelf and slope, Petrie (1975) found that the M_2 component over the slope is dominated by the baroclinic tide of amplitude 5 — 10 cm sec^{-1}. Petrie's observations are consistent with this internal tide being generated by the barotropic tide at the shelf break in 200 m water depth.

Taking into account the appropriate scaling, Fofonoff and Webster (1971) and Webster (1969) found little significant vertical variation to the internal wave field (excluding, once again, inertial and tidal frequencies). An example of this scaling is shown in Figure 3.27. However, there are systematic spatial changes in both the tidal (discussed above) and inertial bands (Fig. 3.27), with inertial energy being greatest near the surface and least near the bottom. Pollard and Millard (1970) have demonstrated rather well that motion at the inertial period in the surface mixed layer is directly wind-forced. This inertial energy propagates downwards as freely travelling waves and apparently, through non-linear conversion processes, becomes scattered into the rest of the internal wave spectrum. Hamilton (1982) has found some evidence for increased inertial wave activity near the bottom at the Atlantic 2800 m dumpsite some 2-3 weeks after the passage of Hurricane Belle.

Gulf Stream Rings

Early investigation of cold core rings (Fuglister and Worthington, 1947, and Iselin and Fuglister, 1948) led to the first observed formation of a cold core ring from a meander (Fuglister and Worthington, 1951). Data accumulated over the next two decades provided the basis for the first good descriptions of the distribution, movement, and decay of cold core rings (Fuglister, 1972, 1977; Parker, 1971; Barrett, 1971; Richardson et al., 1973). In the 1970s interest in rings intensified and a series of new experiments were begun, involving diverse measurement techniques such as airborne XBTs, satellite infrared images, SOFAR floats, satellite-tracked drifters, and vertical profilers.

It became clear that rings are a persistent and ubiquitous feature of the Northwest Atlantic Ocean, so that it was possible to deal with their population statistics (Richardson et al., 1978). Scientists from the United States Naval Oceanographic Office (now NAVEASTOCEANCEN in Norfolk) also began to study rings more actively during this time and published data in the *Gulf Stream Monthly Summary*. This publication was replaced by *Gulf Stream* (1975-81) which has since been replaced by *Oceanographic Monthly Summary*. NOAA now produces weekly and even twice weekly charts (called Oceanographic Analyses) which show Gulf Stream and ring positions on the basis of satellites and other data. The history of ring investigations is described in a review article by Richardson (1983c).

Rings form from 70°W eastward (Fig. 3.28). Most rings have been observed in the region 60-70°W with a maximum number near 65°W, suggesting a preferred formation region. Generation is apparently common near the New England Sea-mounts (Fuglister and Worthington, 1951; Richardson, 1983c). The seamounts are also associated with large meanders and a semi-permanent ring-meander. Rings have been observed to both pinch off from this latter structure and also coalesce with it.

Cold Core Rings

Cold core (or cyclonic) rings have a large raised dome in their thermal, salinity, and density fields, extending down near the sea floor. Different water properties (temperature-salinity, temperature-oxygen, etc.) can frequently be

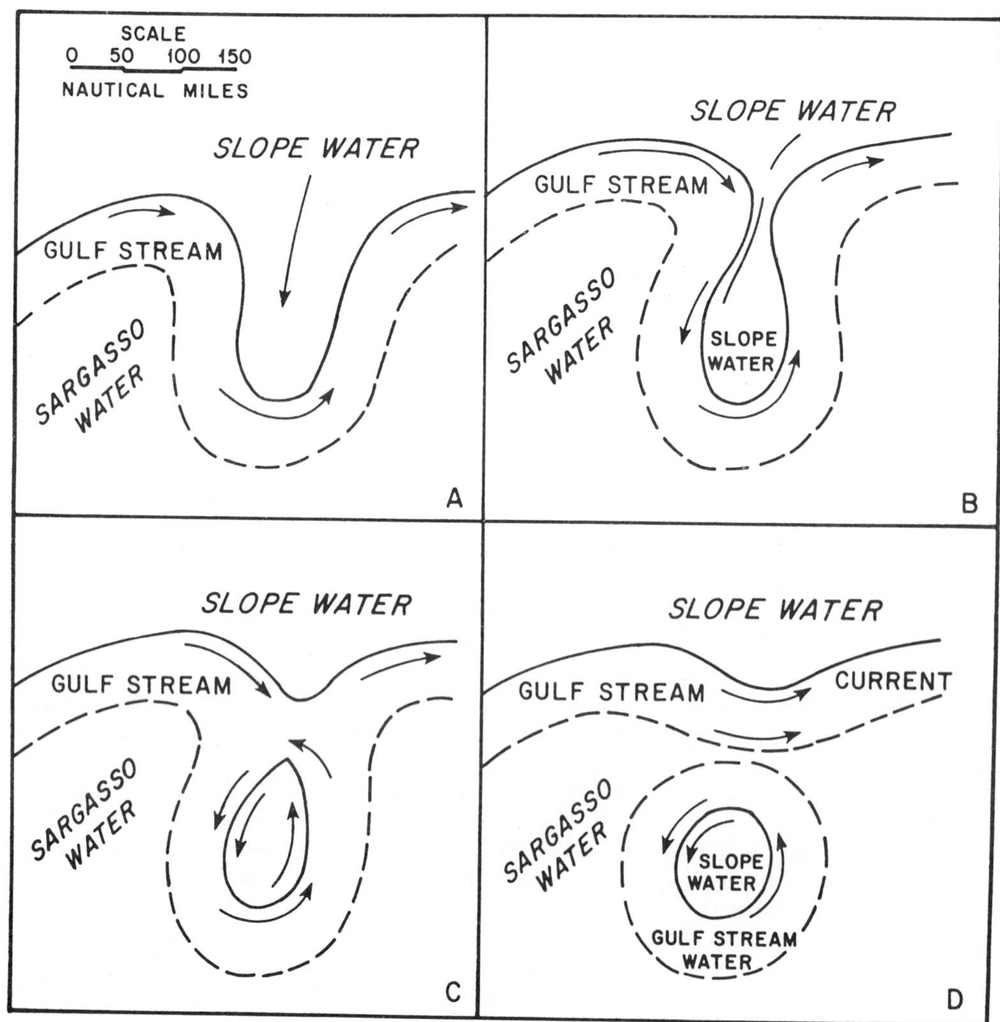

Figure 3.28 Schematic diagram showing genesis of cold water ring.

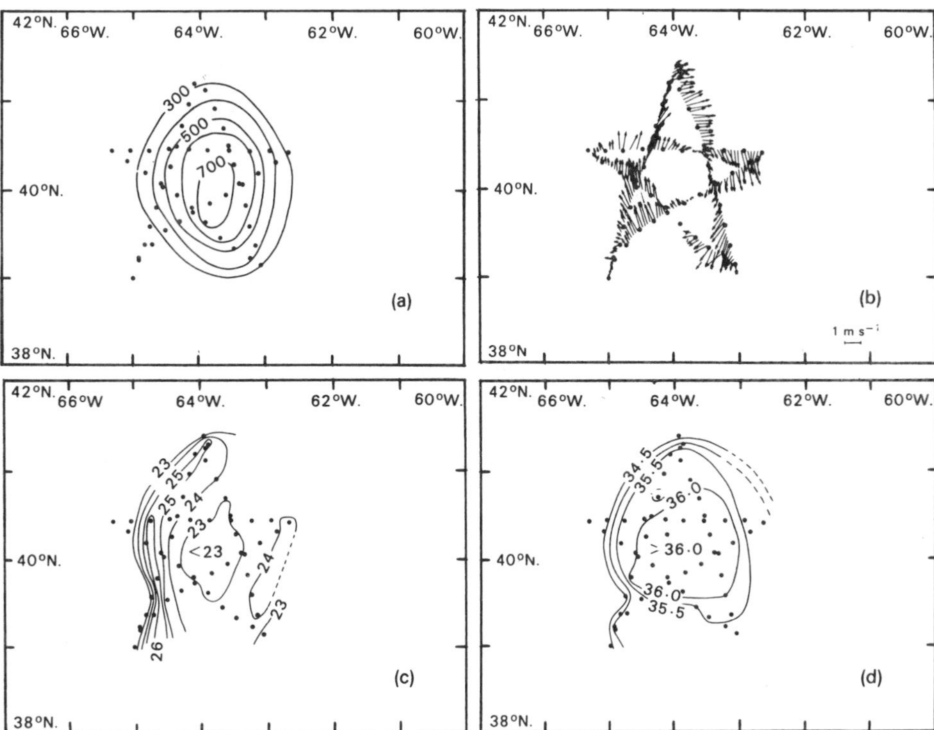

Figure 3.29 The depth of the 10°C isotherm (a), near surface velocities (b), surface temperature (c) and salinity (d) of warm core ring 81-D, 14-16 September 1982. From Joyce et al. (1984).

identified in the upper several hundred meters. Approximately ten cold core rings can co-exist at a single time (west of 55°W). An analysis of hydrographic stations, BTs, XBTs, and satellite infrared images from the period 1932-1976 has resulted in the identification of 225 cold-core ring observations in the area west of 50°W (Parker, 1971; Lai and Richardson, 1977).

Rings usually move westward when they are not touching the Gulf Stream and eastward when they are attached to it (Richardson, 1980; Fuglister, 1972; 1977). The westward movement, sometimes northwest, sometimes southwest, is characterized by a mean speed of 5 cm sec^{-1}, although there are large variations, including periods when rings remain nearly stationary. If rings are attached to and advected parallel to the Stream, their eastward movement can be quite fast: 25-75 cm sec^{-1}. Often such rings split from the Stream after eastward advection and resume westward movement.

Warm Core Rings

Warm core rings form in the Slope Water region between the continental shelf and the Gulf Stream. These rings or eddies are bodies of water 100-200 km in diameter which result from the pinching off of a northward Gulf Stream meander. The Sargasso Sea core of the ring is initially surrounded by a clockwise rotating remnant of the Gulf Stream. Maximum surface currents in the high velocity region of the ring are 50-200 cm sec^{-1}. Knowledge about warm rings has increased in the last decade because they are readily identifiable in satellite infrared imagery. Such imagery and ship of opportunity data have been used by Bisagni (1976), Halliwell and Mooers (1979), and Mizenko and Chamberlin (1979), for example, to characterize the frequency of occurrence and distribution of warm core rings. After formation, rings drift westward in the Slope Water with speeds of 3-5 km/day (Lai and Richardson, 1977). A 5-year summary of ring activity, assembled by B. Christ, A. Friedlander, and J. Fitzgerald of the National Marine Fisheries Service, Atlantic Environmental Group in Rhode Island, and presented at the 1981 ICES statutory meeting in Woods Hole, MA, October 1981, showed 42 rings which were tracked. Once rings move west of the New England Seamounts (into region 4 of the figure), they can have lifetimes of 7 to 8 months (Joyce and Wiebe, 1983).

A major study of warm core rings is now underway which involves a number of different research institutions (EOS, 1982; Joyce and Wiebe, 1983). Hydrographic sections and direct measurements of near surface currents are beginning to yield a picture of the mass, heat, salt, and momentum structure of rings (Joyce et al., 1984; Stalcup et al., 1982). Ring 81D was observed in September-October 1982 near 40°N, 64°W to be elliptically shaped with clockwise currents near the surface of nearly 200 cm sec^{-1} (Fig. 3.29). Ring decay together with interactions with the Gulf Stream can reduce the energy of warm core rings as they move west/southwest towards Cape Hatteras. Nevertheless, they still are the main source of strong current variability in the entire Slope Water region. They can also affect the circulation at the shelf/slope front (EG&G, 1978) and near submarine canyons (Butman et al., 1982) See discussion below. Data suggest that the circulation around rings does not extend into the deep water. Saunders (1971) found little evidence for ring currents at depths below 1000 m.

Within the next 2 years, the physical, biological, and chemical structure of warm core rings should be more clear as a result of the multi-disciplinary warm rings program. This will certainly lead to a better understanding of the effects of these large eddies upon possible offshore minerals exploitation.

Canyons

Along the continental slope between Nova Scotia and Cape Hatteras, the numerous submarine canyons demark regions where physical processes can be altered from those in the surrounding Slope Waters (Shepard and Dill, 1966). Several regional canyon studies have been completed or are nearing completion. These have involved both hydrographic surveys and deployment of moored current meter arrays.

Recent work in Hudson Canyon (Hotchkiss and Wunsch, 1982) has shown energetic, polarized velocity fluctuations associated with surface and internal tides which can be larger near the heads of the canyon than the unpolarized fluctuations in the surrounding waters: large enough to lead to enhanced mixing (Hotchkiss, 1982) and sediment transport. These fluctuating currents of tidal origin were sampled over short periods of time by Shepard et al. (1979) who concluded that strong turbidity currents initiated in canyons were responsible for enough sediment erosion and transport to maintain and modify those canyons. Measurements in Hudson Canyon showed larger fluctuations in the up/ down canyon direction than in the cross-canyon direction (Hotchkiss and Wunsch, 1982). Furthermore, this polarization increased towards the head of the canyon. Similar results were found in a USGS study of Lydonia Canyon off Georges Bank by Butman et al. (1982).

Since surface and internal tides are ubiquitous over the continental shelf and slope, it can be anticipated that these high-frequency fluctuations are important for sedimentation processes in all the canyons along the eastern continental margin of the U.S.

As noted by Hunkins (1983) in a preliminary report to the MMS on physical measurements in Baltimore Canyon, the oscillatory currents will not transport sediments away from the canyon heads; this time-averaged transport will be accomplished by the low frequency currents in the canyons. For this reason, it is necessary to obtain long time series data in canyons before any statements can be made about sediment transport and possible "anomalous" physical processes with time scales from days to months.

It is generally accepted that much of the low-frequency current variability over the continental shelf can be related to passage of synoptic-scale atmospheric events such as winter cyclones and hurricanes (see Beardsley and Boicourt, 1981). Keller et al. (1973) observed unusual levels of suspended sediments in Hudson Canyon following a hurricane. The above time-series measurements of Hotchkiss, Hunkins, and Butman et al. have not yet shown a definitive relationship between passage of storms and low-frequency currents deep in canyons; this is in contrast to findings by Shepard co-workers in west coast canyons. Hotchkiss speculates that the difference may be due to the presence of a narrow, shallow shelf along the west coast while the canyons along the eastern U.S. coast are farther removed from the coast with deeper canyon heads.

In analyzing their moored current meter data in Lydonia Canyon, Butman et al. (1982) find that the dominant source of low-frequency variability is related to passage of warm core Gulf Stream rings, not atmospheric forcing. They found that currents over the slope as well as in the canyon reversed from a southwesterly direction, intensified, and reached speeds as great as 80 cm sec^{-1} to the northeast, coincident with the passage of a warm ring (contrast the monthly mean currents for December, 1980, in Figure 3.30 with those after ring passage in February, 1981). Since ring-related currents attenuate with increasing depth, flow over the bottom of canyons may not be as strongly affected as near the surface. Note that below 800 meters from the surface there appears to be little influence of the ring upon the deep flow in Lydonia Canyon. Hunkins has observed little influence of rings upon the near-bottom currents in Baltimore canyon. Rather he sees a deep down-canyon flow of a few cm sec^{-1} near the canyon head with an up-canyon current near the bottom farther offshore. As these studies come to completion, a better understanding should emerge of the coupling of low frequency currents in canyons to the atmospheric shelf circulation and that of the surrounding Slope Water.

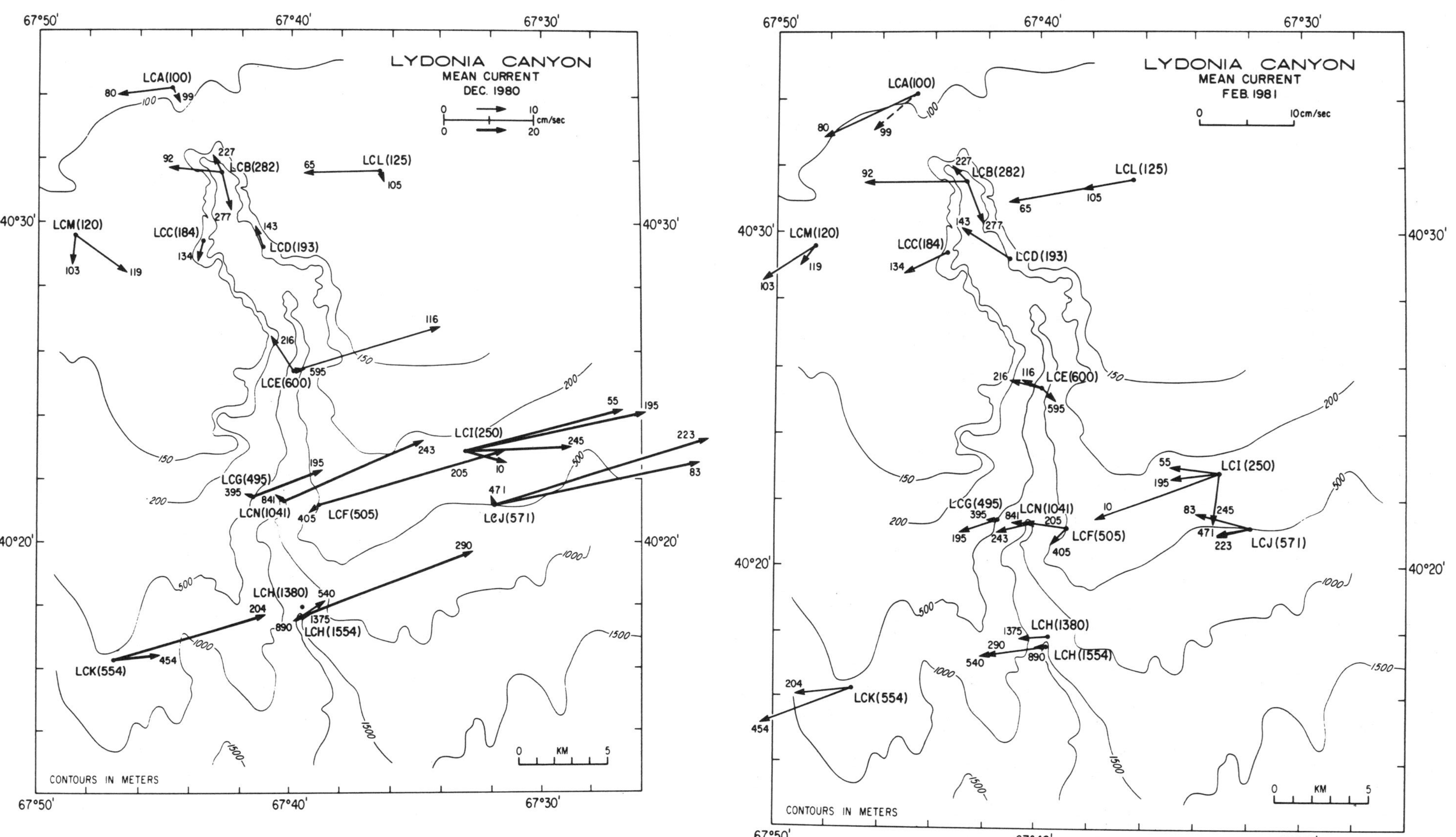

Figure 3.30 Monthly mean currents in Lydonia Canyon in December 1980 (left) and February 1981 (right) (Butman et al., 1982). Instrument depths are next to tips of current vectors.

4 Submarine Geology

B.E. Tucholke

Introduction

The last significant synthesis of the geology of the continental slope and rise off the eastern United States was published by Emery and Uchupi (1972). In the dozen years since their work, a large amount of geological and geophysical data has been acquired, and there have been many new insights into the geologic structure and history of the margin. Within the limited space available, this chapter is intended to address four points: 1) to summarize the kinds and locations of new geological and geophysical data that have become available, 2) to provide an updated overview and summary of the general geologic conditions extant in the region, 3) to identify major uncertainties in our knowledge of the geologic structure and processes along the margin, and 4) to summarize the geologic development of the margin from the time of initial sea-floor spreading in the Middle Jurassic. This work is based on an exhaustive survey of unpublished and "gray" literature as well as published reports, only a small fraction of which can be cited here. Specifically excluded are extended reviews of geophysical potential-field data such as gravity and magnetics. The implications of these data for geologic structure, however, are incorporated in the text. The free-air gravity field recently has been reviewed in detail by Grow et al. (1979b), Bowin et al. (1982), and Ewing (1984a-c), and magnetic anomaly patterns have been mapped most recently by Klitgord and Behrendt (1977) and Zeitz and Gilbert (1980a,b).

This summary focusses on the continental slope and the continental rise above 4000 m, generally within the U.S. exclusive economic zone along the Atlantic margin (Fig. 4.1). Although the continental shelf, lowermost continental rise, and abyssal plains are outside this focus, they are discussed where their features have an important bearing on the geologic development of the continental margin.

Bathymetry/Physiography

Development of Seafloor Maps and Imaging Systems

The first comprehensive maps of submarine topography along the U.S. Atlantic margin were compiled by Veatch and Smith (1939), based on echo-sounding surveys made by the U.S. Coast and Geodetic Survey in the 1930's. The most outstanding accomplishment of this effort was their recognition that the continental slope is heavily dissected by submarine canyons and channels. Veatch and Smith attributed these features to subaerial erosion (a notion now known to be incorrect), principally because subaerial analogs were known and possible submarine mechanisms of erosion were unknown. The earliest, broad-based geological summary of the western North Atlantic margin based on echo-sounding was provided by Heezen et al. (1959). Their definitions still remain the standard descriptions of physiographic provinces, and their physiographic diagram of the seabed (updated several times since) provides a useful generalized depiction of the principal features of the U.S. Atlantic continental margin.

During the following 20 years, precision echosounding profiles were acquired by ships of numerous institutions, and detailed surveys were made over several parts of the continental slope and uppermost rise. Depths in most of these profiles could be read to an accuracy of about one meter. Beam width typically was 30° and greater; hence in areas of irregular topography, side echoes usually masked the true bottom echo beneath the ship. This limitation still exists for conventional echosounding data, and it provides serious limitations for investigation of seafloor topography at large working scales. A large set of echosounding data over the Atlantic margin was summarized in the late 1960's by Pratt (1968) and Uchupi (1965, 1968a). Belding and Holland (1970) also published a bathymetric map of the U.S. Atlantic margin. This compilation includes some industry data not available to Uchupi for his earlier maps.

More recent, generally available bathymetric maps that contain the entire U.S. Atlantic margin included those by Uchupi (1971), U.S. Naval Oceanographic Office (1984), National Ocean Survey (NOS, 1975 to 1981), Perry et al. (1980), and GEBCO (1982). Scales vary from 1:250,000 to 1:10 million, and contour intervals vary from 100 to 1000 m.

The most recent update of bathymetry of the entire U.S. Atlantic margin is that of Shor (1984), Shor and Flood (1984), and Flood and Shor (1984) in the Ocean Margin Drilling (OMD) synthesis (Fig. 4.1). The original maps are at a scale of about 1:1.1 million (Mercator projection, 4" = 1°) and they include an extensive compilation of available published and unpublished data. Contour interval is 100 m.

In recent years, the narrow-beam echosounder (14° beam width) has been used more frequently (Fig. 4.2). Such data are valuable in areas of irregular topography such as the continental slope. Unfortuately even these narrow-beam soundings insonify an area having a diameter equal to up to 0.25 times water depth, so that side echoes remain a prob-

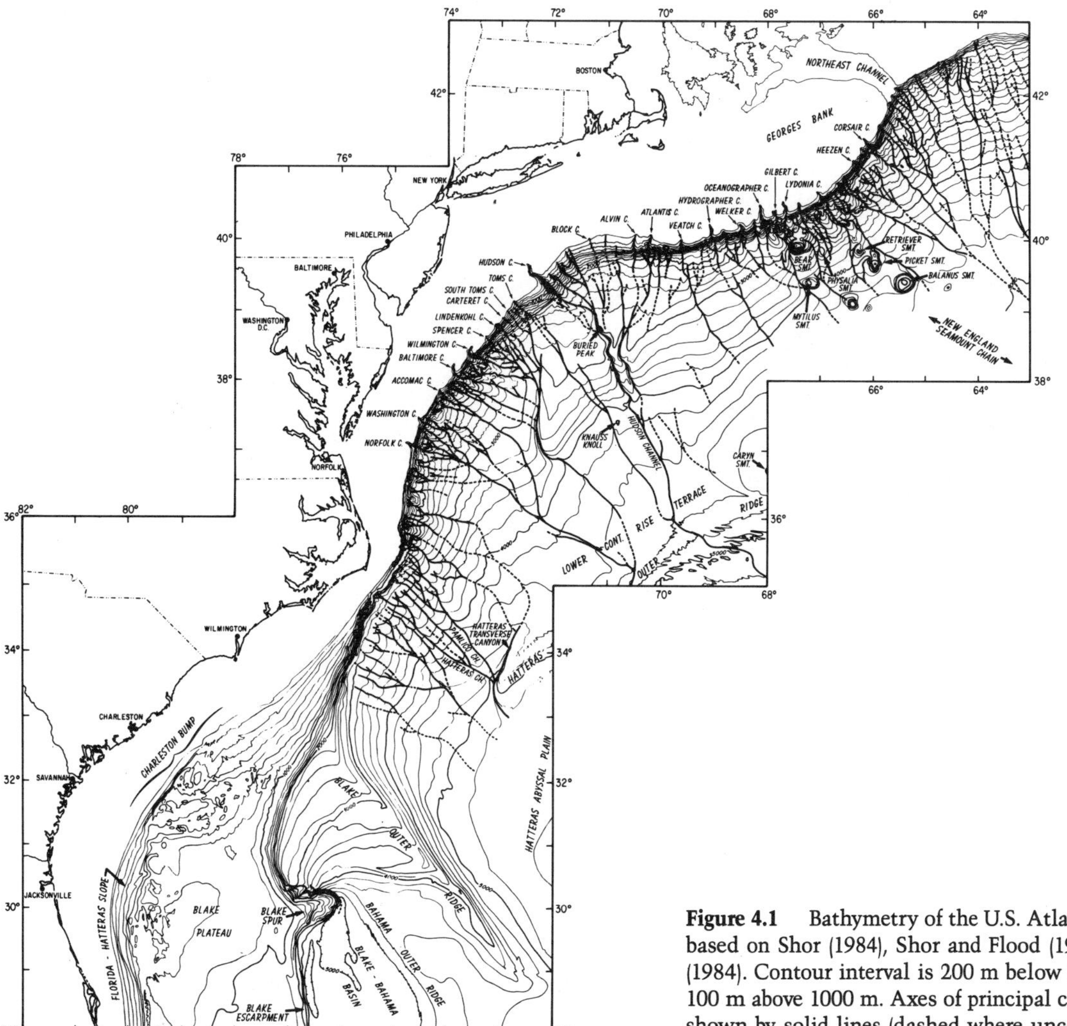

Figure 4.1 Bathymetry of the U.S. Atlantic continental margin, based on Shor (1984), Shor and Flood (1984) and Flood and Shor (1984). Contour interval is 200 m below 1000 m water depth and 100 m above 1000 m. Axes of principal canyons and channels are shown by solid lines (dashed where uncertain or approximate).

lem on steep slopes. Detailed bathymetric surveys of areas of the continental slope have been made using both conventional and narrow-beam echosounders and some of these surveys extend onto the upper continental rise just below 2000 m (Fig. 4.2). However, no such detailed surveys have been made on the deeper part of the continental rise. These areas in particular will benefit from increased use of multi-beam echo-sounding systems which can insonify a width of seafloor up to 0.8 times water depth and subtend angles of 2⅔° in individual beams.

In addition to technical limitations of echosounding in determining precise seafloor depths, an additional limitation is the incomplete track coverage. In rough seabed areas such as the continental slope, it is not uncommon for trends and junctures of canyons and gullies to be miscontoured, as one can readily see by comparison of recent bathymetric maps. Even closely spaced tracks (1 km) can leave ambiguities. This problem is best resolved by use of side-scan sonar imaging to determine morphologic trends and to guide contouring. Numerous side-scan sonar surveys have been conducted in recent years, principally over the continental slope (Fig. 4.3). The side-scan imaging systems used in these surveys can be roughly grouped into three categories: 1) long-range (>10 km total swath; e.g., Popenoe et al. (1982c), Scanlon (1982a,b), and Twichell and Roberts (1982)), 2) mid-range (2-10 km swath; Popenoe et al. (1981b), Robb et al. (1981c), O'Leary (1982a), and McGregor et al. (1982a,b), Kirby et al. (1982), Robb et al. (1981a), and 3) short-range (<2 km swath; Twichell (1983), Johnson and Lonsdale (1976), and Hollister et al. (1974)).

The next "step up" in resolution is bottom photographic systems which can resolve seafloor features down to less than a millimeter and up to several meters. Unfortunately there is a gap in coverage between conventional photographic systems (meters) and short-range side scan sonar systems (tens of meters). This is partially resolved by photographic systems such as ANGUS (Acoustically Navigated Geological Undersea Surveyor; Phillips et al., 1978; Ballard, 1980) which can photograph large areas and provide mosaics of overlapping photographic coverage. ANGUS has been used on the continental slope near Lydonia and Alvin Canyons off Georges Bank. The photographic sled, "Cheap Tow" (Ryan

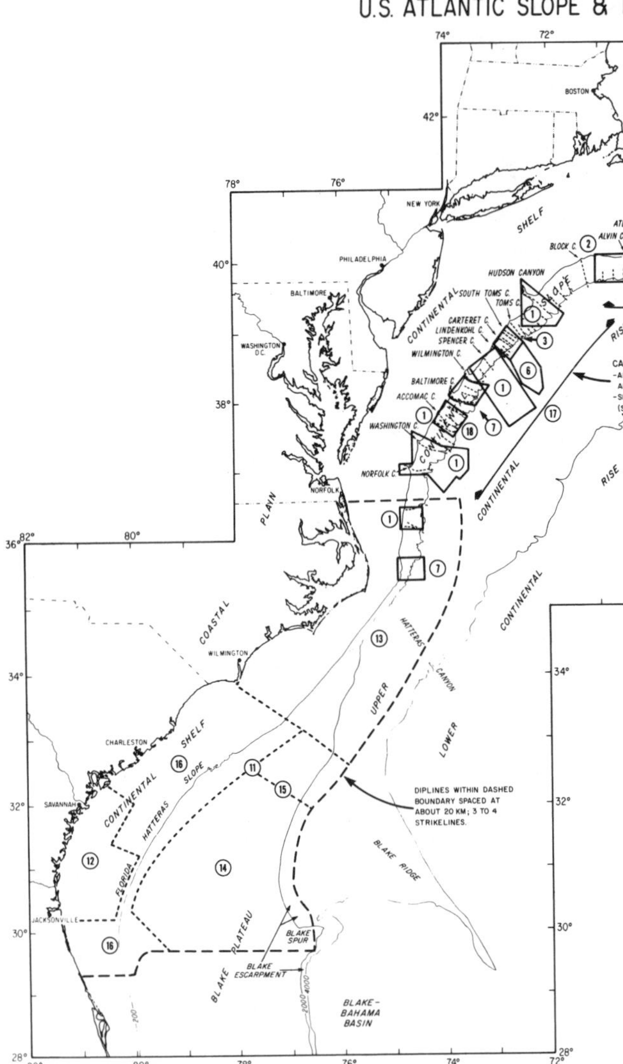

Figure 4.2 Summar of recent detailed seismic and echosounding surveys and compilations along U.S. Atlantic margin. Recent Detailed Bathymetric Surveys and Compilations: 1-Narrow Beam Surveys (0.5 to 1 mile grid) (Nastav et al., 1980); 2-Wide-Angle Echosounding (~1 mile grid) (MacIlvaine, 1973); 3-Wide-Angle Echosounding (0.5 mile grid) (Robb et al., 1981a; Kirby et al., 1982); 4-Wide-Angle Echosounding (J.A. Moody and B. Butman, unpubl. data); 5-Wide-Angle Echosounding (P. Valentine, unpubl. data); 6-Wide-Angle Echosounding (~1 mile grid) (J. Robb, unpubl. data); 7-Wide-Angle Echosounding (~1 mile grid) (B.A. McGregor, unpubl. data); 8-Wide-Angle Echosounding (~½ mile grid) (D. O'Leary and B.A. McGregor, unpubl. data); 9-Wide-Angle Echosounding (~1 mile grid) (N.G. Bailey and J.M. Aaron, unpubl. data); 10-Malahoff et al. (1980); 11-Hurley (1964). *Detailed Seismic Surveys:* 1-Nastav et al. (1980); 3-J. Robb (1980a, b); 6-J. Robb, (unpubl. data); 7-B.A. McGregor (unpubl. data); 8-D. O'Leary and B.A. McGregor (unpubl. data); 9-N.B. Bailey and J.M. Aaron, (unpubl. data). *Additional Seismic Lines:* 10-N.B. Bailey and J.M. Aaron (1982a, b) and J.M. Aaron et al. (1980); 11-Popenoe (1981); 12-Henry et al. (1981); 13-Popenoe et al. (1982c); 14-Pinet et al. (1981) and Popenoe (1980a); 15-Pinet et al. (1981); 16-Popenoe et al. (1981a); Edsall (1978); and Edsall (1980); 17-Robb and Kirby (1980); 18-Malahoff et al. (1980).

et al., 1980), also has been used in Baltimore, Lydonia and Oceanographer Canyons to study biological populations (Hecker et al., 1980); this system optimally provides about 40 percent coverage along track. A new visual survey system is being developed at Woods Hole Oceanographic Institution by R.D. Ballard and colleagues. This "ARGO" system will be able to obtain seafloor images more than an acre in size for overlapping coverage; the system was field-tested in 1984.

Direct observation of the seafloor in submersibles, accompanied by sampling and photography, gives the geologist the closest perspective of seafloor conditions along the margin, and numerous diving campaigns have been conducted in recent years (Fig. 4.4). Although the areal coverage of these observations is small, their resolution of geologic relationships is unequalled by any other survey tool.

Scales of seafloor features on the U.S. Atlantic margin span the entire size range from the sub-millimeter-scale to hundreds of kilometers. These are characterized in Figure 4.5 along with comparable horizontal length scales that can be resolved with various imaging systems. Resolution of vertical scales typically is better by an order of magnitude or more. Individual physiographic provinces (the largest-scale seafloor features) are treated in the ensuing summary, and discussion of seafloor features of smaller length and width scales is integrated therein.

Physiographic Provinces

Continental Slope

The continental slope, extending from the shelf break (60-200 m) to water depths of about 2000 m, is the most prominent physiographic discontinuity along the Atlantic margin and also the most complex. Its present form results from varying contributions of sediment upbuilding and

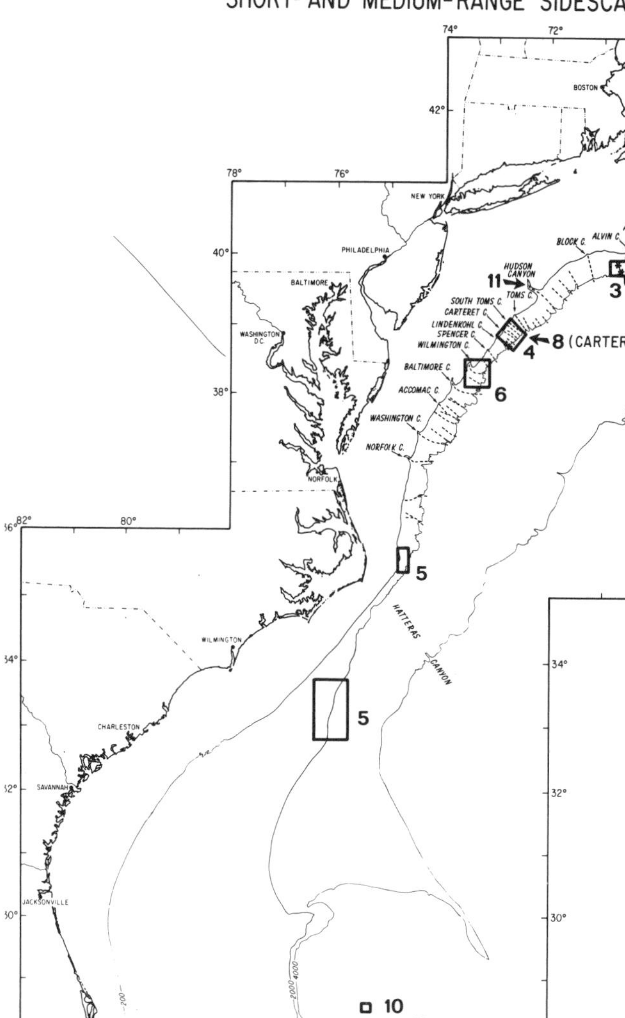

Figure 4.3 Compilation of sidescan-sonar surveys conducted along the U.S. Atlantic slope and rise. GLORIA long-range sidescan- sonar surveys (not shown) cover a 30- to 80-km wide swath along the Blake Escarpment, Blake Spur, western Blake Outer Ridge, and continental slope/uppermost rise from 33° to 42°N (see Scanlon, 1982a, b; Twitchell and Roberts, 1982). *References:* 1. Short-range sidescan (600 m total swath)-D. Twichell, (unpubl. data); 2. Mid-range sidescan (Sea Mark I)-O'Leary (1982). Asterisks show areas of mosaic photographic coverage by Woods Hole Oceanographic Institution's ANGUS survey system; 3. Mid-range sidescan (Sea Mark I)-D. O'Leary and B.A. McGregor (unpubl. data). Asterisks show areas of mosaic photographic coverage by Woods Hole Oceanographic Institution's ANGUS survey system; 4. Mid-range sidescan (Sea Mark I)-Robb et al. (1981c); 5. Mid-range sidescan (Sea Mark I)-Popenoe et al. (1981b) and Popenoe et al. (1982c); 6. McGregor et al. (1982b); 7. Mid-range sidescan (Sea Mark I)-Shell Offshore Inc. (unpubl. data); 8. Mid-range sidescan (Sea Mark I)-Shell Research and Development (unpubl. data); 9. Short-range sidescan, including 125 kHz narrowbeam echosounding, 4 kHz subbottom profiler, and intermittent photographic coverage (Johnson and Lonsdale, 1976); 10. Short-range sidescan, including other data as in (9) above (Hollister et al., 1974a); 11. Short-range sidescan and 3.5 kHz surveys at head of Hudson Canyon (D. Twichell, unpubl. data).

progradation, seafloor erosion, and control by reefs, faults, and diapirs. The slope is distinguished by its *relative* steepness compared with other margin provinces. For purposes of discussion the slope is divided into three segments along the margin, each significantly different in character: 1) Georges Bank to Cape Hatteras, 2) Florida-Hatteras Slope south of Cape Hatteras, and 3) Blake Escarpment (Fig. 4.1).

Slope from Georges Bank to Cape Hatteras

The top of the continental slope in this region is the shelf break, which occurs between 40 and 160 m water depth. Near the shelf break are old shorelines developed during Pleistocene sea-level lowstands; these are Franklin Shore, near 120 m, and Nichols Shore, at 140 m water depth. These shorelines are best defined in seafloor morphology around Hudson Canyon and south to Norfolk Canyon. The base of the continental slope, defined by marked change to a lower seafloor gradient on the upper continental rise, occurs near 2000 m, but varies above or below this level by 200-300 m along strike. Average seafloor gradient of the steepest 600-m depth interval of the slope is about 8°, and it varies from about 4° to 11° (Emery and Uchupi, 1972). Average gradient of the continental slope over its entire depth range is between about 3° and 6° (Heezen et al., 1959). Average width of the slope in this area is about 30 km, but it varies from about 10 km to 50 km. However, submersible observations off Cape Hatteras show that the slope there actually is characterized by nearly vertical scarps separated from one another by more gradual gradients (Milliman et al., 1967).

The continental slope north of Cape Hatteras is abundantly incised with submarine canyons. At least 70 large canyons (most of which are named) and numerous smaller gullies have been identified by bathymetric surveys between Cape Hatteras and the northeast end of Georges Bank (Fig. 4.1). However, even with densely gridded echosounding surveys, side echoes and contouring ambiguities mask the true

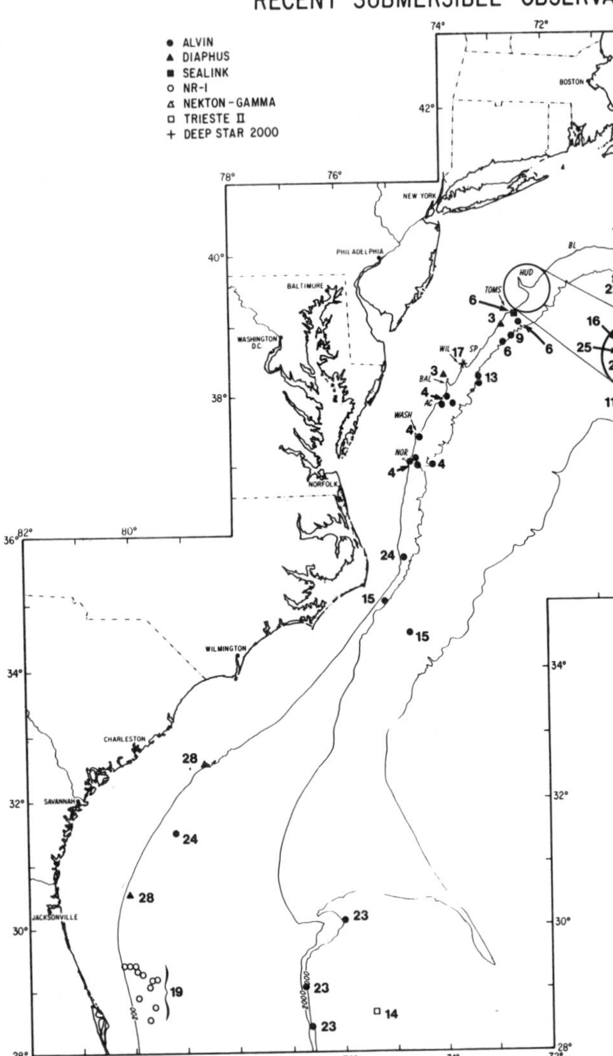

Figure 4.4 Summary of recent submersible observations (≳100m) along the U.S. Atlantic slope and rise. *References:* 1. Ryan et al. (1978); 2. Nastav et al. (1980); 3. Slater et al. (1981); 4. Malahoff et al. (1982), and Malahoff et al. (1980); 5. Hecker and Logan (1981); 6. Hecker et al. (1981); 7. Slater (1981); 8. Valentine et al. (1980); 9. Robb et al. (1983); 10. Valentine et al. (in press); 11. Uchupi et al. (1982b); 12. MacIlvaine (1973), and Ross and MacIlvaine (1978); 13. Stubblefield et al. (1982); 14. Flood (1981); 15. Stanley et al. (1981); 16. Stanley and Freeland (1978); and Knebel (1979); 17. Stanley (1974a); 18. J.R. Uzmann and P. Valentine (unpubl. data); 19. W. Gardner and C.D. Hollister (unpubl. data); 20. Rowe (1972); 21. Heirtzler et al. (1977a), and Houghton et al. (1977); 22. Emery and Uchupi (1972); 23. Gilbert and Dillon (1981); 24. Milliman et al. (1967); 25. C.D. Hollister and A. Malahoff (unpubl. data); 26. Keller et al. (1973); 27. Uzmann et al. (1977); 28. Ball (1978); 29. D. Twichell (unpubl. data). For additional pre-1974 *Alvin* Dives, see: "Deep Submergence Research conducted during the period 16 June 1961 through 31 December 1973": Woods Hole Oceanographic Institution Technical Report 74-60 (1974).

number and form of submarine canyons. An excellent illustration of this fact is provided by recent long-range side-scan sonar (GLORIA II) surveys of the slope between Hudson and Baltimore Canyons (Twichell and Roberts, 1982). In this area alone, more than 50 canyons and gullies were identified in the GLORIA images (Fig. 4.6), and the position of the majority of canyon axes differed by 1-3 km from the axes contoured in the bathymetric data.

Twichell and Roberts' (1982) study brings out several important points that probably apply to most of the slope between Cape Hatteras and Georges Bank:

1. Canyons are not evenly spaced, but spacing tends to decrease with increasing gradient of the continental slope. Canyons are spaced 1.5 to 4 km apart where the gradient is more than 6° (e.g., between Lindenkohl and Wilmington Canyons), they are 2 to 10 km apart where the gradient is 3° to 5° (e.g., Lindenkohl to Mey Canyons), and canyons appear to be absent where gradients are less than 3°.
2. Canyons are "V"-shaped in cross section and often have steep walls with outcropping strata (see for example, McGregor et al., 1979).
3. Canyons are continuous from their heads to the base of the slope, where some stop but others continue as channels onto the continental rise.

Figure 4.5 (Next page, top) Size scale of sea-floor features along the U.S. Atlantic slope and rise, as discussed in text. Coverage and resolution of various seafloor imaging systems are shown at bottom.

Figure 4.6 (Next page, bottom) Interpretation of GLORIA sidescan record of continental slope between Baltimore and Hudson Canyons from Twichell and Roberts, 1982). Note difference of canyon positions in GLORIA data (heavy lines) and as inferred from National Ocean Survey bathymetric maps (light lines). Apparent slump blocks at lower left divert canyons and feeders toward the east in the Baltimore-Wilmington system.

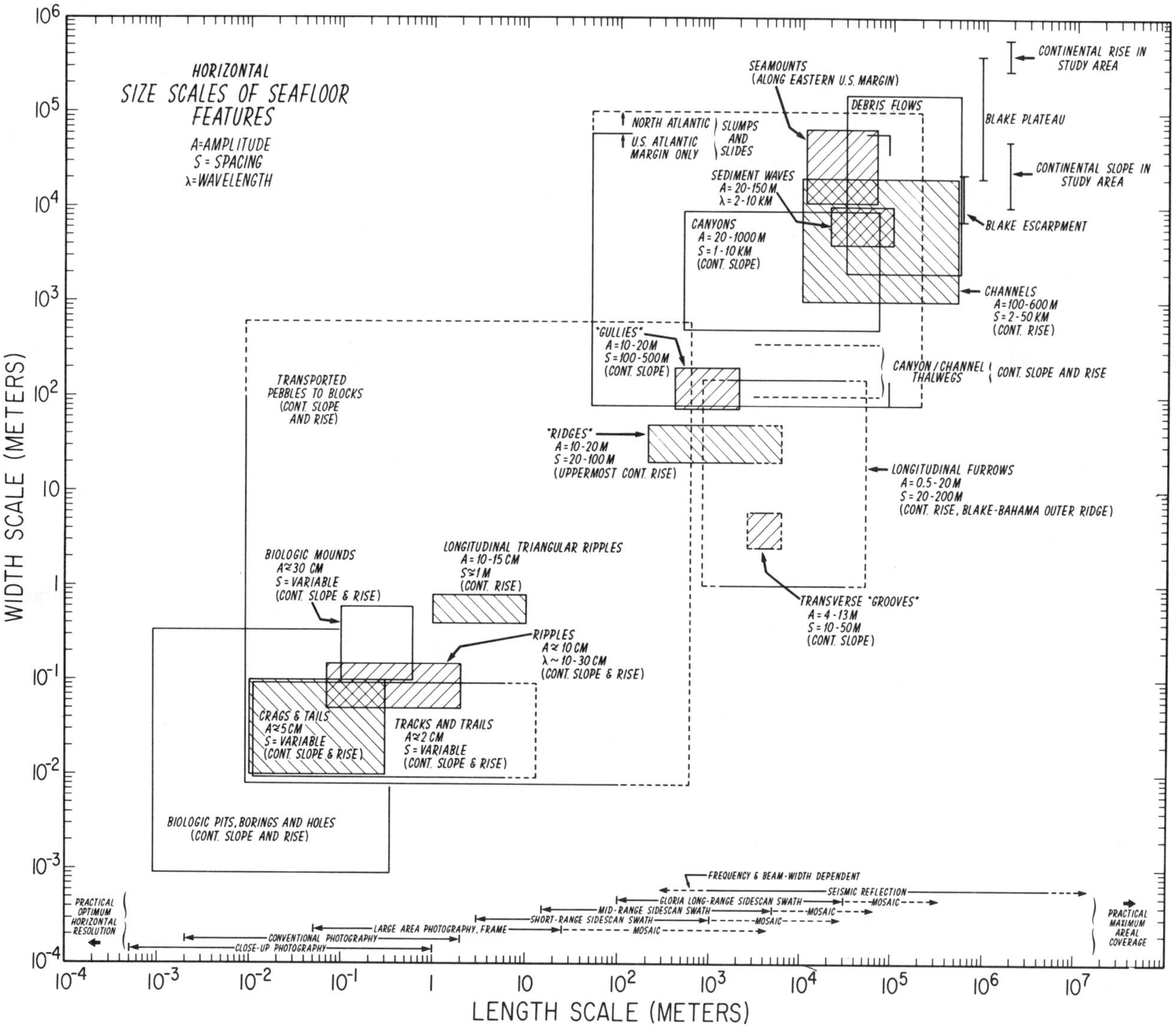

HORIZONTAL
SIZE SCALES OF SEAFLOOR
FEATURES
A=AMPLITUDE
S = SPACING
λ=WAVELENGTH

WIDTH SCALE (METERS)
LENGTH SCALE (METERS)
10^6
10^5
10^4
10^3
10^2
10
1
10^{-1}
10^{-2}
10^{-3}
10^{-4}
10^{-4} 10^{-3} 10^{-2} 10^{-1} 1 10 10^2 10^3 10^4 10^5 10^6 10^7

SEAMOUNTS
(ALONG EASTERN U.S. MARGIN)
DEBRIS FLOWS
NORTH ATLANTIC
U.S. ATLANTIC MARGIN ONLY
SLUMPS AND SLIDES
SEDIMENT WAVES
A= 20-150 M
λ= 2-10 KM
CANYONS
A = 20-1000 M
S = 1-10 KM
(CONT. SLOPE)
CONTINENTAL RISE IN STUDY AREA
BLAKE PLATEAU
CONTINENTAL SLOPE IN STUDY AREA
BLAKE ESCARPMENT
CHANNELS
A=100-600 M
S = 2-50 KM
(CONT. RISE)
"GULLIES"
A=10-20 M
S=100-500 M
(CONT. SLOPE)
CANYON/CHANNEL THALWEGS
CONT. SLOPE AND RISE
"RIDGES"
A=10-20 M
S=20-100 M
(UPPERMOST CONT. RISE)
LONGITUDINAL FURROWS
A= 0.5-20 M
S= 20-200 M
(CONT. RISE, BLAKE-BAHAMA OUTER RIDGE)
TRANSPORTED PEBBLES TO BLOCKS
(CONT. SLOPE AND RISE)
TRANSVERSE "GROOVES"
A = 4-13 M
S = 10-50 M
(CONT. SLOPE)
LONGITUDINAL TRIANGULAR RIPPLES
A = 10-15 CM
S≃1 M
(CONT. RISE)
BIOLOGIC MOUNDS
A≃30 CM
S = VARIABLE
(CONT. SLOPE & RISE)
RIPPLES
A ≃ 10 CM
λ ~ 10-30 CM
(CONT. SLOPE & RISE)
CRAGS & TAILS
A≃5 CM
S = VARIABLE
(CONT. SLOPE & RISE)
TRACKS AND TRAILS
A≃2 CM
S = VARIABLE
(CONT. SLOPE & RISE)
BIOLOGIC PITS, BORINGS AND HOLES
(CONT. SLOPE AND RISE)
FREQUENCY & BEAM-WIDTH DEPENDENT
SEISMIC REFLECTION
GLORIA LONG-RANGE SIDESCAN SWATH
MOSAIC
MID-RANGE SIDESCAN SWATH
MOSAIC
SHORT-RANGE SIDESCAN SWATH
MOSAIC
LARGE AREA PHOTOGRAPHY, FRAME
MOSAIC
CONVENTIONAL PHOTOGRAPHY
CLOSE-UP PHOTOGRAPHY
PRACTICAL OPTIMUM HORIZONTAL RESOLUTION
PRACTICAL MAXIMUM AREAL COVERAGE

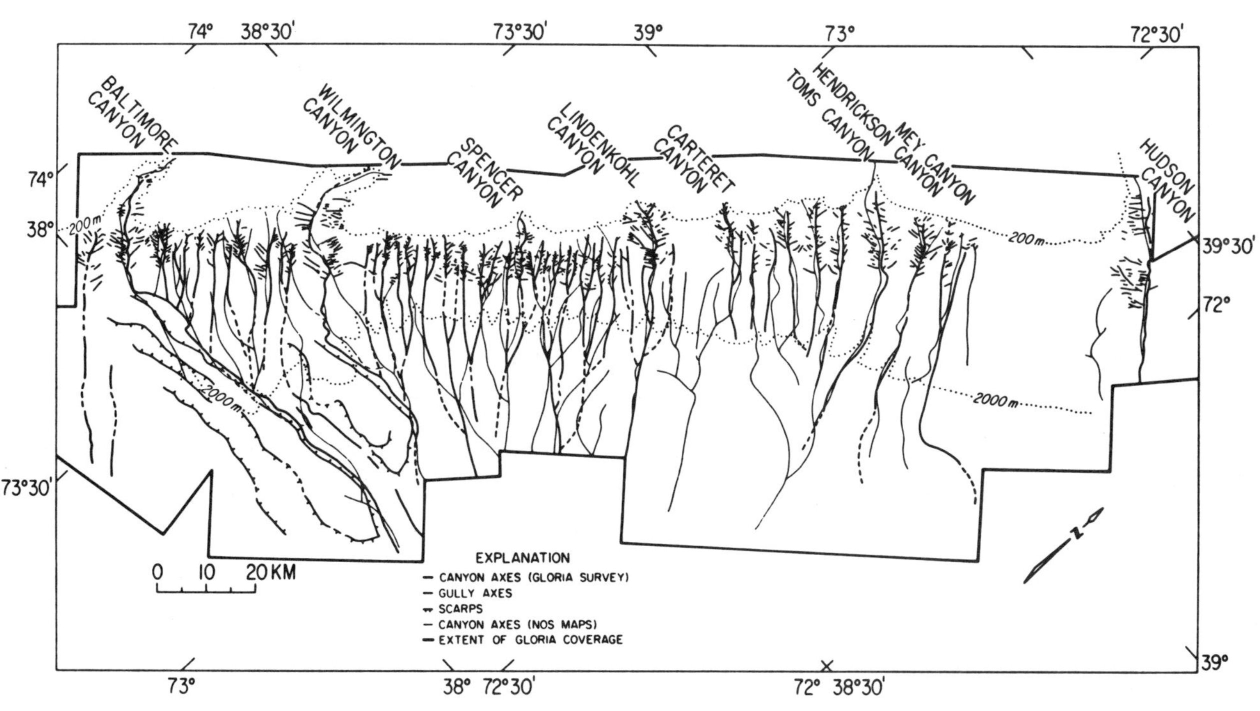

4. Both in general form and in change of form with age, the canyons are remarkably similar to their subaerial counterparts. Numerous tributary gullies intersect each canyon axis at high angles from both sides on the upper slope, creating a pinnate (and locally dendritic) "drainage" pattern. Canyon-wall slopes in this area are 6° to 30° and average 14°. Canyons often merge on the middle slope, lower slope, and upper rise to form a definite dendritic pattern. Gullies are uncommon here, where canyon-wall gradients are 2° to 30° and average 10°. Many canyons have heads well below the shelf break (commonly at 200-700 m water depth); these canyons mostly trend straight downslope and appear to be youthful (Quaternary) features. In contrast, larger and more deeply incised canyons cut into the shelf edge and they are usually more sinuous. These canyons are thought to be significantly older, and there is direct evidence that such canyons have experienced several episodes of filling and re-excavation. Examples include filled channels that have been re-excavated in Wilmington Canyon (Kelling and Stanley, 1970), some of the canyons south of Toms Canyon (Robb et al., 1981d), and canyons south of Baltimore Canyon (McGregor et al., 1979). In Oceanographer, Heezen, and Corsair Canyons, south of Georges Bank, Ryan et al. (1978) made submersible observations that suggest episodes of canyon filling and cutting extending back at least to the Late Cretaceous.
5. The implication is that many, if not all, submarine canyons may first form by mass-wasting processes on the continental slope. However, in several instances a canyon definitely can be traced to a fluvial drainage system via a cross-shelf valley (i.e., Wilmington, Hudson, and Block Island Canyons). In these cases it is possible either that the canyon was developed because of the fluvial link during sea-level lowstands, or that the pre-existing canyon eventually "trapped" the shelf river valley.

Some further details about the intricate morphology of these canyon systems are provided in narrower-swath (5 km) Sea Mark I side-scan images of Wilmington, South Wilmington, and North Heyes Canyons (McGregor et al., 1982b). Again, these features are likely to characterize canyons along much of the rest of the slope:

1. Crescentic scarps appear at the heads of North Heyes and South Wilmington Canyons on the upper continental slope. Thus it appears that these canyons are youthful and are still experiencing headward erosion. However, age of most recent slumping is unknown.
2. Intercanyon ridges are so dissected by gullies feeding into the canyons that gullies from opposite sides of the ridge may cut into one another (Figs. 4.6, 4.7). An arete is a good subaerial analog to many of these ridges. The gullies are 75 to 250 m wide, 10 to 20 m deep, and up to a kilometer or more long. Smaller "chutes" locally feed into the gullies, forming a complex dendritic (as opposed to "simple" pinnate) drainage system.
3. Where Wilmington Canyon crosses the lower slope and upper continental rise it clearly meanders. Meander wavelengths and amplitudes vary from less than a kilometer to more than 2 km. The meanders may result from the older, lower gradient canyon-axis profile. The meandering channel on the upper rise was investigated in submersible dives by Stubblefield et al. (1982), who found a 300 to 500 m wide channel with features similiar to the subaerial fluvial analog. The inner edge of a meander had a gentle slope (usually 5° to 20°, locally 30°), but at the outside of the bend the slopes were 30° to 90°. In addition, depressions paralleling and outside the outer wall of the bend suggest possible local slumping triggered by "bank undercutting".

The morphology of the present continental slope appears to be largely a product of Pleistocene sedimentary processes. These include 1) slope upbuilding and progradation by deltaic sedimentation principally during sea-level low-stands, 2) canyon-cutting by sediment mass movements (slides, debris flows, turbidity currents) during and following lowstands, and 3) sediment slumping (mostly rotation, with limited translation). The relative activity of these processes in the Holocene currently is uncertain.

Florida-Hatteras Slope

South of Cape Hatteras (~34°N) the continental slope has two characteristics (Fig. 4.1): 1) it is relatively smooth along strike and has few of the prominent canyon incisions so prevalent to the north, and 2) it splits into two slopes on either side of the Blake Plateau, one merging with the Blake Escarpment, and one (Florida-Hatteras Slope) continuing nearer the coastline in the more normal position of a continental slope.

The Florida-Hatteras Slope extends from depths of about 60 m down to 700 m at the western edge of the Blake Plateau. South of 31°N the slope is exceptionally smooth and has a uniform seaward slope of about one degree. North of the Charleston Bump, the slope surface is somewhat more irregular and the gradient across it averages less than 1/2° (Fig. 4.1). The shape of the Florida-Hatteras slope results from the interaction of seaward-prograding deltaic wedges with the edge of the Gulf Stream during the Cenozoic (Uchupi, 1967a; Zarudski and Uchupi, 1968; Uchupi and Emery, 1967; Uchupi, 1970). The Charleston Bump is relatively well defined as a delta-like feature. Alternate outbuilding and erosion of the slope appear to have occurred as the Gulf Stream shifted seaward during sea-level lowstands, and landward during sea-level high-stands (Pinet and Popenoe, 1981, 1982). No evidence exists in seismic reflection profiles that faults control the location of the slope as suggested by Sheridan et al. (1965).

Blake Escarpment

The Blake Escarpment extends nearly due north-south south of 29°30′N (Fig. 4.1). Here its slope averages 15°, but in many places the slope is so steep that no echo is returned. Submersible observations at two locations on the escarp-

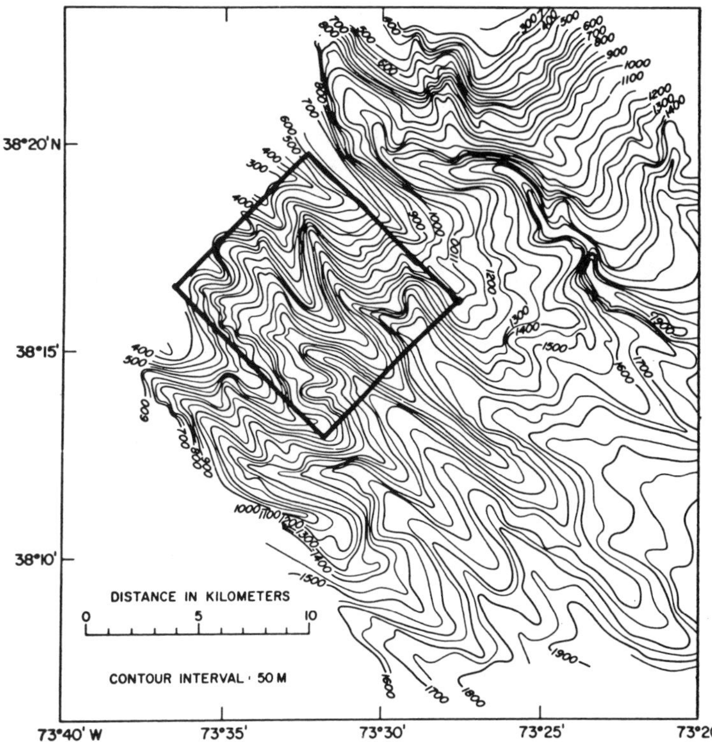

Figure 4.7 Sea Mark I mid-range sidescan sonar image of submarine canyons (North Heyes, South Wilmington), located by box in top figure. Note gullies and chutes that are tributary to canyon axes, and intersection of gullies from adjacent canyons along intercanyon ridge crests (arrows). From McGregor et al. (1982b).

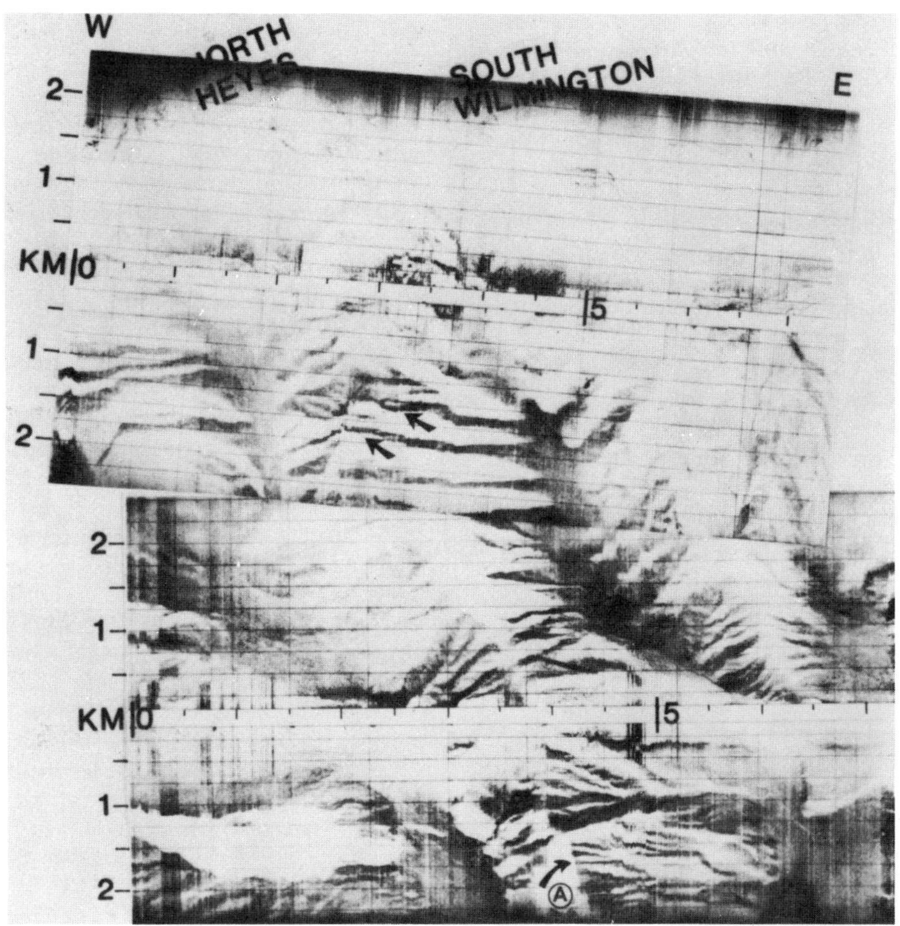

ment and on the Blake Spur show locally vertical scarps (W.P. Dillon and C. Paull, unpubl. data). The top of the escarpment in this area is at 1000 to 1200 m, and the base near 4800 m bounds the western edge of the Blake-Bahama Basin. At 30°N the Blake Spur (Blake Nose) forms a ENE-trending protuberance.

Seafloor across the Blake Spur dips eastward at about one degree to 2400-2600 m, then drops to 4800 m with the same gradient as that of the escarpment to the south. North of the Blake Spur, the escarpment merges with the continental slope. Average slopes between 30°N and 34°N are about 2°, although slopes locally reach 5°-6°. The Blake Escarpment owes its form to the presence of an Upper Jurassic to Lower Cretaceous reef and carbonate-bank system. The face of this edifice has been eroded and corroded by a combination of contour-current erosion, biologic erosion, jointing/spalling, and other mass movements (Paull and Dillon, 1980, 1981; Dillon and Paull, unpubl. data).

Blake Plateau

The Blake Plateau lies between the Florida-Hatteras Slope and the Blake Escarpment. It has a relatively smooth surface at depths of 700 to 1100 m north of 28°N (Emery and Uchupi, 1972) but has terrace-like depth intervals developed. These terraces occur near 800-900 m and 1000-1100 m, (as well as 1200-1220 m south of 28°N). Small terraces near 400 and 600 m can also be observed within the Florida-Hatteras Slope across the Charleston Bump. The terraces in the slope appear to have originated from submarine erosion by the Gulf Stream down to levels of resistant strata.

The only significant irregularities on the surface of the Blake Plateau occur along the western margin; they are elongate depressions, and mounds formed by deep-water corals (Fig. 4.8). The depressions follow the base of the Florida-Hatteras Slope, and they continue northeast along the Charleston Bump to 32°10′N. Seismic reflection profiles (Bunce et al., 1965; J. Ewing et al., 1966; Pinet et al., 1981b) show that the depression walls truncate reflecting interfaces in the adjacent sediments. Pratt and Heezen (1964) and Pratt (1966) concluded that the depressions are produced by erosion at the base of the Gulf Stream (see also Pinet and Popenoe, 1982). However, there is some evidence that seaward escape of groundwater from beneath the Florida-Hatteras slope during Pleistocene sea-level lowstands may have contributed to formation of the depressions by solution and karst development (Emery and Uchupi, 1972). The possibility that "fresh" groundwater is still escaping from some of the depressions is emphasized by the fact that the research submersible *Aluminaut* lost 400 kg of buoyancy as it crossed one depression in a 1966 dive (Manheim, 1967).

The positive-relief features distributed around and among the depressions in this area are deep-water coral mounds or reefs (Stetson et al., 1962, 1969; Pratt 1966, 1968; Ayers and Pilkey, 1981; Pinet et al., 1981a; Popenoe et al., 1981a, 1982). These coralline mounds are still actively accreting, and they vary in height from less than 30 m to more than 150 m (Fig. 4.8).

Figure 4.8 (Next page) Seismic profiles across northern Blake Plateau, located on lower left track chart by bold arrows (from Pinet et al., 1981b; note reefs (top) and erosional depressions (bottom). Inset at lower right shows detailed bathymetry in area of northern profile (from Stetson et al., 1969); note location of known coral mounds (dots) and elongate depressions. →

Continental Rise

The continental rise extends from beyond the northern edge of the study area off Georges Bank southward to 33°N latitude. The Blake Outer Ridge between 29°N and 33°N is discussed separately because, although it is continuous with the continental rise, its origin differs significantly from that of the rise. There is no continental rise present along the Blake Escarpment (Fig. 4.1).

The continental rise begins at the base of the continental slope near 2000 m water depth and extends seaward to a juncture a) with the Hatteras Abyssal Plain (33°N to 35°N, 5450 m to 5200 m), b) with the westernmost Bermuda Rise (35°N to 37°N, 5200 m to 4950 m), and c) with the Sohm Abyssal Plain (north of 37°N, ~5000 m). The seafloor gradient of the continental rise above 4000 m varies from about 0.2° to 1.0° and averages 0.5°. The general morphology varies from gently convex to gently concave. Seaward of the section of the margin between Cape Hatteras and Long Island, the lower continental rise (below 4000 m) contains a terrace extending from about 4000-4200 m to 4600 m with a slope of 0.10° to 0.15°. The seaward edge of the terrace is the Hatteras Outer Ridge. This ridge is formed of current-deposited sediments; it blocked turbidites traversing the continental rise from about late Miocene to late Pleistocene time (Tucholke and Laine, 1983) and caused them to pond and construct the terrace.

The continental rise generally is split into the upper continental rise (~2000 m to 3000 m), the central continental rise (3000 to 4000 m), and the lower continental rise (>4000 m). Because of an almost complete lack of detailed surveys (Fig. 4.2), the morphology of the continental rise is known only from widely spaced, mostly randomly acquired sounding lines. The fact that the continental rise looks less complex than the continental slope in bathymetric maps (Fig. 4.1) is partly a result of inadequate survey data, and partly a reflection of real geomorphic and geologic differences.

The larger (and probably older) canyons that dissect the continental slope continue as deep-sea channels on the upper and central continental rise, and in many locations they cross the lower rise as well (Fig. 4.1). Separate channels commonly coalesce on the rise into a single channel system. This is observed for the Wilmington/Baltimore Canyons, the Norfolk/Washington Canyons, and the Hatteras/Pamlico Canyons. The last group merges with the Hatteras Transverse Canyon near 5000 m on the landward side of the Hatteras Outer Ridge, and it continues from there south and east to the Hatteras Abyssal Plain.

By far the best-developed channels on the continental rise are those with sources at Hudson Canyon and in the Baltimore-Wilmington Canyon group. These incise the cen-

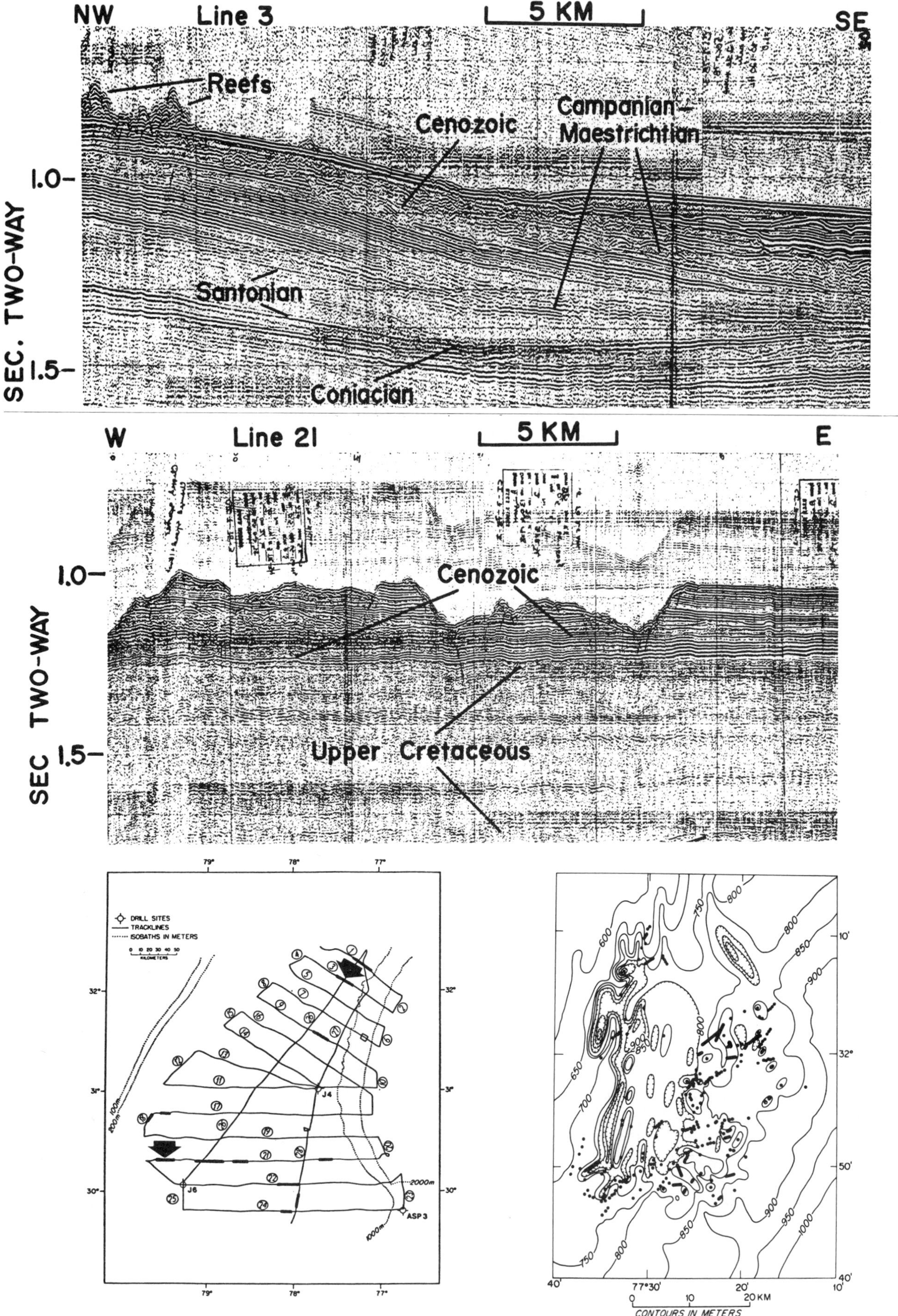
NW
Line 3
5 KM
SE
Reefs
Cenozoic
Campanian–
Maestrichtian
Santonian
Coniacian
SEC. TWO-WAY
1.0–
1.5–
W
Line 21
5 KM
E
Cenozoic
Upper Cretaceous
SEC TWO-WAY
1.0–
1.5–
DRILL SITES
TRACKLINES
ISOBATHS IN METERS
J4
J6
ASP 3
2000m
1000m
100m
200m
79°
78°
77°
32°
31°
30°
10'
32°
50'
40'
77°30'
20'
CONTOURS IN METERS
20 KM

tral continental rise by up to 800 m. Tributaries for the Baltimore-Wilmington Channel extend from just south of Baltimore Canyon as far north as Carteret or South Toms Canyon, and the multiplicity of sources converging at the top of the central rise may explain the strong channel development deeper on the rise. Hudson Channel also appears to tap a number of source canyons, possibly as far east as Atlantis Canyon (Fig. 4.1), but most of these canyons are smaller than those in the Baltimore-Wilmington system. However, Hudson Canyon itself is the best developed canyon on the U.S. Atlantic margin. Sediment load transported down Hudson Canyon and tributary canyons near the front of the ice sheet during Pleistocene lowstands may have equalled or exceeded that of more laterally extensive canyon/channel systems farther south, thus accounting for the strong development of Hudson Channel on the central continental rise.

There is some asymmetry in the development of the right and left banks of Hudson and Baltimore-Wilmington Channels, the right bank (facing downslope) being up to about 170 m higher (Pratt, 1967). This is expected because the Coriolis force causes channel-bank overflow of turbidity currents on the right side of the channel (northern hemisphere, looking downchannel). Ultimately the same effect of differential deposition should force northward migration of the channel, as is suggested in the northward-bowed channel axes across the 3000 m isobath.

Most of the smaller, less developed canyons that incise only the lower part of the continental slope do not appear to retain their expression on the continental rise. They may either end at the base of the continental slope or they may merge to form tributary channels that reach greater depths. Detailed seismic reflection and sidescan sonar surveys are needed to resolve this question.

Meager detailed survey data (Fig. 4.2, 4.3) suggest that the uppermost rise is strongly affected by processes other than just channelling associated with canyons. GLORIA data from the slope and uppermost rise off Baltimore and Wilmington Canyons indicate large east-trending ridges at the base of the slope (Fig. 4.6). These ridges have diverted the Baltimore and Wilmington Channels to the east for distances of about 50 km and 20 km, respectively. The ridge south of Wilmington Channel has been interpreted as a slump block by McGregor and Bennett (1979), and the ridge blocking Baltimore Channel may have a similar origin. McGregor et al. (1982a,b) found that the ridge below Wilmington Channel has a smooth crest and is blanketed by about 100 m of conformable sediments. In contrast, along the south side of the Wilmington ridge a series of small ENE-trending "ridges" occur. These "ridges" are 10 to 20 m high, are spaced 30 to 100 m apart, and consist of clay beds dipping steeply to the northwest (Stubblefield et al., 1982). If the Wilmington ridge is a slump block as surmised by McGregor and Bennett (1979), then these small features may be pressure ridges at the toe of the block.

As noted earlier, Wilmington Channel on this uppermost part of the continental rise has a meandering path. Most likely this results from the reduced seafloor gradient on the rise, and it may be characteristic of other upper-rise channels that lie below well-developed canyons incising the continental slope.

Near 3000 m water depth on the upper continental rise just south of Cape Hatteras (32° to 35°N), salt diapirism has caused the seafloor sediments to be fractured by small faults (Popenoe et al., 1982). At 33°N, two large diapirs in this zone reach nearly to the seafloor; associated salt tectonism created a slump some 60 km wide on the adjacent continental slope (see Fig. 4.16), and the slump scar is clearly defined in GLORIA side scan images.

The central and lower continental rise lie below the immediate zone of influence of mass wasting found near the base of the continental slope. These parts of the rise are generally smoother and are likely to be geologically more uniform. They have been developed dominantly under the combined influence of downslope flow of turbidity currents and the contour-following flow of southward-flowing bottom currents (e.g., Heezen et al., 1966a). The channels and levees developed by turbidity currents have been discussed. In contrast, a general smoothing of the rise probably is effected by the contour-following currents. The best geologic manifestation of the contour currents is developed in the "lower continental rise hills" forming the exposed east flank of the Hatteras Outer Ridge at 4600 to 5200 m water depth. Recent high-quality seismic reflection data clearly show that these hills are migrating sediment (mud) waves (Tucholke and Laine, 1983). No sediment waves are known to occur at the seafloor on shallower parts of the rise, but some moderately developed waves occur in the subsurface (Mountain and Tucholke, 1985).

Blake Outer Ridge

The Blake Outer Ridge extends continuously from the continental rise at 35°N, southeast to 28°N (Fig. 4.1). It narrows and deepens with distance from the continent; the crest is about 2000 m deep at the base of the continental slope and reaches 4000 m at 29°N. The seafloor gradient along the ridge axis is about 0.2°. Southeast of 29°N, the ridge descends more steeply (0.3°) to the edge of the Hatteras Abyssal Plain at 5400 m. The north flank of the outer ridge is remarkably straight, and on a regional scale it is smooth. Except for the continental slope north of Cape Hatteras and the Blake Escarpment, the north flank of the Blake Ridge at 30°N exhibits the steepest slopes along the U.S. Atlantic margin (up to 2.5°). The south side of the outer ridge has much gentler slopes, and a small depositional spur extends west from the ridge crest near 30°30′N. At greater depths (4500 m) to the south, the Blake Outer Ridge merges with the Bahama Outer Ridge. Like the Hatteras Outer Ridge that forms the lower continental rise to the north, the Bahama Outer Ridge exhibits well defined, migrating sediment waves deposited from contour-following bottom currents.

The Blake Outer Ridge is unique in that it is constructed almost entirely of silty clays transported by contour-following bottom currents. The Western Boundary Undercurrent (WBUC) flows southeast along the north flank of the outer ridge. At shallow depths (<4500 m) this current forms a northwestward return flow along the south flank of the outer ridge before turning south along the Blake Escarpment. At greater depths the boundary current follows the bathymetric contours of the Bahama Outer Ridge. Bryan

(1970) modelled the deposition of the Blake Outer Ridge by sediment fallout beneath a zone of interaction between the northward-flowing Gulf Stream and the deeper, equatorward flow of the WBUC. The model is reasonable because the outer ridge is entirely sedimentary in origin. There is no deeper basement structure that could have controlled the depositional locus (Tucholke et al., 1982).

Although the Blake Outer Ridge appears smoothly developed in bathymetric maps (Fig. 4.1), in detail there is considerable seafloor roughness. Seismic profiles show that the north flank of the outer ridge has been strongly eroded by the WBUC (Ewing and Hollister, 1972), and high-resolution 3.5-kHz profiles indicate local slumping in response to this erosion. On the south side of the outer ridge, hyperbolic echoes and echo focussing in 3.5-kHz profiles indicate that contour-parallel seafloor furrows are widespread (Hollister et al., 1974; Tucholke, 1979a).

Seamounts

Seamounts occur in two groups along the U.S. Atlantic margin. One group is at the western end of the New England seamount chain (Fig. 4.1). The westernmost edifice is Bear Seamount with a minimum depth of about 1100 m; its base is buried in sediments of the continental rise below 2500 m. The seamount is flat-topped, but visual observations by K.O. Emery and J.D. Milliman (unpubl. data) from the submersible ALVIN found no evidence of wave-base erosional truncation, only a blanket of calcareous sand. They also observed a nearly vertical basalt slope between 1250 m and 1340 m, although the average gradient of the seamount's flanks is much less.

Mytilus Seamount (minimum depth 2269 m) is not flat-topped but is capped by acoustically transparent sediments that overlie a prominent, flat unconformity and deeper stratified layers (Emery et al., 1970). There is little doubt that the unconformity was eroded at wave base and that the seamount has since subsided by about 2400 m.

Other seamounts in the New England chain above the 4000 m contour of the continental rise include Physalia Seamount (minimum depth 1880 m) and Retriever Seamount (1840 m). A short distance to the east and south are Picket Seamount (1690 m), Balanus Seamount (1470 m), and two smaller peaks with apices at 2670 m and 4050 m. All these seamounts have flank slopes generally exceeding 10°, and all have affected sedimentary patterns near their bases by 1) diverting seafloor channels that cross the continental rise, 2) ponding sediments against their shoreward flanks, or 3) diverting contour-following bottom currents flowing along the rise to form moats and drifts (e.g., Johnson and Lonsdale, 1976).

The second group of seamounts consists of three features mostly buried beneath the continental rise to the southwest of the New England seamount chain. Caryn Seamount (minimum depth 2900 m) on the lower continental rise near 4900 m is the most exposed (Fig. 4.1). To the northwest, Knauss Knoll rises only 900 m above the continental rise near 4000 m seafloor depth. A long sediment drift extends northeast from this peak; Lowrie and Heezen (1967) thought that the sediment lens was a "foredrift" emplaced by the southwesterly flowing WBUC. This is a hydrodynamically untenable conclusion, and its is more likely the drift is a "tail" deposited in response to northeast-directed flow at the base of the Gulf Stream. Nearby current measurements support this conclusion (Chapter 2).

The final "seamount" in this group is not a seamount at all because it is almost entirely buried beneath the continental rise. It occurs beneath the north flank of Hudson Channel where the channel makes a jog to the south (Fig. 4.1); its presence in fact may account for the unusual diversion of the channel. If not exposed along the canyon wall, the peak at least very nearly crops out there. The composition of this buried peak is unknown, but it is assumed to be basaltic.

Seafloor Sediments

Surficial sediments of the U.S. Atlantic continental margin have been studied both in regional and local contexts. More than 6000 sediment samples from the shelf, slope, and parts of the upper continental rise were studied for textural and compositional parameters; results were published in detailed reports by Field and Pilkey (1969), Hathaway (1972), Hollister (1973), Hulsemann (1967), Judd et al. (1970), Milliman (1972), Milliman et al. (1972), Pilkey et al. (1966), Pilkey et al. (1969), Pratt (1968), Ross (1970b), Schlee (1968), Schlee (1973), and Schlee and Pratt (1970). No comparably detailed study has been made for sediments of the continental rise itself. However, other specialized studies of sediments on the slope and rise have been conducted; those of particular interest include Gorsline (1963), Stanley et al. (1981), Stow (1976, 1979), Lambert et al. (1981), McGregor et al. (1979), Bennett et al. (1980), Klasik and Pilkey (1975), Field and Pilkey (1971) and Fritz and Pilkey (1975). The following summary is based largely on Milliman et al. (1972) and is supplemented by other references, particularly for the continental rise.

Texture of Surface Sediments

Most of the continental shelf is covered with sands and locally gravelly sands (Fig. 4.9, top). The finer-grained materials in these sediments have been winnowed out and transported either shoreward into estuaries or off the shelf and into canyons onto the continental slope. The continental slope below the "mud-line" north of Cape Hatteras is mantled mostly by silty clay and clayey silt which locally is sandy (Fig. 4.9). The Florida-Hatteras Slope south of about 31°N is texturally unique in that it is mantled by silty clays, whereas the adjacent continental shelf and the Blake Plateau are covered by sand. Uchupi and Emery (1967) and Hollister (1973) have attributed this textural province to deposition of fine-grained materials from a southward-flowing Gulf Stream counter-current which underlies the western edge of the Gulf Stream. Submersible dives in this area (Fig. 4.4) confirm the general southerly flow with relatively tranquil conditions (C.D. Hollister, pers. comm., 1983). Farther north, the entire northern Blake Plateau and the Charleston Bump are mantled by sands that locally contain gravels. In contrast, the southeastern Blake Plateau ("A" in Fig. 4.9, bottom) is covered by slightly clayey, silty sand and sandy

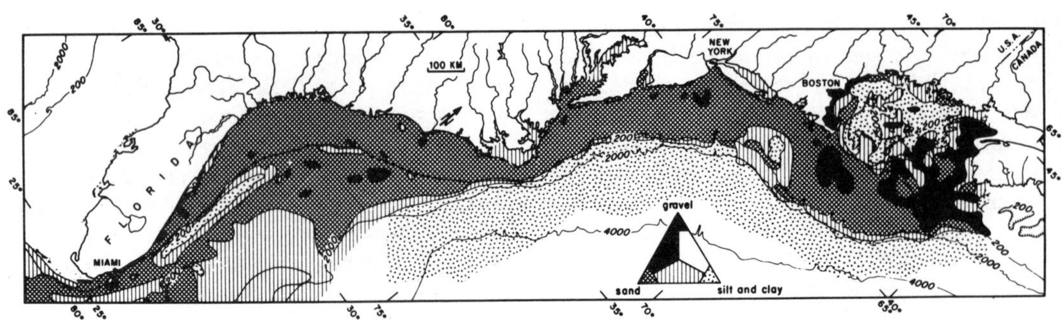

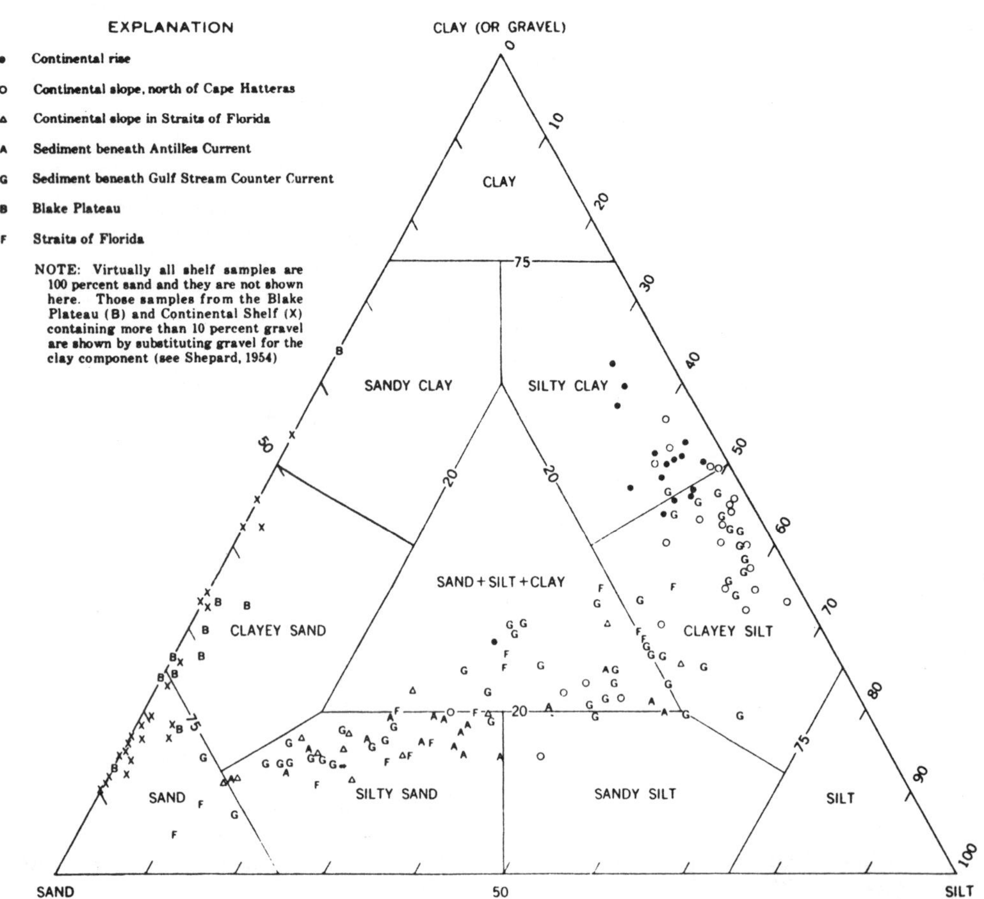

Figure 4.9 Top-Modal grain size of surface sediments along the U.S. Atlantic margin (from Milliman et al., 1972). Bottom-Ternary diagram of sediment type for the slope, rise, and Blake Plateau south of New Jersey (from Hollister, 1973).

silt. These sediments appear to be somewhat winnowed by the Antilles Current sweeping west to northwest across the Plateau to join the Gulf Stream. A similar texture is found along the continental slope extending north to Cape Hatteras. Presumably some of the sand on the slope north of the Blake Plateau is transported there by the Gulf Stream.

On the continental rise, surface sediments are almost invariably silty clays and clayey silts (Fig. 4.9). Mean grain size probably is slightly greater near the major channels crossing the rise (e.g., Hudson Channel, Bennett et al., 1980) because of channel-bank overflow of turbidity currents. Piston cores also show that silty and sandy beds up to a few centimeters thick underlie the finer surficial sediments on the rise (Ericson et al., 1952; Field and Pilkey, 1971).

Carbonate Content

Continental shelf sediments north of Cape Hatteras contain very little carbonate (less than 5 percent), but carbonate contents south of Cape are locally greater than 50 percent (Fig. 4.10). Cape Hatteras is an oceanographic barrier that separates southern, tropical waters from northern temperate waters (Cerame-Vivas and Gray, 1966; Stefansson and Atkinson, 1968), and this is reflected in production of carbonate-secreting organisms. On the Florida-Hatteras shelf and slope the carbonate consists of mollusks, barnacles, coralline algae, oolites, and echinoid fragments (Fig. 4.11); it comprises less than 50 percent of the sediment (mostly less than 25 percent) because of dilution by terrigenous detritus from the adjacent coast. The Blake Plateau has very high carbonate content, consistently more than 50 percent (Fig. 4.10). Within 50-100 km east of the Florida-Hatteras shelf and Charleston Bump, carbonate content on the Plateau exceeds 95 percent (Milliman et al., 1972). Most of this carbonate consists of planktonic foraminifera (Fig. 4.11), but coral fragments are an important component in the vicinity of the deep water mounds and linear depressions beneath the Gulf Stream axis.

Carbonate comprises about 5-20 percent of the continental slope sediments north of Cape Hatteras and consists mostly of planktonic foraminifera, with smaller amounts of echinoids, serpulids, benthonic foraminifera and, in the Fig. 4.9. south, pteropods. On the continental rise the carbonate content averages about 30 percent in a band extending north and northeast beneath the warm Gulf Stream waters (Fig. 4.10). The position of this band is determined mostly by productivity associated with the Gulf Stream, but also by some minor dilution on the landward side, and by dissolution on the deep seafloor southeast of the Gulf Stream. The crest of the Blake Outer Ridge has high carbonate content (50 percent), partly because carbonate swept off the Blake Plateau by the Gulf Stream settles into the Western Boundary Undercurrent and is transported to and deposited on the outer ridge. The decrease in carbonate content with depth on the flanks of the outer ridge is largely a function of dissolution with depth.

Composition of Non-carbonate Sediments

Coarse Fraction

Milliman et al. (1972) studied the 125-250 mm fraction of continental-margin surface sediments to determine sediment provenance and patterns of reworking. Their results, summarized below, are representative of patterns in finer sands and silts, and they also agree with clay-mineral distribution patterns discussed later. Most of the non-carbonate coarse fraction is comprised of quartz and feldspar, although there are local concentrations of glauconite and phosphorite. Heavy minerals (e.g., amphibole, epidote, garnet, staurolite, mica) comprise up to a few percent of the coarse fraction, and siliceous biogenic detritus (radiolarians, diatoms, and spicules) are present only in trace amounts. The ratio of feldspar to feldspar plus quartz (f/f + q) in the 125-250 μm fraction is a useful way of expressing the maturity of a sediment, with high ratios (>0.25, arkosic) being immature and low ratios (<0.05, ortho-quartzitic) indicating maturity. The plot of f/f + q in Figure 4.11 shows a general decrease in feldspar content southward along the margin that reflects the nature of the source areas. Mechanical weathering of Paleozoic and Proterozoic rocks in the north has contributed feldspar-rich sediments to the margin, most prominently during past glacial intervals, while predominantly chemical weathering in the south has preferentially removed unstable feldspars. Four prominent bands of feldspar-rich sediments extend across the shelf and slope between Cape Hatteras to New Jersey and are thought to locate major drainages from glacier-covered terrain during the last glacial interval 15-27,000 yr. B.P. (Fig. 4.11). The feldspar-rich zone along the Florida-Hatteras Slope is not in equilibrium with

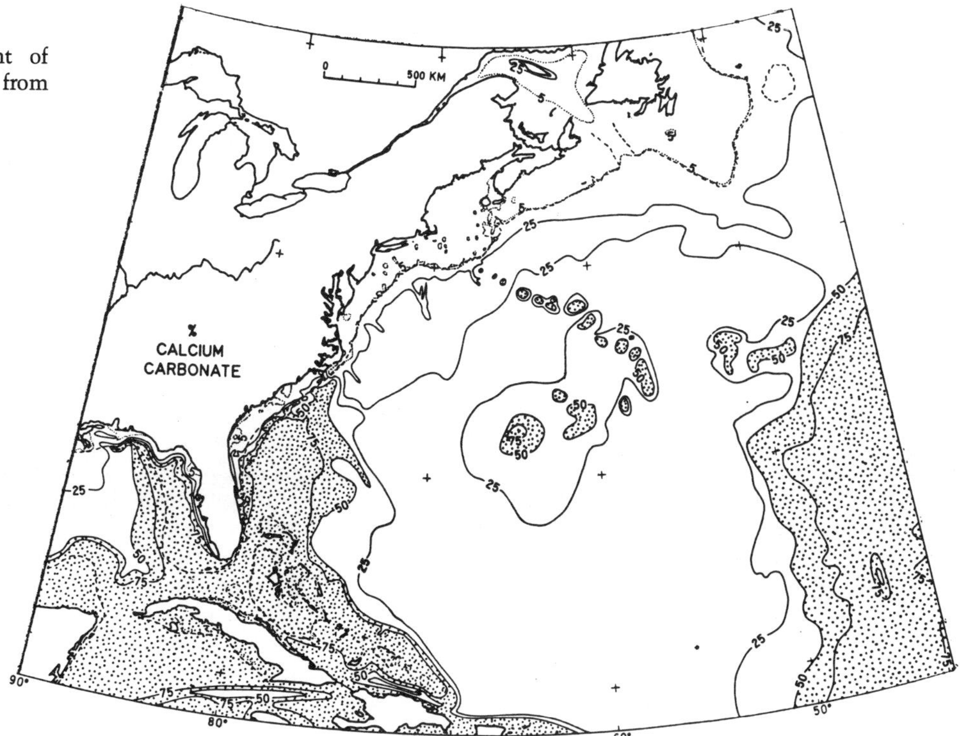

Figure 4.10 Carbonate content of surface sediment (% dry weight), from Emery and Uchupi (1972).

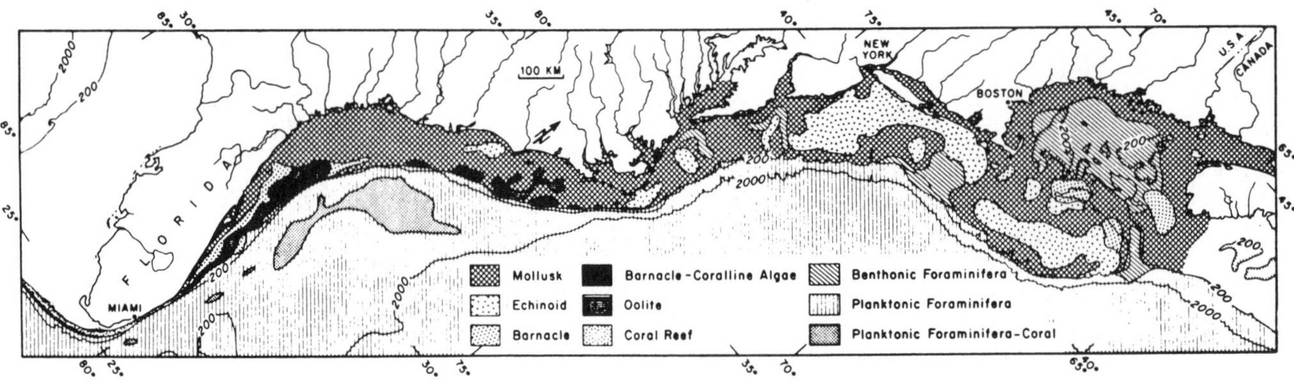

Carbonate assemblages on the continental margin.

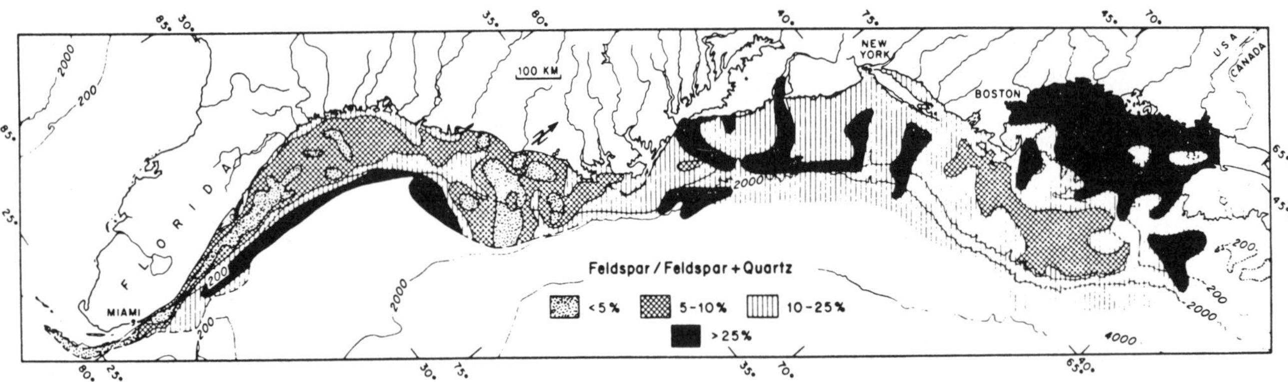

The ratio of feldspar to feldspar + quartz (f/f + q) in the 125 to 250 micron fraction of the surface sediments.

Figure 4.11 Left-composition of the carbonate fraction of surface sediments on the U.S. Atlantic margin. Right-Ratio of feldspar to feldspar-plus-quartz in surface sediments. Both are from Milliman et al. (1972).

Quaternary conditions and probably results from erosion into and exposure of feldspathic Tertiary sediments that blanket the western Blake Plateau.

Glauconite is a locally important mineral on the Florida-Hatteras Slope and the continental slope (Fig. 4.12, left). Glauconites along the Florida-Hatteras Slope north of 28°N and on the Charleston Bump are generally dark green in color, which suggests that they are mineralogically and chemically mature. Like the feldspars in this area, the glauconites are probably weathered from nearby Tertiary outcrops (Milliman, 1972). Dark grains of glauconite farther north on the shelf probably are derived from erosion of Cretaceous and Tertiary coastal-plain sediments; on the adjacent slope they may have the same source or they may be derived from local outcrops of Tertiary strata (Fig. 4.13).

Milliman et al. (1972) also categorized sediment types along the U.S. Atlantic margin according to their overall composition (Fig. 4.12, center). This classification reflects the interaction of sediment source areas, including the contribution of biogenic carbonate from bottom-dwelling and planktonic organisms. Seaward of 2000 m, the continental rise can be classified as a low-carbonate mud, except that muds with high to moderate carbonate contents extend seaward along the axis of the Blake Outer Ridge and northeast beneath the axis of the Gulf Stream.

A "source map" compiled by Milliman et al. (1972) shows that surface sediments have been derived from four major sources: rivers, glaciers, nearby outcrops of older strata, and biogenic productivity (Fig. 4.12, right). "Modern" sediments on the continental slope, the eastern Blake Plateau, and the continental rise are derived principally from two sources: surface-water biogenic productivity, and fine-grained sediment reworked from shelf, slope, and rise areas and redeposited by bottom currents. Rivers draining eastern North America carry little sediment to the continental shelf (Milliman and Meade, 1983), and most of that sediment is trapped in estuaries or coastal marshes (Meade, 1969). Relict sediments, those left from an earlier (glacial) depositional cycle, are concentrated on the continental shelf and consist of fluvial and carbonate detritus and glacial outwash. Fluvial and glacial outwash deposits may have locally reached the upper continental slope north of Cape Hatteras, although the southern limit of grounded glacial ice was about 41°N along a line from Long Island to the northern part of Georges Bank. Residual sediments, having been reworked from nearby outcrops, cover the western Blake Plateau and they

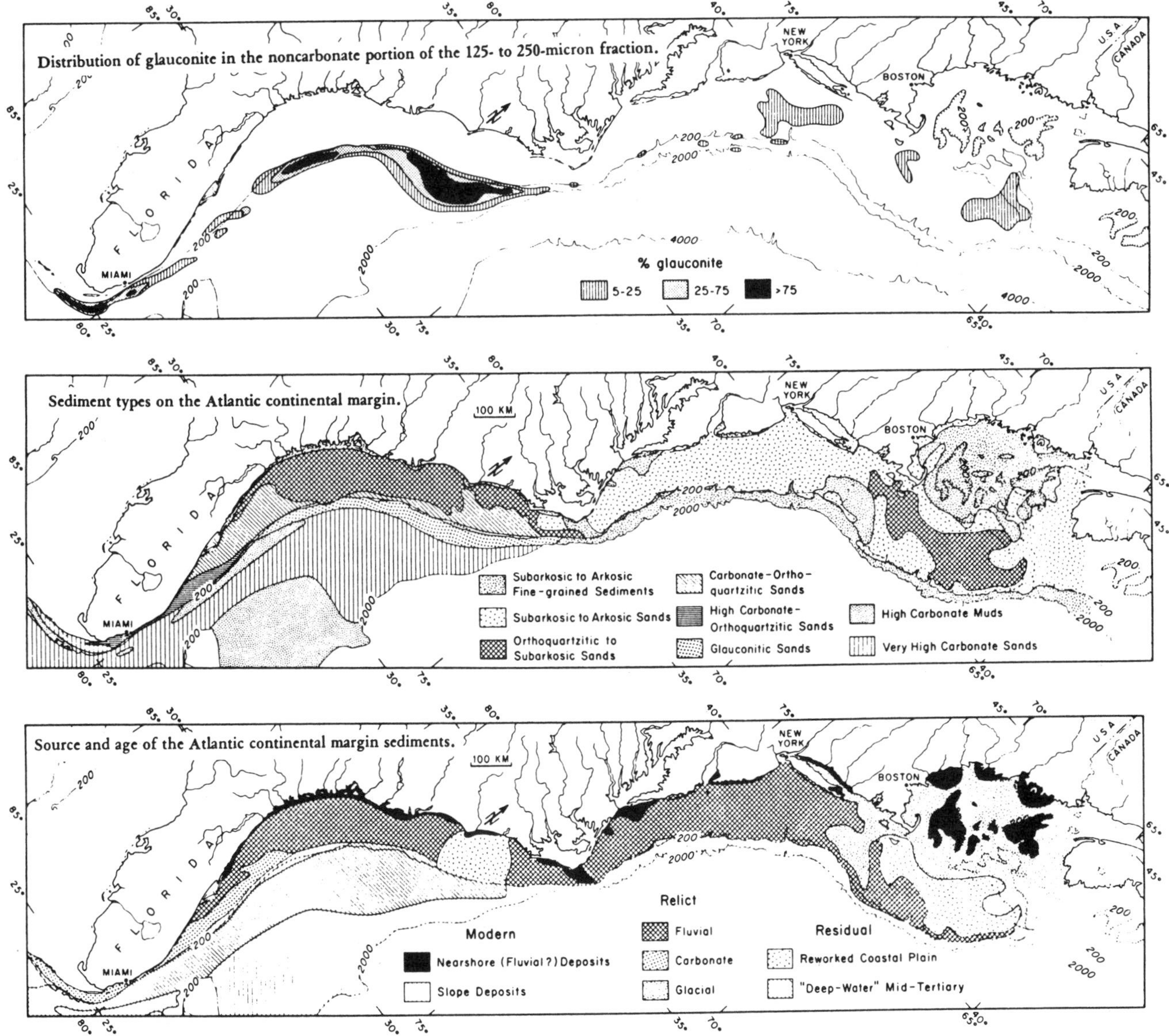

Figure 4.12 Composition and classification of surface sediments along the U.S. Atlantic margin (from Milliman et al., 1972).

may be locally important on the continental slope north of Cape Hatteras.

Mineralogy of the Clay Fraction

Clay mineralogy of U.S. Atlantic margin sediments shows very much the same sediment source and distribution patterns that are observed in the coarse-fraction mineralogy. Studies of the carbonate-free, $<2\ \mu m$ size fraction by Biscaye (1965) and Hathaway (1972) show that there are distinct northern and southern mineral assemblages, divided roughly at Cape Hatteras. The southern assemblage is characterized by kaolinite and montmorillonite derived from low-latitude chemical weathering of exposed continental rocks and sediments (Fig. 4.14). Clay minerals of the Blake Plateau and Florida-Hatteras Slope include 40-50 percent each of kaolinite and montmorillonite, whereas on the continental slope farther north only 10-30 percent kaolinite and 10 percent montmorillonite are present. The same patterns persist offshore onto the continental rise (Biscaye, 1965).

In contrast, mechanical weathering becomes progressively more important north of Cape Hatteras and the clay minerals illite and chlorite dominate there (Fig. 4.15). Amphibole and feldspar are important trace constituents in the $< 2\ \mu m$ fraction of these higher-latitude sediments. On the continental rise illite commonly comprises 50 percent of the clay-size fraction (Biscaye, 1965). Chlorite is present there in concentrations of about 15-20 percent; both it and illite are distributed southward along the continental rise and Blake-Bahama Outer Ridge by the Western Boundary Undercurrent (Tucholke, 1975).

Hathaway (1972) found no evidence that diagenesis was responsible for variations in clay mineralogy. Furthermore, diagenesis seems to be unimportant in components of the coarser fraction of surface sediments along the U.S. Atlantic margin. Diagenetic processes which have affected the subsurface sediments are discussed in later sections.

Organic Matter

The highest organic-carbon contents (locally greater than 2 percent, uniformly greater than 1 percent) occur along the continental slope and uppermost continental rise (Emery and Uchupi, 1972). Upper rise sediments typically contain 0.5 to 1 percent organic carbon, and similar values are found in the fine-grained sediment lens deposited beneath the Gulf Stream countercurrent on the Florida-Hatteras Slope. Sediments on the continental shelf, Blake Plateau, and central and lower rise have less than 0.5 percent organic carbon. Nitrogen follows a similar pattern. The pattern of organic carbon and nitrogen distribution probably is related to grain size, and finer-grained sediments contain higher concentrations of organic material (Emery and Uchupi, 1972). High C and N contents mark locales where fine-grained deposits entrap organic debris, especially beneath the Gulf Stream and beneath productive slope waters. This pattern is accentuated by regeneration (oxidation) of organic matter in the deeper waters over the continental rise and in the turbulently mixed waters of the continental shelf.

Sediment Accumulation Rates

As already discussed, most of the continental shelf is being winnowed by shallow currents and is undergoing net erosion. Much of the inner and mid-shelf sediment appears to be transported landward into estuaries, but a significant fraction of mid- and outer-shelf sediment has been transported seaward during the Holocene epoch to the continental slope and rise. Detailed patterns of sediment accumulation are poorly known along the continental slope, but net Holocene accumulation rates vary from about 10 to 40 cm/1000 yr (Emery and Uchupi, 1972). Embley (1980) determined rates of 10-15 cm/1000 yr for hemipelagic sediments in two cores from the slope south of Baltimore Canyon and a rate of about 15 cm/1000 yr for core material on the slope between Washington and Norfolk Canyons. Emery and Uchupi (1972) have identified an area on the slope immediately east of Cape Hatteras where rates are probably greater than 40 cm/1000 yr, but this is exceptional.

The continental slope also has many areas where net Holocene accumulation rates are zero, or possibly negative. This is immediately made apparent by the common occurrence of pre-Quaternary outcrops along the margin (Fig. 4.13). Most of these exposures occur within submarine canyons and result from long-term erosion, or from cyclic erosion/deposition patterns that have affected at least some canyons since Cretaceous time (Ryan et al., 1978). Other exposures date from early Oligocene time when massive slope erosion was accomplished by bottom currents, and during an ensuing mid-Oligocene sea-level lowstand, as described later in this report. The continued exposure or near exposure of these older beds shows that the Holocene net positive accumulation rates noted above are atypical of Upper Cenozoic sedimentation patterns. Furthermore, in relatively small areas it is entirely possible to find both positive *and* negative rates of Holocene sediment accumulation. Detailed deep-towed high-resolution profiling coupled with accurately located cores could be used to define the local variations in sedimentation patterns. At present no such studies have been attempted off the eastern U.S.

Accumulation rates on the upper continental rise are about the same as the net positive Holocene rates on the continental slope, but rates generally decrease with depth across the continental rise. A C-14 date in a box core from the upper continental rise at 2800 m water depth, east of Accomac Canyon, yielded an accumulation rate of about 7 cm/1000 yr; a rate of 5 cm/1000 yr also has been determined by C-14 for the lower continental rise (4600 m) just north of the Blake Outer Ridge (Embley, 1980). However, rates on the continental rise locally can be higher, for example, next to submarine channels; Klasik and Pilkey (1975) reported C-14-determined accumulation rates of 12 to 13 cm/1000 yr at about 3500 m water depth just south of the Hatteras Canyon system. *Within* submarine channels rates can be zero or negative, as demonstrated by Pliocene and Miocene core samples recovered from Hudson and Wilmington Channels on the central continental rise (Fig. 4.13). On the north flank of the Blake Outer Ridge Klasik and Pilkey (1975) measured a sediment accumulation rate of only 2 cm/1000 yr at depths of about 3500 m. This rate is in accord with conditions of slow net accumulation or erosion inferred from 3.5 kHz profiles in the area, as noted earlier.

Few data are available from the Blake Plateau, and accumulation rates can only be deduced from seismic-reflection observations of sediment distribution. Most of the western Plateau beneath the Gulf Stream, especially at the base of the Florida-Hatteras slope and across the Charleston Bump, probably exhibits zero or net negative sediment accumulation rates (Figs. 4.8, 4.13). The fine-grained sediments beneath the Gulf Stream countercurrent on the Florida-Hatteras slope, however, probably are slowly accumulating. Common recovery of pre-Quaternary sediments in piston cores from the eastern Blake Plateau (Fig. 4.13) also indicates near-zero accumulation rates. The Blake Escarpment is undergoing continued erosion, and sediments as old as Cretaceous are exposed (e.g., Paull and Dillon, 1980).

Dynamic Sedimentary Processes

Gravity-Controlled Mass Movements

Introduction

The fine-grained nature of slope and rise sediments plus the locally moderate to steep gradients are conducive to downslope sediment movement. Slumps, slides, debris flows and

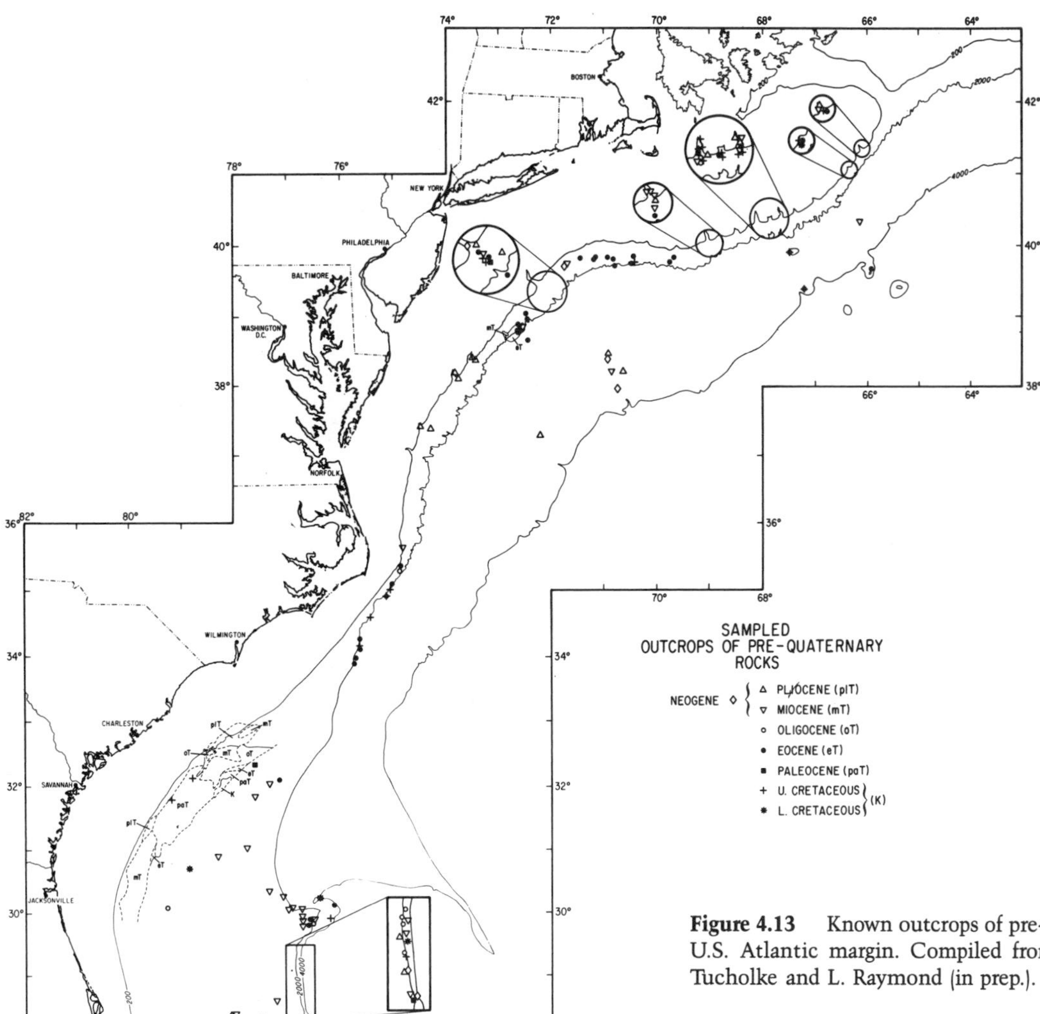

Figure 4.13 Known outcrops of pre-Quaternary rocks along the U.S. Atlantic margin. Compiled from numerous sources by B. Tucholke and L. Raymond (in prep.).

turbidity currents are here considered members of a continuous spectrum of gravity-controlled sedimentary processes. It is very likely that slumps and slides trigger debris flows and turbidity currents. It is also likely that there is a gradational change between debris flows and turbidity currents in many, if not all, instances.

Slumps are defined as mass movements where a "block" of sediments fails en masse but undergoes mostly rotational movement with little downslope translation. The upper perimeter of the slump is marked by a fault or slump scar, and there is little internal deformation within the block. The thickness of the slump block usually represents a small (e.g. 5-10 percent) fraction of the block's lateral dimensions; lateral dimensions can range from tens of meters to tens of kilometers and occasionally are greater.

Slides are mass movements where the block undergoes significant translation but mostly retains its character as a "block". Internal deformation can range from slight (e.g., mostly along the subjacent glide plane) to intense. Slide thicknesses in relation to lateral dimensions are typically less than in slumps. The lateral dimensions involved are comparable to those of slumps but the large slides rarely retain their identity as coherent blocks.

Debris flows consist of centimeter- to meter-scale sedimentary clasts contained within a quasi-fluid flow of finer matrix debris. Debris flows probably are created by the degeneration of slumps and slides, or at their toes, and they can travel hundreds of kilometers over slopes of less than ½° (Embley and Jacobi, 1977; Embley, 1980).

Turbidity currents are fluid density flows that transport usually gravel-sized and finer particles. With the exception of small indurated sediment clasts, most material in a turbidity current consists of disaggregated grains. Turbidity currents can travel more than a thousand kilometers over slopes less than 0.05° (as on abyssal plains).

Figure 4.16 summarizes the major mass movements and associated geologic features known to be present along the U.S. Atlantic continental margin. Features on this map are generalized in some places because of their small scale or close spacing. It should also be kept in mind that the apparent variation in frequency of mass movements along the margin is not entirely clearcut. Data density is highly variable, and interpretations can vary significantly between investigators. Given these caveats, the following discussion summarizes the overall picture of mass movements along the margin.

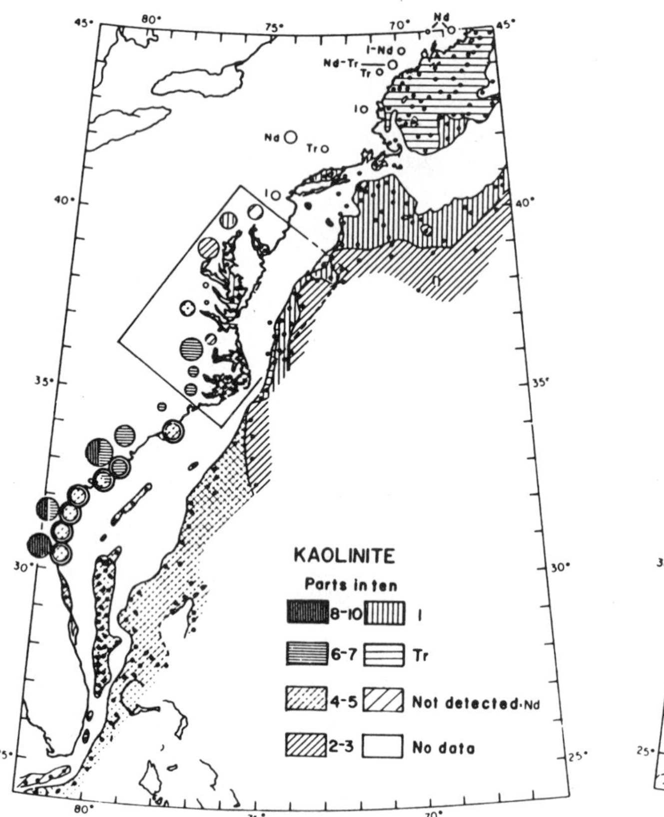

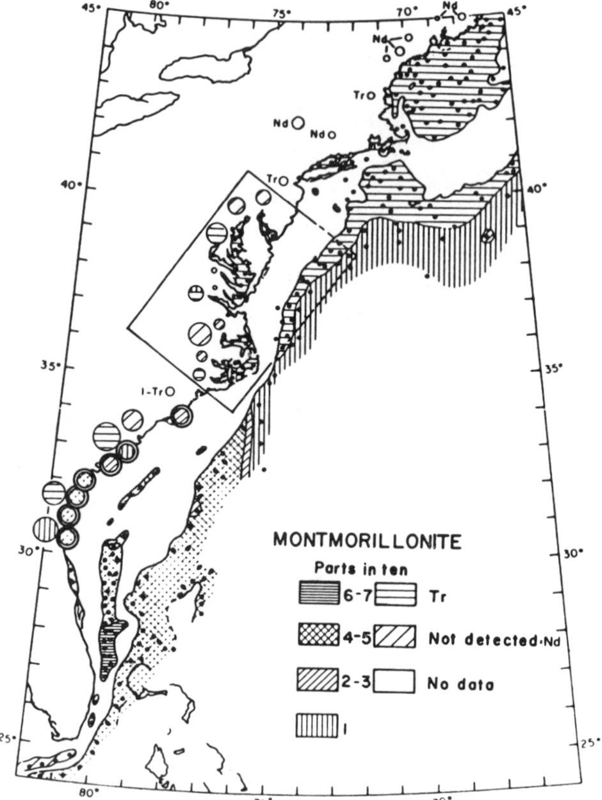

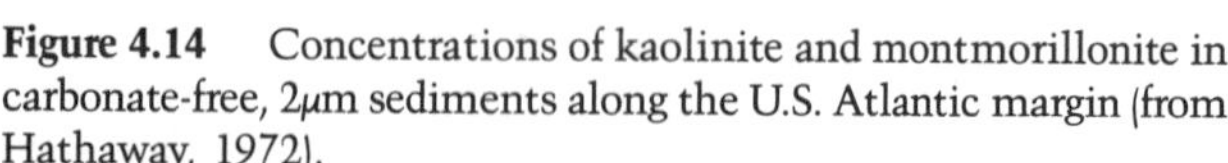

Figure 4.14 Concentrations of kaolinite and montmorillonite in carbonate-free, 2μm sediments along the U.S. Atlantic margin (from Hathaway, 1972).

Slumps and Slides

Known slumps and slides are concentrated almost exclusively along the continental slope and uppermost continental rise. Along the northern part of the Blake Escarpment and the continental slope lying north of the Blake Spur, Pinet et al. (1981a) mapped a series of faults. One scarp or set of scarps just above 2000 m marks a landward limit of apparent slump and slide masses (see location 2, Fig. 4.16). Most of these masses are characterized by chaotic internal structure. Some are mantled by conformable sediments; in others, older beds appear to be exposed, and mass movement forming these structures may be more recent. However the age of the mass movements, and the seaward limit of the slump/slide masses, are not known.

At the northernmost end of the Blake Plateau are numerous seaward-dipping faults. The faults themselves are not associated with slumps and slides, but in many instances buried collapse structures are present along the faults. These structures seem to be restricted to the lower Miocene section; presumably karst solution (sink holes) occurred in the underlying Eocene and Oligocene limestones (Popenoe et al., 1982). Small slump scars occur seaward of the faults near 2000 m on the steeper continental slope. Farther seaward beneath the upper continental rise, at least 28 diapiric structures (probably salt) have been identified in seismic profiles. It is likely that the faulting on the northern Blake Plateau and the associated slumping on the continental slope are the result of salt withdrawal from this region (see Sylwester et al., 1979). At 33°N, a set of clearly defined slump scarps surrounding four diapirs has been identified in GLORIA side-scan profiles (Popenoe et al., 1982).

Along the Florida-Hatteras Slope there is very little evidence of slumps and slides. Ayers and Pilkey (1981) and Popenoe et al. (1981a) have identified one possible slump mass at 32°N (Fig. 4.16). Numerous small faults are present along and within 50-100 km seaward of the Florida-Hatteras Slope. It is possible that intense erosion at the base of the Gulf Stream has removed slumped debris in this area.

Very little is known about possible faulting and slump/slide activity south of 30°N on the Blake Plateau. The fact that the plateau is relatively level and presumably stable suggests a lack of significant mass movement. However, as discussed later, evaporites may be deeply buried beneath the plateau, and their movement could generate instabilities in the shallower sedimentary section.

The Blake Escarpment south of 30°N is precipitous, and seismic reflection data have done little to resolve the structure there. However, recent submersible dives by W.P.

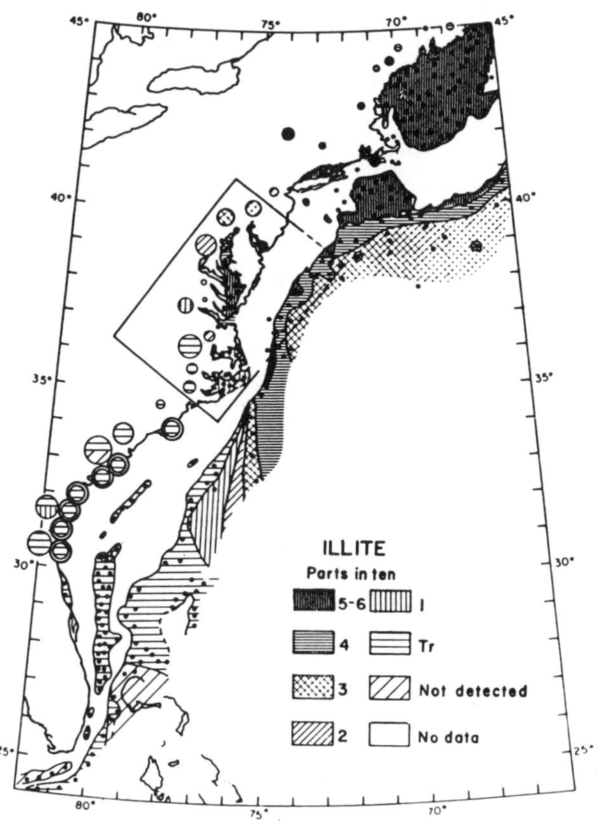

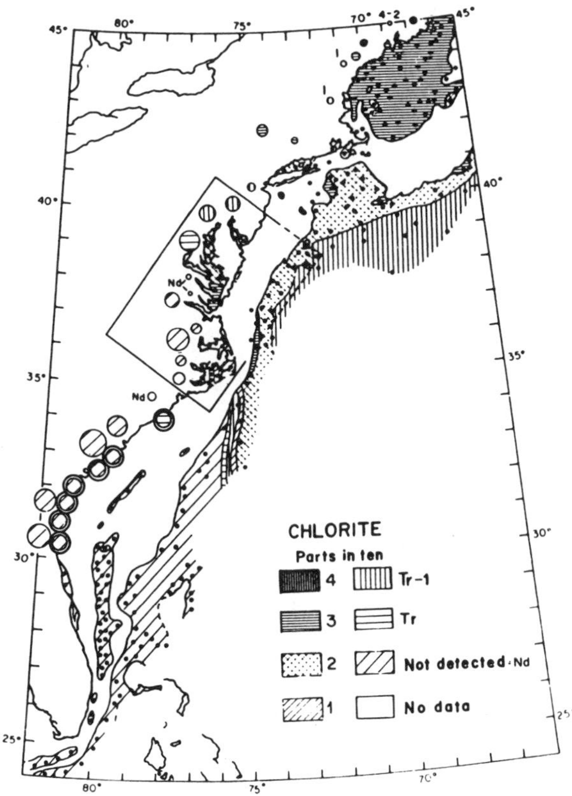

Figure 4.15 Concentrations of illite and chlorite in carbonate-free, 2μm sediments along the U.S. Atlantic margin (from Hathaway, 1972).

Dillon and C. Paull (unpubl. data; Fig. 4.4) identified locally vertical slopes with jointing, faulting, and biologic erosion. Lithified Cretaceous back-reef sediments are exposed, indicating a long history of erosion of this escarpment by mass wasting.

Between 35°N and 38°N there appears to be a moderate amount of down-slope mass movement, increasing northward (Fig. 4.16). Carpenter (1981b) documented several slide areas within proposed lease blocks there, and Knebel and Carson (1979) noted several small slides in their more regional survey of the slope. Abundant examples of mass movement also are documented in seismic reflection data along the continental slope between Baltimore and Accomac Canyons by Malahoff et al. (1980, 1982) and Embley (1982). Malahoff et al. (1982) made ALVIN submersible dives in and around Norfolk Canyon, where they found evidence of fairly recent small slides and slumps (similar observations were made in Washington Canyon). In contrast, in their intercanyon dives (north of Norfolk Canyon and between Accomac and Baltimore Canyons) they observed no outcrops, even on slopes of 30°. Examination from ALVIN of a prominent, seismically defined slide scar (15°-20° slope) and glide plane (4° slope) provided no evidence of recent instability, even within several small, local, slope-restabilization slumps that had formed along the scarp. However, they did find cobbles and large slabs of "beach-rock" that presumably had moved seaward several kilometers (along 2.5° slopes) from the shelf break, north of Accomac Canyon.

Bunn and McGregor (1980) and Popenoe et al. (1982) mapped a significant slump scar at 36°20′N. The position of the actual slump mass and its seaward extent are unknown, but this mass movement is the likely cause for a widespread debris flow observed farther seaward (Fig. 4.16).

The continental slope from about 38°N north to Hudson Canyon has the most intense apparent mass movement, perhaps because it is the most thoroughly surveyed area. The two largest features of possible slump/slide origin, at the base of the slope below Baltimore Canyon and below Wilmington Canyon (Figs. 4.6, 4.16), divert the canyon channels toward the east on the upper continental rise (Twichell and Roberts, 1982). Support for a slump/slide origin is found in the documentation by McGregor et al. (1982b) and by Stubblefield et al. (1982) of features that resemble pressure ridges at the base of the Wilmington block. Submersible dives on these ridges showed them to be free of recent sediment drape, but it is not clear that this indicates recent movement. McGregor and Bennett (1979) had postulated an early Pleistocene age for the activity because they were unable to identify a slump scar in seismic data.

Numerous other slump scars, slumps, and slides have

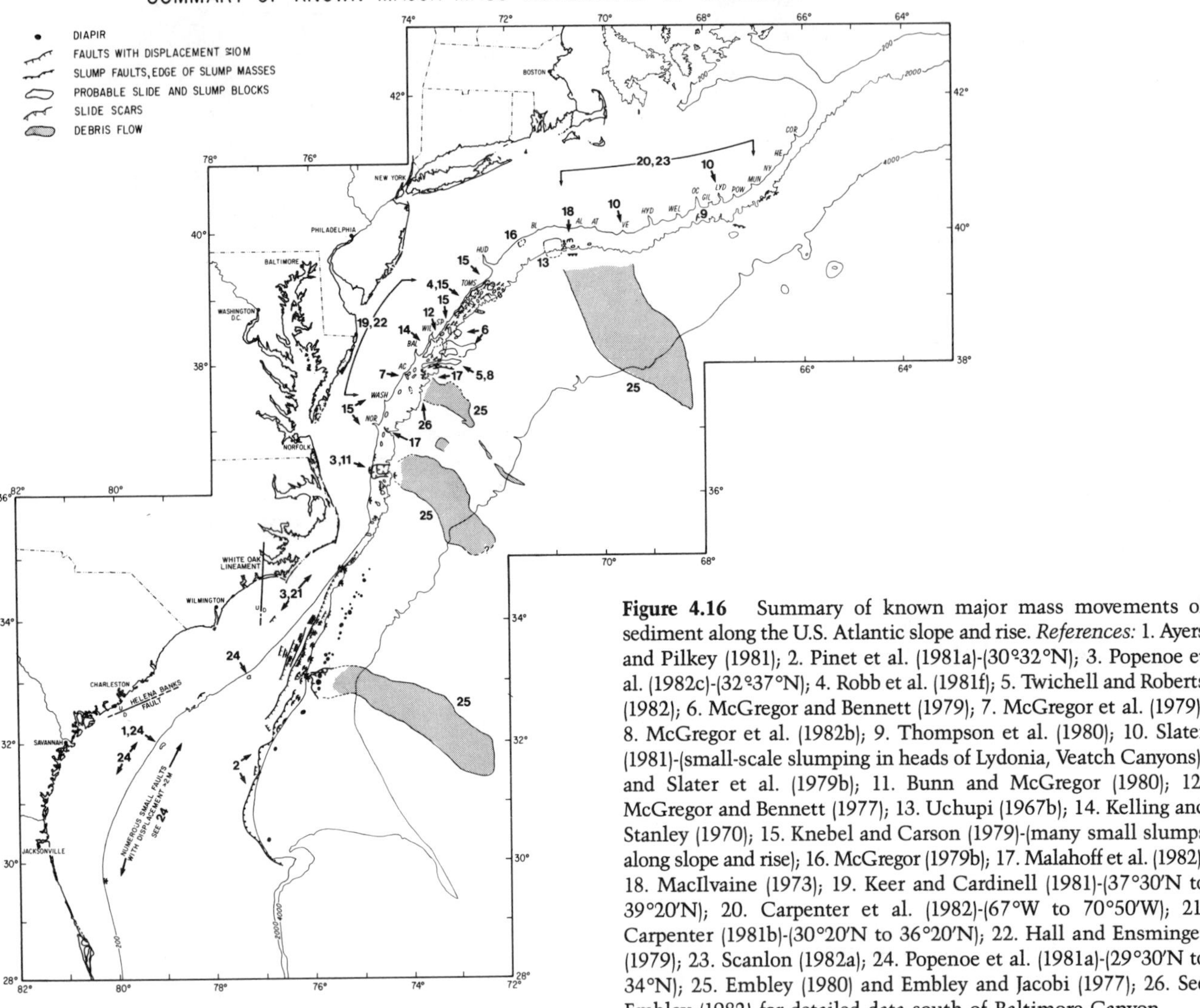

Figure 4.16 Summary of known major mass movements of sediment along the U.S. Atlantic slope and rise. *References:* 1. Ayers and Pilkey (1981); 2. Pinet et al. (1981a)-(30°32°N); 3. Popenoe et al. (1982c)-(32°37°N); 4. Robb et al. (1981f); 5. Twichell and Roberts (1982); 6. McGregor and Bennett (1979); 7. McGregor et al. (1979); 8. McGregor et al. (1982b); 9. Thompson et al. (1980); 10. Slater (1981)-(small-scale slumping in heads of Lydonia, Veatch Canyons), and Slater et al. (1979b); 11. Bunn and McGregor (1980); 12. McGregor and Bennett (1977); 13. Uchupi (1967b); 14. Kelling and Stanley (1970); 15. Knebel and Carson (1979)-(many small slumps along slope and rise); 16. McGregor (1979b); 17. Malahoff et al. (1982); 18. MacIlvaine (1973); 19. Keer and Cardinell (1981)-(37°30′N to 39°20′N); 20. Carpenter et al. (1982)-(67°W to 70°50′W); 21. Carpenter (1981b)-(30°20′N to 36°20′N); 22. Hall and Ensminger (1979); 23. Scanlon (1982a); 24. Popenoe et al. (1981a)-(29°30′N to 34°N); 25. Embley (1980) and Embley and Jacobi (1977); 26. See Embley (1982) for detailed data south of Baltimore Canyon.

been mapped on the continental slope between Baltimore Canyon and Hudson Canyon (Fig. 4.16 and references therein). In addition, faults seem to be a common feature of the slope in that region, especially north of 38°30′N. Keer and Cardinell (1981) noted that the slope commonly has a blocky texture with shallow faulting and mass movements (slumps, slides). It is not presently known whether this activity is recent; like the area studied by Malahoff et al. (1982) to the south, the mass-movement may *appear* to be recent in "low-resolution" seismic data while no recent activity is observable at the visual-observation scale.

The area immediately northeast of Hudson Canyon appears to have very little slump or slide activity, but it must be noted that virtually no studies have been conducted there. The slightly lower gradient of the continental slope northeast of Hudson Canyon and between Norfolk and Washington Canyons could explain the reduced occurrence of significant mass movement in those areas. However, it is also possible that Pleistocene "delta" building in these areas, either because of depositional structure or the nature of the sediments, pre-disposes the areas to be less susceptible to mass movements.

The continental slope east of 71°W appears to have been affected relatively little by sediment mass movements (Fig. 4.16). The most active area is centered on 71°W. Here Uchupi (1967b) reported a massive slump, but the extent and morphology of the feature are poorly defined. Just to the east, MacIlvaine (1973) showed clear evidence of small slide scars, and Carpenter et al. (1982) reported the occurrence of five separate slides. Scanlon (1982a,b) noted the presence of apparent slide scars along most canyon walls, facing inward toward the canyons. She also reported slide scars a) just north of Bear Seamount on the upper continental rise, b) between Munson and Nygren Canyons on the uppermost rise, and c) just west of Alvin Canyon on the uppermost rise. The age of these mass movements and the area of seafloor affected are unknown.

In summary, it is clear that slumps and slides have been

an important agent in sediment mass movement along the entire continental slope. Exceptions appear to be the continental slope south of Georges Bank and possibly just northeast of Hudson Canyon, where only limited mass movement has occurred. There is very little documentation to determine whether slumping activity has occurred on the continental rise, although the relatively steeper gradient of the uppermost rise makes it a likely candidate for such instability. Certainly the upper rise has been affected by slumping and sliding to the extent that it is a catchment area for sediment mass movements that originated on the continental slope. Although seismic reflection and side-scan sonar data give the strong impression of very recent mass movement on the slope, there is as yet little direct (e.g., visual, sample) evidence to support this contention. Evaluation by submersible observation, large-area mosaic photography, and judicious coring are necessary to resolve the question.

Debris Flows

Debris flows have only in recent years been recognized as a widespread sedimentary process in ocean basins, principally through the work of R.W. Embley and R.D. Jacobi. Debris flows generally are recognized in seismic reflection profiles as acoustically transparent (or chaotic), structureless sediment lenses (Fig. 4.17).

Four major debris flows have been documented on the U.S. Atlantic margin (Fig. 4.16). The southernmost very clearly has its origin at the slide scar around the salt domes east of the northern end of the Blake Plateau. It extends more than 350 km downslope onto the Hatteras Abyssal Plain at 5400 m water depth, and it covers more than 25,000 km² of seafloor. The highly variable acoustic character of the deposit (Fig. 4.17, top) probably reflects differences in the rheology and hydrodynamic regime of the flow at different locations along the flow path (Embley, 1980). Cores taken from this deposit contain inclined and convoluted beds, and sediment clasts.

A second debris flow crosses the continental rise at 36°N and appears to have originated from the large slide on the continental slope east of Albemarle Sound (Figs. 4.16-4.18). The flow covers more than 11,000 km² and it may reach the Hatteras Transverse Canyon below 4300 m water depth.

Debris-flow deposits also extend from the continental slope across the upper and central continental rise between Accomac and Baltimore Canyons (Figs. 4.16-4.18). Slide scars along the walls of canyons in the continental slope affect the uppermost 50 m of the sedimentary section; the displaced sediment was funnelled down the canyons onto the continental rise. The main debris flow at 37°30′N covers 4000 km² and reaches a depth of 3100 m. Two deeper tongues of unstratified sediment also have been identified (Figs. 4.16-4.18). Embley (1980) postulated that these may be deposited from a smaller flow that was triggered on the upper continental rise when the main flow loaded the rise.

Embley (1980) reported that cores taken from the northern part of the main debris flow (37°30′N) contained a surfi-

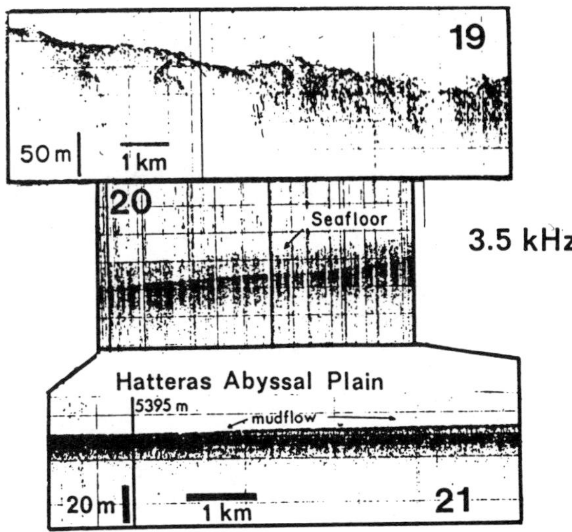

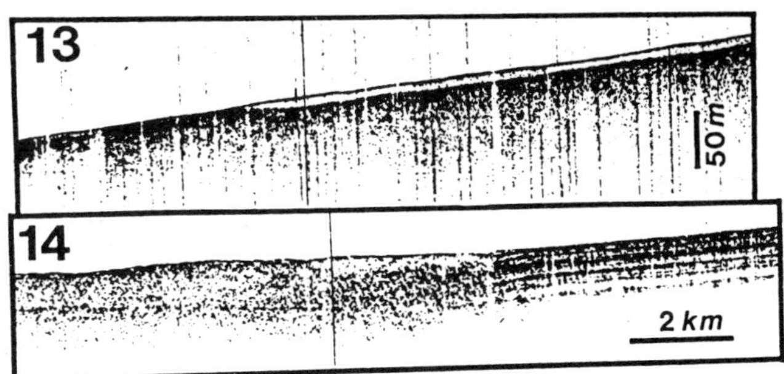

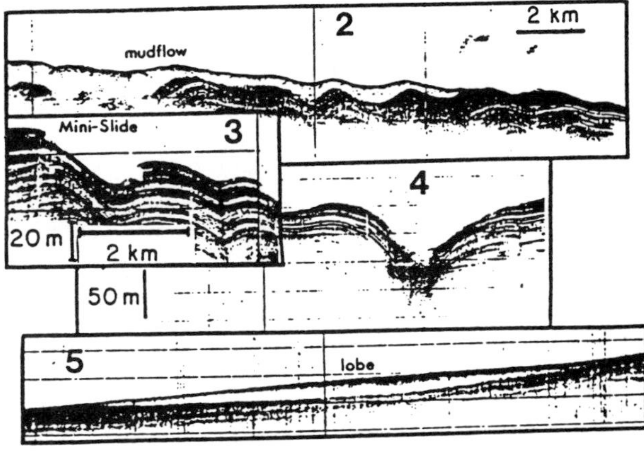

Figure 4.17 3.5 kHz-profiles across debris flows on the U.S. Atlantic continental rise: (from Embley, 1980). Top-east of northern Blake Plateau (Profiles 19-21); Middle-south of Cape Cod (Profiles 13, 14); Bottom-central U.S. Atlantic continental rise (Profiles 2-5; located in Figure 4.18). Profile 3 crosses a small slide scar seaward of Baltimore-Accomac debris flow, and profile 4 contains a small debris flow in a channel below the same feature. See Figure 4.16 for locations of major debris flows.

cial, hemipelagic grayish-brown foraminiferal clay 40 to 120 cm thick overlying an olive-gray clay containing displaced sediment clasts. The base of the hemipelagic clay was radiocarbon-dated at 5200 yr. B.P., 7285 yr. B.P., 10,080 yr. B.P. and 6680 yr. B.P. in four separate cores. In one box-core the matrix of the underlying debris flow was dated at 11,820 and 11,860 yr. B.P. at two different levels, and a clast was dated at >40,000 yr. B.P. Thus the slump/slide/debris flow that originated on the slope between Accomac and Baltimore Canyons appears to have occurred in the latest Pleistocene to early Holocene (>5000 yr. B.P.), but it incorporated debris (clasts) from sediments dating well back into the Pleistocene. Near the southern edge of this same debris flow, Doyle et al. (1979) reported a piston core with "matrix" dates of 25,515 to 26,465 yr. B.P. It therefore seems likely that at least one other separate (and in this case, older) event originated south of Accomac Canyon and contributed to this debris flow.

Older debris flows also are identified buried beneath a veneer of stratified sediments along this part of the continental rise (Fig. 4.18). Embley (1980) estimated that 40 percent of the continental rise may be affected by such (probably Pleistocene) events.

South of Cape Cod, another large slide and debris flow zone has been reported by Embley (1980), but it has not been mapped in detail. It probably is related to the slump/slide activity on the adjacent continental slope (Figs. 4.16-4.17). Cores from this area contain an apparently undisturbed hemipelagic sedimentary section that overlies a unit containing mass-flow structures.

Other, minor debris flows occur along the uppermost continental rise and the continental slope. One deposit was reported by Robb et al. (1981b) on the uppermost rise south of Toms Canyon. Knebel and Carson (1979) noted another deposit, partly slide and partly unstratified debris flow, on the middle continental slope between Hudson and Toms Canyons (Fig. 4.16). Malahoff et al. (1982) reported unstratified, probably debris-flow deposits in canyon axes south of Baltimore Canyon. It is likely that such deposits are a common phenomena in canyons and channels all along the margin.

Turbidity Currents

Turbidity currents are considered to be an end-member of the spectrum of mass movements, and they probably are

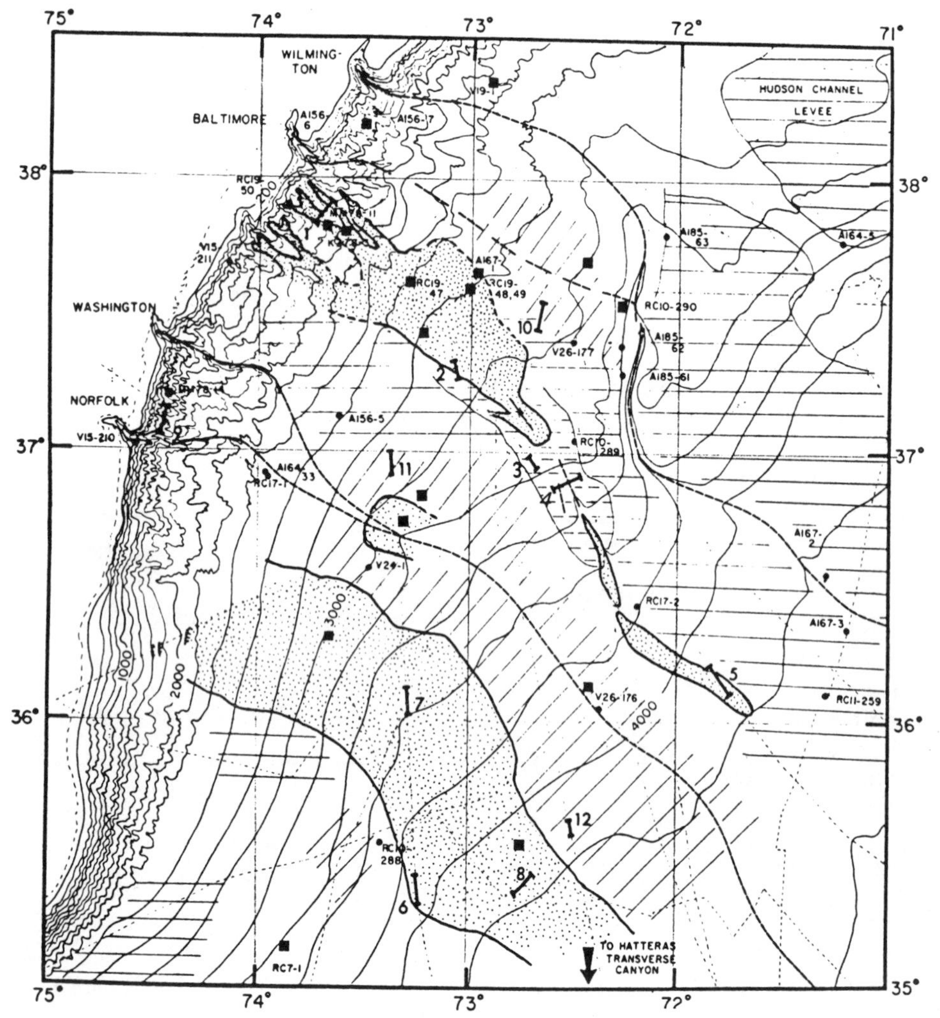

Figure 4.18 Debris flows (stippled), buried debris flows (diagonal lines), and conformable, acoustically laminated deposits (horizontal lines) on continental rise. Black boxes locate cores containing debris-flow structures. From Embley (1980).

generated at the seaward edge of many slumps, slides, and debris flows. Certainly on the scale of seismic-reflection observation, some debris flows grade continuously into flat, widespread, acoustically laminated beds that would normally be considered turbidites (see Fig. 4.17, profile 21).

Turbidity currents have never been observed directly, but their effects have been estimated from submarine cable breaks. The classic example occurred southwest of the Grand Banks in 1929 when several submarine cables were broken by the shock of an earthquake. Heezen and Ewing (1952) noted a series of delayed cable breaks downslope, and the time lag increased with distance. They explained these events as the triggering of a submarine slide that changed downslope to a turbidity current, and they calculated the velocity at greater than 80 km hr^{-1}. The volume of sediment involved was thought to be about 100 km^3 and it appears to have been dispersed over an area of 100,000 km^2.

The continental rise is generally considered to be the depositional product of turbidity currents (e.g., Stanley et al., 1971). In the turbidity-current model the rise consists of a series of abyssal deep-sea fans that overlap extensively to form the relatively uniform sedimentary prism. The basic fan morphology consists of an upper fan with deeply incised canyons (comparable to the incised lower continental slope and upper rise), a middle fan (suprafan) that is convex upward and has relatively well defined channels (middle continental rise), and a lower fan (lower continental rise) that has a fairly low gradient and is crisscrossed by shallow dispersive channels (e.g., Normark, 1970). Turbidity currents traversing the fan (rise) deposit a large volume of sediments on the central fan by channel-bank overflow (often forming levees) and then disperse in progressively smaller distributary channels across the lower fan (rise) and onto the abyssal plain. Turbidity currents debouching on the lowermost fan often deposit coarse (sandy, gravelly), highly reflective sediments. Graded sand and silt beds (turbidites) commonly have been recovered across the continental rise within and near these channels (Ericson et al., 1952; Fritz and Pilkey, 1975). In general, the thickness and grain size of coarse turbiditic beds decrease across the abyssal plain with distance from the lower fan.

This generalized model in some ways fits observations on parts of the U.S. Atlantic continental rise. Cleary et al. (1977) studied the morphology and sedimentology of the continental rise, and they concluded that turbidity currents reach the deep basin principally through three channel systems: Hatteras, Wilmington and Hudson. The situation is complicated by presence of the Hatteras Outer Ridge at the base of the continental rise because this ridge blocked direct turbidite dispersal to the Hatteras Abyssal Plain. Thus the Wilmington Channel has developed its lower fan where it debouches onto the lower continental rise terrace (Fig. 4.1). The situation is similar for Hudson Channel, but both channels feed turbidites across the terrace and the eastern flank of the Hatteras Outer Ridge to the abyssal plain. The Hatteras channel system also is blocked by the ridge but turns southward in the Hatteras Transverse Canyon, in a morphologic setting comparable to the middle fan. The lower fan for this system is formed at the foot of the continental rise along the edge of the Hatteras Abyssal Plain.

There are some problems with the view that continental rise construction is simply a result of coalescing fans. First is the fact that the overall rise morphology does not show the lobate pattern that would be expected. Second, other processes that probably are extremely important in rise construction are ignored in this model. One such process is debris flows, which as noted earlier may affect 40 percent of the continental rise (Embley, 1980). Sediment clasts and conglomeratic muds are known in numerous cores across the rise (Cleary and Connolly, 1974; Embley, 1980), and it is not unlikely that such deposits are widespread but basically unrecognized in many more cores. With respect to debris flows, it is noteworthy that all four major flows observed across the surface of the continental rise lie *between* the three major canyon/channel systems. Thus it is possible that debris flows originating within the reach of tributary systems of the major channels are confined to those channels; there they ultimately degrade into turbidity currents which in turn flow onto and beyond the lower continental rise.

A second important process that shapes the continental rise, and that sometimes is ignored or downplayed in comparison to turbidity currents (Stanley et al., 1971), is contour-following bottom currents. These currents have a significant effect on depositional patterns, as discussed in a subsequent section.

Finally, it should be recognized that not all turbidity currents traverse the entire continental rise to the abyssal plain. The numerous smaller slumps and debris flows distributed along the continental slope each have the potential for generating small turbidity currents. Some of these are channelized on the continental rise; there they probably help to fill the channels rather than simply pass through them. Other turbidity currents probably disperse across the upper and central rise in mostly unchannelized "sheet" flows, thus contributing to construction of the continental rise in a way that is significantly different from the simple fan model.

Other Processes

Other gravity-controlled mass movements of sediment are known to exist, but in general they are either not well documented or their degree of importance is poorly understood. One example is sand flows or sand falls. When shelf sands are swept into the heads of submarine canyons by waves and currents they can pile up, fail, and flow down-canyon. Dill (1966) documented several instances of such sand flows off the tip of Baja California. He found that flows of 5-8 cm sec^{-1} occurred when the angle of sand repose exceeded 30°. Similar phenomena might occur in the heads of U.S. Atlantic margin canyons where adjacent shelf sands are fed into the canyons by cross-shelf transport.

A possibly related phenomenon off the eastern U.S. is suggested by the occurrence of the so-called "mud line" on the upper continental slope near 300 m (Stanley and Wear, 1978; Stanley and Freeland, 1978). Above this boundary is predominantly sandy sediment and below is largely muddy (silty clay) sediment. Stanley and Wear (1978) explained this basically as a winnowing phenomenon within higher energy environments above ∼300 m. However, McGregor et al. (1979) have reported thick sand layers overlying silty muds to depths of 1000 m on the slope and on ridges south of Baltimore Canyon. Fenner et al. (1971) and Stanley (1974a) reported that shelly sands locally spill into the head of Wilmington Canyon, and along the slope Stanley and Wear (1978) also reported gravelly muddy sands locally deeper than 500 m. Thus it is possible that some kind of gravity or suspension flow has transported these sandy grains downslope from the shelf.

A final example is illustrated by downslope grooves scoured in Eocene chalk that is exposed on the continental slope near 2000 m depth south of Toms Canyon. These features have been termed "furrows" by Robb et al. (1983), but they are distinctly different in shape, origin, and orientation compared to normal abyssal furrows formed by bottom currents (e.g., Hollister et al., 1974). The grooves are observed in the bathymetric map of Kirby et al. (1982), and they are mostly linear, 4 to 13 m deep, 3.5 m wide, and spaced 10 to 50 m apart. Their walls are smooth or "plucked" and have the same kind of character as walls of subterranean caves that are smoothed by phreatic erosion. The grooves commonly have floors of flat, fine-grained sediment. At present there is no satisfactory explanation for the origin of the grooves, although erosion by downslope and/or current-induced upslope sedimentary processes seems likely.

Current Effects on Sedimentation

Currents, whether wind-driven, tidal, or thermohaline, are dealt with in detail in Chapter 2. In the present discussion we focus on the observed geologic effects of such currents.

Continental Slope

Sand waves are well documented across the continental shelf from Florida to Georges Bank (e.g., Uchupi, 1968a; Knebel and Folger, 1976). Long-term photographic tripod and current measurements also document that non-cohesive silt, sand, and sometimes coarser particles are widely transported across the shelf by wind-driven and tidal currents (e.g., Butman et al., 1982). Although bottom-drifters indicate a *net* shoreward transport of sediment on the shelf (Bumpus, 1965; Bumpus and Lauzier, 1965), it is clear that suspended and bedload-transported shelf sediment is at times carried seaward over the shelf break (Stanley and Freeland, 1978; McGregor, 1979b). Furthermore, many canyons cut into the shelf far enough that they can intersect cross-shelf transport of sediment, and the canyons ultimately funnel this sediment downslope.

Depths shallower than the mud-line (250-300 m; Stanley and Freeland, 1978; Stanley and Wear, 1978) are thought to represent a higher energy environment in which finer-grained sediment is winnowed out by currents. In detail, this generalized picture shows some significant variations. For example, within the heads of submarine canyons the mud-line generally is somewhat shallower than along the adjacent continental slope (Stanley and Wear, 1978). Although some sand transported west to southwest across the shelf probably is trapped in the heads of submarine canyons, the sand appears to be diluted by finer sediment also trapped there.

Suspended particulate matter (SPM) concentrations over the continental shelf, shelf break, and upper slope typically range from 100 to more than 1000 μg/liter. The SPM usually consists of material up to intermediate silt-size (44 μ), although fine sand is not uncommonly found (e.g., Lyall et al., 1971). Organic debris can comprise up to 60-70 percent of the particulate matter in mid-to-outer shelf areas (Manheim et al., 1970), and it is locally higher on the slope (Bothner et al., 1981). The terrigenous component of the SPM is mostly material resuspended from the seafloor.

Little detailed information exists about transport of such sediment by currents on the slope, but Hunkins et al. (1983) have reported studies of currents and SPM distribution in and around Baltimore Canyon that may be representative of other canyons. They show that SPM concentrations are higher within the canyon than on the adjacent continental slope. Both currents and SPM resuspension events appear to be focussed within the canyon across the thermocline (250-600 m), and there is relatively little comparable effect on the adjacent slope. The currents have a dominant tidal component, but SPM concentrations usually peak only during the up-canyon flow and only as this up-canyon flow begins to wane. A secondary peak sometimes occurs during the down-canyon flow. SPM concentration can increase 10-to-40-fold from background levels within a few minutes during these periodic fluctuations; it then drops sharply to 5-10 times the background level before decreasing back to background level over a period of several hours. Although the current speeds and SPM resuspension events follow tidal cycles, it is not yet known whether purely tidal flows are responsible for resuspension or whether some other mechanism such as internal waves should be invoked.

These results imply a net transport of fine-grained SPM up-canyon, which over the longer term should result in significant deposits of mud in the canyon head, or transport onto the shelf. However, there appears to be a mechanism whereby the SPM is recycled into the outer reaches of the canyon or onto the slope and rise. Hunkins et al. (1983) noted that there is not a coherent response of currents and SPM concentrations along the length and over the entire depth of the canyon; they suggested the presence of "convergence" zones in the canyon which may provide a mechanism for advection of SPM seaward along density (isopycnal) surfaces. Some of this particulate matter probably is redeposited on the slope and in the canyon. However, current measurements on the open slope indicate a net west to southwest flow (e.g., McGregor, 1979a; see Chapter 2), so part of the suspensate probably is advected and deposited in an along-slope direction (Fig. 4.19).

Tidal and internal-wave effects have been studied in other submarine canyons along the U.S. Atlantic margin by Shepard (1975, 1976) and Keller and Shepard (1978), among others. Keller and Shepard concluded that the general net transport is *down*-canyon. It is likely that each submarine canyon has a distinctive pattern of net sediment transport (including focussing of currents and SPM into con-

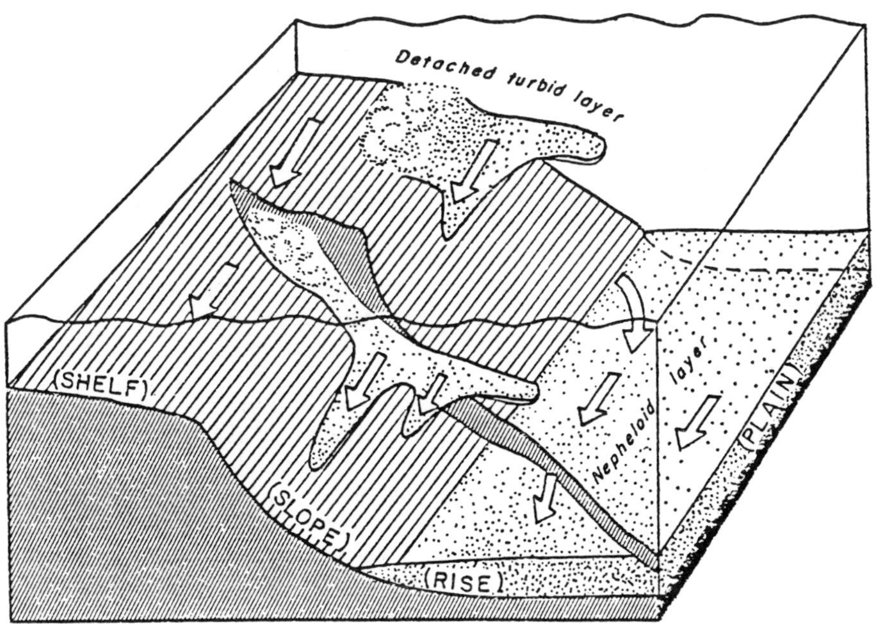

Figure 4.19 Model of turbid water exiting from canyon "convergence" zones along isopycnals, with southwest transport along the continental slope by prevailing currents. Bottom nepheloid layer on continental rise may be a separate phenomenon or may be fed by shallower turbid layers. From W.D. Gardner, *in* Hunkins et al. (1983).

vergence zones) that probably is controlled by the canyon morphology and by the distribution and grain size of erodable material within the confines of the canyon.

The net south to southwest flow along the continental slope appears to have little discernable geologic effect aside from the transport and ultimate deposition of entrained suspended particulate matter. Bottom photographs and visual observations from submersibles rarely show scour or current-produced bedforms in non-canyon areas along the slope. These sediments seem to be cohesive and resistant to current erosion and bedload transport. Emery and Ross (1968), for example, measured instantaneous current speeds of 70 cm sec^{-1} during ALVIN submersible observation at 1300 m water depth south of Cape Cod, but they saw no deformation of the subjacent seafloor sediments. There are virtually no data on the erosion resistance of these slope sediments, but they probably are made more cohesive by some combination of compositional parameters (including high organic carbon content), diagenesis (including biologic reworking), and compaction.

Blake Plateau/Florida-Hatteras Slope

On the northern and central Blake Plateau the geologic effects of the Gulf Stream include surface channelling, scour, and local sand waves (Pinet et al., 1981a; Popenoe et al., 1982). The present axis of the Gulf Stream follows the western margin of the plateau, and it is in this region that strong currents excavate the surface sediments and form deep channels (Figs. 4.8, 4.13). Much of the surface of the Charleston Bump is affected by such erosion. Farther east is a zone of surface scour centered near 31°N, 78°W. This zone is about 120 km long, and 120 km wide at its southern end, but it narrows to the north and is not evident in seismic profiles north of 31°30′N. Bottom sediment in this zone is largely carbonate sand (JOIDES J4 well; Charm et al., 1969). Another, smaller area of current scour is centered near 32°N, 77°30′W, and a small area of sand waves that apparently are migrating to the north is present near 33°N, 77°W (Popenoe et al., 1982).

The Florida-Hatteras Slope, at the edge of a seaward-prograding sediment wedge, is relatively smooth along its length (e.g., Uchupi, 1967a; Zarudski and Uchupi, 1968). In seismic profiles some local scour and truncation of deeper bedding planes are observed, but surface irregularity caused by current erosion is minimal. At the foot of the slope are the linear erosional troughs described earlier in this report (Fig. 4.8). The coral mounds that form positive roughness elements appear to flourish in the zones of turbulent water near these troughs.

The southeastern Blake Plateau has little geologic and geophysical coverage and has not been extensively studied. J. Ewing et al. (1966) showed that the sedimentary blanket is relatively conformable south of the Blake Spur. The Gulf Stream does not affect the seafloor in this area, but it does incise sediment at its western edge nearer the Florida-Hatteras Slope. Relatively thin Quaternary sediment cover across the southeastern Blake Plateau (Fig. 4.13) suggests, however, that the Antilles Current which sweeps west to northwest across the plateau keeps accumulation rates to a minimum.

The Gulf Stream has played an important role in the development of the stratigraphic sequence on the Blake Plateau at least since Eocene time. At JOIDES drill site J4, Eocene and Oligocene calcareous oozes and sands unconformably overlie Paleocene limestones (Charm et al., 1969). The unconformity cutting the limestones is relatively smooth at the northern end of the Blake Plateau but is deeply dissected farther south. Buried cut-and-fill structures related to Cenozoic current erosion by the Gulf Stream (and possibly other shallow currents) are widely distributed across the northern and central Blake Plateau (Pinet et al., 1981a).

Effects of Deep Thermohaline Currents on Continental Rise

A system of contour-following, equatorward-flowing bottom currents, generally termed the "Western Boundary Undercurrent" (WBUC), is well documented along the continental rise of the U.S. Atlantic margin (see Chapter 2). At its shallower edge the current extends along the continental slope at depths less than 1000 m, and it reaches depths greater than 5000 m along the lowermost continental rise. At places across the continental rise north of Cape Hatteras, there are some indications of north to northeast flowing currents. These probably reflect local influences of deep eddies at the base of the Gulf Stream.

The average direction of the abyssal currents appears to be well documented. Although there are a number of direct current measurements (Chapter 2), most of the information on current directions is determined from bedforms observed in numerous bottom photographs. Because the bedforms occur in fine-grained silty clays, at least the larger forms probably require relatively long periods to develop (months to years); thus they average out higher-frequency variations in current direction that are observed in direct current measurements, and they act as vector-averaging current indicators that are representative of longer-term geologic effects. While directions are well determined in bottom photographs, current intensity often is not. For example, as noted earlier, some of the very cohesive silty clays of the continental slope show little deformation beneath currents as fast as 70 cm/s. In contrast, for other areas such as the lower continental rise off Nova Scotia (>4500 m), bottom photographs show unambiguous evidence of very strong bottom currents in the form of crags and tails, scoured seafloor, and ripples. Intermittent current speeds estimated at greater than 50 cm/s from bottom photographs in this area (Tucholke et al., 1979a) were subsequently confirmed by direct current measurements (M. Richardson et al., 1981). The kinds of bedforms observed across the continental rise include textural lineations, crags-and-tails, deflation "flutes", mounds-and-tails, current-elongated biologic mounds, longitudinal ripples, and transverse ripples. Attempts have been made in the past to estimate bottom-current speeds from such features (e.g., Heezen et al., 1966, and Schneider et al., 1967), but generally such estimates are subject to very large uncertainties.

Large-scale current erosion (i.e., detectable in seismic reflection profiles) of the continental rise occurs in two locations (Mountain and Tucholke, 1985). One extends across the northern end of the Hatteras Outer Ridge and onto the continental rise just above 4000 m between Wilmington and Hudson Canyons. The second is centered near 4000 m along the north flank of the Blake Outer Ridge. Deep-sea drilling at Site 104 on this part of the ridge recovered only a thin veneer of Quaternary sediments unconformably overlying upper Miocene silty clays (J. Ewing and Hollister, 1972). 3.5-kHz echograms along the north flank of the ridge show very irregular seafloor with minimal acoustic penetration (Tucholke, 1979a). This echo character is thought to indicate current incision to the level of consolidated sediments, probably with associated downslope mass movement on the relatively steep slope.

Schneider et al. (1967) and Hollister and Heezen (1972) mapped contour-parallel zones of broad hyperbolae from 12-kHz echograms along the continental rise, and Hollister et al. (1976) considered that these might be manifestations of erosion by abyssal currents. More detailed subsequent mapping by Vassallo et al. (1984a, b), however, shows that most of these echo-character zones trend *across* the contours and therefore appear to be related to downslope processes. It is known from side-scan sonar and bottom photographic surveys that many of the zones of hyperbolic echoes, for example on the Blake Outer Ridge (Hollister et al., 1976), are caused by long linear furrows in the seafloor; some of these, especially larger furrows that are tens of meters wide, appear to be erosional in origin (Hollister et al., 1974). Many smaller furrows probably are syndepositional (Tucholke, 1979a).

The Western Boundary Undercurrent also flows southward along the steep Blake Escarpment. Here the WBUC probably does not physically erode the outcropping carbonates, but it may play a role in their chemical corrosion and it may help to support organisms that contribute to biological erosion (W.P. Dillon and C.K. Paull, unpubl. data).

The Western Boundary Undercurrent presently carries SPM concentrations of a few hundred micrograms per liter in the bottom 500 to 1000 m of the water column (Biscaye and Eittreim, 1974, 1977; Eittreim et al., 1975, 1976). This benthic nepheloid layer is best known in vertical profiles of light scattering (nephelometer profiles), which also show a near-surface scattering layer and a clear water minimum at intermediate water depths. By calibrating the nephelometer profiles with directly sampled suspended matter concentrations, and by integrating the light scattering of the benthic nepheloid layer that is in excess of the clear-water-minimum background, Biscaye and Eittreim (1977) were able to determine a "net particulate standing crop" (Fig. 4.20). This net standing crop is thought to represent primarily the interaction of circulating abyssal waters with the ocean bottom. Thus the increased standing crop along the U.S. Atlantic margin presumably results from intermittent resuspension of sediments all along the path of the WBUC. In particular, it is known from detailed seafloor studies that strong, intermittent erosion occurs on the Nova Scotian lower continental rise. This is an area of high abyssal kinetic energy associated with the Gulf Stream; constructive interference of deep Gulf Stream eddies with the mean southwest flow of the WBUC creates "benthic storms" that resuspend sediment and provide large quantities of SPM to the benthic nepheloid layer (Hollister and McCave, 1984). This

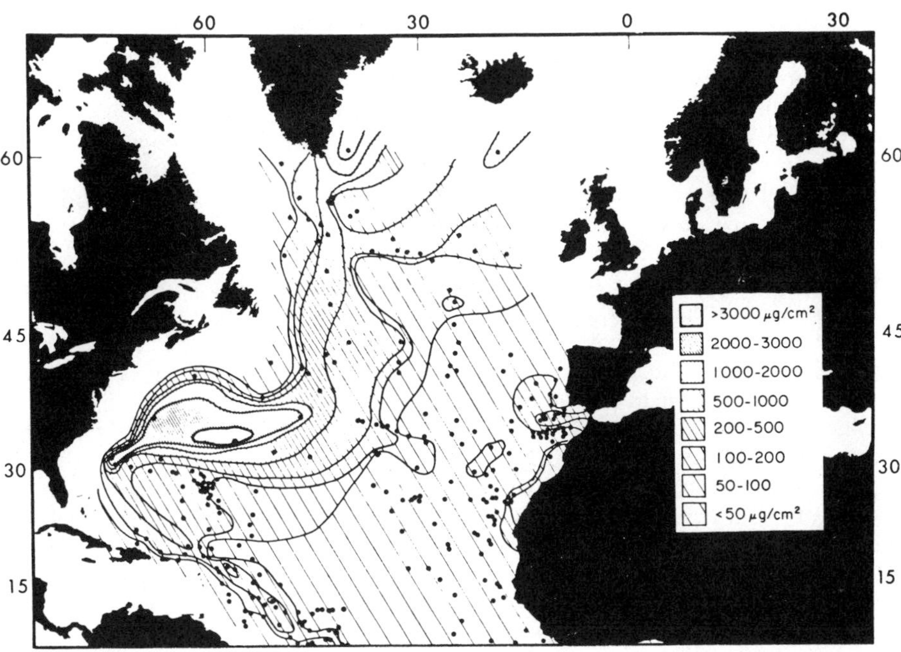

Figure 4.20 Net particulate standing crop in the benthic nepheloid layer of the North Atlantic Ocean (from Biscaye and Eittreim, 1977).

SPM source is apparent in the pattern of the bottom nepheloid layer (Fig. 4.20). Unfortunately, there are relatively few nephelometer stations along the U.S. Atlantic margin and the pattern in Figure 4.20 is generalized. It is possible that significant resuspension is also occurring beneath the Gulf Stream along the northern U.S. continental rise. It is also possible, as discussed earlier, that suspended particulate matter "pumped" from submarine canyons may contribute significantly to the benthic nepheloid layer on the continental rise and in the interior of the ocean basin.

The long-term significance of the WBUC as a geologic agent is best demonstrated by the distribution of sediment tracers carried in the flow. Reddish lutites (clays) derived from the region of the Laurentian Channel have been traced as far south as the Blake Outer Ridge (Hollister, 1967), as have distinctive Upper Carboniferous palynomorphs derived from the same region (Needham et al., 1969). A third tracer is the clay mineral chlorite, derived from high-latitude mechanical weathering of subaerial outcrops. A band of chlorite-enriched sediment can be traced from the shelf, slope and rise off Nova Scotia south along the U.S. Atlantic margin and along the Bahamas (Tucholke, 1975). A progressive southward decrease in chlorite abundance results from dilution by sediment injected laterally from the continental margin.

Excepting direct seafloor erosion, the possible mechanisms of sediment injection into the WBUC include "pumping" from submarine canyons (already discussed); slumps, slides and debris flows; and turbidity currents. The latter two classes of mechanisms are known to be important over longer time scales (i.e., tens of thousands of years), but their significance in the Holocene is uncertain. Studies of stratigraphy and mineralogy of piston cores from the continental rise show that graded beds (turbidites) are common in the immediate vicinity of submarine channels, but away from these areally restricted channels coarse-grained layers are infrequent, usually thin (<2 cm), and ungraded (e.g., Klasik and Pilkey, 1975). Klasik and Pilkey interpreted this to mean that bottom currents have been more important in transporting and depositing sediment to shape the continental rise than turbidity currents have been. The sediment entrained in and later deposited by the contour-following WBUC could either be eroded from upstream seafloor areas or it could be pirated from turbidity currents crossing the continental rise. The latter mechanism is conceptually similar to the pirating of sediment from submarine canyons by the net southwest bottom-current flow along the continental slope (Fig. 4.19).

The Western Boundary Undercurrent exhibits a clear influence on sedimentation and erosion patterns along the continental rise. It currently does not eradicate patterns of cross-slope sedimentation but subdues them and smears them out along bathymetric contours, thus smoothing out the morphology that would be expected if the "fan model" were solely responsible for development of the rise. The relative importance of along-slope (current) versus cross-slope (gravity-controlled) sediment transport has also varied considerably throughout the depositional history of the rise, as discussed in a subsequent section.

Pelagic Sedimentation

Pelagic sediments consist of particles which settle to the seafloor from surface and near-surface waters. They include biogenic detritus, eolian and meteoric dust, and particles rafted along the sea surface by ice and organic debris before being dropped to the seafloor. Of these, only biogenic detritus and ice-rafted debris contribute significantly to the seafloor sediments along the U.S. Atlantic margin.

Cosmic material, most notably small black magnetic iron spherules (cryoconite) and glassy microtektites, have long been known to contribute to deep-sea sediments, but neither has been reported from the submarine sediments of the U.S. Atlantic margin. The contribution by eolian dust is only slightly better known, again because the contribution is so slight that it is strongly diluted by sediment from other sources. Johnson (1979) noted that some Saharan dust probably reaches the U.S. Atlantic margin in the Northeast Trade Wind Belt between 5°N and 35°N; accumulation rates are probably well below 0.4 mm/1000 years. There probably is also some contribution from westerly winds blowing off the northern part of the continental U.S. (Lisitzin, 1972), but this contribution has not been quantified.

The delivery of biogenic detritus to the seafloor is controlled largely by surface circulation and productivity patterns, both of which are described in the Physical Oceanography and Biological Oceanography chapters of this book. However, the material that is preserved in the sedimentary record only partly reflects these patterns because of such factors as dissolution of carbonate and siliceous tests, oxidation of organic detritus, and dilution by terrigenous detritus. Sediment-trap data from the North Atlantic and elsewhere (e.g., Honjo, 1980; Thunnell and Honjo, 1981; Takahashi and Honjo, 1981) indicate that significantly less than 5 percent of the surface production is transported to the deep water. Biogenic contents in slope, rise, and Blake Plateau sediments range from about 20 percent to more than 90 percent; most is calcium carbonate, primarily planktonic foraminifera in the coarse fraction and coccoliths in the fine fraction (Fig. 4.10). Biogenic opal, while not generally measured along the U.S. Atlantic margin, probably represents less than 5 percent of the *carbonate-free* fraction in most areas; radiolarians dominate in the coarse fraction, and diatom frustules and their fragments are prevalent in the fine fraction. In general, biogenic components in surface sediments are much less concentrated than the vertical flux calculated from sediment-trap measurements would imply. Thus dissolution of hard parts and oxidation of organic matter at the seafloor must be important in degradation of pelagic debris.

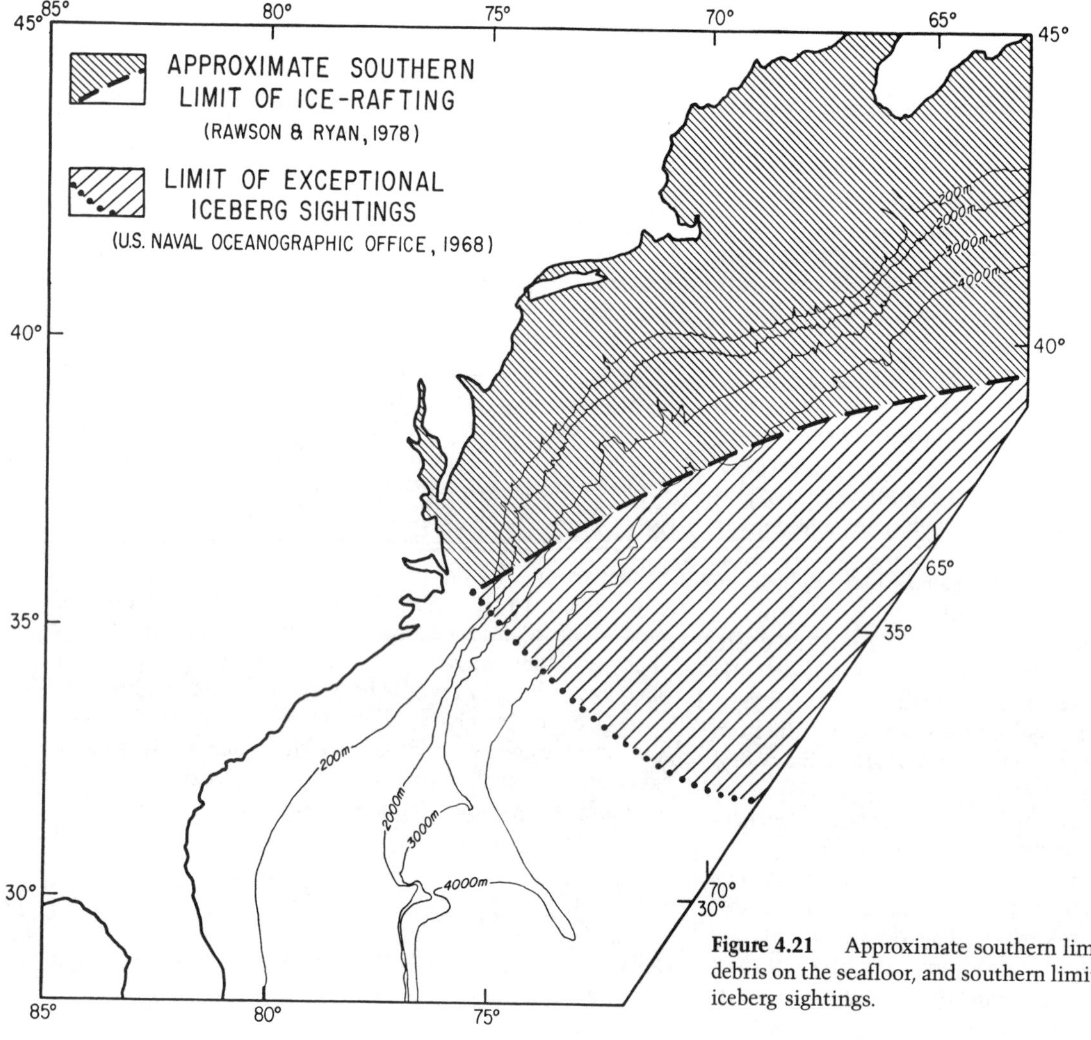

Figure 4.21 Approximate southern limit of significant ice-rafted debris on the seafloor, and southern limit of exceptional historical iceberg sightings.

Ice-rafted detritus (IRD) is delivered to the seafloor off eastern North America by icebergs transported in surface currents. According to models by Ruddiman (1977), during interglacial times these currents carry icebergs west and southwest along the continental margin south of Newfoundland toward Georges Bank, resulting in deposition of IRD as far south as 40°N. During glacial times, IRD probably was deposited as far south as 35°N as ice calved along the New England margin and floated south. Ultimately the southward movement was blocked and the ice was carried eastward by the Gulf Stream, which was shifted to more southerly latitudes during glacial periods. This model generally agrees with the observed distribution of IRD across the western Atlantic seafloor (Fig. 4.21). Bottom photographs show that ice-rafted pebbles, cobbles, and boulders are common along the continental rise and slope as far south as the New England Seamount Chain, and more widely scattered debris is observed south of Georges Bank (Heezen and Hollister, 1971). Unfortunately, no quantitative assessment of IRD has been made along the U.S. Atlantic margin. However, if we extrapolate Ruddiman's (1977) determinations in the subpolar North Atlantic to this area, it is likely that accumulation of ice-rafted sand (>62 μm) and coarser material has exceeded 100 mg/cm^2/1000 yr during interglacial periods, and rates probably were at least several hundred mg/cm^2/1000 yr during glacial periods.

Biologic Reworking of Sediments

Very little is known about sediment reworking and resuspension by benthic organisms except in shallow-water environments. There are two significant exceptions. One includes biologic experiments presently underway as part of the High Energy Benthic Boundary Layer Experiments (HEBBLE) in deep (>4500 m) water on the continental rise off Nova Scotia. Yingst and Aller (1982) have found that the upper 10 cm of sediment in box cores from this area is criss-crossed by tubes and burrows, mostly formed by polychaete worms that numerically dominate the infauna. Most of the infauna appear to be surface-feeding rather than deep-feeding deposit feeders. Numerical densities of macrofauna and meiofauna are about 20 times higher than for comparable groups reported from other deep-sea sediments from similar depths. As noted earlier, this is an area of intermittently strong bottom currents which may support the apparently abnormal population densities by transporting food to the area. Organic carbon content of the surface sediment (0.6 percent) is about twice as high as would be expected in the area. ATP (adenosine triphosphate) and bacterial concentrations also are higher than in other deep-sea localities, suggesting elevated microbial activity. By analogy, we might expect comparably increased biologic activity in other areas swept by strong currents, i.e., in canyons and in places on the U.S. continental slope and rise beneath the Western Boundary Undercurrent.

The second exception is the information derived from visual observations of the seafloor (e.g., submersible dives, bottom photographs) and of burrowing in sediment cores. Cores recovered from the continental slope and rise commonly are laced with burrow mottles produced by mobile benthic organisms, and the organisms themselves are frequently photographed (e.g., Heezen and Hollister, 1971; Rowe and Menzies, 1969; Rowe, 1971b). The organisms appear to be strongly zoned by depth and/or water temperature, although these patterns are modified by the presence of canyons and channels, and by current zonations (Rowe, 1971b).

On outcrops of indurated sediment along the continental slope and on the Blake Escarpment, burrowing organisms very actively erode the substrate. The exact areal exposure of indurated rock in any of these regions is uncertain, but it is clear from the distribution of pre-Quaternary outcrops that the occurrence of such substrates must be widespread (Fig. 4.13). Thompson et al. (1980) estimated that outcrops or contemporary erosional surfaces covered less than 20 percent of the seafloor in Baltimore, Wilmington and Lydonia Canyons. A similar estimate is reasonable for at least some areas of the open slope, such as the areas south of Toms Canyon studied by Kirby et al. (1982). On the Blake Escarpment, exposure of outcrops is the rule, and a relatively small fraction of the seafloor is mantled by recent sediment (W.P. Dillon and C.K. Paull, unpubl. data).

The importance of bioerosion of the continental slope has been documented by many researchers (e.g. Trumbull and McCamis, 1967; Dillon and Zimmerman, 1970; Kelling and Stanley, 1976; Cooper and Uzmann, 1977; Warme et al., 1971, 1978). Valentine et al. (1980) noted that red and/or jonah crabs are especially active on steep outcrops in Oceanographer Canyon. The crabs bore holes 5 to 15 cm in depth and diameter. Lobsters and fish are active in removing loose sediment, especially around talus and ice-rafted blocks. Fish stir up loose sediment in their search for food and possibly for protection. Boring bivalves may be important in creating pock marks on outcrops, and brachiopods and occasional polychaetes have been observed in these cavities (Ryan et al., 1978). Thus it appears that borers and burrowers actively erode consolidated sediments, and their burrows ultimately can cause failure of the sediments. Other surface feeders, particularly fish and infauna, inject fine-grained detritus into the water column where it can be swept away by ambient currents. Even the red crab, *Geryon quinquedens*, has been observed to inject plumes of fine sediment into the bottom boundary layer (Malahoff et al., 1982).

Sediment Mass Properties

Geotechnical Properties

Numerous studies of sediment mass properties (bulk physical properties, geotechnical properties) have been made for areas of the U.S. Atlantic continental slope and rise. Recent overview reports by Keller et al. (1979) and Bennett et al. (1980) are of particular interest. Booth et al. (1983a,b) also reported detailed studies of geotechnical properties of about four dozen cores from the Georges Bank and Baltimore Canyon Trough continental slopes.

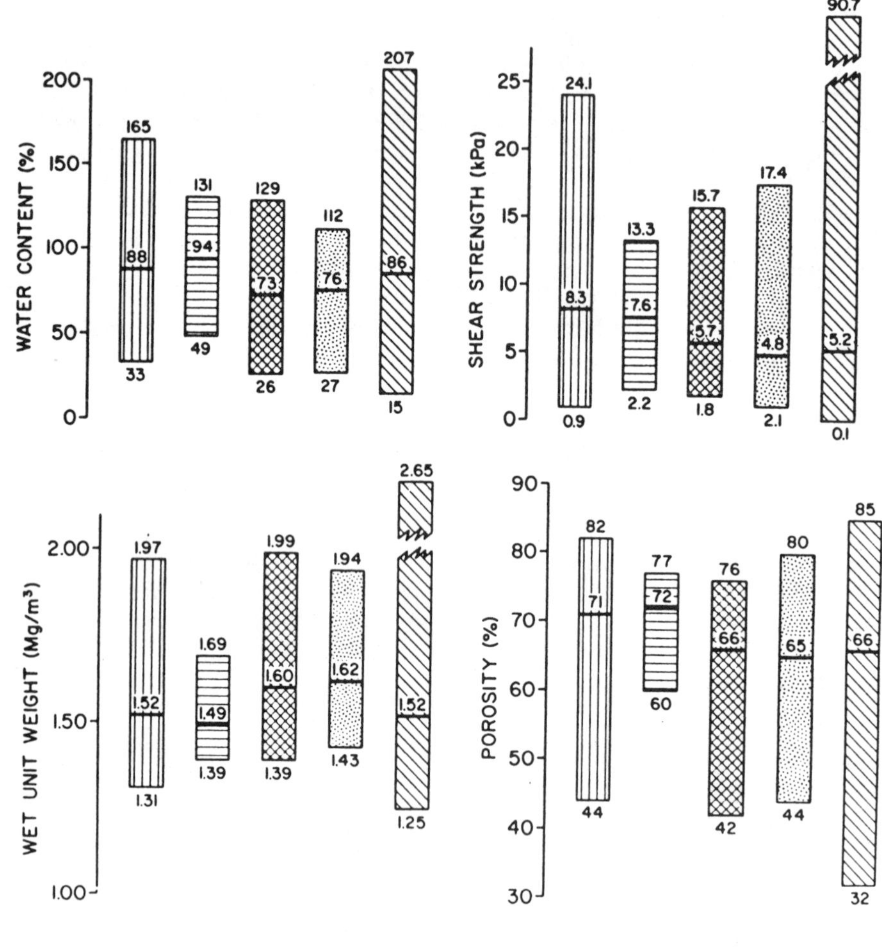

Figure 4.22 Maximum, minimum, and mean values of sediment mass properties for the U.S. Atlantic slope, upper rise (<3000 m), continental rise adjacent to Hudson Channel, and a corridor extending from 3000 m on the rise eastward across the northern Bermuda Rise. Values for the North Atlantic Basin as a whole are based on Horn et al. (1974). From Bennett et al. (1980).

The mass properties of sediments are strongly dependent on texture, composition, and mode of deposition. Thus one can predict that strong variability in mass properties should exist in sediments of the continental slope and across the western Blake Plateau. The continental rise, being affected by contour-following flow of bottom currents, should be relatively uniform, except where perturbed by variable sedimentation patterns around submarine channels or because of downslope sediment movement from the continental slope. This conceptual framework is reasonably accurate in terms of measured sediment mass properties (Fig. 4.22). In water content, shear strength, wet bulk density (wet unit weight) and porosity, the continental slope shows more variability than is present over the continental rise, over the Hudson Channel traversing the rise, or even across a mega-corridor extending seaward from the 3000 m isobath across the northern Bermuda Rise. Only the larger North Atlantic basin as a whole, encompassing a wide variety of sedimentary regimes, is more variable (Horn et al., 1974). According to Bennett et al. (1980), although the shear strengths of slope deposits are highly variable, the mean value (8.3 kPa) is higher than that of all the other provinces. However, the mean shear strength of the upper rise sediments is only slightly lower (7.6 kPa).

Ranges and averages do not provide a totally adequate description of mass properties for sedimentary provinces because important patterns can be missed or misinterpreted; furthermore, there may be aliasing in sample distribution. Keller et al. (1979) summarized mass properties in detail for the continental slope from Cape Hatteras to Georges Bank. While there is significant variability in properties, especially across submarine canyons, there are also well-defined patterns. Mean wet bulk density consistently decreases with water depth across the continental slope, and it is matched by a correlative increasein porosity. Water content also tends to increase with depth across the slope, but there are notable exceptions, including significant variability south of Wilmington Canyon. Keller et al. (1979) also noted that mean shear strengths on the middle and upper continental slope (<7 kPa) are actually *lower* than those on the lower slope and upper rise (7-14 kPa). As noted earlier, this may be a problem of sample aliasing. In general, however, the lower shear strengths seem to correlate with increased proportions of coarse-grained material on the continental slope, increases in mean wet bulk density, and decreases in porosity and water content.

Sensitivity of marine sediments is the ratio of natural to remolded shear strength, and it gives a measure of the strength lost by a sample when it is subjected to a shock or disturbance. Extreme values of sensitivity for continental slope sediments range from 1 (insensitive) to 12 (slightly quick), but mean sensitivities range from 2 to 8, and values of 2 to 4 predominate (Keller et al., 1979). Slope deposits thus are very similar in mean sensitivity to near-surface Atlantic basin sediments, which have a mean sensitivity of 4 (Keller and Bennett, 1970).

It is noteworthy that the surficial sediments on the upper continental rise show the greatest consistency in mass properties, even though they might be expected to show variability because of sediment mass movement from the adjacent continental slope. The relative uniformity of properties could be a result of the low sample density in this area. It also may be that the measurements have sampled a relatively uniform mantle of sediment that has been wafted from the slope and slope canyons and distributed along contours by the ambient southwest-flowing bottom currents.

Bennett et al. (1980) summarized the mass properties of several "typical" cores across the continental margin as a guide to the general geotechnical character of the deposits. It must be realized, of course, that no one core can be considered typical since there is such a high degree of variability within provinces, especially on the continental slope. Several noteworthy observations can be made, however. Cores from the open upper continental slope, a slope-ridge between canyons, a slope valley, the upper rise, and near Hudson Channel all show water contents that are generally above the liquid limit. (The liquid limit is the water content limit between semiliquid and plastic states of the sediment; remolding of a sediment with a water content higher than the liquid limit will transform the sediment into a viscous slurry.) Thus these sediments are susceptible to semi-fluid behavior if sufficiently shocked or remolded. Water content generally decreases and shear strength increases with depth in cores, as would be expected. However, some cores (e.g., a slope ridge core from ~1000 m water depth south of Baltimore Canyon) show constant or increased water content with depth in core, and a corresponding constancy or even a decrease in shear strength with depth. Such sediments are *under*consolidated, a state which is generally recognized as a principal condition for slumping.

Localized zones of underconsolidated sediments, principally on valley walls and ridges, have been identified in the Lease Sale 59 area on the continental slope off New Jersey (Keer and Cardinell, 1981) and south of Baltimore Canyon (Bennett et al., 1980) and Toms Canyon (Booth et al., 1981a). In contrast, no such underconsolidated sediments are known to be present on the continental slope adjacent to Georges Bank (Carpenter, et al., 1982; Booth et al., 1981b). These observations tend to agree with the degree of sediment mass movement documented along the margin (Fig. 4.16). It must be kept in mind, however, that there remain significant uncertainties in defining slope stability. Strong limitations exist for determining whether sediments are prone to failure because of uncertainties about 1) *in situ* pore pressures, 2) triggering mechanisms, 3) lateral and vertical extent of underconsolidated materials, and 4) adequacy of mechanical models of slope failure.

Compressional Wave Velocity

Sound velocity in slope and rise sediments has been studied principally using two techniques: wide-angle seismic reflection and refraction, and velocimeter measurements on core samples. The most widely based data set for sound velocity is a compilation of sonobuoy results used by Tucholke et al. (1982) to map sediment thickness in the western Atlantic basin. Velocity determinations in this data set are grouped into morphologic/acoustic provinces, the boundaries of which have been adjusted to minimize the standard deviation among individual determinations (Fig. 4.23). Along the margin there is a distinct break to higher velocities in the Lower Cretaceous sedimentary section at the level of the seismic reflector, Horizon β. This reflector marks the top of Jurassic and Neocomian limestones beneath largely terrigenous Upper Cretaceous and Cenozoic sediments. For strata below Horizon β, it is appropriate to use a separate velocity function that is dependent on overburden thickness (Fig. 4.23).

Mountain (1981) found little difference between the velocity/depth equations determined in this manner and velocities determined along the USGS/IPOD multichannel seismic reflection line extending southeast from Cape Hatteras. Egelson (1981) made a more detailed study of seismic interval velocities of the continental rise from Cape Hatteras to Cape Cod, based on multichannel seismic reflection profiles. He found 1) a general decrease in interval velocities in a seaward direction, and 2) a velocity decrease from southwest to northeast in Miocene and younger sediments. Both trends can be explained by changes in depth of burial and lithology.

The technique of measuring sound velocity on piston-core samples has not been widely used along the slope and rise. Sound velocities (corrected to *in situ* conditions) measured on 8- to 14-m long piston cores from the lower continental rise and Blake Outer Ridge average about 1500 m/s, with prominent velocity spikes up to 1800 m/s in thin sandy-silt beds (B. Tucholke, unpubl. data). Ratios of surface-sediment to bottom-water velocities typically are 0.96 to 0.99, comparable to values determined on the adjacent Hatteras Abyssal Plain (Tucholke, 1980).

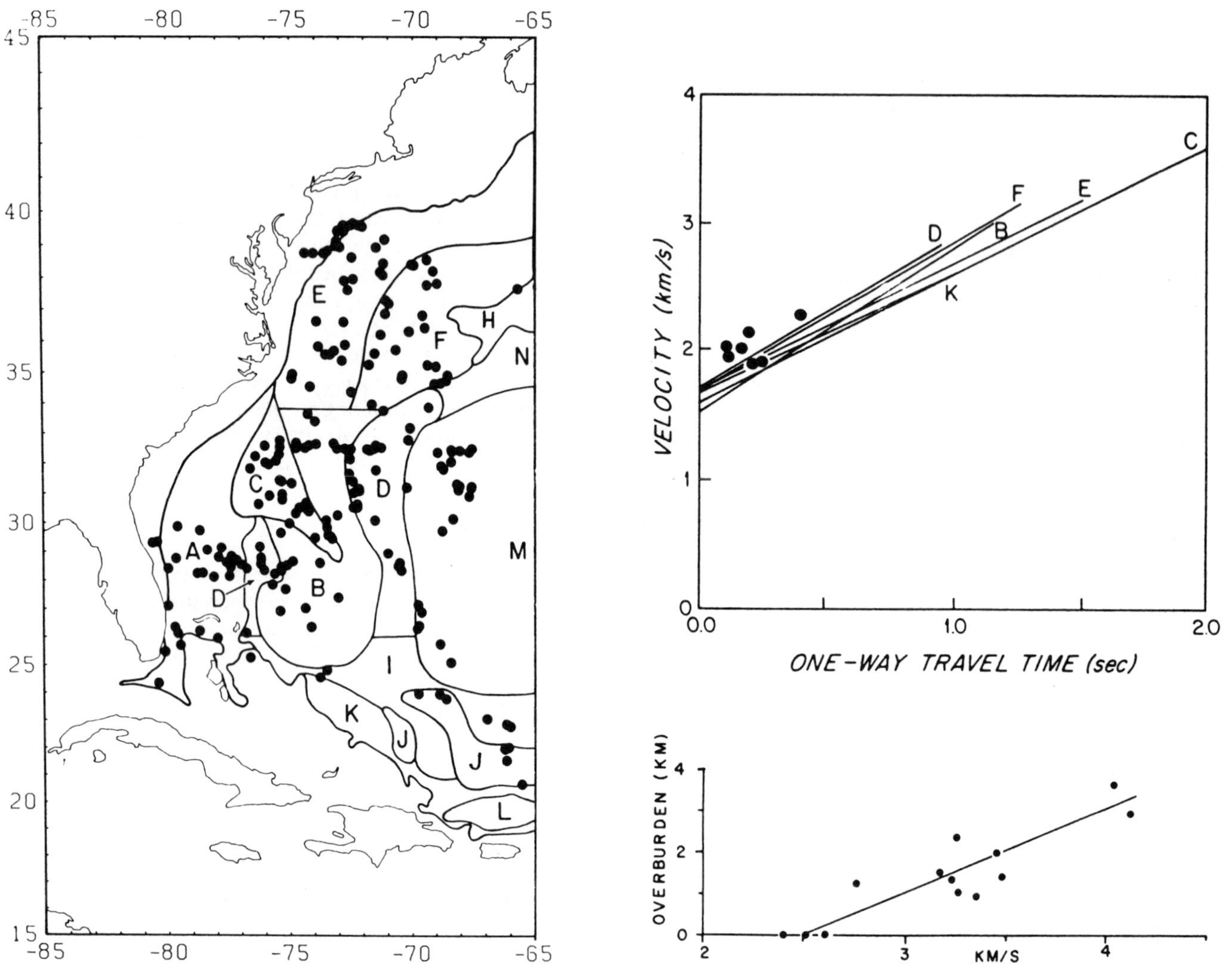

Area	V (km/s)	Standard Deviation S (km/s)	V_0* (km/s)	K** (km/s²)	Std. Error of Estimate ΔV (km/s)	Correlation Coefficient r	Number of Determinations n	Reflection Time† T (sec)
A	2.14	0.21					8	<1.0
B			1.53	1.30	0.18	0.83	39	2.6
C			1.60	1.00	0.16	0.89	69	4.0
D			1.73	1.16	0.15	0.69	88	1.8
E			1.69	0.98	0.12	0.94	51	3.0
F			1.68	1.16	0.24	0.74	50	2.5
G			1.62	1.19	0.19	0.88	55	3.0
H			1.49	1.43	0.23	0.93	62	3.0
I	2.03	0.23					8	<1.0
J	1.88	0.20					7	<1.0
K			1.69	0.91	0.17	0.65	16	1.8
L	2.26	0.16					8	<1.0
M	2.00	0.20					27	<1.0
N-Q	2.00‡							<1.0

* V_0 is the intercept velocity at t=0, where t is one-way vertical travel time
** K is a constant
† Maximum depth, in seconds two-way travel time, to which the functions can be confidently applied.
‡ Assumed velocity; data are insufficient or poorly sampled.

Figure 4.23 *Upper left* - Provinces (lettered) for which velocity-depth functions have been derived from sonobuoy profiles. *Top Right:* Velocity (V) = $V_0+Kt\pm\Delta V$, where V_0 is the intercept velocity at t=0, K is a constant (acceleration, km/s²), t is the *one- way* vertical travel time, and ΔV is the standard error of estimate; values for various provinces are given in the table at top. *Center right* - Velocity-regression lines for provinces; dots are averages for entire sediment column where low data density or thin sediments precluded derivation of a regression line. *Bottom* - Dependence of velocity on overburden thickness for sediments below seismic Horizon *β*. All data from Tucholke et al. (1982).

Geologic Structure of the Margin

Basement

Basement beneath the continental slope and rise typically is 8 to 10 km below sea level (Fig. 4.24), but depths as great as 11 km occur beneath the outer slope and uppermost rise east of Georgia and Delaware. Basement at this location presumably is normal tholeiitic oceanic crust just seaward of the ocean-continent boundary (Grow et al., 1979; Klitgord and Grow, 1980). An apparent basement ridge (depths of 7 to 8 km) beneath the continental slope and beneath the East Coast Magnetic Anomaly (ECMA) is thought to mark the position of the ocean-continent boundary. The existence of this ridge is interpreted from magnetic depth-to-basement estimates by Klitgord and Behrendt (1979), but the actual existence of igneous or metamorphic basement at this level is uncertain. In multichannel seismic reflection profiles the presumed basement ridge lies beneath a seismically chaotic reef or carbonate bank (e.g., Schlee et al., 1979) which masks the underlying basement. There are two partial exceptions (Fig. 4.25). On USGS multichannel line 25 across Baltimore Canyon Trough, basement beneath the continental shelf (top of synrift fault blocks and sediments) can be traced at a depth of 14 to 15 km seaward to a position slightly east of the peak of the ECMA with no indication of a basement ridge (Grow et al., 1983). Also, on USGS line 32 extending southeast from Cape Fear, deep hyperbolic reflections probably originating from the surface of basaltic ocean crust can be traced landward into the east flank of the ECMA at a constant depth of about 11 km. Together, these observations suggest that relatively level basement may extend beneath the East Coast Magnetic Anomaly. Thus the shallow basement ridge inferred from magnetic depth-to-basement estimates may be an artifact of assumptions about basement magnetization.

To the east of the ECMA, the most prominent basement feature occurs in oceanic crust at the Blake Spur Magnetic Anomaly (BSMA). A west-facing basement step or scarp, up to about one kilometer high, extends along the BSMA from the Blake Spur north to at least 36°N (Fig. 4.24). Near the Blake Spur the basement feature is a horst extending more than half a kilometer above the surrounding basement.

The other principal structural trends in oceanic basement are fracture zones that extend from obviously lineated younger crust westward to the ECMA. Unfortunately, most deep-penetration seismic reflection lines across the continental rise are dip lines, so the spacing, size, position, and orientation of fracture zones there are poorly documented. However, largely on the basis of aeromagnetic surveys, Schouten and Klitgord (1977) and Klitgord and Behrendt (1979) have suggested NW-SE-trending fracture zones in this crust (Fig. 4.26). Many of these fracture zones extend into structural offsets bounding major shelf basins west of the ECMA.

Crustal ages below most of the slope and rise are not well constrained because basement has been drilled only along the outer perimeter of the margin at the western edge of the M-Series (Keathley) magnetic anomalies (Figs. 4.24, 4.26). To the west is the Jurassic Magnetic Quiet Zone (JMQZ) which, excepting the Blake Spur Magnetic Anomaly, has only low-amplitude, mostly uncorrelated magnetic anomalies. Magnetic anomaly M-25 occurs beneath the lowermost continental rise just west of DSDP Site 105, and farther south DSDP Site 100 was drilled beside the Bahama Banks very near M-25. Sediments immediately above basement at these sites suggest an early to middle Oxfordian age for anomaly M-25. Depending on the time scale used, the geochronometric age thus could be as young as 145 Ma (e.g., Van Hinte, 1976) or as old as about 161 Ma (e.g., Harland et al., 1982). In the Blake-Bahama Basin, DSDP Site 534 recovered middle Callovian sediments above basaltic crust at magnetic anomaly "M-28" (Sheridan et al., 1982; Bryan et al., 1980). The two geologic time scales cited above give geochronometric ages of about 153 to 166 Ma, respectively, for this crust. This wide variation in geochronometric ages is symptomatic of ambiguities about the relative lengths of Jurassic stages and of radiometric ages (limited in number and some of uncertain quality) that are used to construct the timescale. However, if the basal sediments at DSDP 534 correctly date the underlying basement, then it is at least possible to say that the basaltic crust at the nearby Blake Spur Anomaly is probably lower Callovian.

A complication arises from the fact that the 200-km strip of crust between the BSMA and the ECMA has no counterpart off Africa in the eastern North Atlantic. Vogt (1973) suggested that the mid-oceanic spreading ridge made an eastward jump very early in the opening of the North Atlantic, moving from a median position between North America and Africa (median between the ECMA and BSMA) to a position along the African margin that now is marked by the BSMA. This explanation is still reasonable in the light of more recent magnetic and seismic evidence (e.g., Klitgord and Grow, 1980). Hence, to extrapolate ages from younger crust (anomaly M-25/"M-28") back to the time of initial continental separation, one must consider not only the *total* spreading rate (as opposed to half-rate) west of the BSMA but also the possibility of a significant spreading rate change at the time of the mid-ocean ridge jump.

Probably the best age estimate for initial separation of North America and Africa still is derived from diabase dikes along subaerial regions of the continental margins. Liberian dikes are dated at about 180 Ma (Dalrymple et al., 1975) and diabase intrusions in the Newark-trend basins in Connecticut and Maryland date to about 175 Ma (Sutter and Smith, 1979). The suggested 175-180 Ma age agrees with reasonable extrapolations from the oceanic crustal record.

Sedimentary Structure and Stratigraphy

Sediment Thickness

Total thickness of sediments along the U.S. Atlantic margin is shown in Figure 4.27. The possible basement ridge beneath the continental slope marks the seaward edge of the major shelf basins and the landward edge of the seaward-thinning sediment prism that forms the continental rise. The total volume of sediments on the continental margin is divided subequally between these physiographic prov-

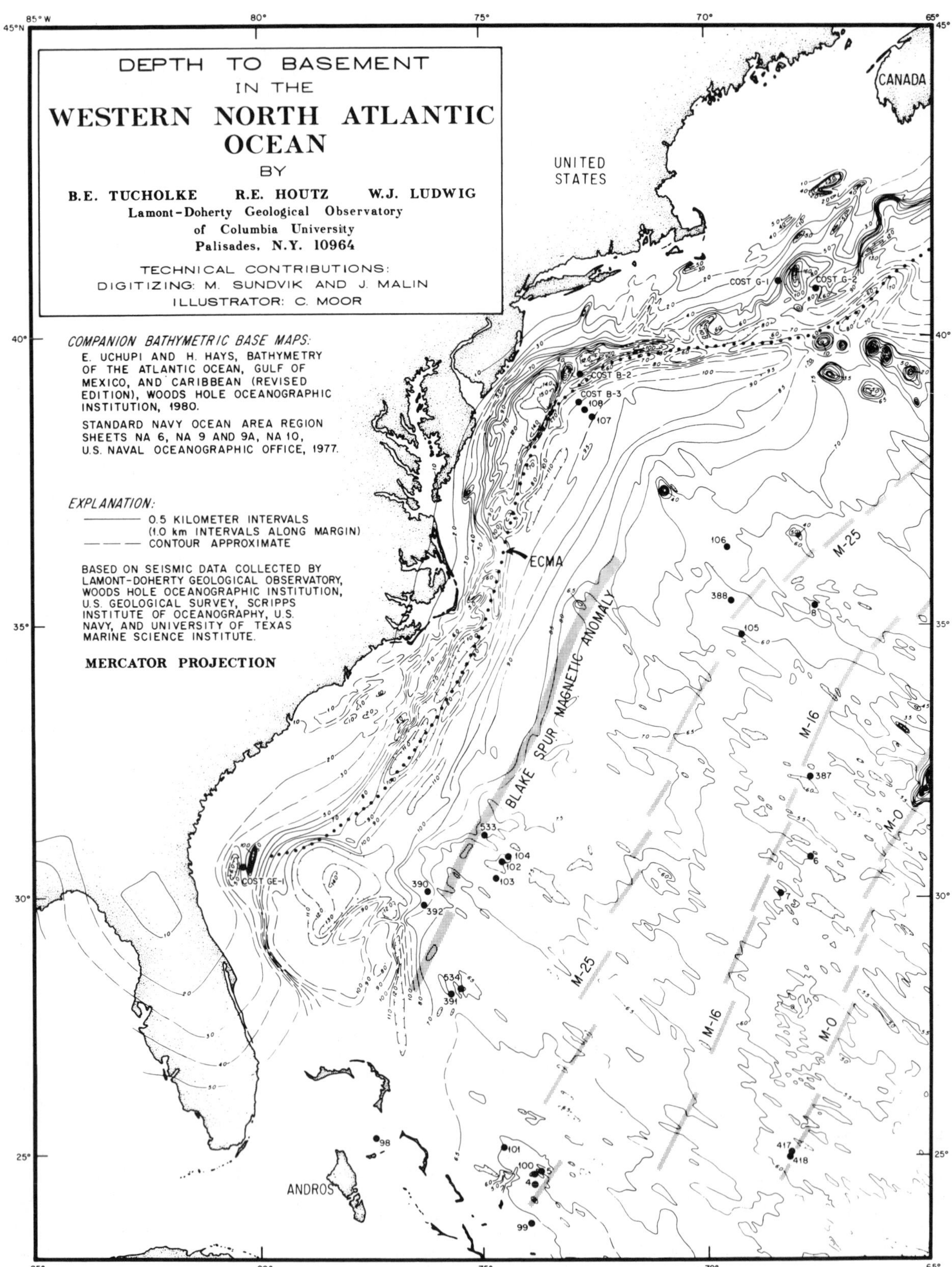

Figure 4.24 Map of depth to basement in kilometers below sea level (adapted from Tucholke et al., 1982). Principal magnetic anomalies and borehole locations are indicated.

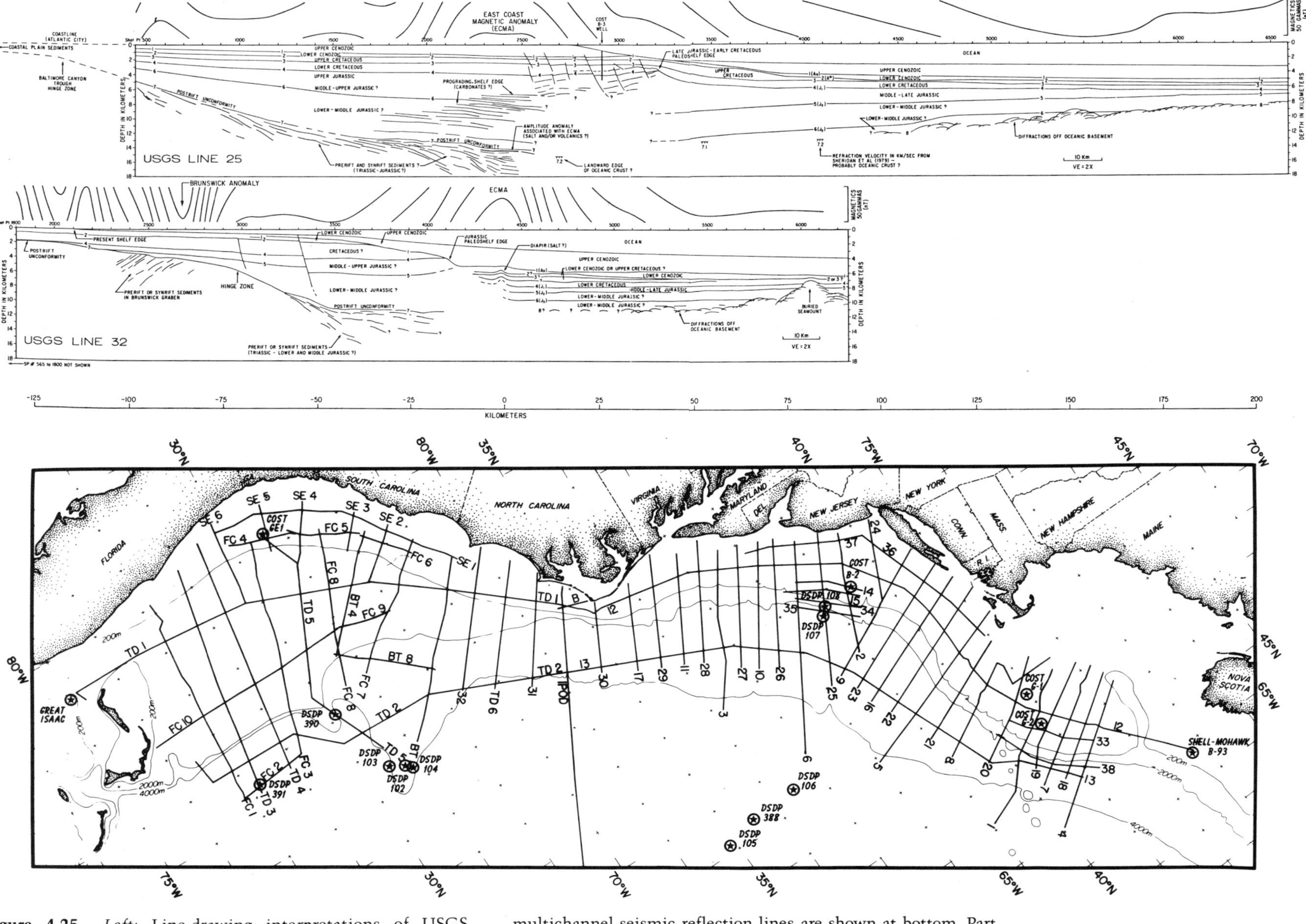

Figure 4.25 *Left:* Line-drawing interpretations of USGS multichannel seismic lines 25 (top) and 32 (center), from Grow et al. (1983). *Right:* Locations of these and other available USGS multichannel seismic reflection lines are shown at bottom. Part of USGS line 25 is depicted in Figure 4.32. Magnetic anomalies are shown along tops of profiles.

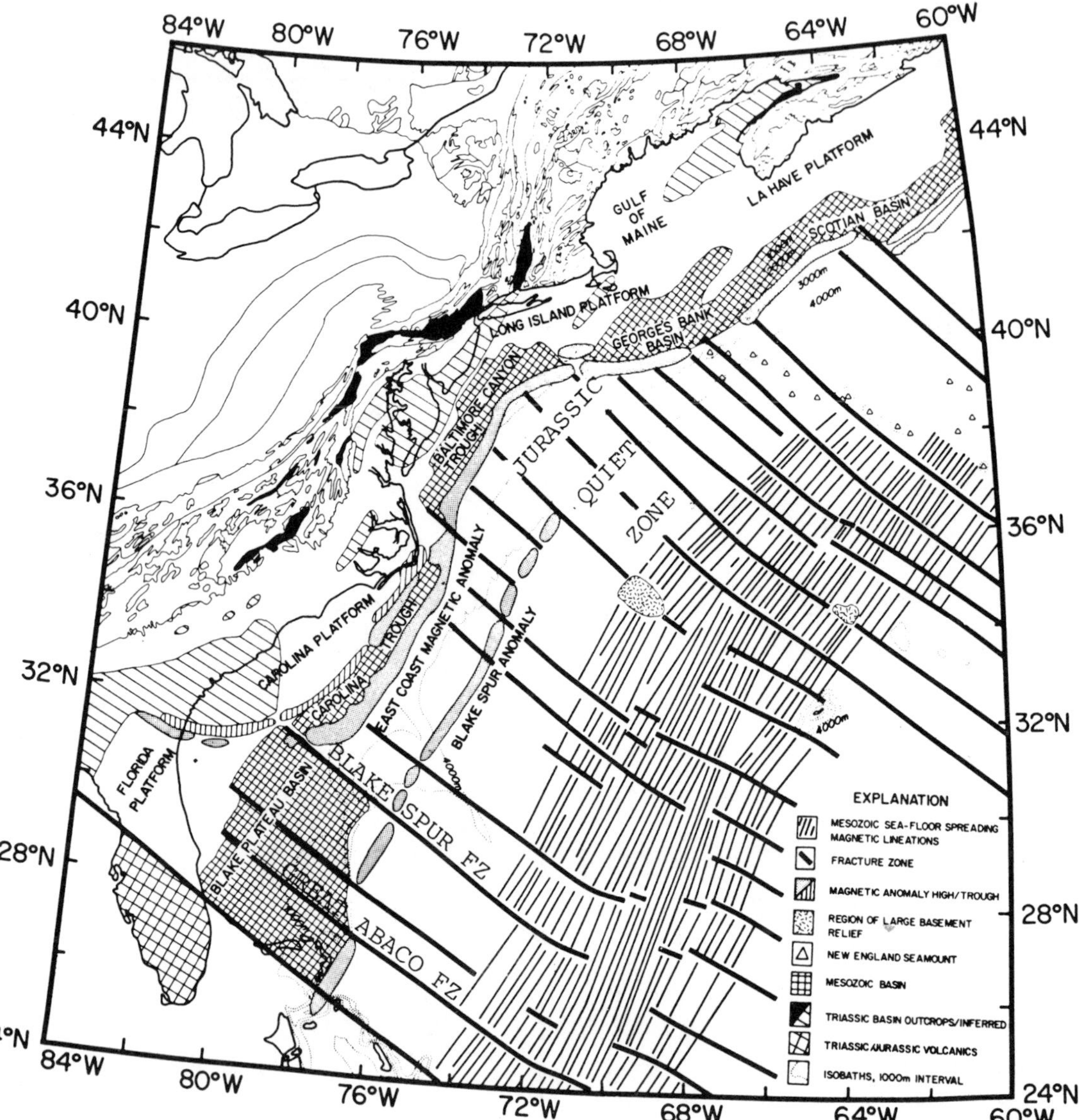

Figure 4.26 Locations of magnetic anomalies (closely-spaced M-Series lineations at right) and known and inferred fracture zones along U.S. Atlantic continental margin (from Klitgord and Behrendt, 1979).

inces. Maximum sediment thickness seaward of the ridge is about 9 km above presumed oceanic crust near the base of the continental slope. On average, up to about 40 percent of the sediments forming the continental rise are of Cenozoic age (Mountain and Tucholke, 1985).

Sedimentary Structure

Primary Structures. The Mesozoic and Cenozoic sediments giving the U.S. Atlantic continental margin its present shape exhibit the entire spectrum of primary depositional structures normally observed in ocean basins. Specific examples are discussed in the ensuing section on stratigraphic development, but a general summary is provided here.

Conformable bedding is most commonly observed in sediments of the continental rise, because this province is most removed from the principal continental source areas of sediments and from the variable controls on input of these sediments to the ocean basin. However, changes in relative sea-level along the margin have strongly affected sediment supply to the rise; during lowstands sedimentary onlaps developed near the base of the continental slope and, more infrequently, downlaps in a seaward direction developed beneath the central and lower continental rise. These relations are more pronounced in the Cenozoic stratigraphic sequence, partially because they are less deeply buried and thus better resolved in seismic reflection profiles. The seismic discordances or discontinuities in contacts between

Figure 4.27 *Next page:* Total sediment thickness in kilometers along the U.S. Atlantic margin (from Tucholke et al., 1982). Borehole locations are shown by filled circles. →

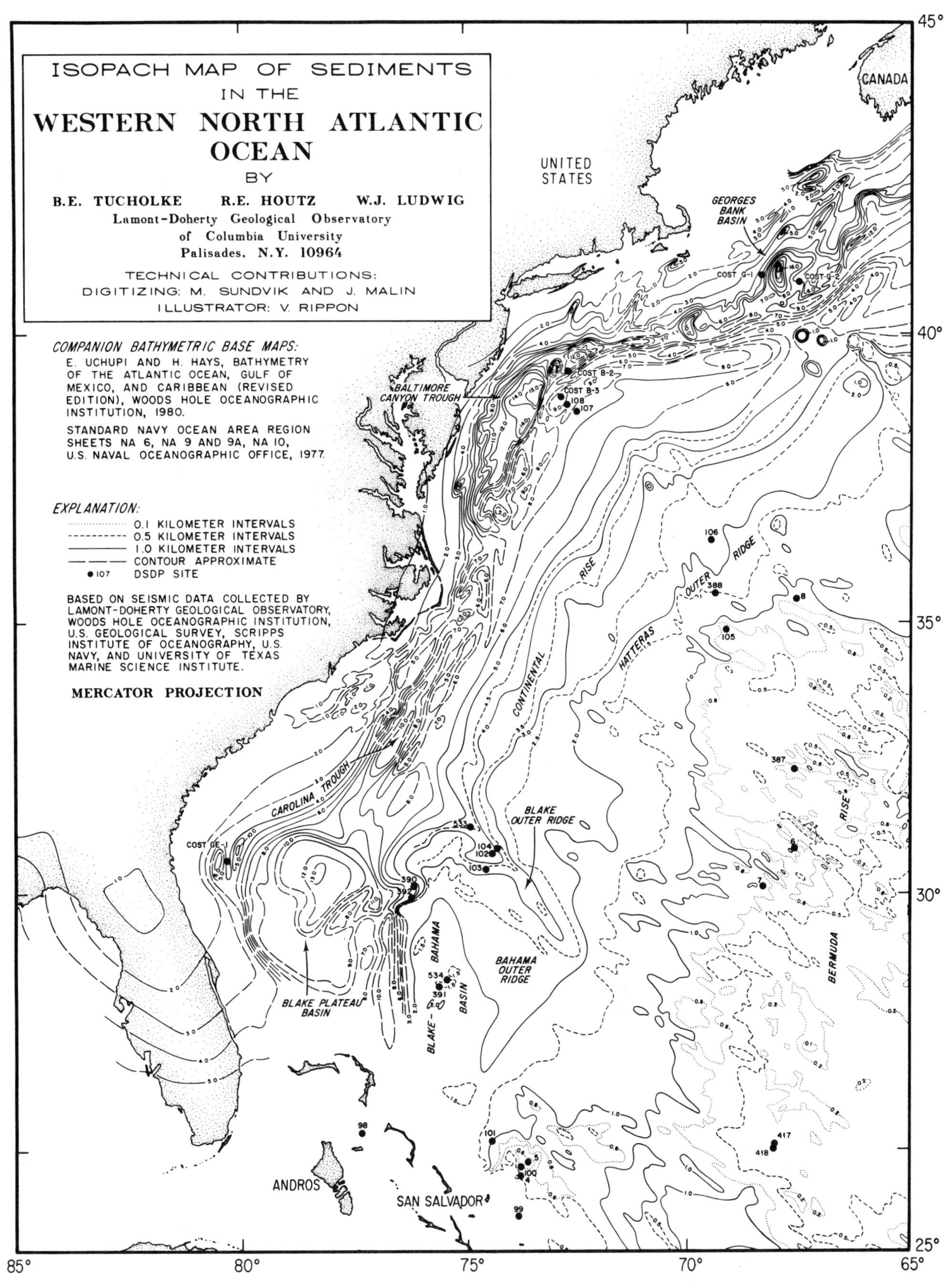
ISOPACH MAP OF SEDIMENTS
IN THE
WESTERN NORTH ATLANTIC
OCEAN
BY
B.E. TUCHOLKE R.E. HOUTZ W.J. LUDWIG
Lamont-Doherty Geological Observatory
of Columbia University
Palisades, N.Y. 10964
TECHNICAL CONTRIBUTIONS:
DIGITIZING: M. SUNDVIK AND J. MALIN
ILLUSTRATOR: V. RIPPON
COMPANION BATHYMETRIC BASE MAPS:
E. UCHUPI AND H. HAYS, BATHYMETRY OF THE ATLANTIC OCEAN, GULF OF MEXICO, AND CARIBBEAN (REVISED EDITION), WOODS HOLE OCEANOGRAPHIC INSTITUTION, 1980.
STANDARD NAVY OCEAN AREA REGION SHEETS NA 6, NA 9 AND 9A, NA 10, U.S. NAVAL OCEANOGRAPHIC OFFICE, 1977.
EXPLANATION:
0.1 KILOMETER INTERVALS
0.5 KILOMETER INTERVALS
1.0 KILOMETER INTERVALS
CONTOUR APPROXIMATE
107 DSDP SITE
BASED ON SEISMIC DATA COLLECTED BY LAMONT-DOHERTY GEOLOGICAL OBSERVATORY, WOODS HOLE OCEANOGRAPHIC INSTITUTION, U.S. GEOLOGICAL SURVEY, SCRIPPS INSTITUTE OF OCEANOGRAPHY, U.S. NAVY, AND UNIVERSITY OF TEXAS MARINE SCIENCE INSTITUTE.
MERCATOR PROJECTION
UNITED STATES
CANADA
GEORGES BANK BASIN
COST G-1
COST G-2
COST B-2
COST B-3
BALTIMORE CANYON TROUGH
CONTINENTAL RISE
HATTERAS OUTER RIDGE
CAROLINA TROUGH
BLAKE OUTER RIDGE
COST GE-1
BLAKE PLATEAU BASIN
BLAKE - BAHAMA BASIN
BAHAMA OUTER RIDGE
BERMUDA RISE
ANDROS
SAN SALVADOR
45°
40°
35°
30°
25°
85°
80°
75°
70°
65°

these rapidly accumulated sequences sometimes are interpreted as unconformities. However, it must be recognized that the sediment accumulation rates are probably low (tens of meters) between these major depositional pulses; thus conventional reflection profiling, which has a resolution of tens of meters, cannot resolve the difference between continuous slow deposition and a true unconformity at these boundaries.

Another locus of mostly conformable bedding is the Mesozoic sedimentary sequence on the southeastern Blake Plateau (e.g., J. Ewing et al., 1966). These beds thin slowly eastward, reflecting a source of detrital material along the Florida margin. The younger Cenozoic beds commonly show apparent unconformities possibly cut by the Antilles Current. Unconformities cut by the Gulf Stream are especially well developed toward the western and northern margins of the plateau.

Progradational bedding is common beneath the continental shelf and is a response to relative changes of sea-level. Small-scale progradation occurs in Pleistocene sediments at the shelf edge in some locations; widened slope contours and Pleistocene drainage patterns suggested by arkosic sediments on the shelf (Fig. 4.11) indicate that such progradation may be most important around Hudson Canyon and in the Baltimore-Washington Canyon area. As noted earlier, the Florida-Hatteras Slope also is the seaward limit of a prograding sediment wedge whose position probably has been controlled by the Gulf Stream.

In post-Eocene sediments of the continental rise and Blake Ridge (i.e., above seismic Horizon A^u), control of depositional sequences by contour-following bottom currents (the Western Boundary Undercurrent) was important. The most obvious manifestation of this control is abyssal sediment waves that are developed locally beneath the central continental rise and the Blake Outer Ridge (Fig. 4.28). However, by far the best-developed sediment waves occur on the Bahama Outer Ridge (e.g., Markl et al., 1970) and on the Hatteras Outer Ridge (lower continental rise east of Cape Hatteras) where sedimentation was less influenced by downslope processes and proportionately more influenced by currents (Fig. 4.29). In addition to these obvious manifestations of current control, it is likely that sedimentary lenses in the continental rise are elongated parallel to regional contours (Mountain and Tucholke, 1985). As discussed previously, the entire Blake Outer Ridge (Fig. 4.28) is composed of sediments deposited under the control of bottom currents. The spectrum of seismic sequences within this sediment body is representative of the range that can be deposited by abyssal currents.

Carbonate reefs and banks were important during the Mesozoic development of the continental margin. In Jurassic to mid-Cretaceous time they flourished along the outer edges of the continental shelf and the Blake Plateau. They are now observed as deeply buried structures beneath the continental slope, and they crop out along the Blake Escarpment.

Secondary Structures. Shallow faulting and mass wasting along the continental slope have already been discussed (Fig. 4.16). The one major surface fault that is known to persist well into the subsurface is a growth fault along the outer edge of the northernmost Blake Plateau (Sylwester et al., 1979; Dillon et al., 1982). This fault is shown in Figure 4.25, Line 32, at kilometer -50. Numerous other faults affect the subsurface of the shelf and slope, particularly near the buried reef or shelf-edge carbonate bank (Fig. 4.25, Line 25, kilometers 0-40). These mostly affect only the Mesozoic and Lower Cenozoic sedimentary sequence. No significant deep or shallow faulting has been reported in continental rise sediments.

Erosional unconformities, created by currents or by downslope mass-wasting, are widespread in the sedimentary record of the slope and rise. Beneath the continental rise, the most prominent unconformity is marked by seismic Horizon A^u, which represents strong erosion of the margin by abyssal currents probably at the beginning of Oligocene time (Tucholke and Mountain, 1979; Miller and Tucholke, 1983); excavation of the upper rise may have oversteepened the continental slope and triggered slumping there (Fig. 4.28). The continental slope was further eroded during a mid-Oligocene sea-level lowstand. As a result of these events the continental slope retreated as much as 30 km, approximately to its present position (Schlee et al., 1979). The present exposure of Cretaceous and Tertiary strata along the slope (Fig. 4.13) is a direct legacy of this erosion. Other less important unconformities interrupt the sedimentary sequence on the continental rise as discussed later.

As noted earlier, unconformities created by current erosion are most dramatically expressed in Cenozoic sediments on the western part of the Blake Plateau and on the Charleston Bump where the Gulf Stream has been active (Fig. 4.8). The eastern Blake Plateau has unconformities developed to a much lesser extent; these unconformities probably were eroded at the base of the Antilles Current. The most dramatic unconformity in the region is that of the Blake Escarpment. It probably has formed in response to a combination of abyssal current erosion/corrosion, biological erosion, and mass-wasting (Paull and Dillon, 1980, and unpubl. data).

Diapirs and associated deformation strongly affect the sedimentary section of the upper continental rise in a 400-km long zone extending from the northern end of the Blake Plateau to Cape Hatteras (Fig. 4.16; see also kilometer 0, Line 32, Fig. 4.25). These diapirs are thought to be cored by salt, which may extend south beneath the Blake Plateau, although no diapirs have yet been recognized beneath the plateau itself. Salt diapirism also strongly affects the continental slope and rise off Nova Scotia where the seaward edge of the salt front is close to 4000 m water depth (Uchupi and Austin, 1979). This zone of salt diapirism extends at least as far south as the eastern edge of Georges Bank (Fig. 4.30), where faulting has been attributed to salt tectonics by Uchupi et al. (1982a). A possible Mesozoic "reef complex" that is seismically continuous with the zone of salt diapirs also extends south to about 40°N (Fig. 4.30; Austin et al., 1980). It is possible that this structure includes non-diapiric salt. Between Cape Hatteras and 40°N there is no well-documented diapirism beneath the continental slope and rise. However, salt has been drilled at 3800 m in Baltimore Canyon Trough (Oil and Gas Journal, 18 Sept. 1978, p. 72); at this shallow depth the salt must be associated with a diapir.

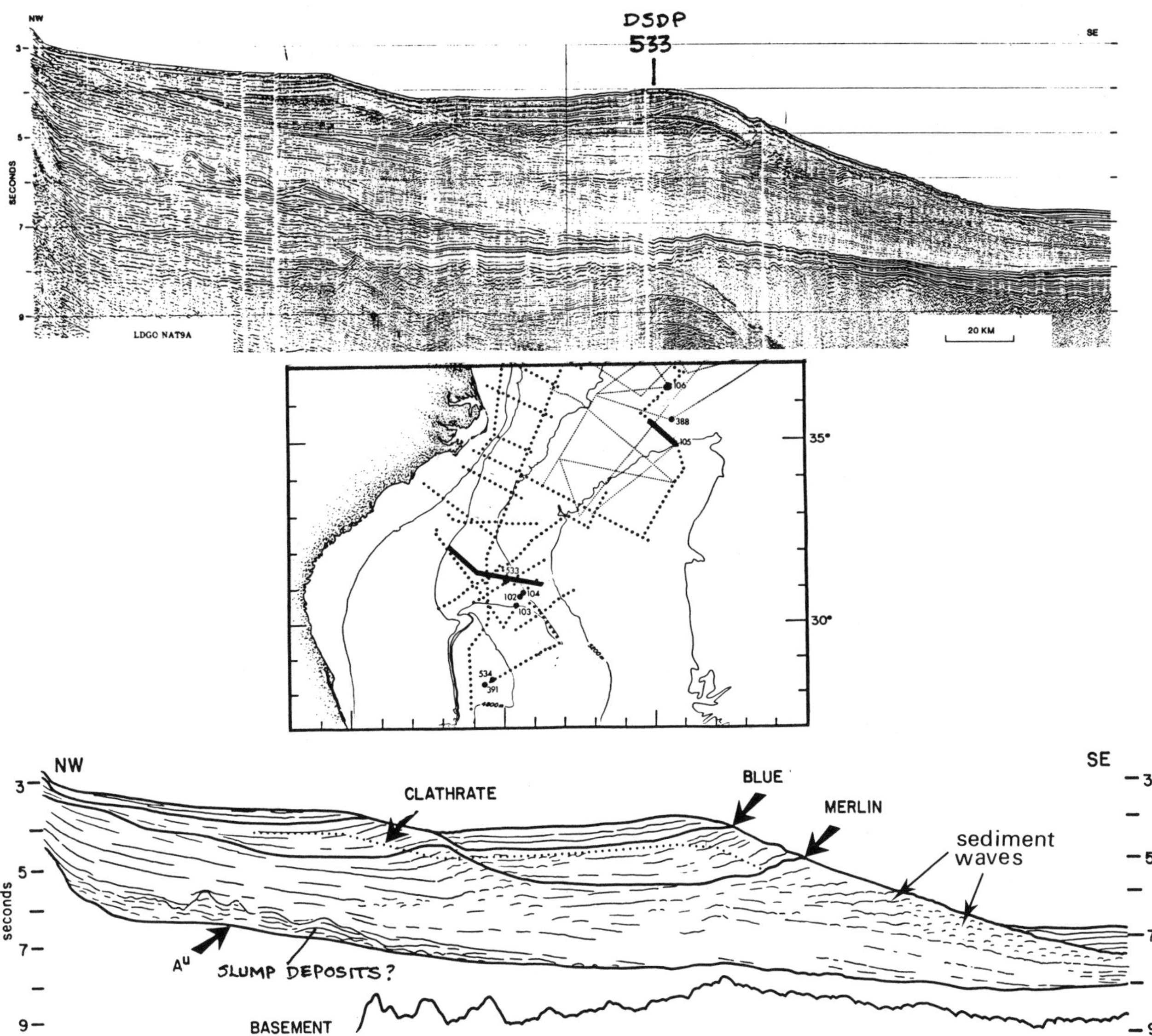

Figure 4.28 Seismic reflection profile (top) and interpretation (bottom) across the Blake Outer Ridge. Note possible slump deposits above Horizon A^u at lower left. Bold southern trackline on map shows location of profile and DSDP drillsites. Adapted from Mountain and Tucholke (1985).

Stratigraphic Synthesis

The continental slope and rise off the eastern United States contain a remarkable record of Atlantic Ocean geologic history extending from the onset of seafloor spreading in the Middle Jurassic upward to the present. The sedimentary record is complex because it has been controlled by interplay of both shallow-marine and deep-marine processes. Near the continental slope, relative sea level, river sediment input, reefs, faulting, slumping, and local diapirism all have exerted direct or indirect control on sedimentation patterns and sediment composition, and none of these was constant along strike. Along the lower continental rise, these processes often were subsidiary to such deep marine controls as abyssal currents and fluctuations in the calcite compensation depth (CCD). Between these areas the central and upper continental rise have felt most strongly the impact of processes related to both shallow and deep marine controls; hence these areas are the most difficult to interpret satisfactorily. Turbidity currents resulting from shallower-water (e.g., sea level) impulses probably have had the most pervasive and geologically long-ranging effects on sedimentary sequences across the entire margin, although in the post-Eocene section they commonly played a role subsidiary to that of contour-following bottom currents.

The data base for examining the geologic history of this area consists almost exclusively of seismic reflection profiles, although limited drilling data are available (e.g., Fig. 4.27). Most available seismic profiles are single-channel, but along the continental slope and uppermost rise, the U.S.

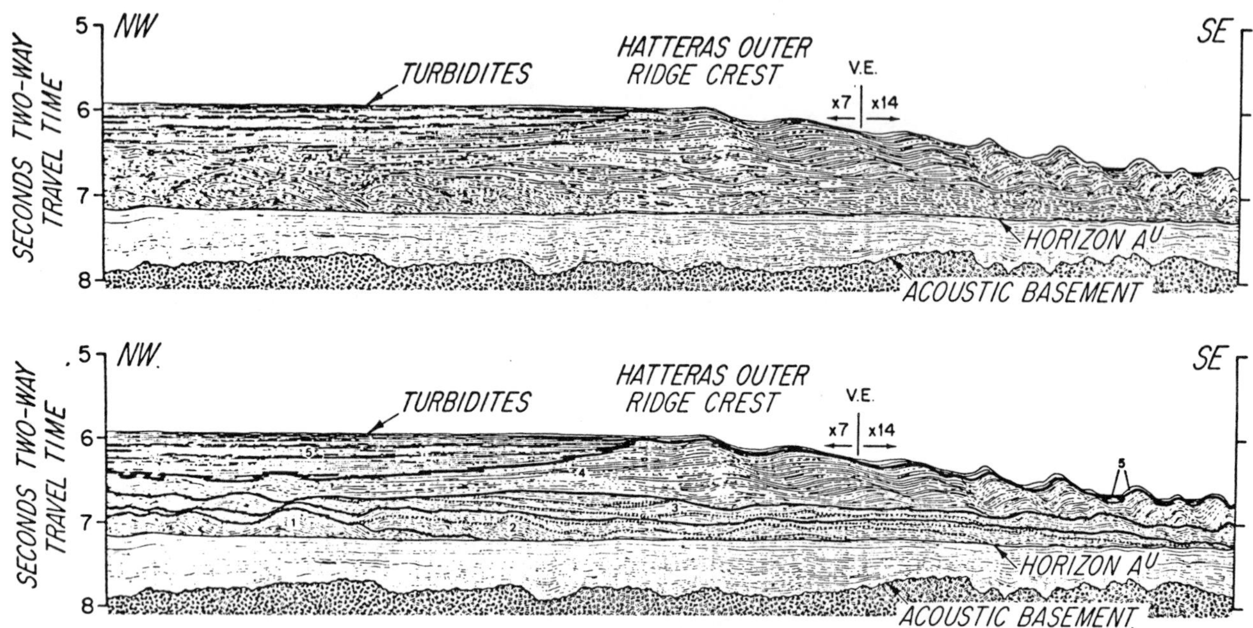

Figure 4.29 Tracing of seismic reflection profile across sediment waves that form the Hatteras Outer Ridge on the lower continental rise. Sedimentary sequences 1 to 4 in lower profile date to the Miocene to mid-Pliocene; sequence 5 is 2.8 m.yr. and younger. From Tucholke and Laine (1983). Bold northern line in Figure 4.28 locates profile.

Geological Survey has obtained a grid of multi-channel seismic (MCS) lines that is very useful (Fig. 4.25). In deeper water, there are several scattered MCS lines, and more detailed MCS surveys have been conducted over the lower continental rise east of Cape Hatteras around DSDP drillsite 603 (Woods Hole Oceanographic Institution, 1980), over the lower continental rise along the USGS-IPOD line (University of Texas Marine Science Institute, 1975), and over the Blake Outer Ridge and in the Blake-Bahama Basin around DSDP drillsites 391 and 534 (Lamont-Doherty Geological Observatory, 1975 and 1977).

There is relatively little continuous corehole sampling of the stratigraphic sequence, especially in its deeper part, but available borehole data still provide significant constraints on the geologic framework (Figs. 4.27, 4.31). Among the boreholes along the deeper part of the margin, DSDP Site 105 provided the first real insight into the nature of the Mesozoic lithofacies that are widespread along the continental rise. However, because the site was drilled on a basement high, it is virtually impossible to correlate the lithofacies with most well-defined seismic sequences in the surrounding areas. The correlations are better resolved in other boreholes along the outer perimeter of the margin sedimentary prism (Sites 5, 100, 101, 387; Fig. 4.27), although it is difficult to trace seismic-sequence boundaries unambiguously into the margin from these sites. DSDP Sites 391 and 534 in the Blake-Bahama Basin recovered sediments as old as Callovian and clarified the stratigraphic sequence in that basin, but extrapolating results and tracing seismic boundaries north across the Blake Outer Ridge to the continental rise still leaves unresolved ambiguities.

DSDP boreholes constrain the Cenozoic sedimentary record about as effectively as they do the Mesozoic record. Sites 105, 106, 388, and 533 (Fig. 4.27) recovered sediments that help document the Neogene development of the continental rise. Site 108 on the upper rise, although it penetrated only 75 m and was poorly cored, provides an important control point for the distribution of Eocene lithofacies. In general, however, the Paleogene sedimentary record beneath the continental rise is almost unsampled and thus is poorly known.

Along the continental slope, the most important control point is the COST B-3 well where the post-Middle Jurassic section was sampled. Several AMCOR sites along the slope (Hathaway et al., 1976; Poppe, 1981), as well as cores and dredges on outcrop areas (Fig. 4.13) help document Cenozoic facies. DSDP Sites 390 and 392 on the Blake Spur (Fig. 4.27, 4.31) recovered shallow-water Cretaceous and Paleogene facies on the Blake Plateau, but these sequences cannot be correlated easily northward along the margin. The COST GE-1, B-2, G-1 and G-2 wells on the shelf are invaluable in interpreting shelf, slope, and rise depositional history (Fig. 4.31).

The sediment composition and patterns of sediment dispersal to and along the slope and rise varied significantly in time. The principal factors that controlled sedimentary patterns have been discussed above; changes in the relative influence of each of these controls created depositional sequences that are widely recognized in the rock and seismic stratigraphic record. The depositional sequences have been outlined as reasonably well-defined seismic stratigraphic intervals by several authors (Fig. 4.31). The ensuing discussion summarizes the sedimentary development of the continental slope and rise in successive time slices and refers to the seismic intervals and boundaries outlined as the "North American Basin Standard" in Figure 4.31.

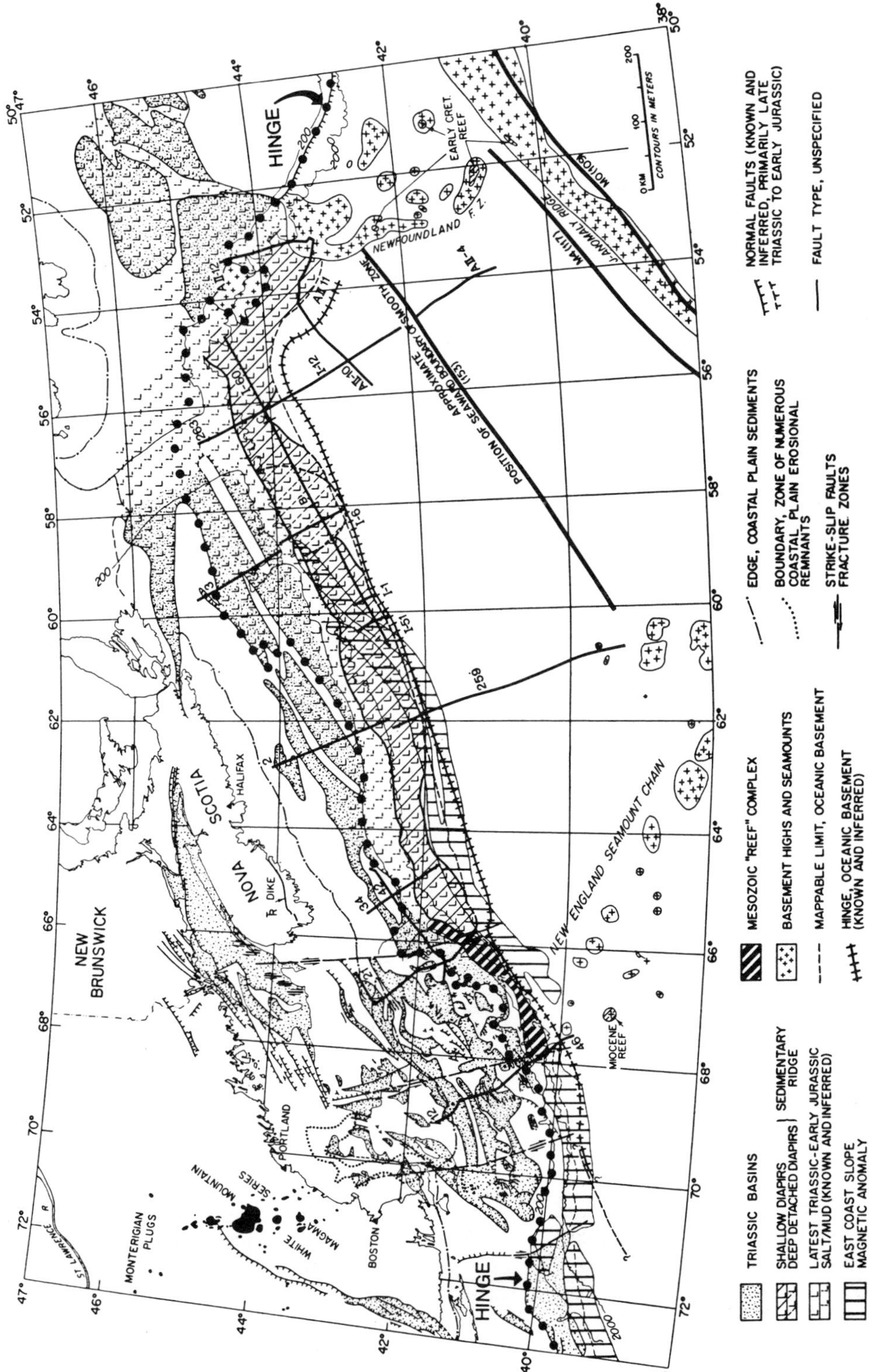

Figure 4.30 Structural framework off the northeastern United States, from Uchupi and Austin (1979). Note the diapiric salt ridge (L-pattern) extending west to Georges Bank.

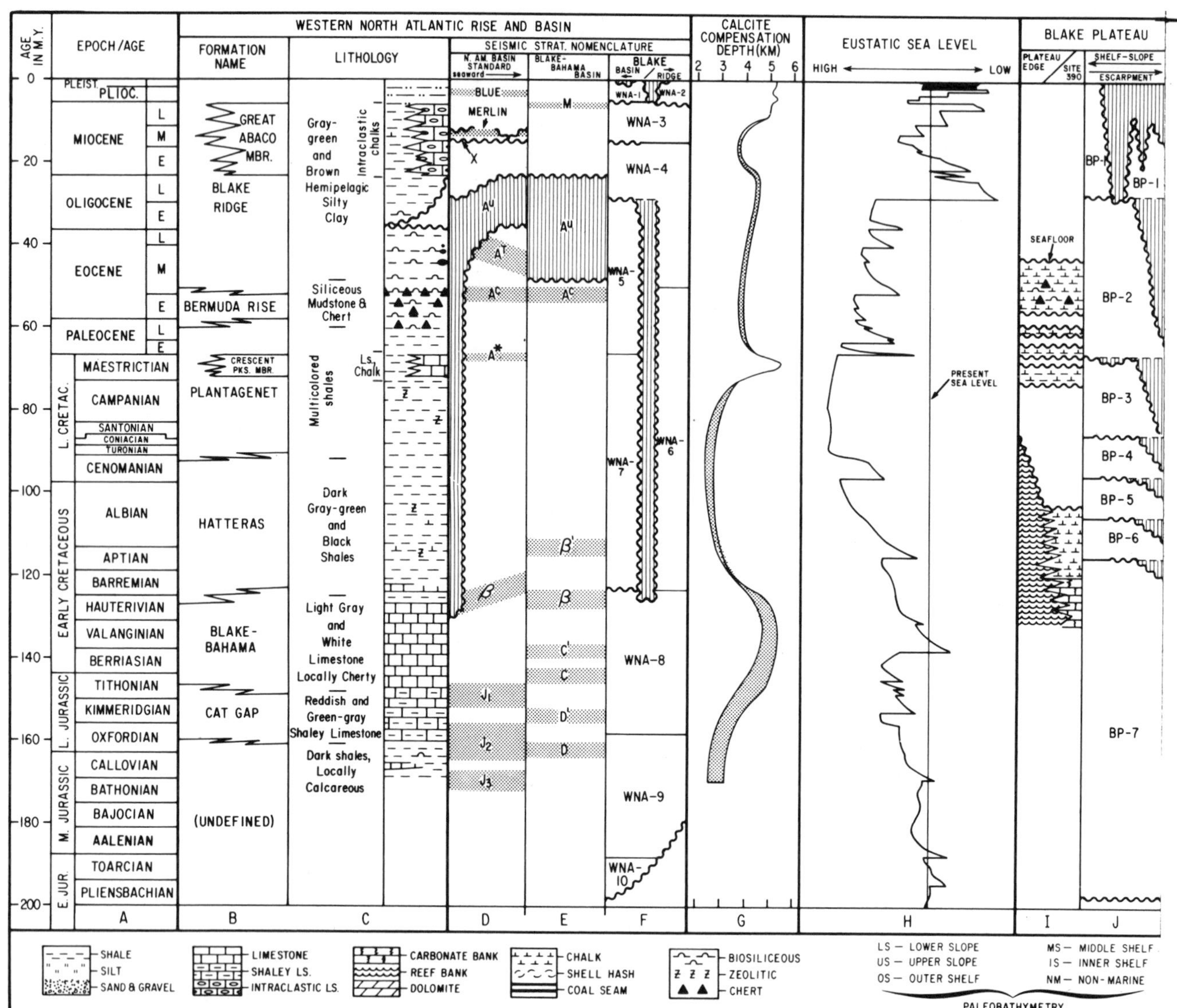

Figure 4.31 Summary of lithofacies, seismic stratigraphic nomenclature, calcite compensation depth, eustatic sea level and paleo-bathymetry versus age for key areas of the U.S. Atlantic continental rise, slope and shelf (from Mountain and Tucholke, 1985). *Data sources:* A- Harland et al. (1982) time scale for Mesozoic, Berggren et al. (1984a,b) for Cenozoic; B- Jansa et al. (1979); C- Jansa et al. (1979) for post-Callovian, and Sheridan et al. (1982) for Callovian; D- Horizons Blue and Merlin: Mountain (1983), Horizon X: Tucholke and Laine (1983), Horizons A^u, A^T, A^c, and β: Tucholke and Mountain (1979), Horizon J_1, J_2, and J_3: Klitgord and Grow (1980)

Middle Jurassic. Presuming an age of about 175 Ma for the earliest seafloor spreading along the continental margin, and using the Harland et al. (1982) geologic time scale, Bajocian to Bathonian sediments overlie the oldest oceanic crust. There is some indirect evidence that the total thickness of Middle Jurassic sediments on this crust is small. First is the fact that Bajocian and Bathonian sediments on the continental shelf (e.g., the COST G-2 well at the outer edge of Georges Bank Basin) are largely limestones and dolomites, indicating carbonate-platform deposition. Most detrital, terrigenous sediments were deposited farther landward (e.g., COST G-1 well, Fig. 4.31), and they rarely passed the shelf edge to the deep basin. A similar situation probably prevailed along the margin as far south as the Bahamas (Jansa, 1981), but this inference is based on extrapolation from very limited well and dredge data.

A second indication that Middle Jurassic sediments overlying oceanic crust may be thin appears when we examine the presumed spreading-ridge jump which occurred at the time that the Blake Spur Anomaly was formed. Jurassic sediments consistently and progressively onlap basement from the continent edge eastward across the abandoned spreading axis to the Blake Spur Magnetic Anomaly (Klitgord and Grow, 1980). These sediments show no disruption or change in accumulation pattern across the abandoned ridge, so very little sediments could have accumulated before the ridge jump. Similarly, there is no known disruption on the African margin; although a

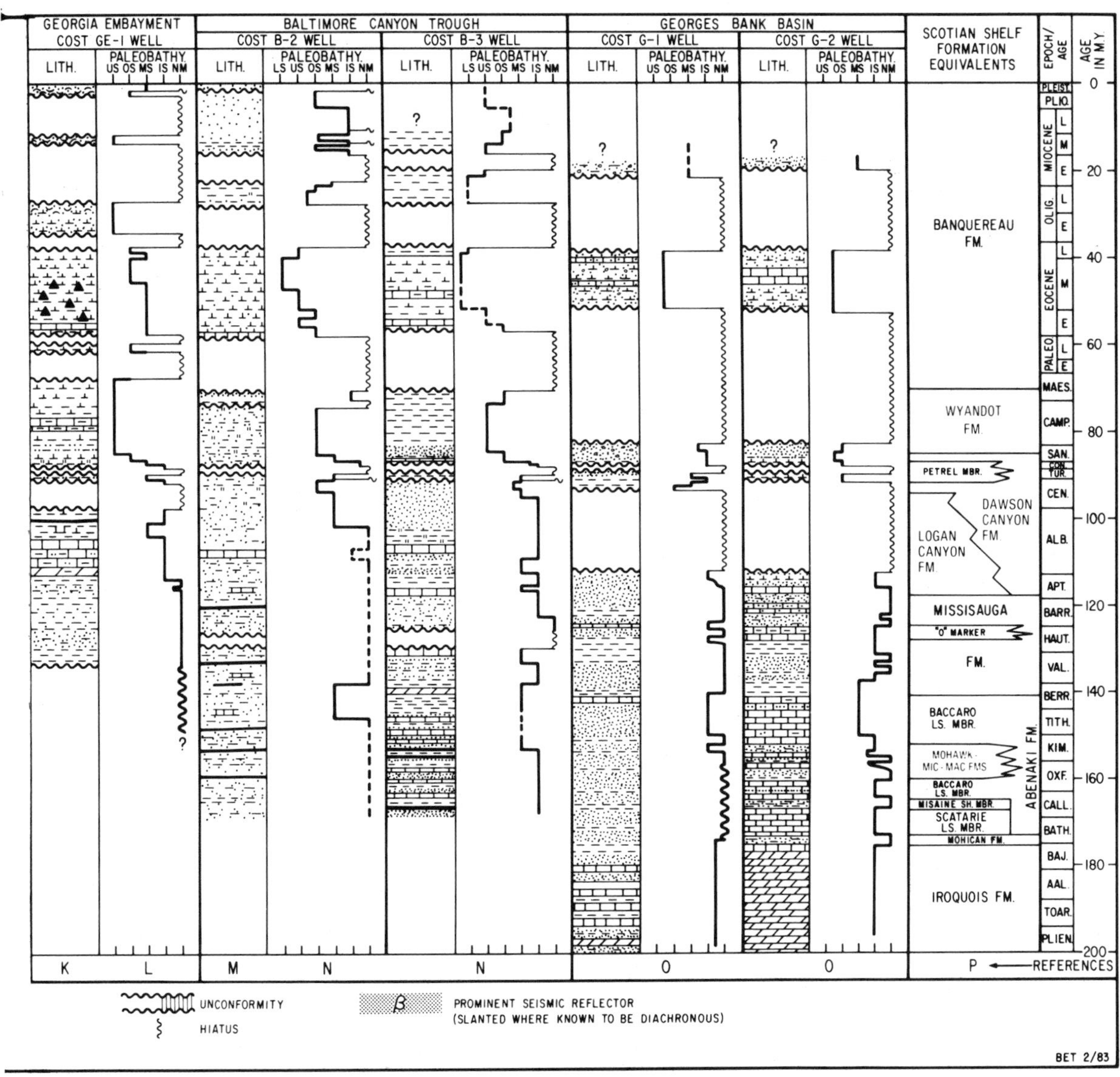

and data in this paper; E- Sheridan et al. (1982); F- Shipley et al. (1978); G- CCD curve based on Tucholke and Vogt (1979) and modified with data from Jansa et al. (1979) and Berger and von Rad (1972); H- Vail et al. (1977) for Cretaceous, and Vail and Todd (1981) for Jurassic; I- Derived from DSDP Site 390 and interpretation of adjacent seismic profiles by Benson et al. (1978a,c); J- Shipley et al. (1978); K- Rhodehamel (1979); L- Poag and Hall (1979); M- Rhodehamel (1977); L- Poag (1981b), Eocene lithofacies inferred from nearby AMCOR boreholes, dredges, and cores; P- Modified from Poag (1982a) and based on Jansa and Wade (1975) and Given (1977).

relatively small sediment accumulation along the African margin (to which the spreading axis jumped) could be masked in seismic reflection profiles by basaltic flows, dikes and sills, it is unlikely that a significant sediment prism was present there.

Klitgord and Grow (1980) identified a deeply buried but conspicuously high-amplitude reflector in the inner JMQZ (west of the BSMA) that they termed reflector J_3 (Fig. 4.32). The reflector is observed along most of the U.S. margin and pinches out on oceanic crust about halfway across the inner JMQZ. According to the above ridge-jump model, crust near the pinchout dates from late Bathonian to early Callovian, which gives a maximum age of about 168-172 Ma for reflector J_3 (Fig. 4.31). The smoothness and high reflectivity of the horizon, and the reverberant character of adjacent gently-dipping reflectors, suggest that J_3 marks a significant event of downslope sediment transport. For example, the sea-level lowstand occurring at the end of the Bathonian (Fig. 4.31) may have triggered the erosion of shelf carbonates and their consequent transport to the continental slope and rise via turbidity currents. The presence of such high-impedance carbonates could explain the high reflectivity of reflector J_3. Whether this lowstand is represented in the sedimentary record of the COST G-1 and G-2 wells is unknown because biostratigraphic age control at these levels is poor.

Another well-developed reflector packet, termed J_2 by Klitgord and Grow (1980), occurs within the Jurassic sedimentary sequence above J_3 (Fig. 4.32). Like J_3, it slopes smoothly and gently seaward but it reaches at least as far seaward as the Blake Spur Magnetic Anomaly where it abuts the west-facing basement step. J_2 has not been traced as a continuous horizon eastward of the basement step, but a similar reflector that probably is J_2 locally extends about a third of the way from the Blake Spur anomaly eastward toward anomaly M-25. This onlap onto oceanic crust suggests a maximum age of middle Callovian (~165-166 Ma), but the reflector could be younger, perhaps as young as early Kimmeridgian.

Depositional patterns during the Middle Jurassic are shown by the map of total sediment thickness between reflector J_2 and basement in Figure 4.33. Strata of this age form a rather uniform sedimentary prism all along the continental margin. No particular point sources of sediments appear to have been present along the shallow carbonate shelf to form distinct deep-sea fans.

Late Jurassic. A prominent seismic horizon, termed J_1 by Klitgord and Grow (1980), overlies Horizon J_2 and usually caps a seismically laminated sequence (Fig. 4.32). J_1 locally has been traced as far eastward as magnetic anomaly M-23. The horizon cannot be traced confidently southward beneath the Blake Outer Ridge, but in stratigraphic position it is roughly equivalent to Horizon C in the Blake-Bahama Basin (Jansa et al., 1979). The best age estimate for this reflector is Upper Jurassic, probably upper Kimmeridgian or Tithonian (Fig. 4.31).

The thickness of Upper Jurassic sediments (J_1-J_2) is depicted in Figure 4.34. It is clear in seismic reflection profiles that shelf-edge reefs and carbonate banks were well developed by this time (Fig. 4.32; Schlee et al., 1979). Along much of the margin this barrier prevented sediment transport from the shelf to the deep basin. However, off Baltimore Canyon Trough, detrital sediments flooding the shelf apparently passed through breaches in the shelf-edge barrier and formed a large deep-sea fan at the base of the slope(Fig. 4.34). A smaller fan probably also developed immediately north of the Blake Spur. Drilling has not penetrated either of these fans. Because the Calcite Compensation Depth (CCD) was relatively deep in the Late Jurassic (Fig. 4.31), the fan sediments probably are finer-grained and more calcareous equivalents

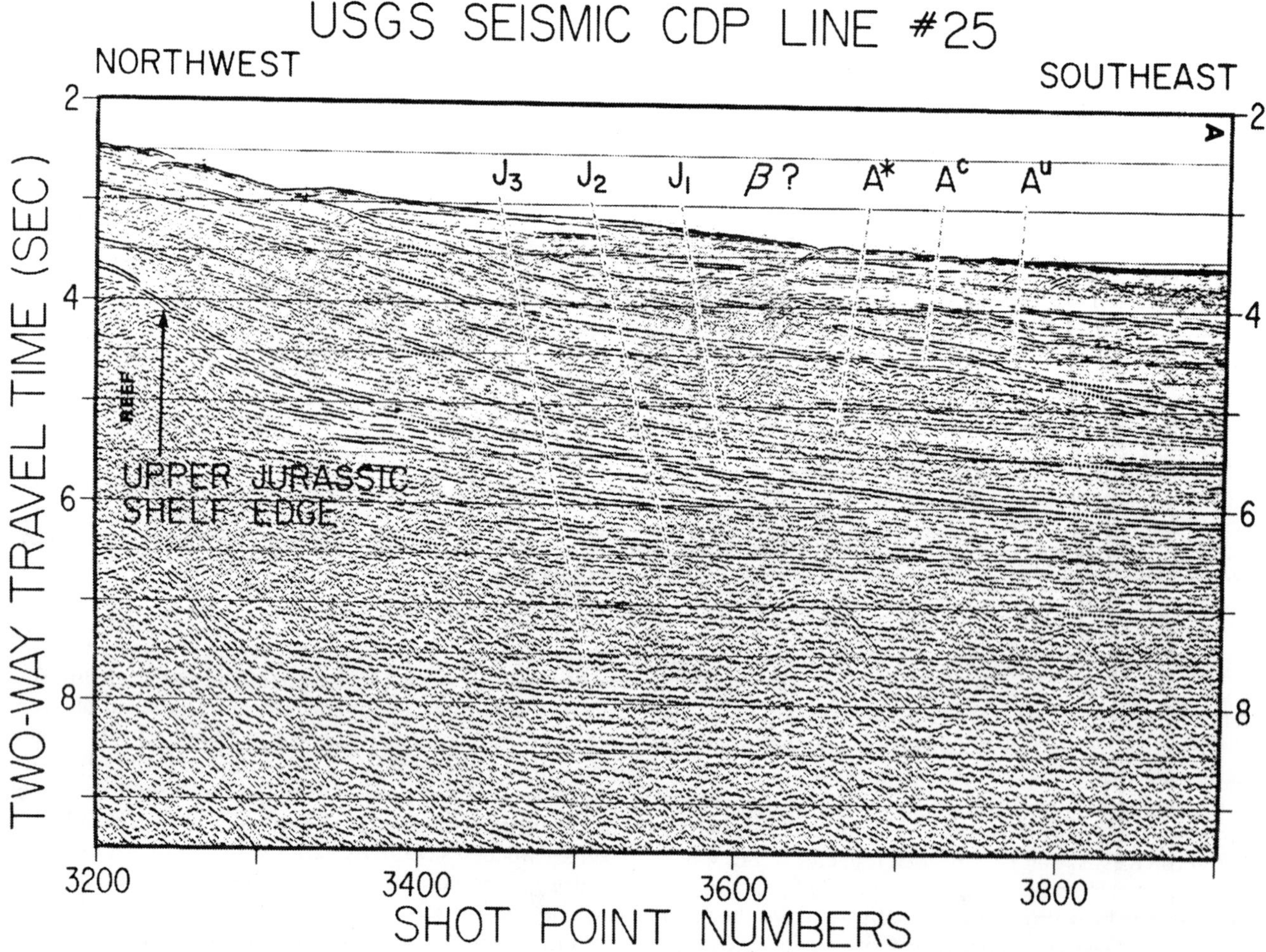

Figure 4.32 Section of USGS multichannel seismic line 25 across the lower continental slope and upper rise seaward of Baltimore Canyon Trough, from Klitgord and Grow (1980). See Figure 4.25 for location. Note identifications of seismic horizons and reef control on Upper Jurassic shelf edge.

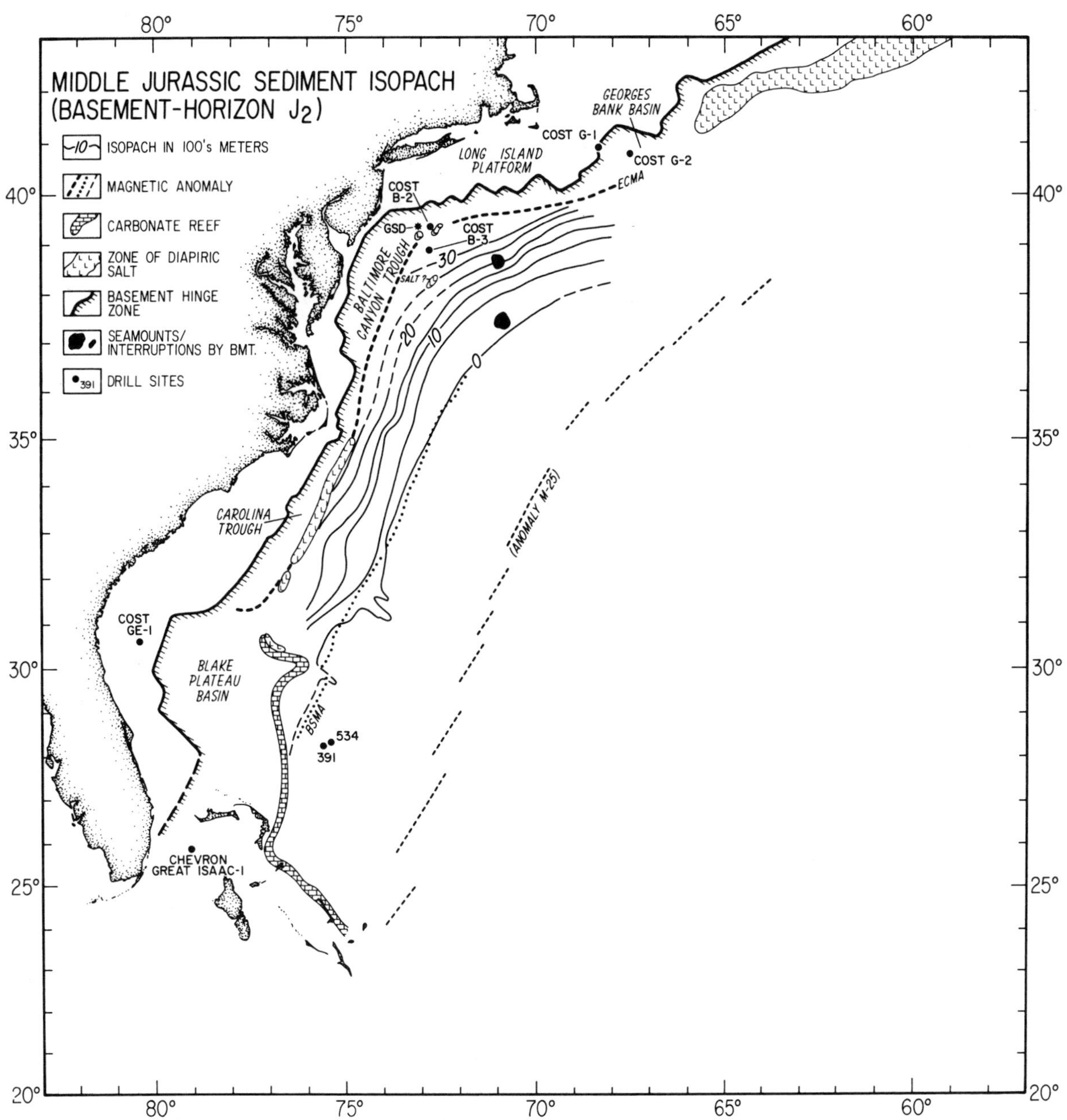

Figure 4.33 Sediment thickness between Horizon J_2 (Callovian/Kimmeridgian) and basaltic basement, from Mountain and Tucholke (1985). ECMA is East Coast Magnetic Anomaly, BSMA is Blake Spur Magnetic Anomaly.

of the sandy shales that were being deposited on the shelf at that time.

Along the outermost edge of the margin where Deep Sea Drilling Project sites have penetrated Upper Jurassic sediments, they consist of reddish and gray-green limestones and shaley limestones. This facies has been termed the Cat Gap Formation by Jansa et al. (1979).

Early Cretaceous. In the deep basin along the outer perimeter of the continental margin, the Lower Cretaceous sedimentary sequence consists of light gray to white limestones (Fig. 4.31), defined as the Blake-Bahama Formation by Jansa et al. (1979). The limestones locally include sandy detrital carbonates and terrigenous debris nearer the continent, for example at DSDP Sites 391 and 534 in the Blake-Bahama Basin. In the Atlantic basin, the top of this limestone sequence is a prominent seismic marker termed Horizon β (Tucholke and Mountain, 1979). However, the reflector usually becomes much less well defined beneath the central and upper continental rise (Fig. 4.32), probably because the highly calcareous sediments that give rise to the strong impedance contrast in the deep basin are diluted by non-calcareous detritus nearer the continental margin.

Lower Cretaceous sediment thickness (Horizon J_1 to β) shows a continuation of the sediment dispersal patterns developed during the Upper Jurassic (Fig. 4.35). Continental-

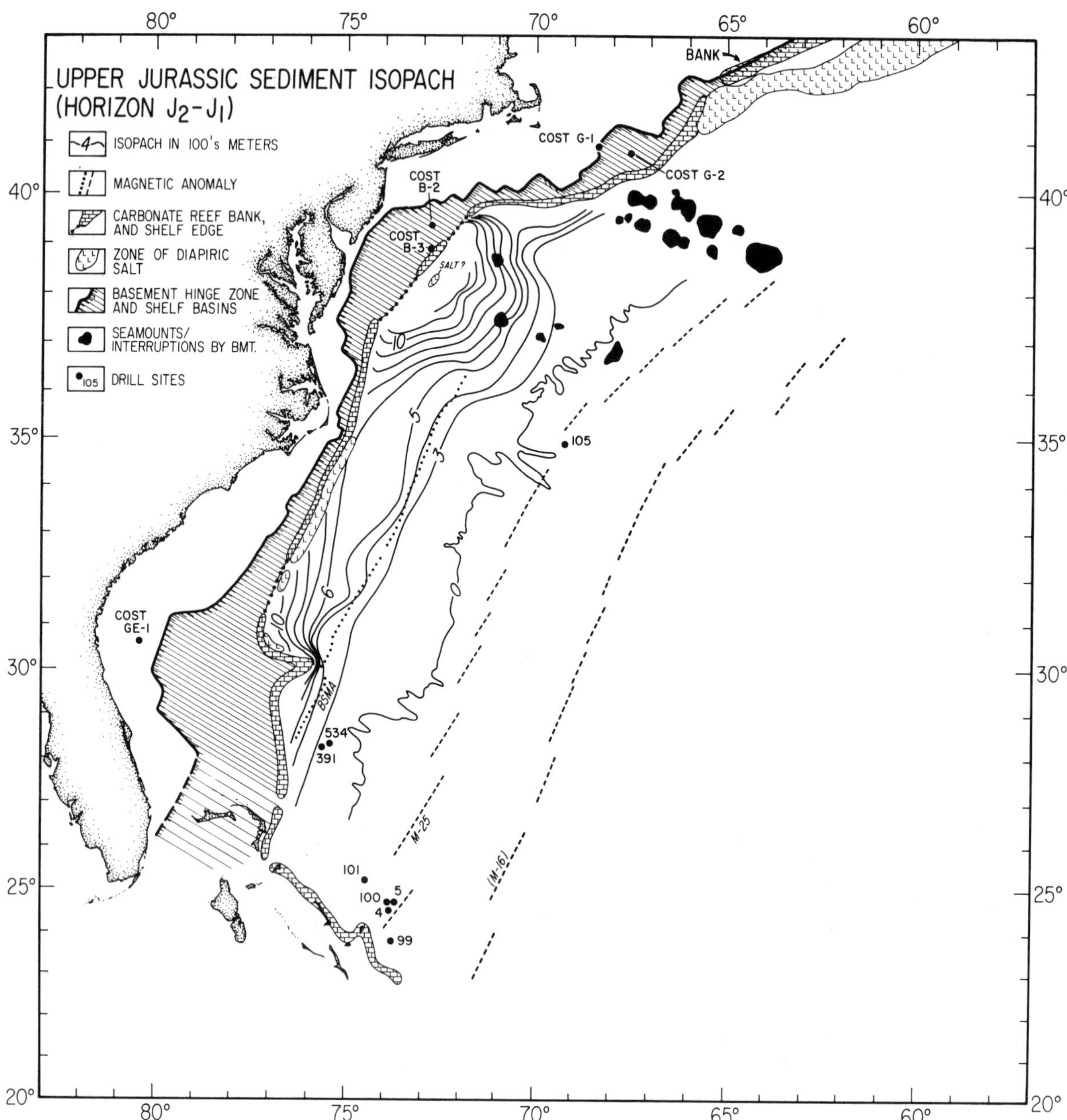

Figure 4.34 Upper Jurassic sediment thickness (Horizons J_1 to J_2). Note carbonate reef and bank buildups at shelf edge. From Mountain and Tucholke (1985).

rise fans were developed seaward of Baltimore Canyon Trough and between the Blake Spur and Cape Hatteras. At least one abyssal fan began to form at the southwestern edge of Georges Bank. Seismic profiles show that the shelf-edge reef system was overstepped by seaward-prograding sediments next to these fans, but it continued to flourish and to block seaward transport off most of the rest of Georges Bank, off Cape Hatteras, and along the outer edge of the Blake Plateau.

Latest Early and Late Cretaceous. Uppermost Lower Cretaceous sediments drilled in the deep basin consist of cyclically deposited black and gray-green shales that accumulated following a sharp rise in the CCD at the end of the Neocomian (Hatteras Fm., Fig. 4.31). The gray-green shales are low in organic carbon and are burrowed by benthic organisms. They apparently were deposited under low-oxygen, but not reducing, conditions. In contrast, the black-shale interbeds are finely laminated, unburrowed, and often high in organic carbon (typically several percent), suggesting deposition under anoxic conditions. Most of the organic carbon is of terrigenous origin, but a prominent organic-carbon spike (up to 15 percent) at the end of the Cenomanian is almost entirely marine carbon (e.g., Tucholke and Vogt, 1979).

Upper Cretaceous basin sediments along the periphery

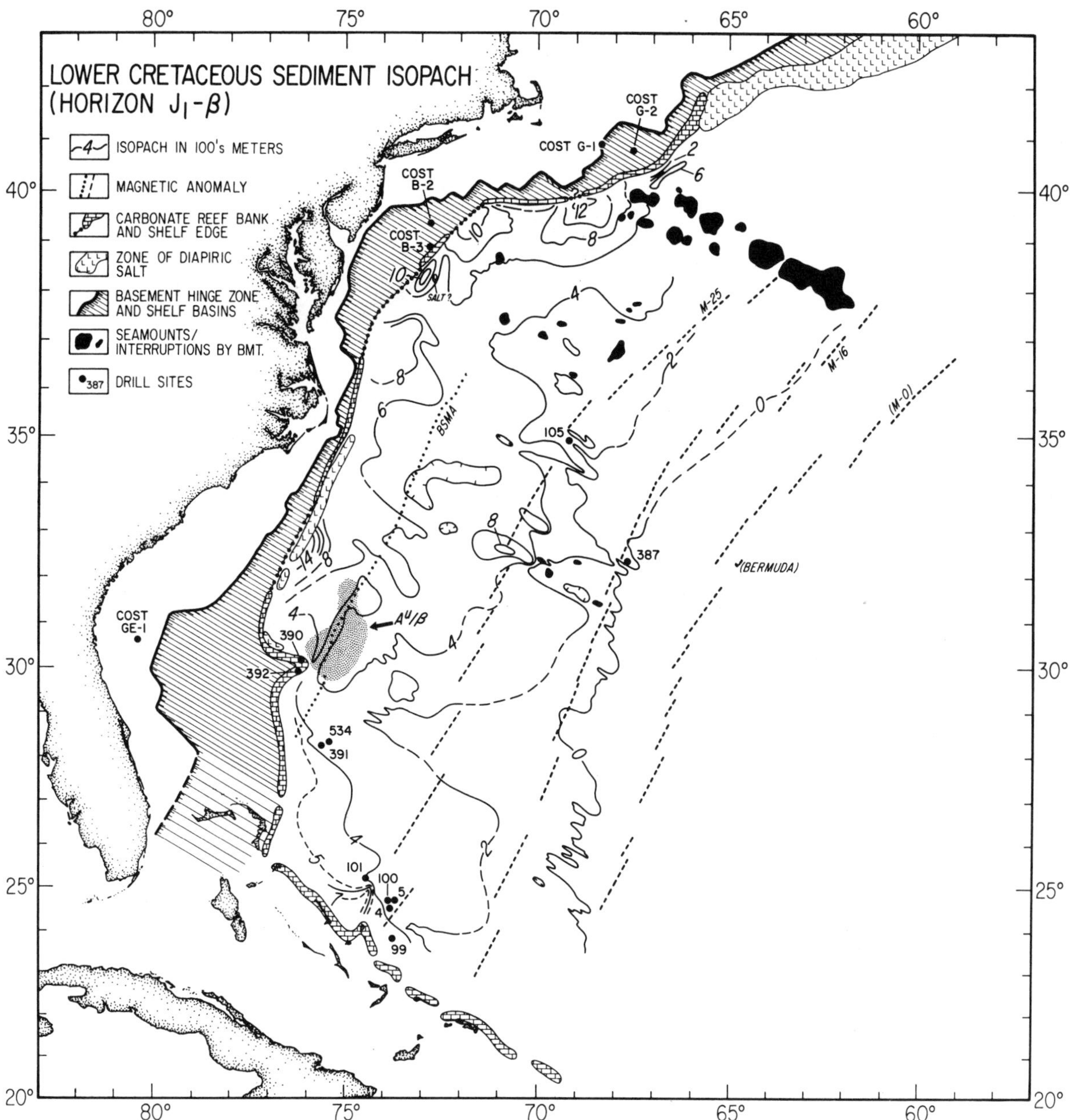

Figure 4.35 Lower Cretaceous sediment thickness (from Mountain and Tucholke, 1985). Shaded area (Au/β) shows where later current erosion cut down to level of Horizon .

of the margin are multicolored shales, deposited at very low rates (~1 m/m.y.) in an oxygenated environment below a shallow CCD. Where drilled at DSDP sites in the basin, these shales are relatively thin (tens of meters).

The latest Early and Late Cretaceous rise in sea level progressively restricted sediment transport beyond the continental shelf, as witnessed in the occurrence of deep-basin multicolored, pelagic shales and in the middle-shelf/outer-shelf/upper-slope paleobathymetric estimates for the COST wells (Fig. 4.31). Intermittent relative lowstands of sea level, however, particularly in the latest Early Cretaceous, triggered erosion of the shelf with consequent sediment transport from the shelf onto the slope and rise.

At the end of the Cretaceous (late Maestrichtian), a sharp depression and rise of the CCD resulted in widespread deposition of a carbonate bed across the basin and margin (Tucholke and Vogt, 1979; Tucholke and Mountain, 1979; Fig. 4.31). These carbonates form the seismic reflector Horizon A*.

The thickness of upper Lower and Upper Cretaceous sediments depicted in Figure 4.36 shows a continuation of fan building on the continental rise seaward of Baltimore Canyon Trough. By this time the reef bank bordering Georges Bank was overstepped by prograding shelf sediments (Poag, 1982), and at least two abyssal fans extended seaward from the base of the continental slope. South of

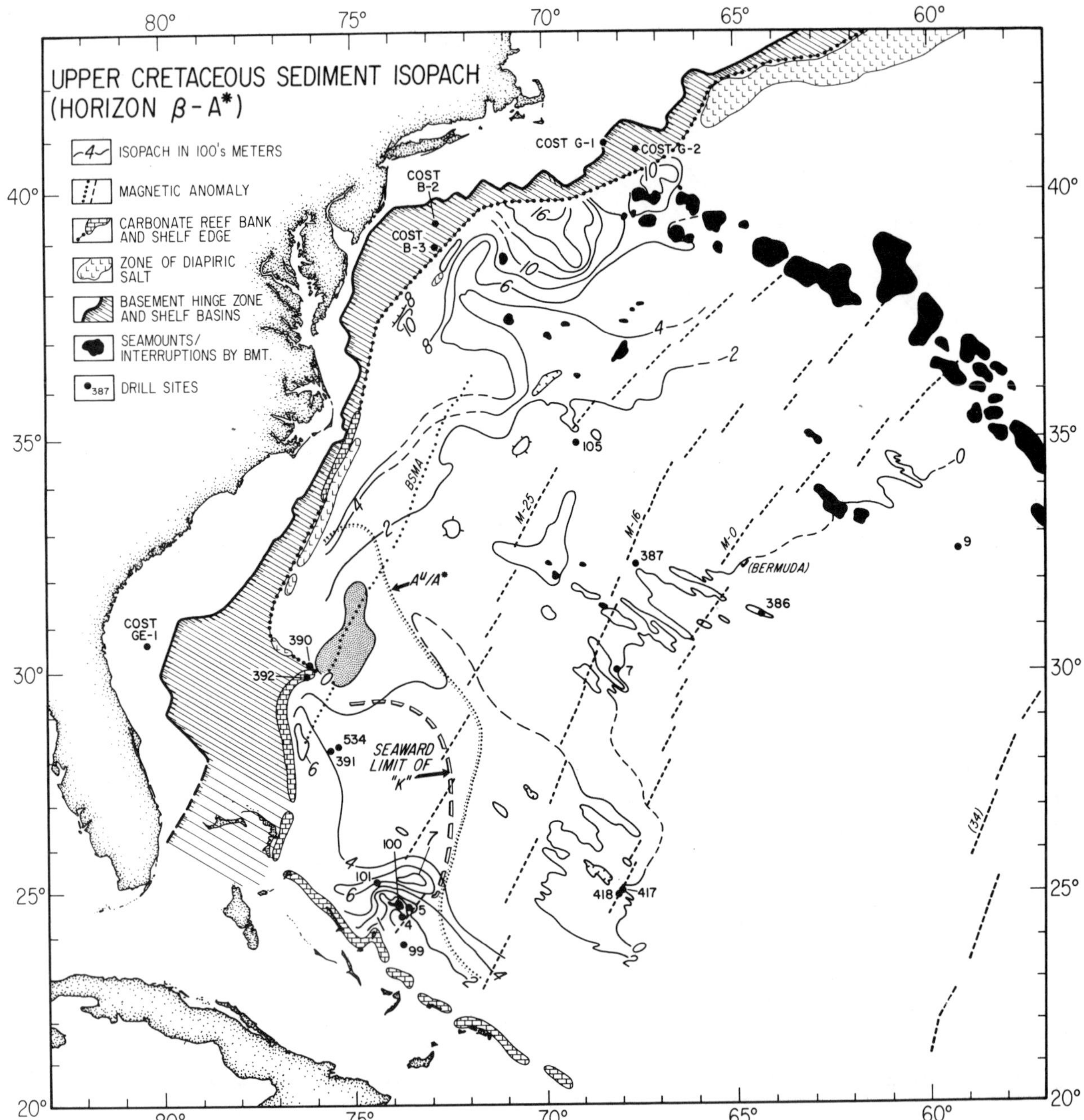

Figure 4.36 Thickness of upper Lower and Upper Cretaceous sediments along the U.S. Atlantic margin (from Mountain and Tucholke, 1985). West of A^u/A^* dotted line, Oligocene bottom-current erosion removed Cretaceous sediments below Horizon A^*. Erosion down to Horizon β occurred in shaded area east of Blake Spur. Seaward limit of reflector "K" defines limits of abyssal sediment fan.

Cape Hatteras massive erosion by abyssal currents in the Oligocene removed much of this stratigraphic sequence, so little is known about later Cretaceous fan development in that area. For example, just north of the Blake Spur and beneath the western end of the Blake Outer Ridge the erosion excavated sediments all the way down to Horizon β (shaded area, Fig. 4.36). However, along the northwestern edge of the Bahama Banks and seaward of the Blake Plateau, the remnants of an abyssal fan are apparent. Great Abaco Canyon and Northeast Providence Channel are two likely conduits that funnelled shelf sediment to this fan. A distinct Cretaceous reflector, termed "K" by Mountain (1981), laps onto Horizon in a seaward direction as indicated in Figure 4.36. This reflector approximately defines the limits of the abyssal fan.

Early Cenozoic. A trend of lowering eustatic sea level and superimposed, intermittent sea-level lowstands began in the Early Cenozoic. Curiously, most shelf sediments, as indicated in COST wells (Fig. 4.31), appear to have been deposited in relatively deep water during this period, and they are highly calcareous and locally siliceous. It is unclear whether the numerous unconformities on the shelf were subaerially eroded during sea-level lowstands or whether other mechanisms (e.g., currents, wave-base erosion) were responsible. In either case, the eroded sediments were transported off shelf to the slope and rise. Highly calcareous and siliceous turbidites originating from the continental margin are known from deep-sea drilling to have been transported far into the deep basin during the Paleocene and Eocene (Tucholke, 1979b). Eocene chalks commonly crop out on the continental slope (Fig. 4.13); their distribution has been studied most extensively by Robb et al. (1981d) on the continental slope south of Toms Canyon.

One characteristic feature of early to middle Eocene deposits is the presence of chert. The chert is usually porcelanitic rather than quartzose, and it formed by diagenetic transformation of biogenic opaline silica (opal A) to cristobalite (opal CT). In the deeper basin, the top of the cherty sediments forms a prominent seismic reflector termed Horizon A^c (Tucholke, 1979b;Tucholke and Mountain, 1979). Along the continental slope and rise, a large part of these sediments was removed by Oligocene bottom-current erosion before the transformation to chert occurred. Still there are siliceous and cherty sediments known to be present locally, for example at DSDP Site 108. Siliceous and cherty Eocene sediments also are present on the Blake Plateau, continental shelf, and in coastal plain deposits (Fig. 4.31).

Very near the Eocene-Oligocene boundary, strong abyssal currents containing waters derived from Arctic sources began to scour the U.S. Atlantic continental slope and rise (Tucholke and Mountain, 1979; Miller and Tucholke, 1983). The currents eroded a profound unconformity all across the continental rise and at least part way up the continental slope. This unconformity is easily identified in seismic reflection profiles and is termed Horizon A^u. As noted earlier, the erosion cut as deeply as Horizons A^* and β south of Cape Hatteras (Fig. 4.36).

The distribution of Paleocene-Eocene sediment thickness in Figure 4.37 is largely a reflection of the scouring and sculpting effects of the strong Oligocene abyssal currents. South of Cape Hatteras there is little Paleocene-Eocene sediment preserved along the deep-water continental margin. Possible remnants of Paleogene abyssal fans are observed seaward of Baltimore Canyon Trough and Georges Bank. They probably formed principally during the Paleocene and middle Eocene sea-level lowstands (Fig. 4.31).

By Paleogene time, shelf-edge reefs survived only along the Bahama Banks. The reef front along the edge of the Blake Escarpment died out during the Late Cretaceous and the entire Blake Plateau slowly subsided and continued to accumulate a cover of calcareous sediments. By the end of the Paleocene the Gulf Stream became active along the western margin of the Blake Plateau (Pinet et al., 1981c; Manheim et al., 1982). During succeeding epochs it shifted alternately seaward during sea-level lowstands and landward during highstands, scouring the Tertiary sedimentary cover across the western part of the Plateau (Fig. 4.8). The Florida-Hatteras Slope probably began to develop once the Gulf Stream provided a barrier to seaward progradation of shelf sediments (Manheim et al., 1982).

Late Cenozoic. The intense erosion of the continental slope and rise during the early Oligocene appears to have oversteepened the slope in some areas, so that slumps, slides, and debris flows locally cover the Horizon A^u unconformity (Fig. 4.28). Furthermore, a major sea-level lowstand in the mid-Oligocene appears to have resulted in erosion of the continental shelf on Georges Bank (COST G-1 and G-2 wells; Fig. 4.31) and in the area of Baltimore Canyon Trough (COST B-2 and B-3 wells); the Southeast Georgia Embayment (COST GE-1 well) was relatively unaffected. Turbidity currents crossing the continental slope during this lowstand, together with additional slumps, slides, and debris flows, further eroded the slope. In total, these events caused erosional retreat of the continental slope for distances of between 5 and 30 km (Fig. 4.37; Schlee et al., 1979). The continental slope never recovered from this erosional cycle and is now positioned well landward of the Upper Jurassic and Cretaceous shelf edge (e.g., Fig. 4.32). This erosional cycle is the principal reason that pre-Quaternary sediments are still exposed across much of the continental slope (Fig. 4.13).

Following the early Oligocene erosional episode, contour-following bottom currents probably began to decrease in intensity (Miller and Tucholke, 1983). However, they continued to exert strong control on depositional patterns, and they molded the later Oligocene slump and lowstand deposits into an elongate sedimentary body along the base of the slope and beneath the continental rise (Fig. 4.38). At the same time, the southerly flowing bottom-current system interacted with the base of the Gulf Stream where the currents crossed one another north of the Blake Spur. Deposition in this zone of interaction is thought to be responsible for the nucleation and growth of the Blake Outer Ridge (Bryan, 1970). About early Miocene time, the bottom-current regime must have changed significantly because the Hatteras Outer Ridge began to form along the edge of the lower continental rise (Fig. 4.29), and the Bahama Outer Ridge began nucleating on the southern edge of the Blake Outer Ridge (Tucholke and Laine, 1983; Mountain and Tucholke, 1985). Both of these outer ridges are comprised of silty, clayey sediments that are molded into well-developed sediment waves. The exact nature of the change in abyssal circulation that triggered construction of these outer ridges presently is uncertain, but a general decrease in current intensity is implied (Mountain and Tucholke, 1985). Abyssal currents have continued to interact with downslope sedimentary processes right up to the present day, as discussed in an earlier section of this report.

Quaternary. Along the U.S. Atlantic margin the Quaternary epoch was characterized by significant fluctuations in relative sea level, locally perhaps as great as 150 m (e.g., Milliman and Emery, 1968), that resulted from episodic continental ice buildup. Along the northern U.S. Atlantic margin, glacial ice reached as far south as Long Island, Martha's Vineyard, Nantucket, and Georges Bank. The lowered sea level and proximity of ice during the glacial episodes had several profound influences on sedimentary environments. First, sediments were transported in pro-

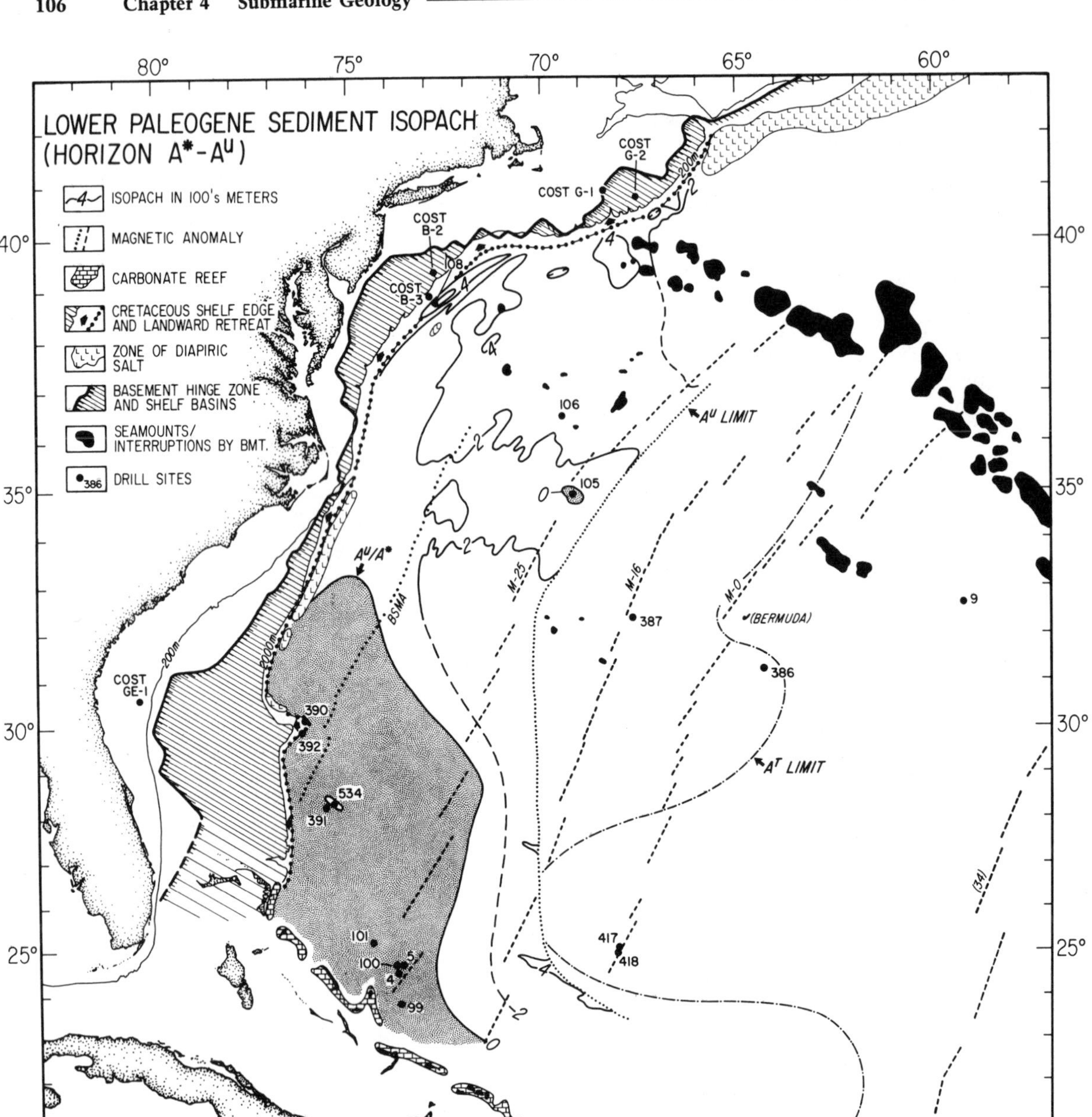

Figure 4.37 Sediment thickness between Horizons A* and A^u (Paleocene-Eocene) along the U.S. Atlantic margin. Stippled area shows where erosion has cut down to and beyond Horizon A*. From Mountain and Tucholke (1985).

glacial stream valleys (e.g., Hudson Valley) across the subaerially exposed continental shelf and deposited directly in submarine canyons and on the continental slope. Rapid sediment accumulation on the slope probably is responsible for the widely observed Quaternary slope instability in this region. Large volumes of sediment also were funnelled across the continental rise to abyssal-plain depocenters, either directly through submarine canyons or after temporary residence on the continental slope (e.g., Laine, 1980). Second, sediment composition and texture were significantly affected. For example, as noted earlier, feldspar-rich sediments north of Cape Hatteras appear to define major drainage patterns from glacier-covered, mechanically weathered terrain (Fig. 4.11; Milliman et al., 1972). Sediment textures, notably in the classes of ice-transported sand, gravel, and coarser debris, also were strongly affected in areas proximal to the ice front (e.g., Schlee and Pratt, 1970). Finally, surface- and bottom-current patterns were changed during glacial periods. Most notably, lowered sea levels displaced the axis of the Gulf Stream to more easterly positions on the Blake Plateau (Pinet et al., 1981c) thus shifting the loci of erosional zones and depocenters. In general, the high-frequency sedimentary cycles characterizing the Quaternary are atypical of the longer-term geologic development of the U.S. Atlantic slope and rise, yet they have imposed a strong signal. We presently

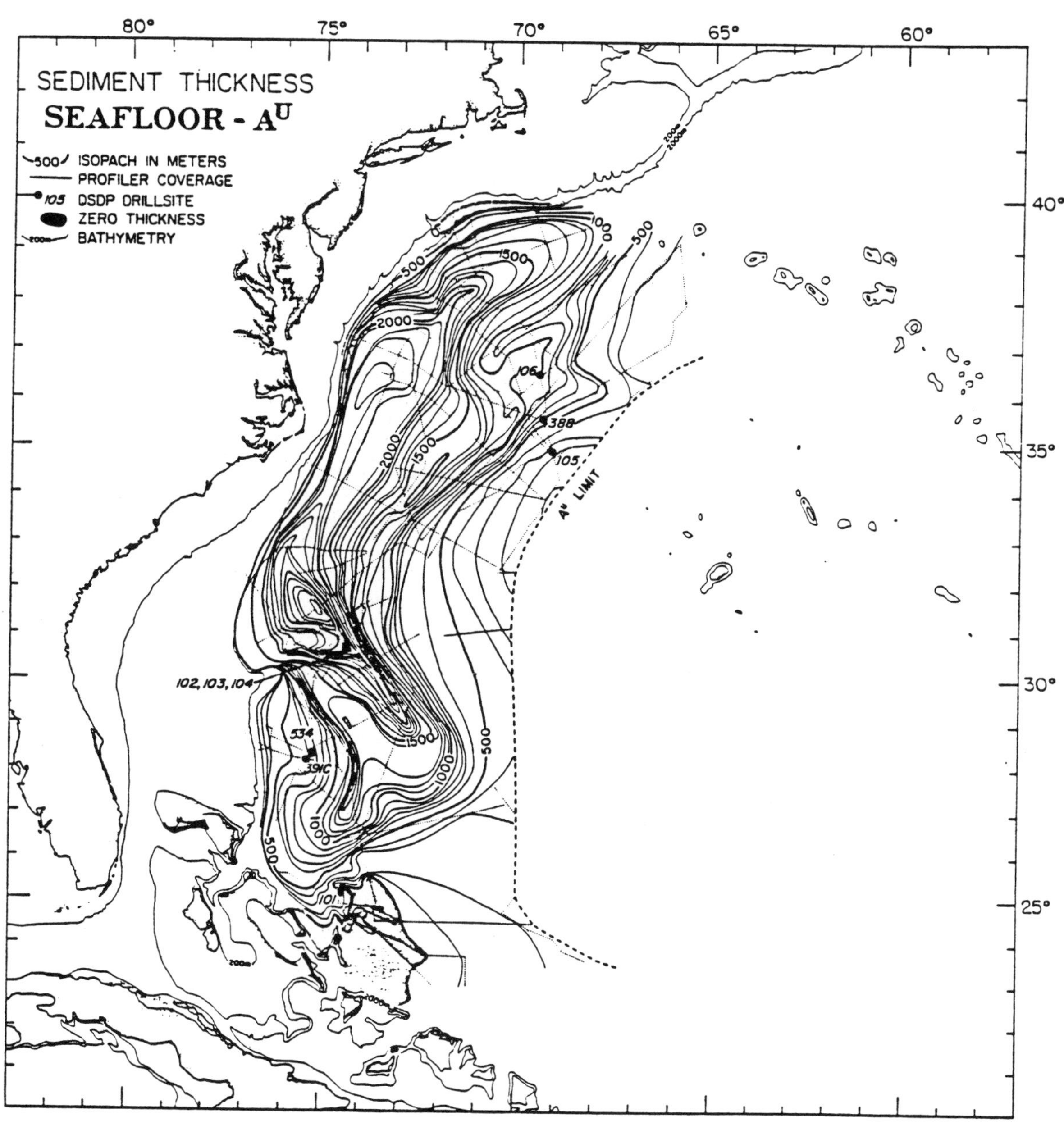

Figure 4.38 Upper Cenozoic sediment thickness (Oligocene-Recent) along the U.S. Atlantic margin. From Mountain (1981).

have a poor understanding of whether present sea-floor sedimentary patterns are in dynamic equilibrium with modern conditions or whether they are still adjusting to conditions imposed during the last glacial interval, 15-27,000 yr. B.P. These questions need to be better resolved before we can predict the likelihood of such dynamic readjustments as sediment mass movement.

Seismicity

Seismicity of the submarine part of the U.S. Atlantic margin generally is not well known. Smith (1960) compiled the seismicity of the northeastern United States and eastern Canada for the period 1534 to 1959, and he noted several offshore events of magnitude five or greater. The post-1927 epicenters are based on instrumental data but their locations are poorly determined (> ±20km). Two events of magnitude 4 to 5 were located on the slope and upper rise off Georges Bank. Between 1962 and 1977 the National Oceanic and Atmospheric Administration logged 5 earthquakes that occurred on the lowermost slope and upper rise of the Baltimore Canyon Trough area (Fig. 4.39). One event of magnitude <4.5 occurred on the continental shelf off South Carolina.

Fletcher et al. (1978) studied the distribution of epicenters and travel-time residuals in the eastern continental United States. They showed that seismic activity, particularly large earthquakes, tends to occur along possible structural trends that extend into major offshore fracture zones. Small circles about the pole of early Atlantic opening appear to fit both the off-shore fracture-zone trends and the onshore seismic zones. It is possible that these are zones of crustal weakness; however, the correlation is not yet well documented and geologic explanations for such possible seismic zonation presently are inadequate.

Heat Flow

Heat flow values along the U.S. Atlantic margin are shown in Figure 4.40. Most values are in the range 40-50 milliwatts/m^2 (0.96-1.20 Heat Flow Units [HFU]), although values as low as 30-40 milliwatts/m^2 and as high as 60-70 milliwatts/m^2 are observed. Sediment thermal conductivities are mostly in the range 0.8 to 1.0 Watts/m – °C, but several measurements of significantly higher conductivity (up to 1.2 Watts/m – °C) have been made on the lower continental rise (Fig. 4.40).

Very few measurements of heat flow and thermal conductivity have been made on the U.S. Atlantic slope and rise since the compilation by Jessop et al. (1976). One exception is a program of heat-flow and pore-pressure stations conducted on the continental slope just north of Wilmington Canyon by R.W. Embley (NOAA/NOS) and M. Hobart (L-DGO). These data currently are being analyzed and prepared for publication (R.W. Embley, pers. comm.).

Economic Resources

Gas Hydrates and Gas

Gas hydrates are ice-like crystalline solids formed by the physical bonding of gas molecules within a lattice of water molecules. Under high pressures, gas hydrates are stable at temperatures well above the freezing point of water. Pressure/temperature conditions over most of the seafloor are conducive to the formation of gas hydrates, but the gas concentration must exceed that necessary for complete saturation of the porewater before gas hydrates will form (e.g., Claypool and Kaplan, 1974; Miller, 1974).

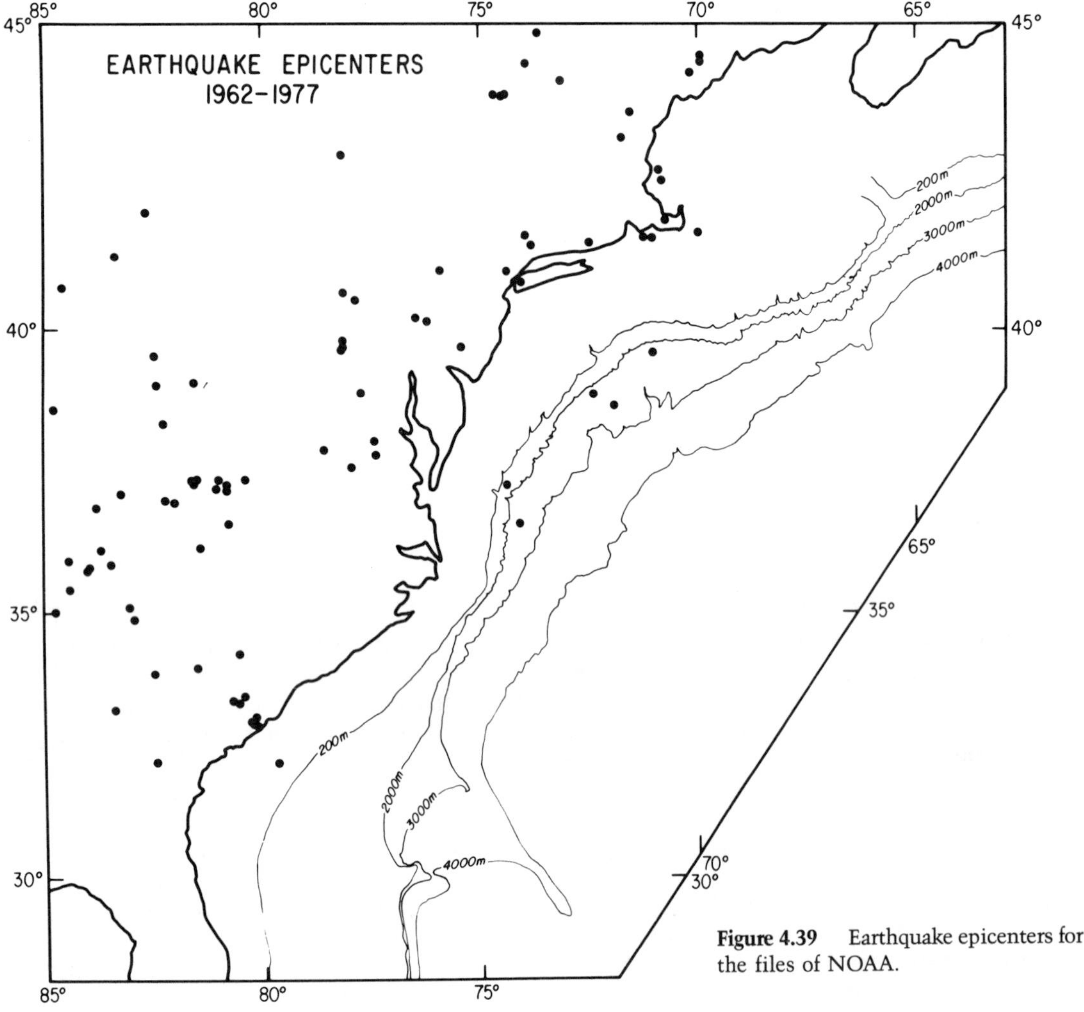

Figure 4.39 Earthquake epicenters for the period 1962-1977, from the files of NOAA.

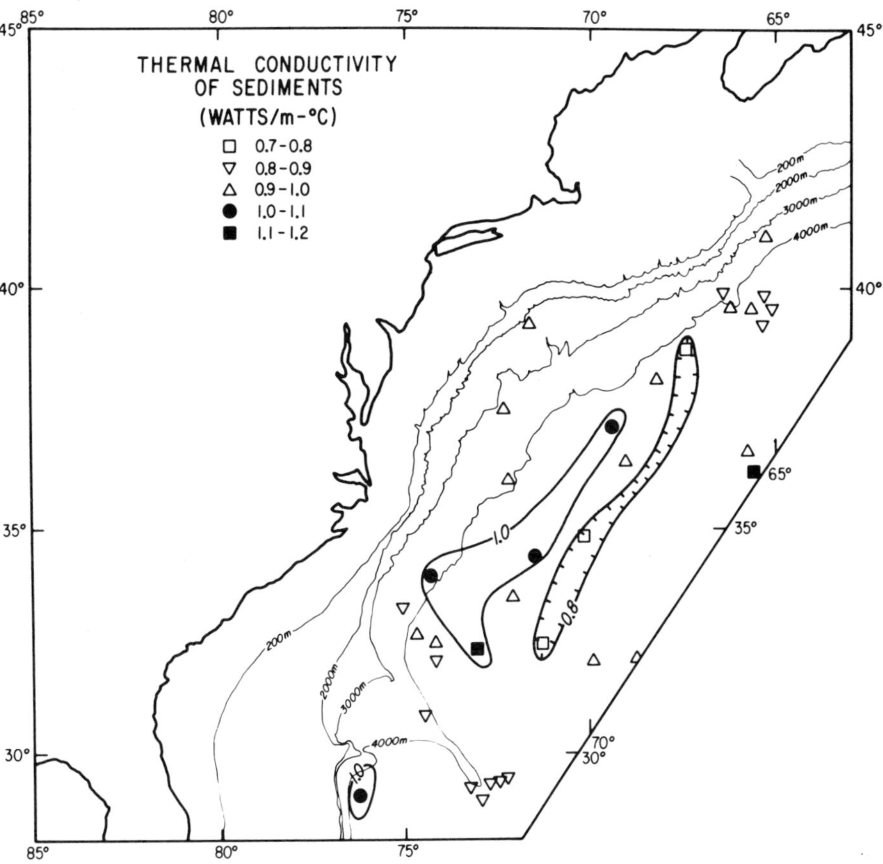

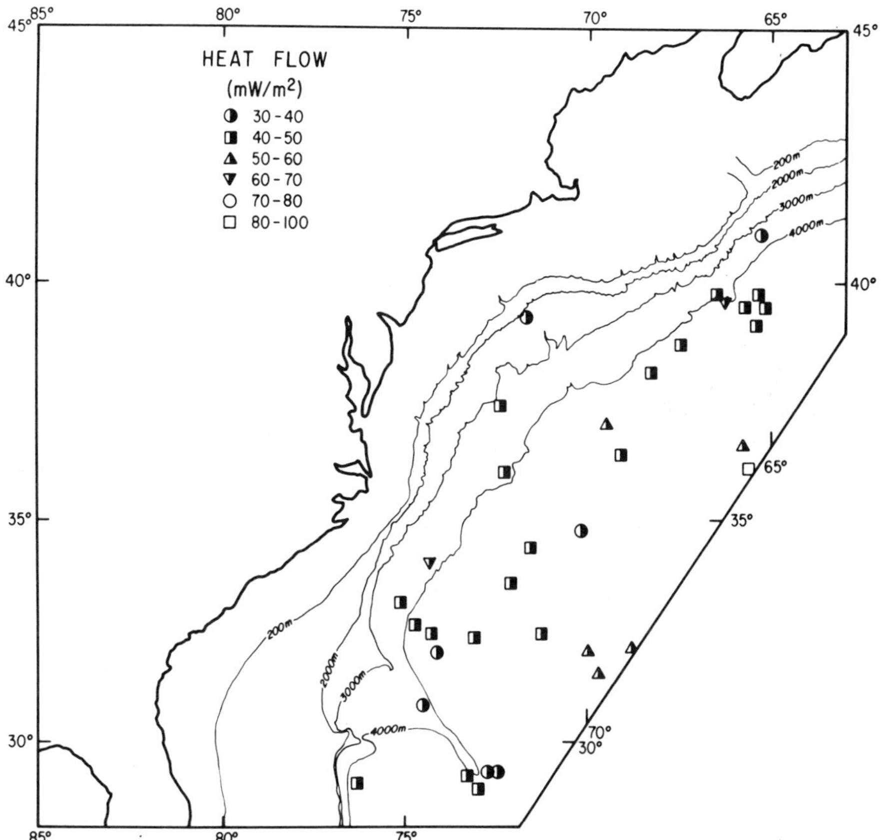

Figure 4.40 Top-Measured values of heat-flow along the U.S. Atlantic slope and rise (Units = milliwatts per square meter). Bottom-Measured values of thermal conductivity along the U.S. Atlantic slope and rise. Based on Jessup et al. (1976).

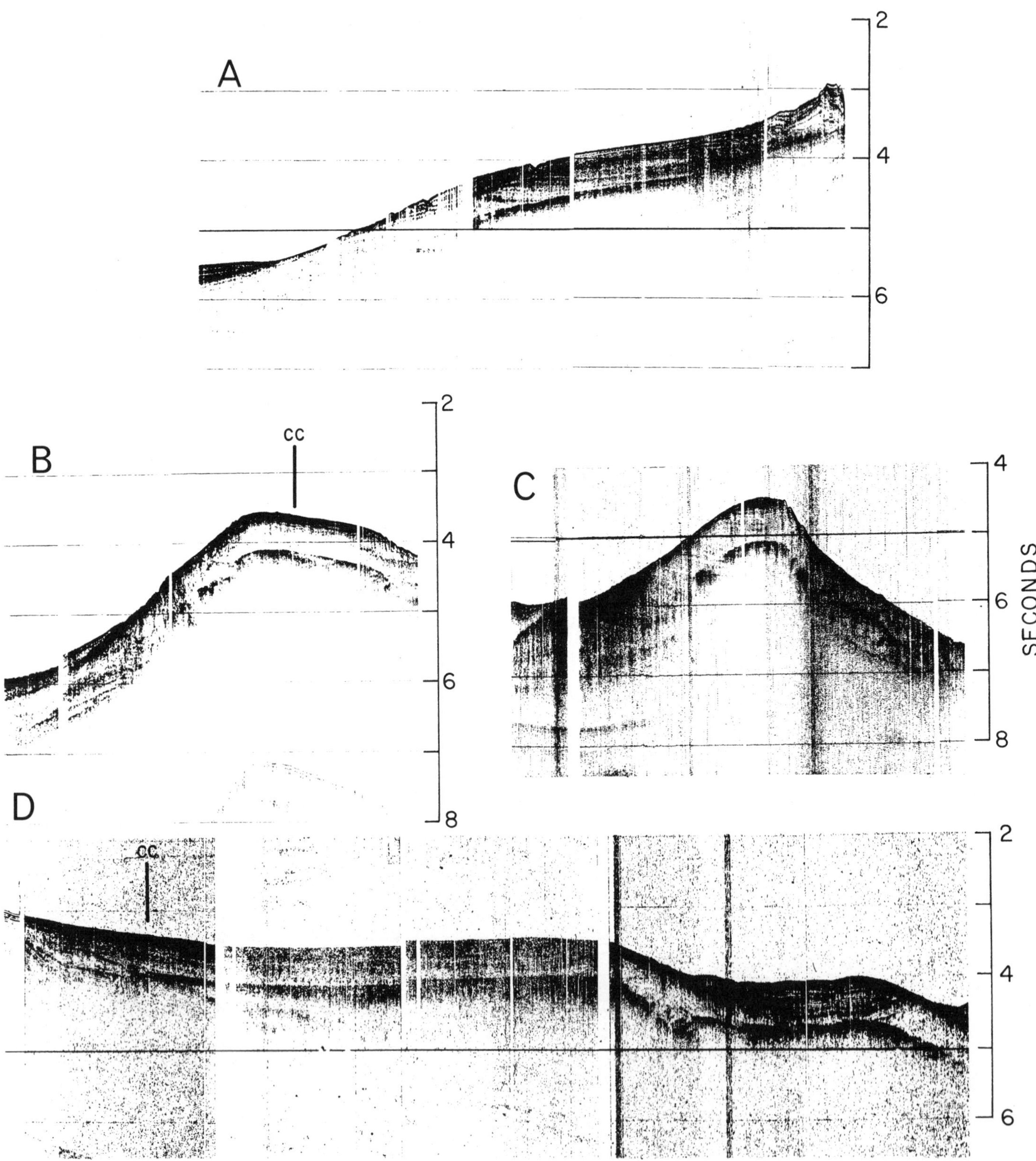

Figure 4.41 Seismic reflection profiles showing development of BSR under the continental rise near Hudson Canyon (A) and within the Blake Outer Ridge (B, C, D). After Tucholke et al. (1977).

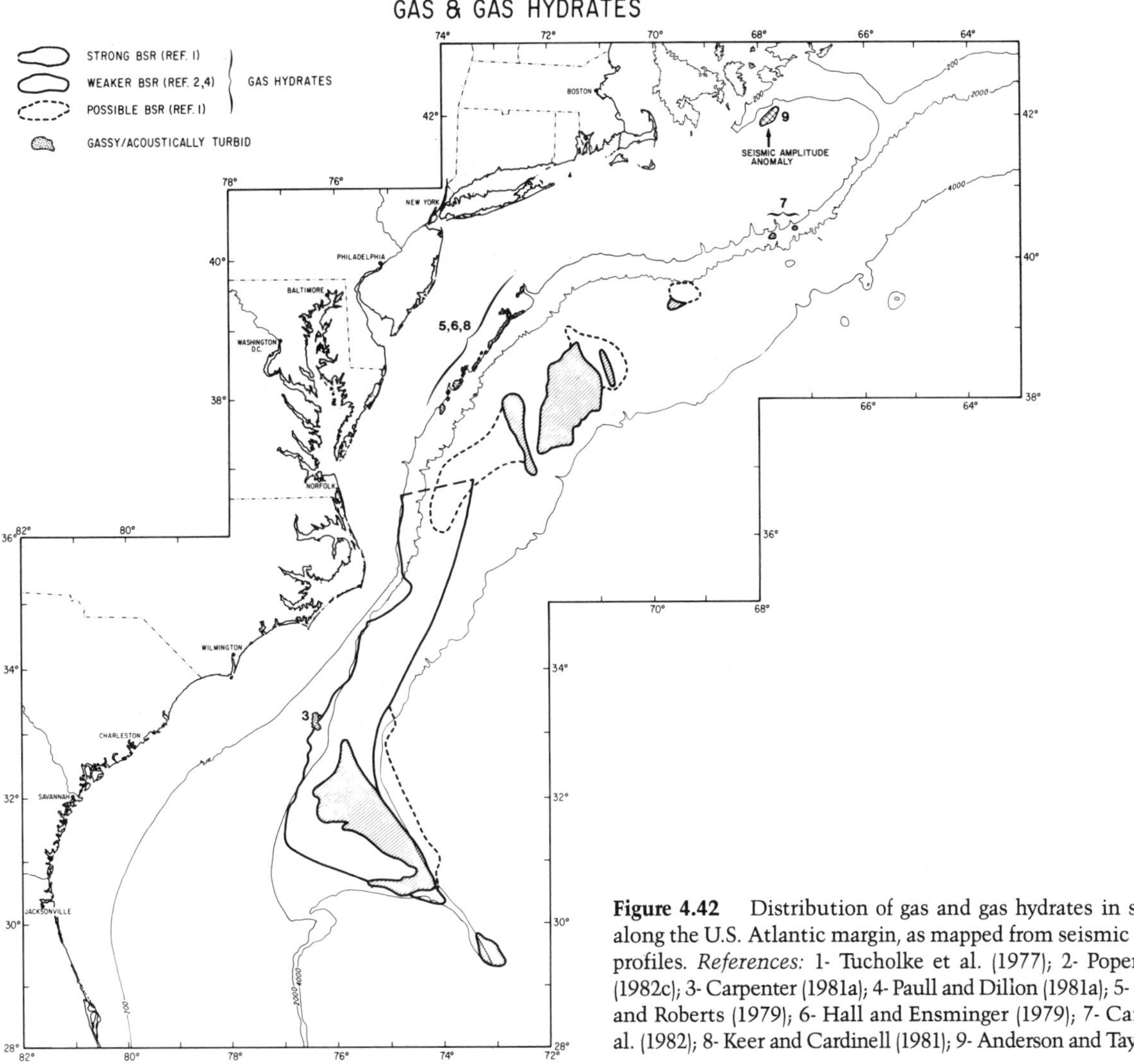

Figure 4.42 Distribution of gas and gas hydrates in sediments along the U.S. Atlantic margin, as mapped from seismic reflection profiles. *References:* 1- Tucholke et al. (1977); 2- Popenoe et al. (1982c); 3- Carpenter (1981a); 4- Paull and Dillon (1981a); 5- Carpenter and Roberts (1979); 6- Hall and Ensminger (1979); 7- Carpenter et al. (1982); 8- Keer and Cardinell (1981); 9- Anderson and Taylor (1981).

Starting within the hydrate stability field at the seafloor and proceeding into the sediment column, pressure and temperature both increase until the phase boundary between hydrate + water and gas + water is reached. Hydrate is not stable in the higher temperature regime below this boundary. Sediments containing gas hydrates are known to have elevated seismic velocities (Stoll et al., 1971). Thus the phase boundary between hydrate-containing sediment and gaseous sediment at the base of the hydrate stability zone is likely to be a boundary of high impedance contrast. It should appear as a strong reflector in seismic reflection records and it should generally follow the contours of the seafloor. Furthermore, because the higher-impedance sediment overlies the phase boundary, the reflector should show a phase reversal.

Bottom-simulating reflectors (BSR's) having exactly these characteristics have been mapped beneath the crest of the Blake Outer Ridge and beneath the central continental rise (Tucholke et al., 1977; Shipley et al.. 1979; Figs. 4.41, 4.42). In the upper part of the sediment column, these two areas both show normal bedding planes that dip landward (or toward the crest of the Blake Outer Ridge), so that gas migrating updip along permeable beds is trapped at the base of the gas hydrate layer. The gas accumulation immediately below the hydrated sediment appears to give rise to the exceptionally high-amplitude BSR's mapped by Tucholke et al. (1977) from single-channel seismic reflection profiles.

Several other workers subsequently identified lower-amplitude BSR's in high-resolution single-channel and multichannel seismic data, extending the areas first mapped and identifying new areas where gas hydrates may be present (Fig. 4.42). Other "bright spots" that may be either gas pockets or possibly related to gas hydrates have been identified along the continental slope of Baltimore Canyon Trough and around Georges Bank.

Gas hydrates were a specific target for deep-sea drilling at DSDP Site 533 at the crest of the Blake Outer Ridge (Fig. 4.28). Visual evidence, geochemical measurements, and the gas-pressure-release history of sediments cored in a pressure-core-barrel at this site all indicate that hydrate is present in the sediment (Sheridan et al., 1982). Gas hydrates also may have been cored in DSDP Sites 102, 103, and 104 on

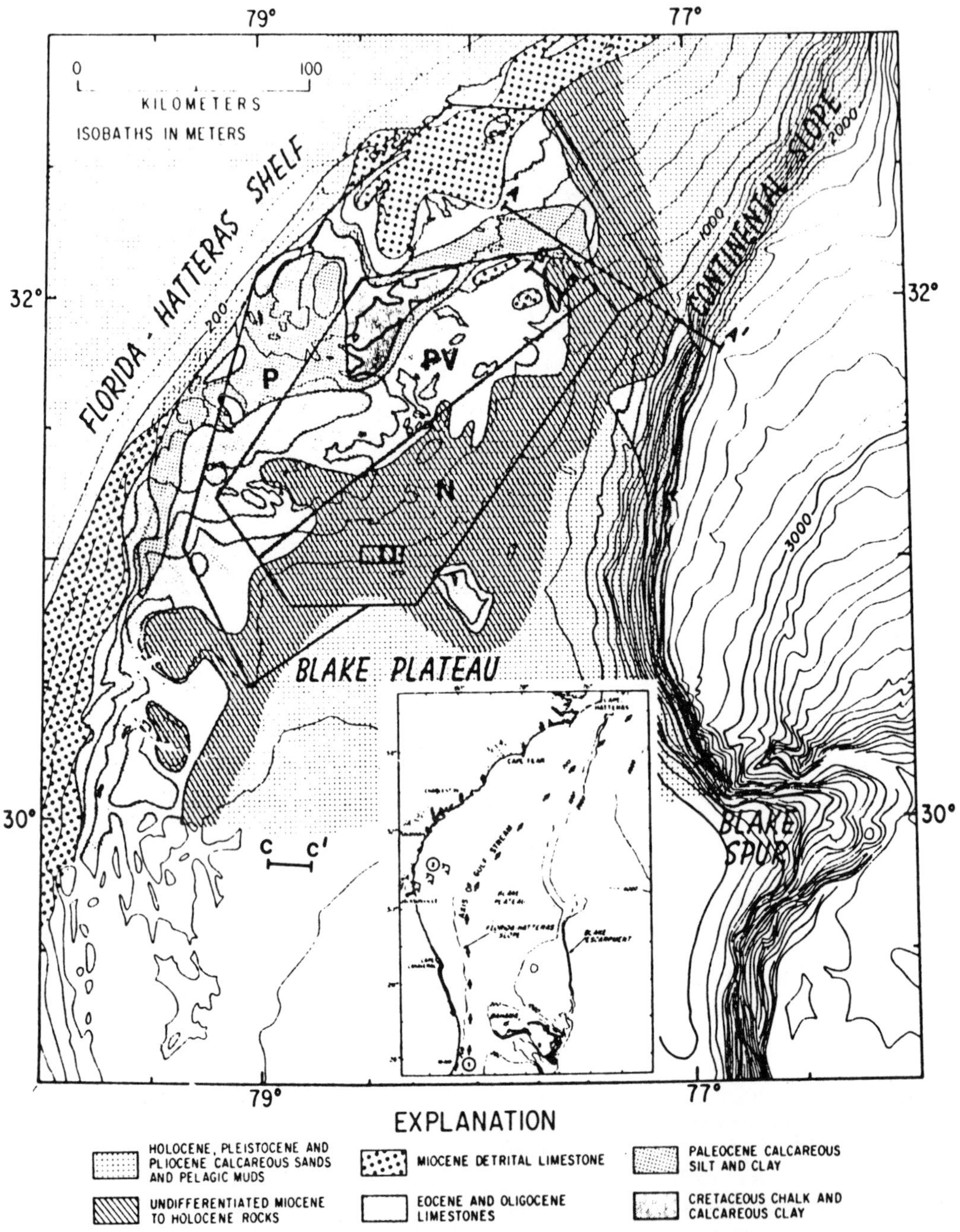

Figure 4.43 Tentative resource tracts of phosphorite and manganese oxides on the northern Blake Plateau (Charleston Bump). N = Mn nodules, PV = Mn pavement, P = phosphorite. From Manheim et al. (1982).

the crest of the Blake Outer Ridge (Hollister, Ewing et al., 1972), although the possible significance of the BSR was not fully realized at that time. The cores recovered gassy sediments, but no visual indications of solid hydrate were recorded. However, siderite and other diagenetic carbonate minerals found near the level of the BSR suggest that chemical reactions occur near the gas-hydrate phase boundary (Lancelot and Ewing, 1972).

It is possible that gas hydrates play a role in sediment mass movement along the continental margin (Carpenter, 1981a; McIver, 1982). If the hydrate forms an impermeable seal to gas migration, gas may accumulate below the hydrate, causing overpressured conditions that can lead to failure. In addition, if the equilibrium conditions at the base of the hydrate zone are disturbed (for example by sea-level fluctuations or rapid sedimentation/erosion), the hydrate may decompose. The constituent gas and water occupy a larger volume than the hydrate. Thus the pressure below the hydrated sediment would increase and could lead to lifting, breaching, or large-scale failure of the overlying sediment column (McIver, 1982).

Oil

The U.S. Atlantic shelf and upper slope currently are being actively explored for hydrocarbon resources. To date, more than 30 wildcat wells have been drilled in the Baltimore Canyon Trough and Georges Bank areas, but no commercially producible hydrocarbons have been encountered.

A likely target for future exploration is the Jurassic-Cretaceous reef trend beneath the continental slope (e.g., Fig. 4.34). Fore-reef talus seaward of this structure interfingers in places with the upper Lower Cretaceous carbon-rich black shales beneath the outer slope and upper rise. If the thermal and pressure history of these shales has been appropriate for hydrocarbon formation, the hydrocarbons could have migrated into the adjacent forereef and reef facies which are potential reservoir beds.

Phosphorite and Manganese

The principal concentrations of phosphorite and manganese oxide deposits along the U.S. Atlantic margin occur on the northern Blake Plateau (Fig. 4.43). Manheim et al. (1982), in a survey of the historical data in this region, noted that an area of more than 14,000 km^2 contains major concentrations of manganese nodules and/or manganese-phosphorite pavement. They suggested that the phosphorite deposits were related to upwelling and high productivity along the landward side of the Gulf Stream. Phosphate-enriched fecal material settles to the seafloor, where chemical and biochemical reactions in the sediments produce apatite pellets and phosphatic gravels. These muds are reworked by marine currents (e.g., Gulf Stream) during sea-level transgressions, and by nearshore processes during regressions, so that the phosphorites are concentrated as lag deposits. On the Charleston Bump the interaction of the Gulf Stream and a flow through the Suwanee Straits across Florida (during sea-level highstands) has produced, reworked, and concentrated the phosphorite deposits. The prolonged exposure of phosphorites at the seafloor also makes them prone to act as nuclei for manganese oxide precipitation. Furthermore, where continuous zones of phosphorite gravel are exposed, they often become cemented by calcite to form pavement sheets.

Extensive, continuous manganese oxide pavements of the Blake Plateau seem to be concentrated on outcrops of Eocene-age rocks between 500 m and 650 m water depth (Manheim et al., 1982). The thicknesses of the pavements and the composition of deeper sections are poorly known, given the present sample coverage. Manganese nodules are more common where they overlie Miocene exposures at 675 m to 1050 m, but slabs and intermittent pavement occur locally to 750 m. Below 1000 m, nodules and crusts are more earthy, friable, and have higher copper contents than in shallower areas.

The manganese nodules on the northern Blake Plateau have the following average concentrations of metals: Mn-16%, Fe-10%, Ni-0.6%, Co-0.3%, and Cu-0.1%. In pavements, concentrations of all metals but Fe are lower. The nodules contain some potentially valuable minor constituents, notably Pt and Mo (Manheim et al., 1982). The valuable metal content of the Blake Plateau deposits is lower than that of prime nodules in the Pacific, but the proximity of the deposits and the fact that they are in U.S. territorial waters makes them a potential economic resource.

Phosphatic sand and lesser concentrations of Mn nodules also occur on the southern Blake Plateau. Phosphatic detritus is widely distributed along the U.S. Atlantic margin and is easily accessible in mostly subsurface deposits of the southern states' coastal plain (e.g., Emery and Uchupi, 1972).

Manganese pavements are relatively common on the New England Seamounts along the U.S. margin (Houghton et al., 1977), but composition of these deposits is not well documented.

Sand and Gravel

Sand and gravel are important economic resources along the U.S. Atlantic margin. These deposits are heavily concentrated on the continental shelf (Fig. 4.9), but there are no significant off-shelf deposits on the continental slope or rise.

Acknowledgments

This summary account of the geology of the U.S. Atlantic margin was supported by the Minerals Management Service, U.S. Department of the Interior. I thank E. Uchupi and J.D. Milliman for reviewing the manuscript and making useful comments. J. Zwinakis drafted many of the figures and P. Barrows typed the text. Contribution No. 5975 of Woods Hole Oceanographic Institution.

5 Marine Chemistry

R.B. Gagosian, J.W. Farrington, H.D. Livingston, and F.L. Sayles

In order to understand more fully the chemistry in the ACSAR area of interest in this report, the subjects of the chemistry of nutrients, trace metals, radionuclides and hydrocarbons will be discussed separately. In each of these subject areas a review of the dissolved and particulate matter chemical concentrations in seawater is followed by a discussion of the studies conducted in organisms and in sediments. The review of sediment studies is divided into research conducted in pore waters and in the solid sediment matrix. Because of the generally small number of applicable studies in the ACSAR area there is considerable discussion of specific data sets.

Nutrient Chemistry

F.L. Sayles

Water Column Distribution

Dissolved Constituents

The distribution of nutrients in the water column is largely a function of the water mass type and biological interactions. For most of the water column, concentrations are closely linked to water mass type and are consistent over the entire area. Shallow waters, on the other hand, exhibit relatively large deviations due to biological activity. Therefore shelf water, which is present through continual exchange with Slope Water, is the most variable water type present in the slope-rise area in terms of nutrient concentrations (see Kester and Courant, 1971, for a review). Distributions of NO_3^- and PO_4^{3-} in shelf waters reflect seasonal changes in vertical mixing and biological productivity. South of Long Island in winter, surface waters (0-50 m) are well mixed. In spring thermohaline stratification and increased biological activity produce depletions of NO_3^- and PO_4^{3-} in the shallower (upper 30 m) waters and enrichments in deeper water. By September NO_3^- is completely depleted and PO_4^{3-} severely so (.2 to .4 μM). The onset of winter brings enhanced vertical mixing and return of nutrients to shallower waters (NO_3^- concentrations >5 μM) and repeat of the cycle. Sections off the Carolinas (Stefansson et al., 1971; Stefansson and Atkinson, 1971) have demonstrated that water mass mixing also influences nutrient distributions, reflecting impingement of the Gulf Stream, river runoff, and Caribbean water in this area. In short, nutrient concentrations in shelf waters reflect a number of processes which are variable through time. However, concentrations of nutrients, particularly NO_3^- and SiO_2, in shelf waters are generally low relative to the other water masses of the slope-rise region.

The data presently available in the ACSAR area (Fig. 5.1) show rather regular distributions of nutrients. Surface concentrations of NO_3^- are low and show seasonal influences, especially nearshore. Peterson's (1976) data show surface (0-50 m) concentrations of NO_3^- at 1 μM during May 1974 sampling. Similar values are reported in the NOAA (1977) report for August 1976. PO_4^{3-} concentrations are also low (<1 μM, often <0.2 μM). The NH_4^+ analyses reported for the 1976 cruise are generally very low (<0.1 μM) but on occasion the data are noisy, quite possibly reflecting contamination rather than real variation. The NH_4^+ data reported for 1974 are 5 to 10 times the concentrations reported for the 1976 cruise and should be considered questionable. Below the surface water, PO_4^{3-} and especially, NO_3^- concentrations rise rapidly to a maximum; the depth of this maximum in the 1974 data (Peterson, 1976) is almost invariantly at 300 m, occasionally 500 m. Values at the maximum vary somewhat but NO_3^- is usually 22 (± 2) μM. PO_4^{3-} concentrations rise to about 2 μM ($\pm$ about .1) at about the same depth as the NO_3^- maximum. The TTO data (PACODF, 1981) yield similar results for stations in 4000 m of water and less, (i.e., NO_3^- and PO_4^{3-} maxima at $\sim$300 m). Concentrations of NO_3^- in the TTO sections are slightly higher (23-25 μM) at the maximum, and PO_4^{3-} concentrations are substantially lower (1.4 to 1.5 μM) at the maximum and extremely consistent relative to the data reported by Peterson (1976). Deeper, further offshore, TTO stations indicate a substantial increase in the depth of the NO_3^- and PO_4^{3-} maximum (water depth >4000 m) to about 1000 m. Below the maximum both NO_3^- and PO_4^{3-} are quite constant at $\sim$18 μM and 1.2 μM, respectively. Silica parallels NO_3^- and PO_4^{3-} in most respects, exhibiting a maximum at similar depth. The SiO_2 maximum is, however, more poorly developed. Values rise from <5 μM in shallow waters to maximum values of $\sim$15 μM at the NO_3^- and PO_4^{3-} maximum, decrease slightly, then increase gradually to the bottom to concentrations of $\sim$36 μM.

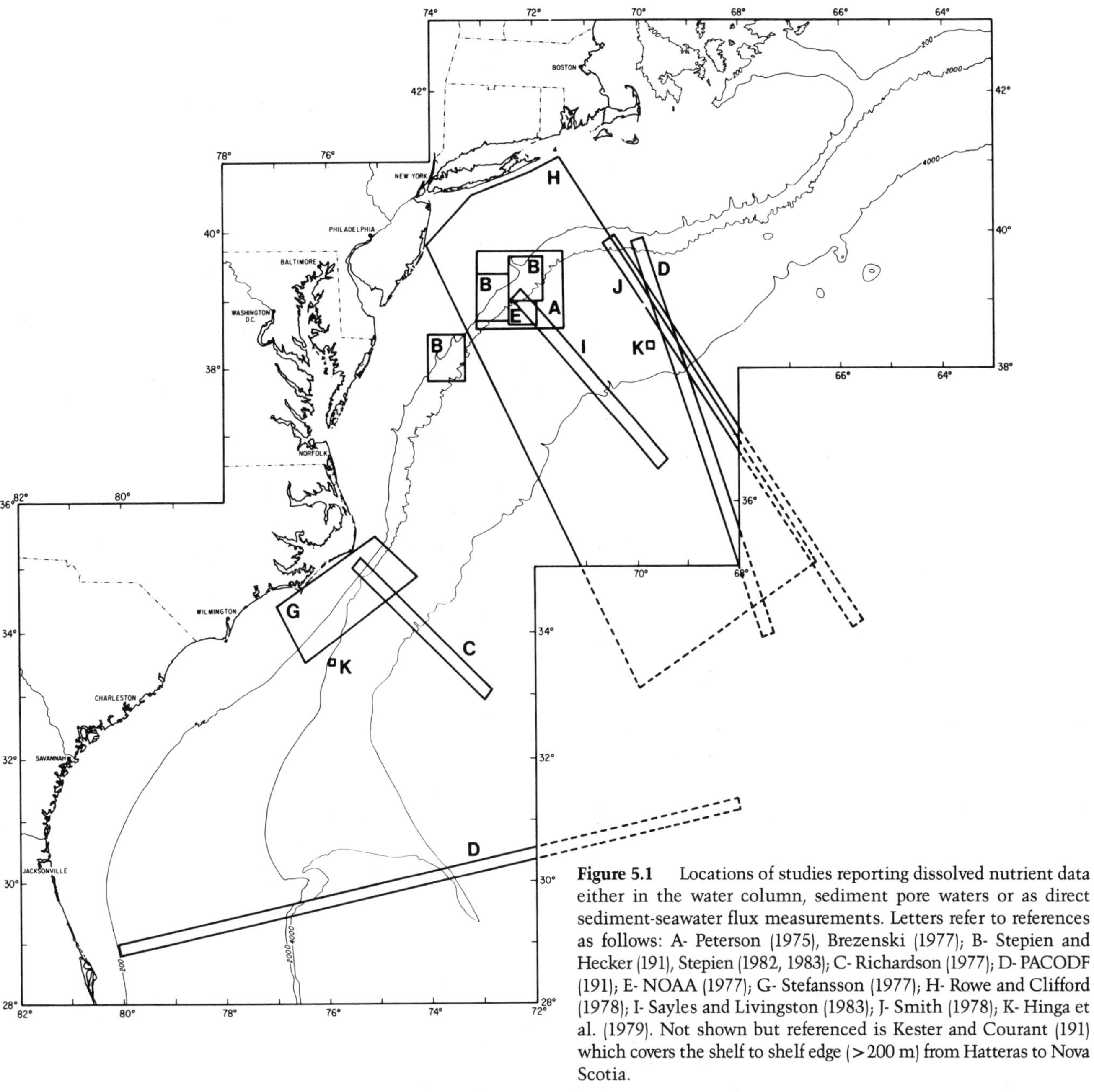

Figure 5.1 Locations of studies reporting dissolved nutrient data either in the water column, sediment pore waters or as direct sediment-seawater flux measurements. Letters refer to references as follows: A- Peterson (1975), Brezenski (1977); B- Stepien and Hecker (191), Stepien (1982, 1983); C- Richardson (1977); D- PACODF (191); E- NOAA (1977); G- Stefansson (1977); H- Rowe and Clifford (1978); I- Sayles and Livingston (1983); J- Smith (1978); K- Hinga et al. (1979). Not shown but referenced is Kester and Courant (191) which covers the shelf to shelf edge (>200 m) from Hatteras to Nova Scotia.

Particulate Matter and Nutrients

A major process in nutrient cycling in the oceans is the transport of C, N and P in particulate organic matter. While C, N and P are not strictly nutrients, they are "potential" nutrients and the exchange between dissolved and particulate C N P pools is well documented. The depletion of soluble nutrients in shallow water (<30 m) and enrichment in deeper waters noted by Kester and Courant (1971) is a typical example. Data on carbon and nitrogen in particulate matter, as an important part of nutrient cycles, are presented in the sections on hydrocarbons and organic matter. Cycling of nutrients and its influence upon biology is discussed in Chapter 6 of this book.

Sediment Distributions

Interstitial Waters. While the general characteristics of nutrient chemistry in the interstitial waters of marine sediments are reasonably well known, to our knowledge only three studies have reported nutrient profiles on the Atlantic slope and rise. Rowe and Clifford (1978) determined NH_4^+, NO_3^-, NO_2^-, and PO_4^{3-} in a large number of cores on the shelf, slope, and rise off the mid-Atlantic states over a depth range of 20 to 5000 m. The NO_3^- and NH_4^+ concentrations show extreme variations from station to station and, occasionally, within a given station. Values of both NH_4^+ and NO_3^- at times exceed 400 μM, while values in excess of 200 μM are common. At some stations, values of 20-40 μM are characteristic. NO_3^- production is linked to O_2 diffusion into the sediments. The relation between O_2 and NO_3^- is defined by diffusion coefficients and the stoichiometry of reactions. In the absence of O_2, NO_3^- is consumed, not produced. Values of 400 μM are about one order of magnitude above the theoretical limit of NO_3^- concentration for known bottom water conditions. Further, the profiles reported are an order of magnitude or more higher than any reported elsewhere for suboxic diagenesis of NH_4^+ or NO_3^- (see references cited above) as well as in this area itself (see below). Because so many of the values are impossibly high, all of the reported concentrations should be considered questionable. The PO_4^- data are closer to the range reported in previous studies of sediment pore waters. However, there is frequently a great deal of scatter in the data (± 1-2 μM), whereas a steady increase would be expected (and has been reported) as a result of continued breakdown of organic matter during diagenesis. Consequently, the PO_4^{3-} data should also be considered sufficiently suspect to preclude their quantitative use.

Pore water NO_3^- profiles are reported by Sayles and Livingston (1983) for a series of stations between Bermuda and Cape May. In the deeper stations, NO_3^- rises to maximum values of 30-40 μM and then decreases with depth to zero. Shallpwer stations do not exhibit maxima, with concentrations decreasing to zero from the sediment surface. These profiles are typical of the range of behaviors generally exhibited by deep sea sediments (e.g., Emerson et al., 1980, 1982; Grundmanis and Murray, 1982), with a general increase in reducing conditions with decreasing water column depth. The production of NO_3^- (to produce a maximum) is indicative of oxidation of organic matter; NO_3^- depletions reflect the utilization of NO_3^- as an electron acceptor in the further oxidation of organic carbon.

Analyses of SiO_2, NO_3^- + NO_2^-, and PO_4^{3-} in pore water from a core at 3500 m are reported by Hinga et al. (1979). The data, similar to those reported by Sayles and Livingston (1983), exhibit an NO_3^- (+ NO_2^-) maximum in the upper few centimeters. Concentrations decrease to a few μM at ~8 cm. PO_4^{3-} concentrations increase to ~8 μM and remain constant in the lowermost sections (6-8 cm).

The gradients between the pore water and overlying sediment imply a flux across the water-sediment interface. Since NO_3^- is not utilized in the presence of O_2 (Cline and Richards, 1970), there should be a zone of finite thickness where O_2 is consumed and NO_3^- produced. Production must result in some enrichment and a flux of NO_3^- out of the sediments, however small. Only under circumstances where the overlying water becomes anoxic should denitrification proceed at the interface; this is not believed to occur on the slope-rise area. This, however, is not entirely consistent with direct flux measurements (see below). The fluxes of NO_3^- and PO_4^{3-}, coupled with reactions at the interface, are responsible for the bulk of the remineralization of particulate C, N, and P removed from the euphotic zone by primary production and subsequent settling at depths of less than about 2000 m (Hinga et al., 1979).

Solid "Nutrients" in Sediments. The inverse of the oxidative remineralization of organic matter discussed above is the burial of C, N, and P in organic matter. Discussion of C, N, and P in the sediments, that which has escaped diagenetic reaction is included in the section on organic geochemistry.

Processes Influencing Nutrient Distributions

A number of processes acting on different spatial and temporal scales exert primary influence upon the nutrient concentrations in the area of interest. Over most of the water column advective processes dominate. In surface waters biogenic cycling often dominates. Particle settling removes nutrients in the form of particulate C N P from the biologically active zone. Finally, the particulate products of near-surface biological activity collect at the sediment surface where further biological activity utilizes reduced carbon as an energy source. Sediment metabolism can be described in two steps: reactions at or near the sediment-water interface, and diagenetic reactions within the sediments.

The integrated influence of benthic metabolism is embodied in the fluxes of nutrients and O_2 across the sediment-water interface. This flux is composed of the two components noted above: fluxes due to reactions "at" the interface, and sediment metabolism occurring below the interface. Direct measurements of fluxes by the emplacement of benthic chambers (cf. Smith and Teal, 1972; Pamatmat, 1973) provide estimates of total integrated metabolism on the seafloor. The first reports of chamber measurements of O_2 utilization in the slope-rise area are those of Smith and Teal (1973) for a single station at 1850 m off New Jersey. Smith (1978) has summarized other measurements in the slope-rise area (Fig. 5.1). Oxygen uptake shows a sharp drop (nearly two orders of magnitude) in passing from the shelf environment to the slope-rise environment. Oxygen uptake decreases from ~20 x 10^{-6} M cm^2 yr^{-1} at 1850 m to 1.5 x 10^{-6} M cm^2 yr^{-1} at 4830 m. Hinga et al. (1979) report an O_2 uptake measurement of 70 x 10^{-6} M cm^2 yr^{-1} at 1345 m. This value is higher than Smith's, but given the rapid change in O_2 metabolism between shelf and slope, it is not inconsistent.

Smith (1978) detected no "chemical demand" for O_2 (i.e., O_2 uptake due to reaction within the sediment). On the slope and rise, no chemical demand was detected, indicating that reactions "in" sediments do not contribute significantly to overall metabolism. This conclusion is at variance with other studies (e.g., Murray and Grundmanis, 1980). Sayles (1981) estimated aerobic oxidation from the dissolution flux of $CaCO_3$ from sediments. The estimates, for depths in the North Atlantic between 4800 and 3200 m, are comparable to Smith's overall estimates. Since chamber measurements include sediment demand, this is a critical uncertainty.

While O_2 demand is a good measure of benthic metabolism, it does not *a priori* translate to nutrient fluxes despite the plethora of models attempting to relate the two. Smith et al. (1978) report benthic fluxes of NH_4^+, NO_3^-,

NO_2^-, and PO_4^{3-} for two of the stations described in Smith (1978). For NO_3^- and PO_4^{3-} the fluxes are into the sediment; for NH_4^+ the flux is out of the sediment. Given that O_2 is present, the data imply that reactions at the interface are complex and that a number of different reactions (nitrification as well as denitrification) occur simultaneously. This pattern of complexity is confirmed by the more extensive measurements of Smith and Horrigan (1983) in the Pacific.

As noted above, only three studies of nutrient concentrations are reported for the area of interest. The huge NO_3^- concentrations reported by Rowe and Clifford (1978) require large scale fluxes of nutrients into the interface region or into the bottom water. For reasons discussed above, these NO_3^-, NH_4^+, and PO_4^{3-} data are highly questionable. The data of Sayles and Livingston (1983) and Hinga et al. (1979) can be used for estimating fluxes due to diagenesis. This should be done with caution, however, for as pointed out above, direct measurements of NO_3^- fluxes often are variable even on a spatial scale of a meter. Thus despite the consistency of most pore water profiles (cf. Sayles, 1981; Emerson, 1980; Murray and Grundmanis, 1980), the relationship of sediment fluxes to fluxes to the overlying seawater may be greatly complicated by reactions at the interface.

As noted earlier, nutrient concentrations in the slope-rise area (and others) largely reflect water mass. The exception to this is primarily in the surface water where biological activity dominates. To a first approximation, then, the advection of nutrients into the area reflects water mass exchange. This aspect of the slope area is described under physical oceanography. The nutrient profiles available for the slope-rise area are reviewed earlier in this chapter; the temperature and salinity characteristics are also given in the references cited. These data, combined with the water mass exchange characteristics described in Chapter 2, can be utilized to present a reasonably complete picture of advective processes as they influence nutrient distributions.

Trace Elements

F.L. Sayles

Trace Element Distributions in Seawater

Dissolved Trace Metal Concentrations

Most direct determinations of dissolved trace metals in the ACSAR area came from studies of Deepwater Dumpsite 106 (DWD106). Additional, non-systematic, studies of various trace metals are also available. Analyses of water masses advected into the ACSAR area of interest greatly extend the amount of available data, providing a useful adjunct to the specific determinations.

Analyses of seawater samples of trace metals at DWD106 are concentrated on the dumpsite area, but extend north and south as well as to depths of slightly less than 3000 m (see Fig. 5.1). Analyses of Hg, Cd, Zn, Cu, Mn, and Pb are tabulated by Brezenski (1975) for a 1974 cruise, with no interpretation included. Analyses from the same and an additional (1976) cruise are reported by Hausknecht (1977). Analyses of samples on a second 1976 cruise are reported by Kester et al. (1977); they report values of Pb, Cd and Cu that are "similar to other oceanic" regimes and are representative of "background" concentrations in this area. The concentrations are up to an order of magnitude lower than those reported by Brezenski (1975) and Hausknecht (1977). Reporting on subsequent (Pb, Cd, Cu) data, Kester and Hausknecht (1981) conclude that all of the earlier data are suspect. The basis for this conclusion, along with average values, is summarized in Table 5.1. Kester et al. (1981) report additional measurements of dissolved Fe, Cd, Cu, Pb, and Zn. These data include a "control eddy" outside the dump area as well as the monitoring of the dumpsite during dumping operations (Table 5.2). Their data are reported as suspended matter concentrations and total metal concentrations; solution concentrations must be obtained by difference.

Table 5.1 Comparison of 1976 trace metal data from DWD 106, suggesting that earlier data were erroneously high. Data from Kester and Hausknecht (1981), Bender and Gagner (1976), Schoule and Patterson (1981.

Sample set	Depth range	Cd	(μg/1)	Cu	(μg/1)	Pb	(μg/1)
May 1974	Z < 50 m	0.30 ±	0.14 (40)*	0.7 ±	0.4 (40)	3.3 ±	0.9 (40)
	Z > 50 m	0.30 ±	0.14 (56)	0.7 ±	0.3 (56)	3.0 ±	1.2 (56)
February 1976	Z ≤ 50 m	0.44 ±	0.65 (86)	0.4 ±	0.7 (88)	0.7 ±	1.6 (85)
	Z > 50 m	0.56 ±	1.1 (59)	0.6 ±	1.2 (60)	0.5 ±	1.0 (59)
August 1976	Z ≤ 50 m	0.006 ±	0.003 (5)	0.29 ±	0.05 (6)	0.09 ±	0.04 (6)
	Z > 50 m	0.028 ±	0.011 (20)	0.22 ±	0.08 (23)	0.07 ±	0.05 (23)
Analysis of Sea water for dissolved metals	Surface	<	0.010		0.12		----
	NADW (1,800 – 3,700m)		0.025		0.15		----
Typical value for Pacific Ocean water	Surface		----		----	.03 ±	.00

* Number of samples averaged

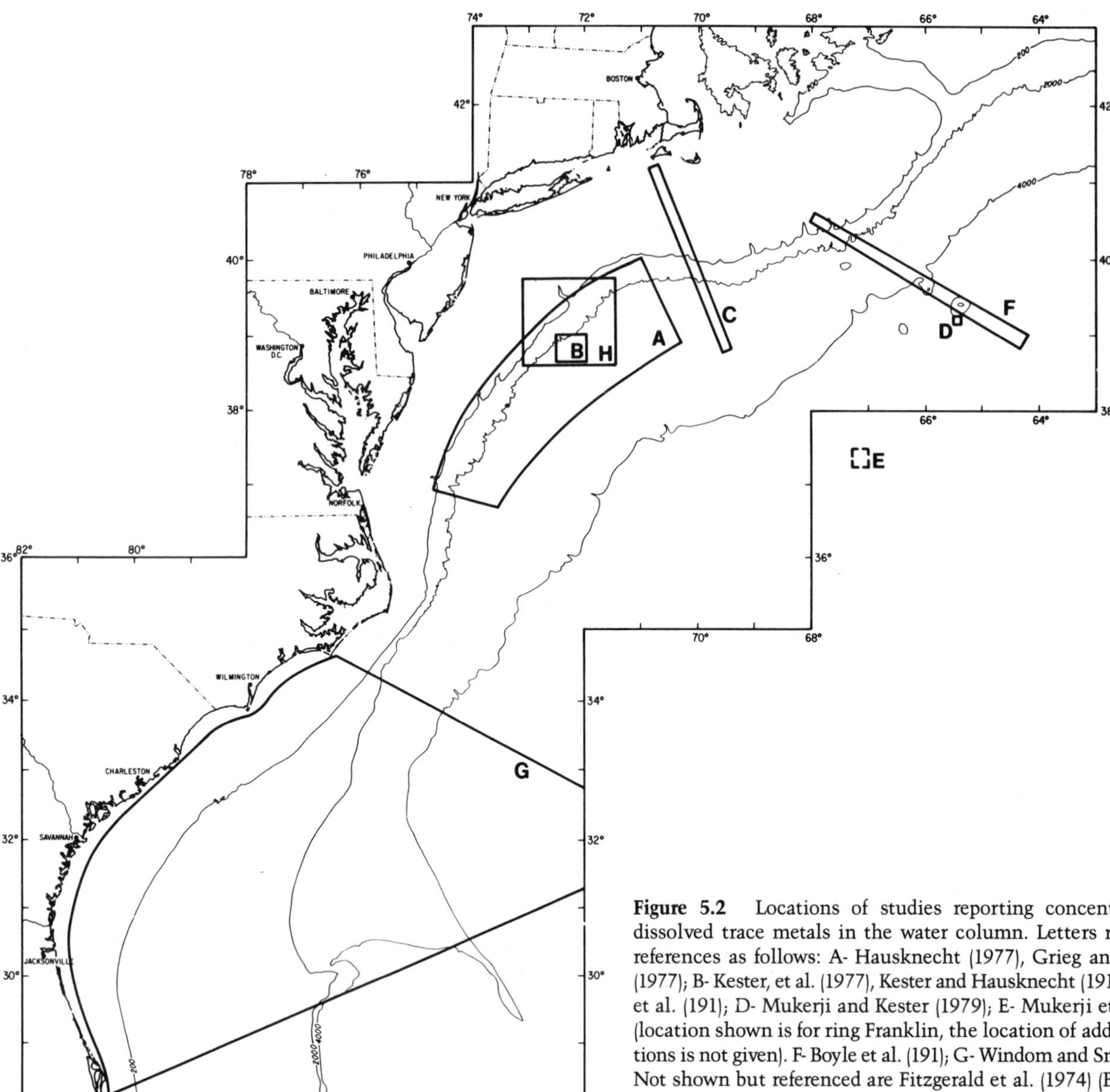

Figure 5.2 Locations of studies reporting concentrations of dissolved trace metals in the water column. Letters refer to the references as follows: A- Hausknecht (1977), Grieg and Wenzloff (1977); B- Kester, et al. (1977), Kester and Hausknecht (191); C- Kester et al. (191); D- Mukerji and Kester (1979); E- Mukerji et al. (1979), (location shown is for ring Franklin, the location of additional stations is not given). F- Boyle et al. (191); G- Windom and Smith (1979); Not shown but referenced are Fitzgerald et al. (1974) (Bermuda to Halifax transect) and Bewers et al. (1976) (Scotian shelf/slope).

Data also have been reported by several authors for studies that have included samples within the geographic area of the slope and rise (Fig. 5.2). These data tend *not* to provide a systematic set of analyses relevant to the slope-rise area per se, as they were not collected with any specific focus in this region.

Analyses of mercury are reported by Fitzgerald et al. (1974) in the northern fringe of the U.S. continental rise on a transect from Halifax to Bermuda. Samples include shelf, slope (Scotian), and deep water samples. All values reported are greater than or equal to 140 $\mu g\ l^{-1}$. Mukerji and Kester (1979) report a single profile taken in the Gulf Stream and covering 4400 m of the 4700-m water column. Their average value of 4.1 ($\pm$1.0) $\mu g\ l^{-1}$ strongly indicates that the samples of Fitzgerald et al. were subject to severe contamination. Mercury is also reported by Mukerji and Kester (1978) and Mukerji et al. (1979) for slope, Sargasso Sea, and Gulf Stream ring waters. A consistent correlation with SiO_2 and hence biological cycling is noted.

Cadmium analyses are reported by several authors for samples from the slope-rise area. Mukerji et al. (1979) report Cd for a variety of water masses, including slope water, Sargasso Sea water and Gulf Stream rings. The DWD106 studies (Kester et al., 1981) also include Cd concentrations. Surface water of the shelf were reported by Boyle et al. (1981). All of these authors note a correlation of Cd to nutrients, in particular NO_3^-. Boyle et al. report Cd concentrations of $< 1\ \mu g\ l^{-1}$ for surface waters of the Sargasso Sea, but 18 $\mu g\ l^{-1}$ in surface waters of the continental shelf. Mukerji et al. (1979) report similar concentration variations for samples of the upper 100 m: 1 $\mu g\ l^{-1}$ in the Sargasso Sea and 14 $\mu g\ l^{-1}$ in slope waters, with values rising to 69 $\mu g\ l^{-1}$ in nu-

Table 5.2 Trace metal concentrations obtained from pump samples (surface waters) and Niskin samples (upper 1000 m of water column) at a control station near the DWD 106 dumpsite. Data from Kester et al. (1981).

Metal	Mean	σ	n
	ng/kg		
	Pump sampler (16 – 21 m)		
Fe	230	60	6
Cu	110	30	6
Cd	2	0.5	6
Pb	110	40	6
Zn*	500	144	5
	Niskin sampler (20 – 1,000 m)		
Fe	670	140	10
Cu	170	64	10
Cd**	13	6	5
Pb***	276	34	5
Zn	4,600	1,960	10

* Value at 16 m, 2,400 ng/kg, was excluded because it is significantly different from other values.
** Upper 100 m only, due to systematic increase in deep waters.
***Upper 100 m only, due to decrease in deep waters.

trient-rich slope water. On the Scotian shelf and slope Bewers et al. (1976) report results consistent with these values including analyses of shelf, slope and North Atlantic Central water (NACW).

Cadmium analyses at depths 100 m in the slope-rise area are reported only by Kester et al. (1981). Cadmium concentrations are shown to increase with depth to values of ~40 $\mu g\ l^{-1}$ at 1000 m, the deepest water sampled. Bender and Gagner's (1976) values of Cd concentrations in the Sargasso and North Atlantic Deep Water (NADW) are somewhat lower than those of Kester et al. (1981), an average of 25 $\mu g\ l^{-1}$. As positive correlation with NO_3^- has been shown by essentially all of the data reported, it is not surprising that somewhat lower concentrations are associated with NADW than with NACW, which is associated with the deep NO maximum.

More data are available on Cu concentrations in seawater than on any other trace metal. Kester et al. (1981) report values to depths of 1000 m in the area of DWD106; Cu at depths of less than 50 m averages 0.29 $\mu g\ l^{-1}$ whereas concentrations at > 50 m average slightly less at 0.22 $\mu g\ l^{-1}$ (Table 5.1). Data reported by Boyle et al. (1981) for shelf surface waters are slightly lower: 0.16 $\mu g\ l^{-1}$. Surface waters for all oceans, including the Sargasso Sea, are significantly lower at about 0.08 $\mu g\ l^{-1}$. Windom and Smith (1979) report a number of analyses of surface waters in the southeastern portion of the slope-rise area that bracket the results of Kester et al. (1981) and Boyle et al. (1981). Windom and Smith (1979) report no obvious systematic areal variation in waters to the east of the Gulf Stream, the Gulf Stream, and the shelf, with Cu concentrations ranging between 0.02 and 0.30 $\mu g\ l^{-1}$. To the north of the ACSAR area, Bewers et al. (1976) report somewhat higher values for water off Nova Scotia: shelf = 0.56 $\mu g\ l^{-1}$, slope = 0.24 $\mu g\ l^{-1}$, NACW = 0.39 $\mu g\ l^{-1}$.

For Fe and Pb, there is only one report (each) relating directly to concentrations in the slope-rise area. For Fe this doubtlessly reflects its ubiquitous presence as a contaminant on ships and in laboratories. Kester et al. (1981) report values for particulate Fe and total Fe collected both by *in situ* pumping (16-21 m) and Niskin (0-1000 m) bottles (table 5.2). The values are comparable and translate to dissolved concentrations of 0.050 to ~0.20 $\mu g\ l^{-1}$. Pb concentrations, reported by Kester and Hausknecht (1981), are for a number of stations over a depth range of 0-2800 m; surface (< 50 m) values at two stations in the DWD106 area average 0.06 and 0.18 $\mu g\ l^{-1}$ deeper waters (> 50 m) at these stations average 0.04 and 0.09 $\mu g\ l^{-1}$ (table 5.1).

Ni concentrations in the specific area of the slope-rise are limited to a single surface water analysis, while no analyses of Mn and Co have been carried out in this area. Boyle et al. (1981) report a Ni concentration of 0.23 $\mu g\ l^{-1}$ for a surface water sample on the shelf south of Cape Cod. This compares to about 0.1 $\mu g\ l^{-1}$ in the Sargasso Sea. Any further estimates for Ni, Co, and Mn concentrations must be inferred from analyses in peripheral areas or from analyses of water masses that are advected into the slope-rise area.

Table 5.3 Calculated speciation of various trace metals in the ACSAR area. From Kester and Hausknecht (1981). Percentage of total metal in indicated dissolved species.

Species	Hg			Pb			Cu			Cd		
pH:	7.60	8.10	8.40	7.60	8.10	8.40	7.60	8.10	8.40	7.60	8.10	8.40
M^{2+} (free)	<<0.1	<<0.1	<<0.1	3.3	1.9	1.3	1.0	0.7	0.4	4.1	3.8	3.4
MOH^+	<<0.1	<<0.1	<<0.1	10.3	18.8	25.0	1.5	3.3	3.4	4.0	11.4	20.4
$M(OH)_2$	<<0.1	<<0.	<<0.	<0.1	0.2	0.5	4.3	29.5	60.8	<0.1	<0.1	0.6
$M(SO_4)^\circ$	<<0.1	<<0.1	<<0.1	1.1	0.6	0.4	0.1	<0.1	<0.1	0.5	0.5	0.4
M – organic	<0.1	<0.1	<0.1	<0.1	<0.1	<0.1	81.9	56.6	29.3	<0.1	<0.1	<0.1
MCl^+	<<0.1	<<0.1	<<0.1	14.6	8.4	5.6	7.6	5.2	2.7	62.7	57.5	51.3
MCl_2	2.9	2.9	2.9	23.8	13.8	9.2	2.0	1.4	0.7	22.2	20.3	18.2
MCl_3^-	16.3	16.3	16.3	16.3	9.4	6.3	– –	– –	– –	5.9	5.4	4.9
MCl_4^{2-}	80.8	80.8	80.8	– –	– –	– –	– –	– –	– –	– –	– –	– –
$MHCO_3^+$	– –	– –	– –	1.6	0.8	0.5	– –	– –	– –	0.2	0.2	0.1
$MCO3^\circ$	– –	– –	– –	28.9	46.0	51.3	1.6	3.1	2.7	0.3	0.9	1.3

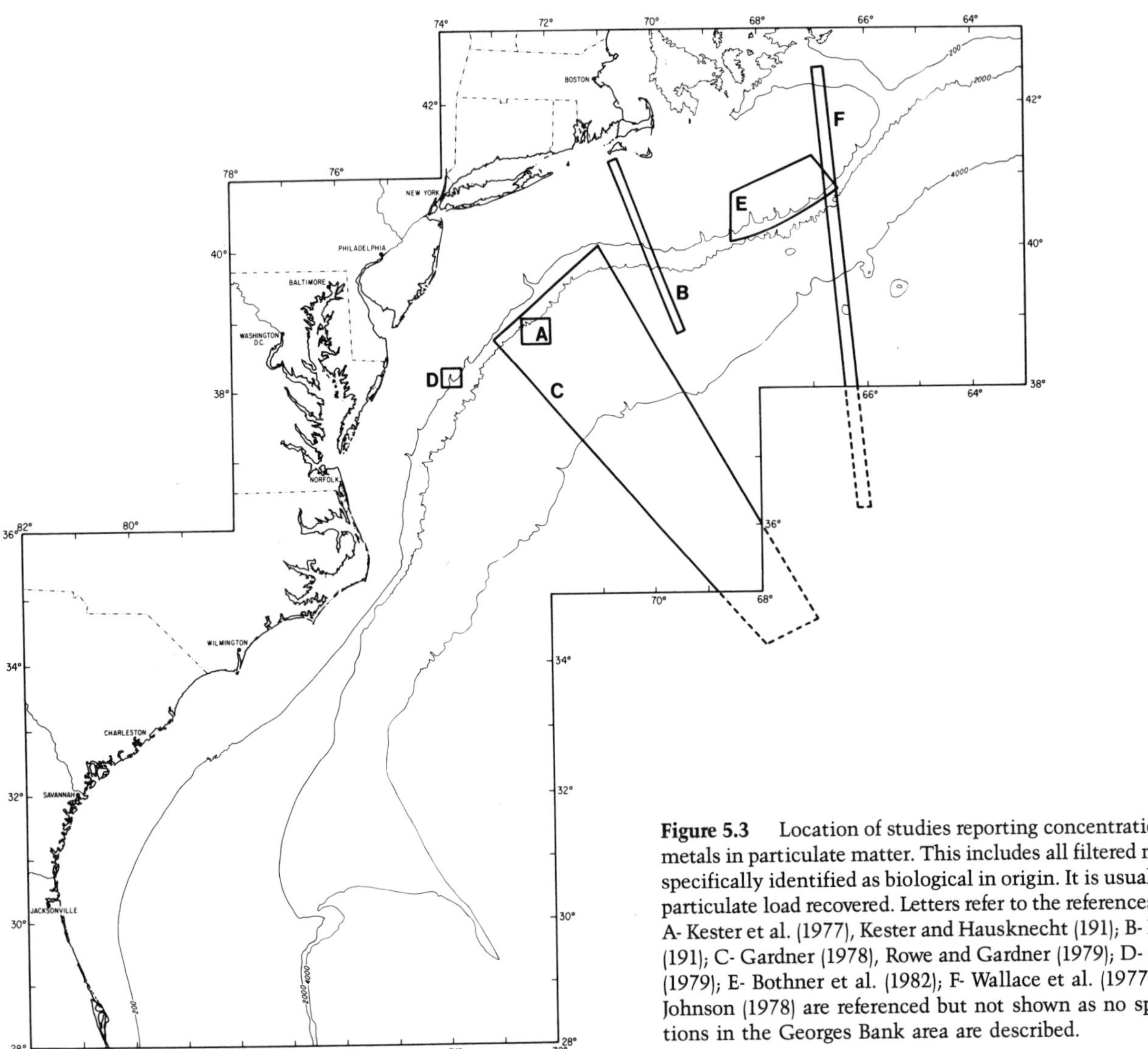

Figure 5.3 Location of studies reporting concentrations of trace metals in particulate matter. This includes all filtered material *not* specifically identified as biological in origin. It is usually the total particulate load recovered. Letters refer to the references as follows: A- Kester et al. (1977), Kester and Hausknecht (191); B- Kester et al. (191); C- Gardner (1978), Rowe and Gardner (1979); D- Gibbs et al. (1979); E- Bothner et al. (1982); F- Wallace et al. (1977); Stick and Johnson (1978) are referenced but not shown as no specific locations in the Georges Bank area are described.

On the Scotian shelf and slope Bewers et al. (1976) report Ni concentrations of 0.17 to 0.25 $\mu g\ l^{-1}$ in NACW, shelf, and slope waters values quite comparable to that of Boyle et al. (1981). Mn in Sargasso surface water is reported by Bender et al. (1977) as 0.14 $\mu g\ l^{-1}$; NADW is reported to be slightly lower at 0.10 $\mu g\ l^{-1}$. Off the Scotian shelf Bewers et al. (1976) report values of 0.33 $\mu g\ l^{-1}$ in shelf water and 0.08 $\mu g\ l^{-1}$ in slope water. Cobalt analyses of water are limited to a single report, that of Bewers et al. (1976): shelf = 0.013, slope = 0.019, NACW = 0.027 $\mu g\ l^{-1}$.

Efforts to describe the distribution of metals between various species within the slope-rise area are largely limited to the model of Kester and Hausknecht (1981). This study has attempted to characterize the speciation of four trace metals (Cu, Cd, Hg, and Pb) on the basis of thermochemical considerations. The model was calculated for a salinity of 34.5%, a value that is a rough median found in the area of DWD106. In addition to inorganic speciation, the model incorporates organic speciation at an assumed organic ligand concentration of 1.5 x 10^{-5} M. The results of the Kester and Hausknecht calculations are summarized in Table 5.3 for a range of pH values.

Suspended Particulate Matter Concentrations

One of the major sources of information on trace metals in particulate matter of the slope-rise area is the DWD106 study (Fig. 5.3). As discussed above, early cruise data are questionable. Later data reported by Kester et al. (1977) include background ranges; Cu = 10 to 40 $\mu g\ l^{-1}$, Cd = 0.2 to 1.0 ng l^{-1}, and Pb = ~3 to 25 ng l^{-1} (table 5.4). Cadmium shows a tendency to be enriched in waters above 100 m relative to greater depths; Cu exhibits a similar but less well developed trend. Kester et al. (1981) tabulate and discuss total and particulate Fe, Cu, Cd, Pb and Zn from the DWD106 area. Included in these studies are disposal events as well as a "control eddy" to establish background levels.

In surface water particulate matter, Wallace et al. (1977) reported a strong correlation of Al, Fe and, to a lesser extent, Mn. They conclude that this correlation is a reflection of

Table 5.4a/b Near surface profiles of particulate and total trace elements concentrates, total suspended matter and disolved nutrients. Data from Kester et al. (1981).

Table 5.4a

Depth (m)	Temp. (°C)	Salinity (‰)	TSM (μg/1)	Particulate Hg (ng/1)	Particulate (Fe (μg/1)	Particulate Cd (ng/1)	Particulate Cu (ng/1)	Particulate Zn (μg/1)	Particulate Pb (ng/1)
10	24.05	36.20							
20	21.80	36.15							
20	21.80	36.15	100		0.084	0.54	16	0.21	5.8
30	19.60	36.50							
30			110		0.060	0.14	12	0.17	5.6
50	18.65	36.55							
50			160		0.164	0.64	9	0.64	17.0
60	18.6	36.55							
60			180		0.198	0.22	31	0.50	6.9
80	18.3	36.5							
80			140		0.169	0.29	37	0.32	18.8

Depth (m)	PO_4 (μM/kg)	Si (μM/kg)	TOC (mg/1)	POC (μg/1)	pH	Total Fe (μg/kg)	Total Cd (μg/kg)	Total Cu (μg/kg)	Total Zn (μg/kg)	Total Pb (μg/kg)
10			0.96		8.11					
20					8.14					
20	0.136	1.17				0.60	0.018	0.17	6.26	0.29
30			0.89		8.32					
30	0.195	1.46			*	*	*	*	*	
50			0.68		8.15					
50	0.110	1.17				0.64	0.005	0.27	6.38	0.29
60					8.22					
60	0.059	1.11				0.76	0.016	0.21	6.10	0.31
80			1.26		8.17					
80	0.146	2.24				0.53	0.019	0.14	6.87	0.27

Table 5.4b

Depth (m)	Temp. (°C)	Salinity (‰)	TSM (μg/1)	Particulate Hg (ng/1)	Particulate (Fe (μg/1)	Particulate Cd (ng/1)	Particulate Cu (ng/1)	Particulate Zn (μg/1)	Particulate Pb (ng/1)
16					0.333	0.21	40	1.3	10.9
17					0.207	0.19	71	0.49	10.0
18			30		0.054	0.08	18	0.20	4.9
19			180		0.429	0.12	65	0.58	23.0
20			140		0.184	0.10	27	0.17	13.5
21			130		0.249	0.19	24	0.22	2.6

Depth (m)	PO_4 (μM/kg)	Si (μM/kg)	TOC (mg/1)	POC (μg/1)	pH	Total Fe (μg/kg)	Total Cd (μg/kg)	Total Cu (μg/kg)	Total Zn (μg/kg)	Total Pb (μg/kg)
16	0.250	2.43			8.17	0.27	0.002	0.15	2.4	0.17
17	0.244	2.24	1.09		8.17	0.24	0.002	0.11	0.62	0.15
18	0.230	2.51o	0.90		8.19	0.18	0.002	0.07	0.53	0.12
19	0.224	2.24			8.24	0.13	0.002	0.07	0.34	0.08
20	0.244	2.12		13.1	8.18	0.28	0.003	0.14	0.65	0.07
21	0.204	2.24	1.22		8.20	0.27	0.002	0.11	0.36	0.07

*Contaminated data.

the association of these elements with clay minerals. Nickel and Cr do not exhibit readily definable associations showing some affinities for both clay minerals and organic matter. Copper, Zn, Pb, and Cd exhibit a high degree of association with organic matter. A comparison of the chemistry of surface particulates with average North Atlantic surface sediment (from the literature) indicates substantial differences in element ratios. The authors concluded that < 30% of the

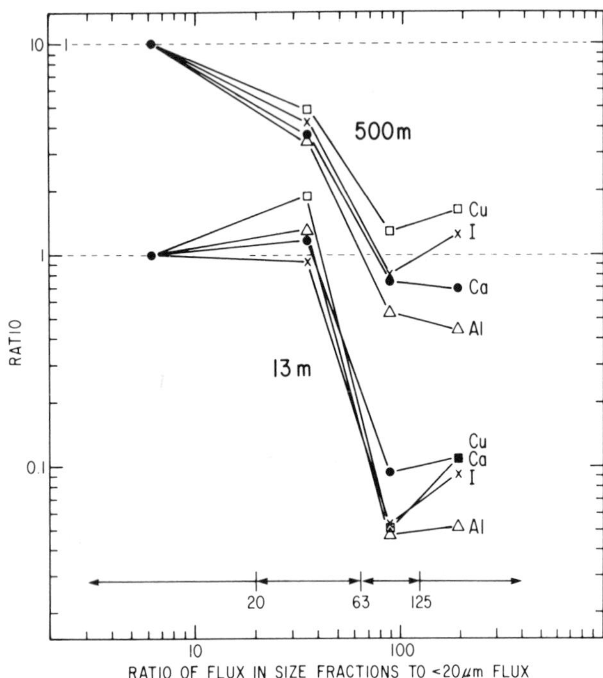

Figure 5.4 Ratio of the flux of several elements in various size locations relative to the 20μm fraction. Data are determined from sediment trap data from traps deployed at two heights above the bottom (500 m and 13 m). Figure from Gardner (1977).

Ni and Cr are associated with the clay minerals and that >90% of the Cu, Zn, Pb, and Cd are associated with particulate organic matter. In attempting to elucidate the sources of particulate trace metals, Wallace et al. (1977) estimate that atmospheric input to the sea surface is very similar to trace metal removal from the mixed layer, implying an atmospheric source for the trace metals.

An extensive study of particulate matter off the mid-Atlantic coast was carried out and reported by Gardner (1978) and Rowe and Gardner (1979). These studies are more comprehensive and coherent than any of the others encountered. Gardner studied particulate fluxes with sediment traps and particulate concen- trations with Niskin casts as well as comparing results with sediment cores. Sediment trap data were collected at DWD106 and a location termed DOS No. 2 (the Deep Ocean Station No. 2 of the Woods Hole Oceanographic Institution), at heights approximately 15, 100 and 500 m above bottom at both stations. Chemical analyses for Ba, Ti, Sr, Mn, Cu, V, Al, Mg, Ca and I were carried out on bulk sediment trap material, size-fractionated trap material (<20, 20-63, 63-125, >125 μm), suspended particulate matter from Niskin bottles (SPM), and surface sediments. As a consequence, Gardner was able to calculate elemental fluxes for the above components and determine the systematics of sedimenting materials.

According to Gardner, 83-96 percent of the total flux is in the <63 μm fraction. The fine fraction flux dominates for all elements. This is apparent in a graphical comparison of flux vs. grain size (Fig. 5.4). The concentrations of almost all elements measured are greater in the fine fractions than in the coarser fractions, with the possible exception of Cu and I (Fig. 5.5). Copper, Sr and the organic-related elements are more abundant in the trap material than in the Niskin samples. This reflects the fact that organic constituents are more abundant in the coarser fractions than in the fine fractions. Gardner attributes this to association of organic material with coarse skeletal fragments rather than fecal material. Gardner reports strong compositional gradients in the near bottom samples and interprets this as being indicative of rapid large scale reaction in the benthic boundary layer.

A small amount of data on particulate matter is also given in scattered abstracts (Fig. 5.3). Generally these reports do not include quantitative data on trace metal compositions. Stick and Johnson (1978) report studies of particulate matter in slope waters of the Georges Bank area and Scotian Shelf. Their analyses include Cu and Zn and focus on estimates of transport of these two elements onto the shelf. Gibbs et al. (1979) describe investigations of particulate matter at the shelf break on the mid-Atlantic coast. They pre-

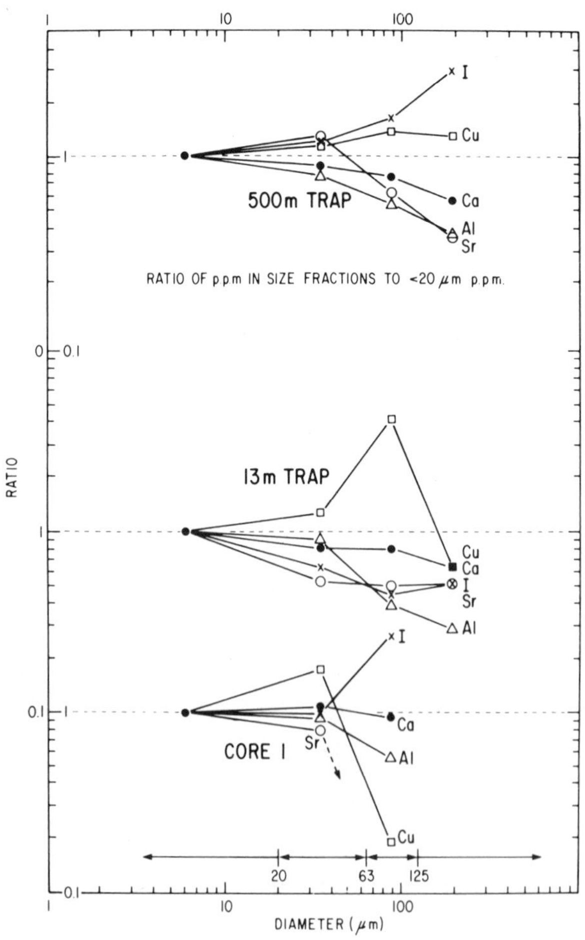

Figure 5.5 Ratio of the concentration of several elements in various size fractions relative to the <2μm fraction. Data are for the same material as that of Figure 5.4. In addition to two traps (500 and 13 m above the bottom) analyses from a core is included. Figure from Gardner (1977).

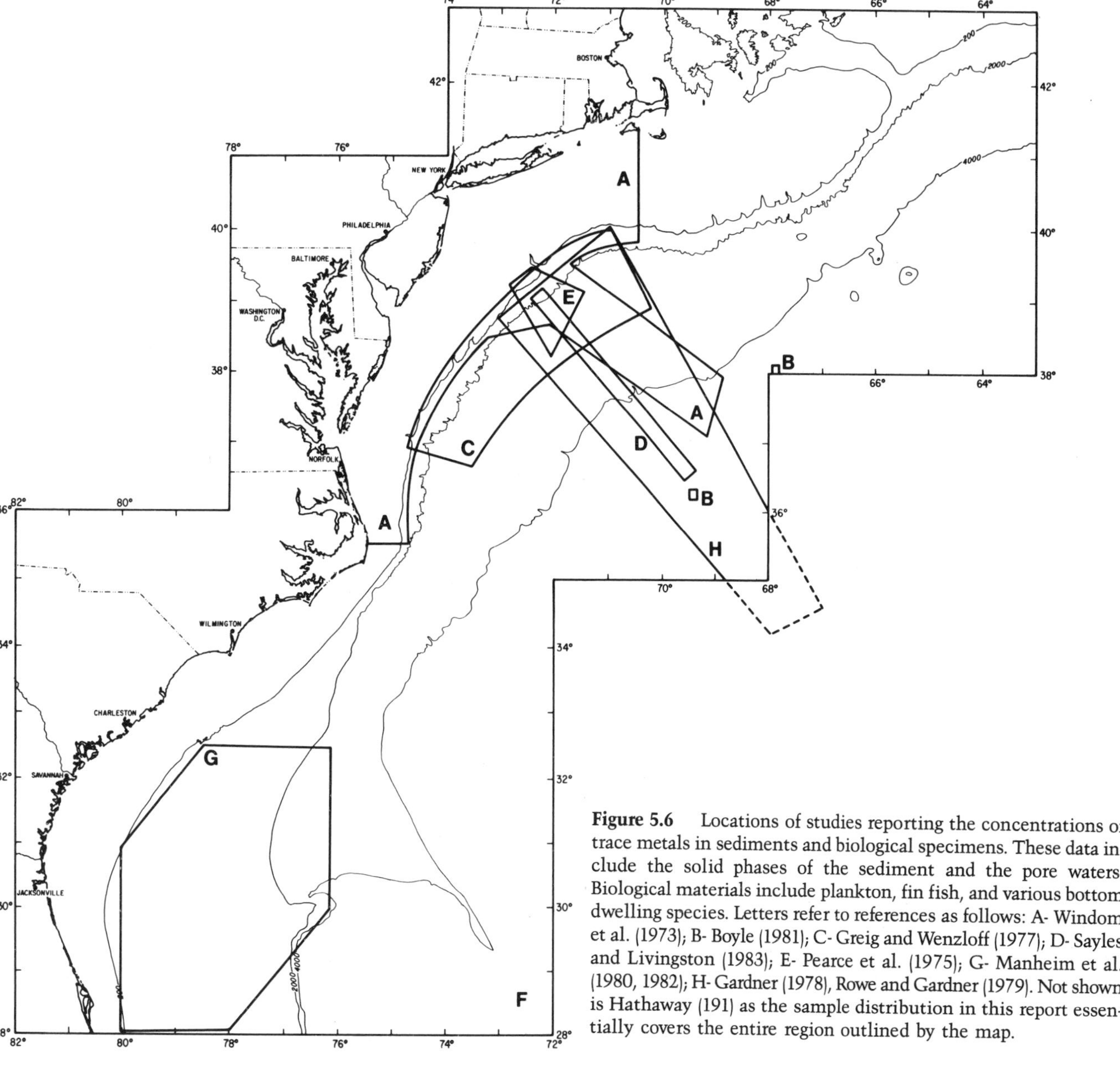

Figure 5.6 Locations of studies reporting the concentrations of trace metals in sediments and biological specimens. These data include the solid phases of the sediment and the pore waters. Biological materials include plankton, fin fish, and various bottom dwelling species. Letters refer to references as follows: A- Windom et al. (1973); B- Boyle (1981); C- Greig and Wenzloff (1977); D- Sayles and Livingston (1983); E- Pearce et al. (1975); G- Manheim et al. (1980, 1982); H- Gardner (1978), Rowe and Gardner (1979). Not shown is Hathaway (191) as the sample distribution in this report essentially covers the entire region outlined by the map.

sent no chemical analyses but do discuss size distributions in the turbidity maximum.

A number of sediment trap samples on the continental shelf have also been analyzed for trace metals. Bothner et al. (1982) report analyses of sediment trap matter in the Georges Bank area, including one trap in Lydonia Canyon at 1380 m. Since exchange between shelf and slope is inevitable, their results are relevant to the slope studies discussed here. By the same token, the results reported by Betzer (1978) for studies in the area of the New York Bight should be noted. Betzer collected particulate matter from a large number of sediment trap deployments as well as suspended samples and reports both flux and trace metal composition data, including chemical partitioning into weak-acid-soluble and refractory fractions.

Biogenic Trace Metals

Few reports discuss the association of trace elements with biogenic materials in the slope-rise area. Windom et al. (1973) report Hg in "plankton" on the continental shelf of the southeastern U.S. (Fig. 5.6), with some analyses beyond the shelf break. The general trend of the distribution is a decrease with distance from the source of Hg pollution. The lowest values reported are found beyond the shelf break and are generally $\leq 0.2\ \mu g\ l^{-1}$ (dry wt.). Boyle (1981) has studied the Cd, Zn, Cu, and Ba content of carefully cleaned foraminifera tests from core tops of the North Atlantic; two of the cores studied fall within the confines of the ACSAR area. Greig et al. (1977) report a large number of analyses of several trace metals in various organs of fin fish in the area of DWD106. Their data include Cd, Cu, Hg, Mn, Pb and Zn. Pearce et al. (1975) also report trace metal analyses for fish and include benthic organisms.

Trace Elements In Sediments

Pore Water Trace Elements

While the general relation of pore water trace element chemistry has been reasonably well worked out, only one such study of sediments has been done within the ACSAR area. Sayles and Livingston (1983) report profiles of Fe and Mn in a transect of the slope and rise of the mid-Atlantic states. The data, along with NO_3^- concentrations, have been used to depict redox conditions in the sediments. The Mn profiles generally depict increasingly reducing environments with decreasing water depth. This is a reflection of increased sedimentation rate, particularly of organic carbon. Mn^{+2} enrichments are observed only at depths in excess of 30 cm in the more oxidized cores, but are found at 3-5 cm in the more reduced cores. Conditions sufficient for the reduction of Fe^{3+} to Fe^{2+} were observed at only one station, on the mid-slope.

Solid Phase Trace Elements

Three primary sources of chemical analyses of sediments exist for the slope-rise area (see Fig. 5.6). All three, however, deal with specific geographic areas, thus meaning that the general characteristics of the trace element (and major components) composition of the continental rise and slope are not well documented.

A large number of analyses were carried out in conjunction with the DWD106 studies. Pearce et al. (1975) reported analyses of surface sediment samples extending from the slope break to in excess of 2000 m depth. Concentrations of Cr, Cu, Ni, Pb, and Zn were tabulated with little discussion; moreover, the data were accompanied by little ancillary information, making interpretation of distributions difficult. Concentrations exhibit little variation in most samples and concentrations fall in the ranges (ppm): Cr = 23 to 28, Cu = 20 to 35, Ni = 18 to 34, Pb = 24 to 32, Zn = 52 to 64. A few samples exhibit markedly lower concentrations for all trace metals. One of these is a shelf sample with concentrations 10-20% of the average, presumably reflecting the often noted inverse relationship between metals and grain size in which sediments (cf. Bothner et al., 1982). Grain size data are included for deeper sediments, and there appears to be little correlation between grain size and metal content. Indeed, the coarsest sample (with 2 to 5 times as much sand as the average) shows no significant decrease in trace metal concentration. Even though the analyses were carried out in the area of DWD106, the authors comment that they see no evidence of metal contamination.

Greig and Wenzloff (1977) presented extensive analyses of sediments from the DWD106 area and its environs. They report concentrations of Cd, Cr, Cu, Ni, Pb, and Zn over approximately 200 miles of the slope and rise around DWD106. As with other data from these studies, no ancillary information, in particular major element chemistry, is given. As a result no interpretation of causative relation is possible. The authors note that the concentrations observed are similar to those of Pearce et al. (1975). As with the Pearce data, variations in concentration are limited. The shallowest samples (≤ 180 m) stand out with concentrations on the order of 10-20% that of deeper samples, again, presumably, reflecting increasing content of coarse, low- metal, detritus on the shelf.

Analyses of sediment samples throughout the slope-rise area are tabulated by Hathaway (1971). This is unquestionably the largest sample/analytical compilation for the east coast slope and rise. The samples are almost exclusively surface sediment. The wide range of analyses includes trace and minor elements, major elements, grain size, organic carbon and nitrogen, $CaCO_3$ content, mineralogy and lithologic descriptions. The breadth of the data, plus their wide geographic base, are valuable for interpreting genetic relationships and causative factors in the distribution of the trace metals. However, while the geographic base is large, there is an obvious bias in the samples analyzed. The great bulk of the analyses are either of shelf samples or, where deeper than the shelf, from the Blake Plateau where a distinct focus on phosphorite nodules and pavements is apparent. The actual number of analyses of more "typical" slope-rise sediments is quite limited. Nonetheless, the inclusion of a wide range of ancillary data makes this compilation a valuable source as regards processes influencing trace metal distribution in the study area.

Relatively specialized but extensive sets of analyses are available for the Blake Plateau area (Fig. 5.6). The Blake Plateau contains extensive deposits of Mn and Fe phosphates and oxides (Manheim et al. 1980). Studies in this area have focussed upon assessing the potential for economic recovery of phosphorites for their trace metal content. Compositional data are reported by Manheim et al. (1980), and a large number of analyses are included in Hathaway (1971). In addition to first transition series elements, Manheim et al. (1980) report lanthanide series (REE) contents of both nodules and pavements. Prospecting has been carried out by Deep Sea Ventures, Inc., in the area, and a large number of analyses have been done in conjunction with these studies.

According to a comprehensive review and compilation of the Blake Plateau data by Manheim et al. (1982), phosphorite dominates the shallower depths (500-750 m) while Mn oxides as crusts and nodules occur at greater depths. Reported average compositions are (in percent):

	Mn	Ni	Co	Cu	Fe
Crust	9.2	0.38	0.14	0.027	—
Nodules	16	0.6	0.3	0.1	10

The range of trace metals reported for Deep Sea Ventures nodule analyses is (in percent):

Ni = 0.4 to 0.95
Co = 0.25 to 0.40
Cu = 0.08 to 0.15.

The authors interpret the concentrations as resulting from two processes. The occurrence of phosphorite is linked to high productivity due to the upwelling of nutrient-rich bottom water. Growth rates are, as with most deep sea nodules, exceedingly slow. Consequently the observed concentrations in the pavements and nodules are thought to result from non-deposition or active winnowing of fine grained materials, leaving the nodules and crusts as lag deposits.

Two additional sources of chemical analyses of sediments for the slope/rise area exist. Gardner (1978) reported analyses of the surface sediment of two cores collected in conjunction with sediment trap studies; Rowe and Gardner (1979) reported ancillary data on these two cores. Sayles and Livingston (1983) report the distribution of MnO_2 in cores as an indicator of redox conditions and the diagenetic remobilization and precipitation of Mn in sediments of the rise and slope.

Several studies of sedimentary chemistry fall outside the geographic bounds of this study but nonetheless deserve mention. The break between shelf and slope is an arbitrary one and exchange across it must occur. The Middle Atlantic Outer Continental Shelf Studies (Harris et al., 1978) carried out by the Virginia Institute of Marine Science report chemical and trace element studies for the shelf. Included are partitions into dilute-acid-leachable and total metal concentrations. These studies show an increase in metal content seaward across the shelf. This is presumed generally to reflect the correlation of metals with the fine grain size fraction and the increase of the latter with increasing depth. Experiments on acid leaching indicate 25-35% of the Pb, Zn, and Cu is extractable whereas 8-15% of the V, Ni, Fe, and Cr is leachable. Exchange across the shelf-slope break will inevitably lead to the occurrence of the fine-grained shelf materials in slope sediment.

Studies on Georges Bank are also relevant to the slope-rise area. Bothner et al. (1982) report a large number of trace metal analyses for the Georges Bank area. Included in their analyses are grain size measurements. Chemical analyses include both bulk sediment and size-fractionated ($< 62\ \mu m$) samples. Exchange between shelf and slope will lead to the occurrence of fine-grained shelf materials in slope sediments. The seaward increase in metal content is correlated with increase in the abundance of the fine-grained component.

Radionuclides

H.D. Livingston

Artificial Radionuclides

Input To The Study Area

The ACSAR region has received some input of radionuclides from accidents, but the major input has been from the fallout of fission and activation products introduced to the atmosphere from above-ground nuclear weapons tests. In broad terms the input of artificial radionuclides from this source may be described as having taken place in two episodes. The first, approximately 25% of the total, took place in the first few years after the series of tests, during the interval 1952-1958. The second, about 75% of the total, took place shortly after the 1961-1962 series of tests. Since that time, the overall input has been very small, representing fallout from countries that did not sign the atmospheric nuclear test ban treaty. The variation with time of these inputs to the northern end of the study region is well represented by a detailed record of deposition of ^{90}Sr from fallout delivered to a site in New York City. This record (Fig. 5.7) can be taken to be representative of the direct atmospheric delivery of this radionuclide to the offshore areas north of Washington, D.C. The input of other nuclides produced in weapons tests can be assumed to be proportional to their production relative to ^{90}Sr. Direct input to the central and southern areas of the ACSAR area would have taken place at rates diminishing from north to south — as fallout was at a maximum in mid-latitudes in both hemispheres. Some indication of the magnitude of this latitudinal difference in input magnitude is inferred from the record of fallout plutonium by 10° latitude bands recorded in soils collected in 1970-71 (Hardy et al., 1973).

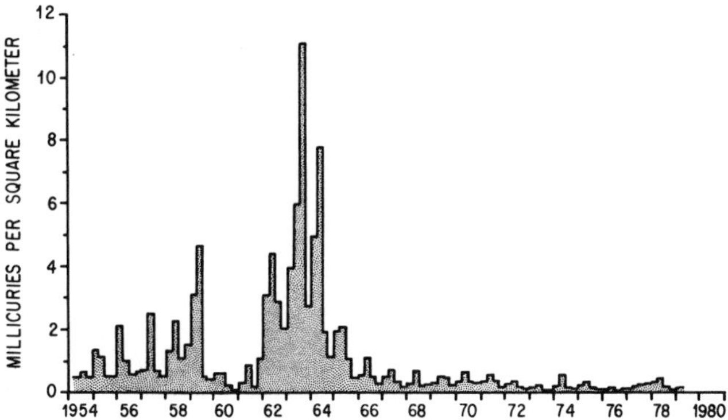

Figure 5.7 Quarterly deposition of strontium - 90. New York City. After Toonkel (1980).

In addition to this direct input to the region, there have been further inputs of fallout delivered to adjacent oceanic areas which have been advected into the region. Against these various input terms there has been removal by the following mechanisms: (1) radioactive decay, (2) advection out of the region, (3) removal of particle-active nuclides to underlying sediments.

Compared to radionuclide fallout from nuclear weapons testing, the input of artificial radionuclides from other sources has been relatively small. These sources include fallout to the region of ^{238}Pu introduced to the atmosphere from the 1964 satellite failure in the Southern Hemisphere (Hardy et al., 1972), the accidental loss in 1962 of the U.S. nuclear submarine *Thresher* (U.S. Department of the Navy, 1982) and the disposal of low-level nuclear waste. The delivery of ^{238}Pu to the region followed a latitudinal pattern similar to that noted above for fallout. The *Thresher* sinking and the two disposal sites represent highly localized sources of artificial radionuclides in the region and, so far as studies to date have shown, no evidence has been discovered for other than extremely localized dispersal of any released radioactivity.

Soluble Nuclides. Those radionuclides whose oceanic behavior is such that they exhibit long residence times in ocean waters and trace advective and diffusive processes in ocean water columns are frequently described as "soluble" or conservative. Of the longer-lived radionuclides produced in nuclear weapons testing, ^{90}Sr, ^{137}Cs, ^{3}H and ^{14}C are generally considered to be water tracers. ^{137}Cs exhibits some affinity for particle association when it comes into contact with sediments or suspended particulates but its dominant behavior is as a water tracer in solution. Likewise, bomb ^{14}C is involved in biological cycles which remove and transfer carbon from surface waters to mid-depths or to ocean sediments but this does not significantly alter the distribution of this nuclide in the water column over short (10^{1}-10^{3} year) timescales.

Because most North Atlantic studies of these nuclides have focused on distributions in open ocean waters (Ostlund et al., 1976, 1974; Kupferman et al., 1979; Bowen and Roether, 1973, 1974), the waters of the ACSAR area have received little specific attention. The time history of tritium and ^{90}Sr in North Atlantic surface water (Dreisigacker and Roether, 1978) has been reviewed (and for tritium reconstructed) and serves to describe the concentrations obtained in the study area from 1952 through 1974. The most comprehensive series of measurements of these tracers were made as part of the GEOSECS program in the North Atlantic in 1972. The results for ^{14}C and ^{3}H have been reported by Ostlund et al. (1974, 1976) whereas those of ^{137}Cs and ^{90}Sr (as well as from other cruises at the same time period) have been presented by Kupferman et al. (1979). These data have been used extensively to test models of large-scale oceanic circulation (Rooth and Ostlund, 1972) with little specific relevance to the area of this report. However, a variety of measurements of the distribution of both artificial and natural tracers was reported for the second GEOSECS intercalibration and test station at a position (35°46′N, 67°59′W) which is not far east of the 4000 m contour of the study area (Roether and Munnich, 1972; Trier et al., 1972; Chung et al., 1974).

One study (Bowen et al., 1974) showed that a distinction can be made between shelf waters and open ocean waters on the basis of their $^{137}Cs/^{90}Sr$ ratio. Open ocean water is characterized by a ratio of 1.44 whereas coastal waters have lower ratios, around 1, presumably related primarily to freshwater input of ^{90}Sr. Clearly these differences can be used in studies of exchange of inshore and offshore waters in slope waters.

The southward movement of the Western Boundary Undercurrent has been followed through measurements of soluble artificial radionuclides, especially tritium, in the core of the current (Jenkins and Rhines, 1980). These authors postulate a transit time of 15 years for water to reach 30°N from the far northern ocean surface. Dilution of the tracer over this transit is estimated to be about 10-fold.

Reactive Nuclides. This group of artificial radionuclides includes:

1. *Fission product* lanthanides - such as ^{144}Ce, ^{147}Pm, ^{155}Eu
2. *Activation products* - such as ^{55}Fe, ^{60}Co, ^{63}Ni
3. *Transuranic elements* - such as Pu and Am.

The characteristic geochemical behavior of this group of nuclides is their removal from the water in association with sinking particles and transfer to marine sediments. Removal rate is determined by reactivity with respect to scavenging of the stable elemental form of a given nuclide (where they exist). Very few water column studies of these nuclides have been made in the ACSAR area. Santschi et al. (1980) describe a decreasing gradient of Pu concentrations from open ocean surface water to nearshore surface water in the New York Bight area. This gradient is similar to that found for reactive naturally occurring nuclides such as $^{228}Th/^{228}Ra$ (Santschi, 1980) or ^{210}Pb (Bacon et al., 1976), and seems to be a consequence of enhanced removal of particle active species in coastal, shelf and slope waters by a combination of removal processes, such as association with resuspended or riverine particulates, biological scavenging associated with primary productivity, and direct removal on contact with sediments at the interface with the water column. These workers also find that a high proportion of the total water concentration of Pu in shallow coastal waters is particulate. Again this is a phenomenon paralleled by reactive nuclides of the natural Th and U series. Some evidence of Pu removal in oceanic water columns offshore of the study area appears in a report by Cochran and Livingston (1983) on Pu distribution relative to ^{137}Cs in water columns above the East Hatteras Abyssal Plain. A more general review of the oceanic behavior of reactive artificial nuclides appears in a paper by Livingston and Jenkins (1982) and references therein.

Once delivered to marine sediments, reactive artificial radionuclides are exposed by physical mixing by bioturbation, burial in areas of high sedimentation and diagenesis within sediment interstitial waters. The former process seems to control the distributions of plutonium in most regions in the ocean (Bowen et al., 1976) and is a dominant mechanism controlling the distributions of Pu, ^{137}Cs and ^{55}Fe in the Atlantic continental rise (Livingston and Bowen, 1979; Bowen and Livingston, 1981; Cochran and Livingston, 1983; DeVito, 1981). There has been much discussion in the literature on the question of mobility of Pu in marine sediments after deposition, and ongoing studies are directed toward these and related questions. One long-term study started in 1983 is the SEEP program (Shelf Edge Exchange Program) supported by the U.S. Department of Energy. Within the framework of this program, reactive radionuclide scavenging in the slope region and subsequent sedimentary diagenetic behavior are subjects of intensive study.

Site Specific Studies

Within the study region there are three sites which have received an input of artificial radionuclides as a result of either an accident or planned disposal of radioactive waste (low level). These are the accidental loss of the nuclear submarine U.S. N/S *Thresher* in 1962, and the dumping of low-level radioactive waste at two sites from 1951 to 1962.

Thresher Site. In 1962 the nuclear submarine *Thresher* sank in a water depth of 2590 m off the Gulf of Maine, at a position about 41°45'N and 65°00'W. Because of the sensitivity regarding details of the accident, little information has been available until recently. Early in 1983, the Navy published a draft environmental impact statement of the possible use of the deep ocean for disposal of the hulls and reactor pressure vessels from decommissioned nuclear power submarines. To support their contention of the minimal radiological impact from such an operation, they cite the environmental studies carried out at the *Thresher* site. These studies have not revealed radioactivity levels in the sediment, biota or water in the immediate vicinity of the sunken submarine which are significantly greater than natural levels. Potentially there exist the suite of fission product radionuclides present in the submarine reactor core and fuel rods as well as the activation products in construction material of the vessel surrounding the reactor core. The absence of released radioactivity at the site is consistent with the design characteristics claimed for submarine reactors and for corrosion rates of structural materials.

Atlantic Low Level Radioactive Waste Dumpsites. Packaged low level radioactive waste was dumped at two locations on the lower continental slope east of the Delaware-Maryland coast during 1951-1962. The shallower site (2800 m) is centered at 38°30'N, 72°06'W and occupies an area of 256 km² (Dyer, 1976). It is about 15 km SE of Deepwater Dumpsite 106 which is discussed elsewhere in this report. The deeper site (at 3800 m) is located down the slope at a position about 37°50'N, 70°35'W. Some details of numbers of drums of waste dumped at each site and the amounts of radioactivity involved are given in Table 5.5 and in the report by Dyer (1976). Four times more drums were dumped at the deep-water site, but the total activity dumped at the 2800-m site was between one and two orders of magnitude higher than at the 3800-m site. In fact, almost half of the total radioactivity dumped at the 2800-m site was contained in the pressure vessel of the reactor of the U.S. N/S *Seawolf* (33,000 Curies). Detailed information about the composition and form of the radionuclides dumped at the two sites does not appear in any report, but such information would be extremely valuable for any assessment of the radiological impact of contained radioactivity at each site. For example, in the more than twenty years since the cessation of dumping, much of the radioactivity may have been reduced to very much lower levels by radioactive decay of relatively short-lived species. But without such knowledge no such predictions are possible.

Technological obstacles involved in any program of surveying the status of such dumped materials are substantial, or have been until recently. A 256 km² area with 14,300 drums dumped corresponds to a density of one drum per 2000 m². In reality, many drums would be in clusters so that the areas in the dumpsite without drums have to be exceedingly large. It is therefore not surprising that early (1961) attempts to photograph the drums failed to find any — despite the taking of 11,000 underwater photographs (Dyer, 1976). Likewise, we have not found any reports of the sighting of the *Seawolf* reactor pressure vessel. It is unclear whether this is a failure of technology or of Navy sensitivity toward release of their own studies.

Table 5.5 Summary of U.S. Sea Disposal Operations, 1951-1967 (Atlantic Ocean).

Year	Approximate Location	No. Containers All Types	Estimated Activity at Time of Packaging (Ci)
1951 – 1958	42°25'N 70°35'W	4,008	2,440
	36°56'N 74°23'W	432	6.5
	Midocean	97	<0.1
	37°50'N 70°35'W 38°30'N 72°06'W	23,000	8,000
1959 – 1960	38°30'N 72°06'W	5,800	68,500 *
between	36°44'N 45°00'W	228	456
and	36°50'N 74°23'W	204	24.5
	Midocean	22	0.1
1961 – 1962	36°56'N 74°23'W	137	15.6
1963 – 1964	36°56'N 74°23'W	58	5.3
1965	36°56'N 74°23'W	6	4.3
1967	36°56'N 74°23'W	6	30.5
Totals		33,998	79,482.9

* Includes pressure vessel of *Seawolf* reactor – estimated 33,000 Ci of induced activity.

In recent years, technological advances have resulted in greater success in attempts to monitor the condition of the dumped wastes. These include the sitings and photography of dumped drums by both manned and unmanned submersibles (Dyer, 1976), retrieval of sediment cores from areas adjacent to drums (Dyer, 1976; Dayal et al., 1979; Bowen and Livingston, 1981) and the recovery of an 80-gallon canister for corrosion studies at Brookhaven National Laboratory (Columbo et al., 1980). The Environmental Protection Agency Office of Radiation Programs has supported much of the research effort at this site and should be a source of future information. Such studies as the evaluation of the sediment characteristics at the site in regard to their affinity for waste nuclide retention would be included in the research (Neiheisel, 1979).

Because of the proximity of the shallow dumpsite to the Deepwater Dumpsite 106 (DWD106) used for industrial waste disposal, many of the basic site characterization studies of the latter have direct bearing on the former and vice-versa. In this respect an extremely useful bibliographic review of these Atlantic Ocean Disposal Sites was prepared for the Sandia National Laboratories for an assessment of their suitability with respect to the marine disposal of large structures (Jackson, 1982).

Measurements of artificial radionuclides in sediments (Bowen and Livingston, 1981; Dayal et al., 1979) and biota (Schell, 1980) in and adjacent to the ACSAR area have been made to evaluate the extent to which leakage and movement of dumped radioactivity has taken place. Radionuclides included ^{134}Cs, ^{137}Cs, ^{60}Co, $^{239,240}Pu$, ^{238}Pu, ^{241}Am, ^{242}Cm and ^{244}Cm. The overall impression from these studies was that, at least in the areas covered, leakage was relatively minor and confined to the immediate area surrounding the point of release. It should not be overlooked that the effort made thus far has only been able to survey a miniscule fraction of the area which has received the waste radioactivity.

A further difficulty of artificial radioactivity in such a disposal environment is that the distributions and concentrations of some radionuclides in the dumped waste must be determined against a "background" which includes oceanic input from fallout of debris from atmospheric nuclear weapons tests. As shown by Bowen and Livingston (1981), the fallout component is not uniform but considerably variable reflecting the complexity of the processes involved in the delivery and subsequent redistribution in the sediments and biota of the area. Comparison of several criteria may allow a distinction between the two sources. These include such properties as interisotopic ratios, sediment nuclide inventories and distribution patterns, the presence of nuclides absent in fallout, and unusual concentrations.

Naturally Occurring Radionuclides

There are a variety of naturally occurring radionuclides involved in oceanic processes which are present in the ACSAR region. They include:

1. primordial nuclides such as ^{40}K, ^{87}Rb, and nuclides of the natural Th and U decay series
2. cosmogenic radionuclides produced in the upper atmosphere by the interaction of cosmic rays with light elements of the atmosphere. This group includes nuclides such as ^{7}Be, ^{3}H and ^{14}C.

In contrast to man-made radionuclides, the concentrations and distributions of naturally occurring radionuclides are controlled by steady state processes at or tending toward equilibrium. Thus the present-day patterns have evolved over geologic time in the ocean in response to their chemistries, biogeochemical processes in the oceans and their radiological decay characteristics. The concentrations of various nuclides generally characteristic of ocean water and sediment are tabulated in the report of Joseph et al. (1971). The behavior of a given radionuclide is controlled by its elemental chemistry. Thus a spectrum of behavior relative to scavenging is exhibited ranging from highly soluble nuclides, such as ^{40}K, to highly reactive ones, such as the isotopes of Th.

The naturally occurring U and Th decay series, headed by ^{238}U, ^{232}Th, and ^{235}U, have been widely used as oceanic tracers because of the oceanic disequilibria between individual members of each series. Because some members are mostly soluble in their oceanic behavior (^{238}U, ^{234}U, ^{235}U, ^{226}Ra, ^{228}Ra and ^{222}Rn) while others are reactive (^{232}Th, ^{234}Th, ^{230}Th, ^{228}Th, ^{231}Pa, ^{210}Pb and ^{210}Po), disequilibria are established in different marine compartments. The extent of these disequilibria provides tools to study the rates of a variety of marine processes.

As with the oceanic behavior of the artificial radionuclides, a comprehensive review of these nuclides and their marine chemistry and biology is outside the scope of this review. One recent review by Cochran (1982) can serve as a reference to general oceanographic studies of the chemistry of these nuclides.

Atmospheric Input

There are no specific studies of the direct input of atmospherically derived natural radionuclides to the study area. As with the input of fallout isotopes, reliance must be placed on studies of input in adjacent areas — either on land or in coastal areas. For example, Krishnaswami et al. (1980) have used atmospherically derived ^{7}Be, ^{210}Pb, and Pu to determine particle mixing rates and accumulation of sediments in marine (Long Island Sound) and freshwater sediments (Lake Whitney, Connecticut) in the coastal area close to the north end of the study region.

Water Column Studies

Sources from Atmosphere, Rivers, and Sediments. As noted above, atmosphere sources include cosmogenic nuclides and ^{210}Pb produced by outgassing of ^{222}Rn from its ^{226}Ra parent in soils and sediments of the continents. Rivers represent sources of soluble natural nuclides to the oceans (e.g., U and ^{40}K) and of insoluble nuclides associated with particulate material. Marine sediments represent both a source and a sink for naturally occurring nuclides. ^{226}Ra and ^{228}Ra produced by Th parent decay in sediments diffuse out of sediments into overlying waters. This process is especially important in shallow coastal areas and deep ocean areas where the dispersal of the diffused species can be traced by its radioactive decay constant. Although its half-life is short (3.8 days), ^{222}Rn diffusion into bottom water following its production by ^{226}Ra decay can be used to study vertical mixing. Some studies of these nuclides and processes have taken place in the study area. Li et al. (1977) studied the flux of ^{228}Ra in shelf and slope areas of the New York Bight and off Cape Hatteras, and they concluded that the sediment flux of ^{226}Ra to the surface ocean was of similar magnitude as the flux from abyssal sediments to the deep ocean. Biscaye et al. (1980) and Carson et al. (1979) report on some of the sources from underlying sediments and controls on vertical mixing of ^{222}Rn in shelf and slope waters of the New York Bight. A general description of the open ocean distribution of ^{226}Ra has been reported by Broecker et al. (1976) based on GEOSECS Atlantic Ocean stations. One of these, an intercalibration station, is located near the eastern boundary of the study region and is representative of the near margin oceanic water column ^{226}Ra distribution (Chung et al., 1974).

Advection and Diffusion of Chemically Inert Radionuclides. Strictly speaking there are few truly chemically inert naturally occurring radionuclides. Most of the metals, such as Th, Pb, Bi, Pa, are particle active and participate more in processes controlled by interaction with particulate phases of the water column. Even ^{40}K and U, which are conservatively distributed in the ocean and vary with salinity, have some chemical reactivity. U is incorporated into calcium carbonate and ^{40}K can enter into ion exchange reactions in clay minerals. So the advection/diffusion behavior of these elements occurs on long timescales reflecting their long oceanic residence times.

^{222}Rn is the only natural radionuclide which is truly chemically inert. In surface waters ^{222}Rn is deficient with respect to its ^{226}Ra parent as it diffuses into the atmosphere (Cochran, 1982) and has found use as a tracer for study of gas exchange of the surface ocean with the atmosphere. As noted above, there is a flux of ^{222}Rn to bottom waters by diffusion from the underlying sediments. This process has been used to determine vertical mixing rates in the deep sea, and its use has been recently reviewed (Cochran, 1982). Although both processes undoubtedly are active in the study area, the only applications within the study area are those reported by Biscaye et al. (1980) and Carson et al. (1979) for the ^{222}Rn diffusion studies in bottom waters of the shelf and slope of the New York Bight.

Particle Fluxes of Reactive Nuclides. Direct measurement of reactive nuclide transport to ocean sediments can be achieved through analysis of material collected in sediment traps. Such studies as reported by Brewer et al. (1980) in the open equatorial North Atlantic have been used to calculate isotopic fluxes of the reactive natural series nuclides. No such measurements have been made in the study area.

Indirect measurements the flux of reactive nuclides can be derived from water column measurements of their deficiencies with respect to their radioactive parents. For example, the deficiency of ^{210}Pb relative to ^{226}Ra in the deep ocean may reflect scavenging at the boundaries (Bacon et al., 1976). Similarly, the extent of the deficiency of ^{228}Th relative to its ^{228}Ra parent can be used as a measure of the residence time of Th in a given water mass. Broecker et al. (1973) found an average activity ratio for $^{228}Th/^{228}Ra$ in the World Ocean of 0.21 (some of their measurements were in surface waters of the study region). They suggested that the calculated removal time for Th of 0.7 years could provide a basis to predicted the rate of removal of reactive pollutants released to the surface ocean from coastal areas. Feeley et al. (1980) found that off New England higher ratios in Slope Water than in shelf waters implied longer residence times (3-4 months in Slope Water as against 1 month in shelf waters). When compared with the 0.7-year average for World Ocean surface water this shows that scavenging of reactive radionuclides (and by analogy reactive nonradioactive pollutants) is much more rapid in shelf and slope areas, presumably because of relatively high particle fluxes.

Sediments

Tracers of Sedimentation, Mixing and Diagenetic Processes. Because of their chemical reactivity and range of half-lives, nuclides of the natural series have been used widely to study these processes. Depending on the nature of a sediment and the half-life of the nuclide, its distribution in a sediment core may reflect 1) its mixing profile by sediment in-fauna, 2) radioactive decay after delivery, 3) the effects of diagenetic processes within the sediment, or 4) a combination of all or some of the first three. Biological mixing and sedimentation rates are derivable when 1) and 2) are the sole or controlling features.

^{234}Th ($T\ 1/2 = 24$ days) in excess of its ^{238}U parent has been used to evaluate short term particle reworking and diagenetic rates in sediments of Long Island Sound (Aller and Cochran, 1976). ^{7}Be and ^{210}Pb have been used to derive mixing and accumulation rates in sediments in the same area (Krishnaswami et al., 1980). It seems clear that in continental shelf and slope sediments, biological mixing is a dominant mechanism controlling the distribution of nuclides with half-lives less than 100 years. ^{14}C should determine accumulation rates within marine sediments because of its relatively long half life, but recent ^{14}C input from atmospheric weapons testing has obscured the record of cosmogenic ^{14}C and made determination of sediment accumulation rate more difficult. Instead, bomb ^{14}C can be used a tracer of particle mixing and organic material diagenesis in the biologically mixed zone of sediments. Problems of this sort in dating fine-grained sediments on the continental shelf south of Cape Cod have been addressed by Bothner and Bacon (in press). Biological mixing rates derived from ^{228}Th, ^{210}Pb, and $^{239,\ 240}Pu$ distributions(Santschi et al., 1980) were used to delineate the depth below which biological mixing ceased.

There have been comparatively few studies using natural series radionuclides as tracers in slope sediments of the study area. DeVito (1981) used ^{210}Pb (as well as bomb $^{239,\ 240}Pu$ and ^{137}Cs) to study sediment accumulation and mixing in cores from one location on the North Carolina continental slope. Brownawell and Sayles (WHOI) have unpublished data for ^{210}Pb (and other radionuclides) in a series of cores across the continental slope of the Mid-Atlantic region of the study area. The U.S. Department of Energy is currently supporting a 10-year multi-disciplinary program of study of exchange processes across the continental shelf-slope boundary in the northern region of the study area. Included in this program will be a number of studies of sediment mixing, accumulation and diagenesis in several series of cores taken in lines across the slope. These studies should represent a substantial increase to the data base for these kinds of sediment/radionuclide interaction processesin the Atlantic slope region.

Interstitial Water. To date there have been few studies of natural series radionuclides in interstitial water to parallel those of major ions, trace elements, nutrients, organics etc. Most sediment studies have focussed on the distribution of nuclides within solid phases of the sediments. Presumably this has been a consequence of the analytical difficulties in making measurements with the small amounts of water generally recoverable from marine sediments — especially by *in situ* sampling techniques. Recent work has demonstrated that it is possible to obtain several liters of interstitial waters from 2 cm horizons of sediments collected in shallow coastal waters and to make measurements of their natural or artificial radionuclide concentrations (Sholkovitz et al., 1983). An extension of this kind of approach to the sediments of the study area could provide valuable data which speak to the nature of the diagenesis of natural series radionuclides in the sediments of the U.S. Atlantic Continental Rise.

The natural series nuclides for which fluxes to the overlying water have been established are ^{222}Rn, ^{226}Ra, and ^{228}Ra.

^{222}Rn, produced by decay from its parent ^{226}Ra, has been studied diffusing out of New York Bight shelf and slope sediments (Biscaye et al., 1980; Carson et al., 1979). Kadko (1980) noted that deep sea sediments have higher Rn fluxes than do shallow, rapidly accumulating sediments, because of their elevated ^{230}Th and ^{226}Ra concentrations. In studies of the Hudson River estuary, it has been argued that surficial stirring can produce enhanced Rn fluxes to bottom waters (Hammond et al., 1977). Similarly, Martens et al. (1980) have argued that the enhanced Rn fluxes which they found from sediments of Cape Lookout Bight (North Carolina) can be explained by a biogenic formation of methane bubble tubes.

^{226}Ra is produced within marine sediments through decay of its ^{230}Th parent. The flux of this nuclide to bottom water is responsible for oceanic water column distributions noted in an earlier section. An account of the factors which affect the deep-sea sediment flux of ^{228}Ra is given by Cochran (1980, 1982). Li et al. (1977) showed that near-shore sediment fluxes of ^{226}Ra are of a magnitude large enough to be a major influence in the input of ^{226}Ra to waters of the surface ocean. ^{228}Ra is produced within the sediment-water column through decay of its ^{232}Th parent. A review of its geochemistry has been given recently by Cochran (1982). Its relatively short (5.8 year) half-life restricts the time and, hence, distance over which it can be used to trace water mixing after it diffuses into overlying water from marine sediments.

Organism Uptake

We are not aware of any reported measurements of natural series nuclides in either planktonic or benthic organisms collected in the study area. There is no question that uptake of these nuclides takes place in the biota, but clearly it is not an area which has received much attention. Cherry and Shannon (1974) have reviewed the literature on the uptake of alpha-emitting radionuclides by marine organisms. The nuclide which exhibits the greatest biological affinity in terms of its biological concentration enhancement is most likely ^{210}Po. Cochran (1982) reviews several papers on the subject and cites ^{210}Po excesses (relative to its ^{210}Pb parent) in phytoplankton and zooplankton. A recent paper on the natural radiation dose to marine organisms (Cherry and Heyraud, 1982) concluded that mesopelagic ocean biota receive relatively large doses of radiation, especially from ^{210}Po. This fact has been used to support certain practices associated with ongoing or projected plans for disposal of radioactive materials in the ocean.

Hydrocarbons

J.W. Farrington

Monitoring and research concerned with hydrocarbons in segments of the Western North Atlantic ecosystem have focussed primarily on estuarine, near-shore, and continental shelf locations. Initial measurements during the early 1970s documented the presence of elevated concentrations of fossil fuel compounds in surface sediments and organisms of these areas compared to a few measurements in open ocean areas including the ACSAR area. Surveys of input sources of fossil fuel compounds to the marine environment during the early 1970s were the first efforts to determine absolute and relative magnitudes. These were reviewed and summarized by the National Academy of Sciences report (1975), which confirmed the need for emphasis on the nearshore measurements since a majority of chronic inputs were in those areas. The emphasis of initial Outer Continental Shelf (OCS) leasing activities and accompanying environmental studies programs on the continental shelf resulted in most of the funding and hence most of the research and monitoring being focussed in the shelf areas during the period 1975 to the present. Thus, the meager literature from the ACSAR area provides a stark contrast to the more voluminous data for the estuarine, nearshore, and continental shelf areas. Nevertheless, this information coupled with studies of hydrocarbons in the continental shelf areas and studies of other organic compounds in the continental slope areas provide some guidance to the inputs, distributions, and fates of hydrocarbons in continental slope areas.

We will review first the general knowledge of hydrocarbons in the marine environment in order to provide a framework for presentation of the data of the study area.

Sources of Hydrocarbons

Sources of hydrocarbons in the marine environment have been discussed in several reviews (Farrington and Meyers, 1975; NAS, 1975; Clark and Brown, 1976; Whittle et al., 1977; Neff, 1979; Geyer, 1980; Wakeham and Farrington, 1980). One important aspect of environmental quality studies related to petroleum exploration and production activities is to distinguish various sources of hydrocarbon input to the region in question. The various criteria for distinguishing sources of input are discussed in NAS (1975), Wakeham and Farrington (1980), Mackenzie et al. (1982), and Farrington (1980). The various sources of input are as follows.

Biological Sources

All organisms examined to date have the capability to biosynthesize or accumulate hydrocarbons from their food and habitat. These hydrocarbons are usually of simple structure. Hydrocarbons originating in land biota are often transported to the marine environment by various means as will be reviewed later in this section.

Geochemical Sources

Hydrocarbons are often produced by transformations of precursor molecules, especially in surface sediments during early diagenesis of organic matter in aquatic and terrestrial ecosystems (Hunt, 1979; Tissot and Welte, 1978; Mackenzie et al., 1982).

Ancient sediment outcrops contain hydrocarbons deposited in these sediments or produced over long periods of diagenesis (Hunt, 1979; Tissot and Welte, 1978; Mackenzie et al., 1982). The exposure of these sediments to subareal or subaqueous weathering contributes hydrocarbons to rivers and directly to the marine environment. If the organic matter of the ancient sediment is sufficiently mature in the sequence of diagenesis and mutagenesis towards petroleum or coal production, then distinguishing these inputs from recent contemporary petroleum inputs to sediments will be very difficult at low concentrations of compounds.

Natural oil seeps contribute petroleum hydrocarbons to the marine environment. There are no known seeps in the ACSAR study region. A few have been reported in nearby deeper water areas by Wilson et al. (1974) but the evidence is not discussed. A curious series of water samples near the Caribbean with high contentrations of hydrocarbons has been interpreted as a transient subsurface seep to the south (Harvey et al., 1979). Natural forest and grass fires are thought to have contributed hydrocarbons — especially polynuclear aromatic hydrocarbons — to ecosystems since before man's farming and industrial activities.

Anthropogenic Sources

There are several routes for the input of petroleum (Table 5.6). The suite of compounds coming from incomplete combustion of fossil fuels (coal, gas, petroleum) contains many of the same compounds as found in petroleum. Coal spills must also be kept in mind especially when considering analyses of sediments for low concentrations of petroleum hydrocarbons. The composition of sediment extracts containing coal can be quite similar to that expected for petroleum (Farrington, 1980; Tripp et al., 1981; Prahl, 1982).

Table 5.6 Estimates for Petroleum Hydrocarbon Input to the Oceans (from NAS, 1975)

Source	NAS Workshop (1973) (mta)*
Marine transportation	2.133
Offshore oil production	0.08
Coastal oil refineries	0.2
Industrial waste	0.3
Municipal waste	0.3
Urban runoff	0.3
River runoff**	1.6
Subtotal	4.913
Natural seeps	0.6
Atmospheric rainout***	0.6
Total	6.113

*Millions of tons per annum.
**PHC input from recreational boating assumed to be incorporated in the river runoff value.
***Based upon assumed 10% return from the atmosphere.

Distributions of Hydrocarbons

Gases: C_1-C_4

Atmospheric and Water Column. There are apparently no published measurements of methane (C_1) or the other hydrocarbon gases in the atmosphere and water column of the ACSAR area since 1976. There are probably several industry-underway sniffer measurements but these are not available to us. Earlier data reported by Swinnerton and Lamontagne (1974) bracket the study region on a transect off the Chesapeake Bay (Table 5.7).

Biota. Gas measurements in biota are rarely made because water solubilities, exchange to the atmosphere, and the biochemistry of the majority of marine organisms make it unlikely that appreciable quantities of methane will accumulate in pelagic or benthic organisms of the continental slope area with the exception of microorganisms.

Sediments. Gas distributions have been reported in a number of deeper (>1 km depth) sediment samples from wells along the Atlantic continental shelf (Hathaway et al., 1979). Sources of methane in surface sediments appear to be mainly biological while sources in deeper sediments could be both biological and petrogenic (Hathaway et al., 1979). Concentrations range from non-detectable to 4.0 x 10^5 ppm by volume.

In summary the distributions of gases in the atmosphere and water column of the ACSAR region are not well known in terms of spatial and temporal variability. Extrapolations from measurements and research from other areas of the North Atlantic, however, should provide reasonable clues (Scranton and Brewer, 1977; Lamontagne et al., 1973; Scranton, 1977; Ehhalt, 1978; and references cited therein). Particular attention should be focussed on possible inputs of methane from patches or areas of temporarily high biological productivity and lower oxygen on the water column as possible sources of temporary high methane concentrations (Scranton and Brewer, 1977). The reports by Hathaway et al. (1979) and Hunt and Whelan (1979) on C_1 to C_4 hydrocarbons in deeper sediments of the Blake-Bahama Basin provide information about the distributions and amounts of these compounds in sediments.

The main sources of methane input to the region based on extrapolation from other areas appear to be early diagenetic processes (Sansone and Martens, 1982) and petrogenic sources (Hathaway et al., 1979; Hunt and Whelan, 1978). The other gases, such as ethane and propane, are thought to originate mostly from diagenetic processes deep in the sediments or from anthropogenic activities. However, recent re-evaluations based on new data from recently deposited surface sediments (Hunt and Whelan, 1979) suggest early diagenetic biological and chemical reactions as the source.

It suffices to state that low concentrations of methane and other volatiles can have the following possible sources in the region: biogenic in the water column and sediments, petrogenic from seepage from deep sediments or exposed ancient sediment outcrops, and advection from nearshore areas where anthropogenic activities release these compounds to the oceans. Removal processes active on these compounds based on general knowledge extrapolated to this area are: biological consumption, advection out of region, and loss to the atmosphere (Scranton and Brewer, 1977; Sansone and Martens, 1982).

Table 5.7 Light Hydrocarbons in Surface Waters, Atlantic Ocean (see Swinnerton and Lamontagne, 1974).

	Concentration Units nanoliters/liter							
Descriptive Location, No. of Samples ( ) and Date	CH_4	C_2H_6	C_2H_4	C_3H_8	C_2H_6	Iso and $n-C_4H_{10}$	Cl no.	Coordinates
50. Lower Sargasso Sea (5) 5/71	44	0.30	6.8	0.10	2.0	Trace	0.6	30°14′N 70°09′W
51. Transit – Sargasso to Trinidad (6) 5/71	40	0.20	6.4	0.09	2.0	Trace	0.5	26°43′N 68°03′W
52. Transit – Sargasso to Trinidad (7) 5/71	38	0.13	6.5	0.06	1.5	Trace	0.4	22°49′N 66°47′W
53. Transit – Sargasso to Trinidad (5) 5/71	39	0.12	5.1	0.07	1.3	Trace	0.4	18°22′N 63°24′W
54. East of Trinidad (4) 5/71	38	0.26	2.7	0.18	1.3	Trace	0.6	11°28′N 60°29′W
55. Transit – Bermuda to Norfolk (1) 12/69	65	Trace	11.0	Trace	0.6	Trace	0.4	32°14′N 64°37′W
56. Transit – Bermuda to Norfolk (1) 12/69	46	Trace	3.5	Trace	0.3	Trace	0.3	35°12′N 67°53′W
57. E. of Norfolk, Va. Shelf (1) 12/69	280	1.0	3.3	0.10	0.3	Trace	2.7	36°36′N 74°41′W
58. Trinidad Shelf (2) 5/71	108	0.40	2.7	0.10	0.9	Trace	1.1	11°28′N 60°22′W
59. N. of Lesser Antilles Isls. (1) 4/69	36	Trace	3.4	Trace	0.7	Trace	0.2	17°50′N 61°30′W
60. N. of Lesser Antilles Isls. (2) 4/69	57	Trace	2.4	Trace	0.8	Trace	0.4	16°50′N 60°54′W
61. E. of Lesser Antilles Isls. (1) 4/69	42	Trace	2.3	Trace	0.5	Trace	0.3	16°08′N 60°26′W
62. E. of Lesser Antilles Isls. (1) 4/69	40	Trace	2.1	Trace	0.6	0.05	0.3	15°21′N 59°55′W
63. Near Barbados (1) 4/69	31	Trace	2.4	Trace	0.6	Trace	0.2	13°13′N 59°07′W
64. R. of Trinidad (1) 4/69	33	Trace	1.1	Trace	0.4	Trace	0.2	10°38′N 60°05′W
65. E. of Trinidad (1) 4/69	39	Trace	0.7	Trace	Trace	Trace	0.3	10°38′N 60°05′W
66. E. of Trinidad (8) 4/69	36	Trace	1.8	Trace	Trace	Trace	0.2	10°38′N 60°05′W
67. Mid Atlantic (1) 5/66	40	Trace	9.9	0.20	0.6	Trace	0.5	52°35′N 20°09′W
68. E. of N. Carolina Coast (1) 6/68	37	1.0	3.7	Trace	0.5	Trace	0.9	33°40′N 74°47′W
69. E. of S. Carolina Coast (1) 6/68	37	1.1	3.6	0.10	0.4	Trace	1.1	32°58′N 74°52′W
70. E. of Northern Florida (1) 6/68	62	1.0	6.8	0.10	0.9	Trace	1.2	30°08′N 75°18′W
71. E. of Southern Florida (1) 6/68	34	1.1	3.0	Trace	0.8	Trace	1.0	26°35′N 74°43′W
72. Gulf Stream – Miami (10) 1972	59	0.60	6.0	0.70	1.5	0.80	1.5	Jan. thru June
73. E. of Bahaman Islands (4) 7/73	36	0.21	6.9	0.08	2.2	0.05	0.5	25°20′N 75°11′W
74. Exuma Sound (1) 6/68	45	1.0	3.8	0.10	1.0	Trace	1.1	24°00′N 75°30′W
75. Grand Caicos Isl. (1) 6/68	48	1.0	3.0	Trace	0.8	Trace	1.0	22°44′N 71°43′W
76. Exuma Sound (3) 7/73	39	0.40	7.7	0.09	2.3	Trace	0.6	24°33′N 76°14′W
77. Miami – Nearshore (10) 1972	190	1.5	16.0	0.94	4.7	2.20	3.2	Jan. thru June
78. Miami – Nearshore (9) 6/72	280	1.4	26.0	1.20	4.8	4.40	4.0	24 – hr sampling period
79. Miami – Dockside (4) 1972	1,300	2.6	30.0	2.8	11.0	4.70	13	Jan. thru June
80. N.E. of Windward Passage (1) 7/73	39	0.21	10.0	0.16	2.2	Trace	0.6	21°58′N 72°52′W
81. S.E. of Chesapeake Bay Entrance (1) 6/68	96	1.5	7.2	0.50	2.9	Trace	2.1	36°30′N 75°43′W

Volatile Compounds: C_5-C_{10}

The distinction between the analytical chemical operational definition of "volatile" compounds of C_5-C_{10} range and higher molecular weight compounds (greater than C_{10} molecular weight) results in an overlap in data reported in the literature for the C_{10} to C_{15} molecular weight range. Some of these M_2/M_4 will be discussed in the section on heavier molecular weight compounds.

There are no published data for volatile compounds in the ACSAR area; this includes atmosphere, water column, sediments and biota. This is not surprising since analyses for these groups of compounds in seawater have only been widely undertaken since 1976. General discussions of the biogeochemical cycles of these compounds are provided by Wakeham et al. (1982), Gschwend et al. (1982), Sauer (1981), and Mantoura et al. (1982). Several compounds, especially the n-alkanes, branched alkanes, and alkenes, have mixed biological, geochemical and anthropogenic sources. Aromatic compounds and several of the cycloalkanes have mainly mixed geochemical and anthropogenic sources. Removal processes for the region are the same as for the lighter gases.

Heavier Molecular Weight Compounds

Atmosphere. Duce and Gagosian (1982) have recently reviewed distributions of n-C_{10} to n-C_{30} alkanes in the atmosphere and estimates of inputs to the oceans. They point out that there are few data for n-alkanes on a global basis and no reliable data for aromatic compounds in the atmosphere over the ocean. None of the data they reviewed are within the study region. Unpublished data from Bermuda for one sampling period for May to December, 1973, are reported by Duce et al. (1974); the value is 2-4 μg of n-C_{14} to n-C_{32} hydrocarbons/SCM (standard cubic meter of air). Duce and Gagosian (1982) conclude that rain scavenging of particulate n-alkanes appears to be the primary transport path to the ocean. Since there are measurements in surface sediments of the region (see below) which implicate atmospheric inputs as a possible source to the overlying water column, and since petroleum inputs to the ocean via the atmospheric route have been considered a major input for several years (NAS, 1975; NOAA, 1978), the lack of reliable, more extensive data for hydrocarbons in the atmosphere and for atmospheric inputs is a major gap in our knowledge of the ACSAR region. The existing sampling and analytical technology certainly permit such measurements (Gagosian et al., 1982; Simoneit and Mazurek, 1981).

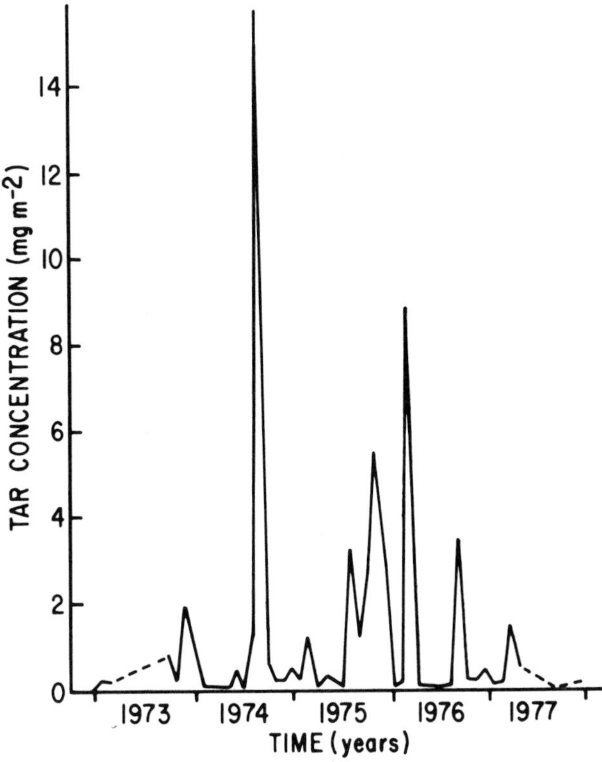

Figure 5.8 Monthly variation of the mean tar concentration (mg/m²) off New Jersey (5° square 116/3). Levy et al. (1981).

Seawater: Pelagic Tar. Pelagic tar has three possible sources: weathered oil from oil seeps (Geyer and Giammano, 1980; Spies et al., 1980), cleaning discharges or bilge discharges from routine tanker operations (NAS, 1975; Morris et al., 1976), and weathered spilled oil from tanker accidents or exploration and production activities (NAS, 1975). Although there is some controversy, it appears that much of the pelagic tar in the western North Atlantic is from tanker discharges associated with ballast and tank cleaning operations (Morris et al., 1976; Mommessin and Raia, 1974; McGowan et al., 1974). However, care must be taken (and has been taken in the references cited herein) not to classify all small nondescript organic blobs collected in neuston tows as tar of petroleum origin. Domestic and industrial waste products, such as plastic spheres carried out to sea as flotsam and dumped from ships at sea, also are present in neuston tows (e.g. see Mommessin and Raia, 1974; Carpenter, 1976; Colton et al., 1974).

The Marine Pollution (Petroleum) Monitoring Pilot Project (MAPMOPP) of the Joint International Oceanographic Commission/World Meteorological Organization has recently published the results of the program between 1975 and 1978 (Levy et al., 1981). This report contains data on visual slick sightings, pelagic tar concentrations (mg/m²) as determined by neuston tows and subsequent gravimetric determinations, and dissolved/dispersed hydrocarbons determined by U.V.-fluorescence of hexane extracts. These data demonstrate that substantial measurable quantities of pelagic tar are found in these areas (Figs. 5.8 and 5.9). The data are similar to the 1971-1974 data set for the western North Atlantic reported by Levy and Walton (1976). A more recent compilation of data for the North Atlantic has been presented by Stoner et al. (in prep.) which shows similar concentrations of pelagic tar in several transects through the study region (Fig. 5.9).

Pelagic tar comes in several shapes, sizes, colors and textures (Butler et al., 1973a). Tar particles are mostly dark brown or black, cm to tens of cm in diameter, with consistencies ranging from hard rubber to lubricating grease. Considering physical chemical evaporation, the profiles of hydrocarbon components in tar determined by gas chromatography, and previous work in the early 1970s, Butler (1975) suggested that the average residence time of pelagic tar in surface waters of the Sargasso Sea adjacent to the ACSAR area is on the order of one year. He also suggested that the numerous tar particles in the size range of .1 to 10 cm³ were formed within days to a few years from larger crude oil masses of the order of 1,000 cm³ median size.

Small tar particles distributed throughout the water column also have been reported by Morris et al. (1976) and Zsolnay (1977) for the Sargasso Sea adjacent to the ACSAR region. The size of these tar "specks" averaged around 3 x 10^{-6} cm³ (Zsolnay, 1977) to a range of <20 μm (Morris et al., 1976). It has been suggested in both papers that these small "flakes" or "specks" of tar in the water column originate from disaggregation or disintegration of tar particles. Concentrations of these filterable tar specks range from not detected ($<.01$ μg/l) to 30 μg/l with mean values in the 0.1 to 10 μg/l range depending on water depth (Morris et al., 1976).

In terms of present state-of-the-art (high resolution glass capillary gas chromatography mass spectrometry analyses)

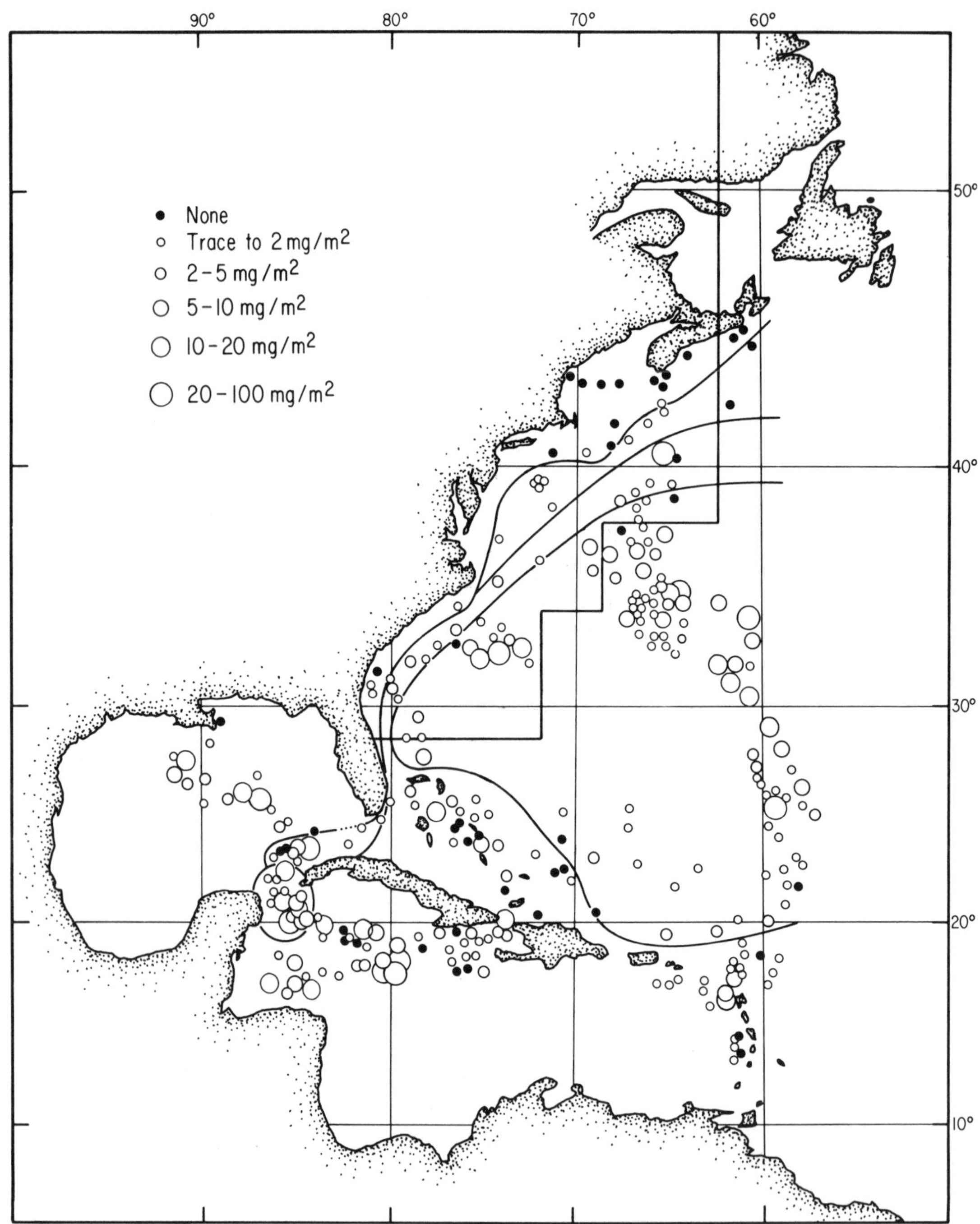

Figure 5.9 Geographical distribution of tar ball concentrations in western North Atlantic. (After Stoner et al. (1983).

pelagic tars are not well characterized chemically. There is enough information from low resolution gas chromatography (e.g. Morris et al., 1976; Butler, 1975; Mommessin and Raia, 1974) and column chromatography separations, followed by compound type mass spectrometry, to establish that most pelagic tar resembles a weathered and partially biodegraded crude oil. The exception is the frequent presence of wax inclusions resulting from precipitation of n-paraffins, presumably onto the sides of cargo tanks in oil tankers (Butler et al., 1973b). Pelagic tar in the ACSAR region has several fates: (1) disaggregation into smaller and smaller particles; (2) loss of components to evaporation and dissolution — especially the lower molecular weight compounds; (3) biodegradation; (4) direct deposition to sediments or incorporation into larger fecal particles by zooplankton and small fish and deposition (Morris et al., 1976) to sediments; (5) advection out of the region. Other than the estimate of an average residence time with a wide uncertainty of months to years by Butler (1975), and the hypothesis that disaggregation is an important process (Butler, 1975; Morris et al., 1976), few quantitative data are available concerning the fate of tar in this region.

Dissolved and Dispersed Compounds In The Water

The terms dissolved and dispersed compounds have been adopted because several studies have analyzed water samples without filtration to remove particulate matter. The term "dissolved" usually applies to a sample filtered through 0.4 to 1 μm glass fiber filters. This situation is the result of the existence of a continuous spectrum of organic matter in seawater from truly dissolved to very large particles. Separation into size classes is based primarily on what is technologically easily attainable coupled with research requirements. The importance of this issue becomes obvious when anyone attempts to model the biogeochemical behavior of hydrocarbons or other compounds in the marine environment since diffusion, air-sea exchange and biological uptake (as a few examples) can operate at different rates depending on the physical-chemical form of the compound.

Many of the data for dissolved/dispersed hydrocarbons in seawater come from the surface water surveys by the MAPMOPP program, by measurement of the fluorescence of aromatic hydrocarbons and then converting to concentrations of hydrocarbons. There has been and continues to be considerable controversy about how much credence should be given to data obtained by this technique. It appears to be a reasonable survey technique, but must be supplemented by more detailed and specific methods (IOC, 1982; Farrington, 1982; among others).

The MAPMOPP dissolved-dispersed hydrocarbon data from U.V.-fluorescence estimates in the ACSAR study region are sparse (Fig. 5.10). Gordon et al. (1974) report similar concentrations on a transect from Nova Scotia to Bermuda. Monaghan et al. (1974) and Brown and Hoffman (1976) report more dissolved/ dispersed hydrocarbon data for the Western North Atlantic. The concentration ranges are on the order of .1-10 μg/1 using solvent extraction and column chromatography techniques followed by quantification of hydrocarbons by infrared measurements. Several of these samples were analyzed by low resolution gas chromatography and compound type probe mass spectrometry. Brown and co-workers concluded that most of the hydrocarbons have compositions consistent with a crude oil origin. These data also have been recompiled and succinctly reviewed by Meyers and Gunnerson (1976).

It is instructive to compare the dissolved/dispersed hydrocarbon concentration data with the pelagic tar data of the previously discussed reports. The clear overlap of ranges of concentrations suggests that much of what is measured as dissolved/dispersed petroleum is small pelagic tar specks because most of the water samples were not filtered prior to analysis. Filtration is not a simple procedure because of the problem with adsorption of dissolved compounds on filters. Wade and Quinn (1975) determined that most hydrocarbons in the few water samples they collected in the Sargasso Sea were removed by filtration which would remove both "particulate" hydrocarbons and a portion of dissolved hydrocarbons adsorbed onto filters.

Dissolved/dispersed hydrocarbons are lost to the atmosphere, biodegraded, adsorbed to or incorporated into particulate matter and deposited in sediments, and advected out of the region.

Sediments

There have been more detailed analyses of heavier molecular weight hydrocarbons in the surface sediments of the ACSAR region than for any other component of the ecosystem. Farrington and Tripp (1977) and Farrington et al. (1977) reported on concentrations of alkanes, cycloalkanes and alkenes in samples of the upper 2 to 8 cm of sediments and transects on the slope and rise in and immediately southwest of the Hudson Canyon. These data are given in Tables 5.8 and 5.9. A combination of gas chromatographic analyses results was used to show that no more than 1-10 μg of petroleum hydrocarbon contamination existed per g of dry sediment for continental slope and rise sediments. However, we must caution that very recently deposited material, such as tar specks in fecal material, can be diluted by background material due to sampling of 2 to 8 cm depth which could include material more than a thousand years old, given the sediment accumulation rates in the area. The complex mixture of alkanes, cycloalkanes, and aromatic compounds which are indicative of petroleum contamination could also have been transported to the region from urban air pollution or grass and forest fires on the nearby continent. Early diagenesis of thousand year-old sediments or inputs from nearby outcrops of ancient sediments are also possible sources (Farrington and Tripp, 1977).

Regardless of the possible additional sources of input, it appears that no more than 10 μg/g of petroleum hydrocarbons are present in the upper 2 to 8 cm of sediments. There was one exception. A core from a station in the Hudson Canyon at 986 meters water depth yielded hydrocarbon concentrations in surface sediments of about 50 μg/g dry weight, most of which appeared to be of petroleum type. Farrington et al. (1977) analyzed several sections of this core and found a deeper section (90 to 94 cm) had hydrocarbon concentrations similar to those found in the other slope and rise samples. They suggested that transport of contaminated material from the New York Bight area via the Hudson trough and channel could explain the observed depth distribution in the core and the elevated concentrations.

La Flamme and Hites (1978) analyzed a few of the same sediment samples as Farrington and Tripp (1977) for polynuclear aromatic hydrocarbons (PAH). The qualitative distribution of the parent compound PAH and alkyl-substituted homologs clearly indicated that the predominant source of the PAH was pyrogenic, either combustion of fossil fuels or forest and grass fire inputs. Thus, two specific markers of hydrocarbon inputs from land are present in these surface sediments of the slope: land plant n-alkanes (Farrington and Tripp, 1977), and pyrogenic PAH (La Flamme and Hites, 1978). The mechanism of transport from land to the sediments is not known but must be either a combination of aeolian transport followed by deposition through the water

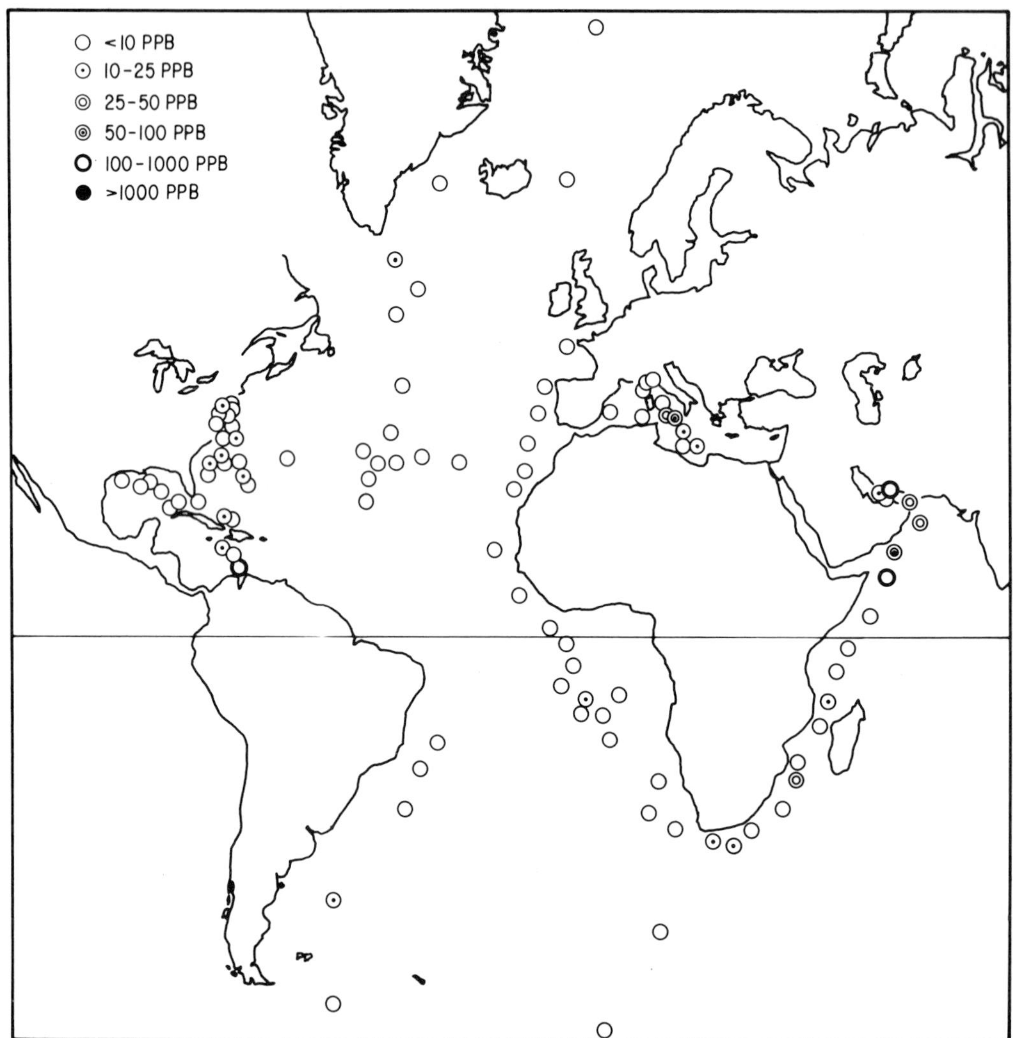

Figure 5.10 Total hydrocarbons in western Atlantic surface water samples. After Monaghan et al. (1974).

column or resuspension of nearshore sediments followed by advection or slumping to the slope region.

The fates of hydrocarbons in sediments of the region include: (1) burial in sediments as a geological deposit, (2) resuspension and release to the water column, or transport out of the area, and (3) biodegradation. None of these can be estimated with any certainty given current data.

Organic Carbon

Hinga et al. (1979) have reviewed the information on the cycle of organic matter in the western North Atlantic including the continental slope and rise areas. They concluded, based on few available measurements, that the vertical flux of organic matter was sufficient to satisfy the oxygen demand of the benthos at 3500 m depth. However, inputs in addition to the vertical flux of organic matter are required to satisfy benthic oxygen demand at shallower depths (600 m and 1300 m depths). A comparison of vertical flux of organic C with benthic oxygen demand and then with accumulation of organic carbon in surface sediments suggests that most of the vertical flux of organic carbon is consumed by the benthos. Gardner (1977) has presented similar results in regard to the flux of particulate orgainc carbon as measured in surface sediments. Deuser and Ross (1980) have provided important data about seasonal influences in the flux of organic matter to the nearby deep Sargasso Sea. Permuzic (1980) and Permuzic et al. (1982) have reviewed the nature and distribution of organic matter in surface sediments of the world's oceans including the ACSAR area.

Biota

We can find only two references to measurements of hydrocarbons in biota of the ACSAR area. Brown and Pancirov (1979) measured polynuclear aromatic hydrocarbons (PAH) in six Baltimore Canyon fish. The low concentrations of PAH in fish (Table 5.10) may signify that the fish were not

Table 5.8 Sampling data for data reported in Table 5.9. (Farrington and Tripp, 1977)

Station	Location	Date	Water Depth (m)	Sampling Gear
Abyssal Plain				
K19 – 4 – 9	30°01′N, 60°00′W	3/29/71	5250	Core
K33 – 2 – 6	32°25′N, 70°13′W	10/2/73	5465	Core
K19 – 5 – 3	33°39′N, 66°33′W	4/8/71	4950	Anchor Dredge
K19 – 5 – 4	35°19.6′N, 69°45.3′W	4/11/71	4900	Anchor Dredge
K19 – 5 – 5	37°44.7′N, 69°45.3′W	4/11/71	3950	Core
K19 – 5 – 6	37°41′N, 70°31′W	4/12/71	3923	Anchor Dredge
Continental Slope				
K19 – 5 – 7	38°10.5′N, 71°25.0′W	4/13/71	2950	Anchor Dredge
K19 – 5 – 9	38°53.5′N, 71°47.8′W	4/14/71	1830	Anchor Dredge
K19 – 5 – 13	39°16.1′N, 72°25′W	4/14/71	190	Grab
Hudson Canyon				
K33 – 2 – 8	38°10′N, 70°50′W	10/5/73	3785	Core
K33 – 2 – 9	39°05′N, 71°29′W	10/5/73	2626	Core
K33 – 2 – 10	39°27.8′N, 72°13.0′W	10/6/73	986	Core
K33 – 2 – 11	39°40′N, 72°29′W	10/6/73	137	Core
Hudson Channel				
K19 – 5 – 15	39°47.4′N, 73°00.1′W	4/16/71	79	Grab
G187DG	40°04.2′N, 73°28.5′W	4/17/72	78	Grab
G187MG	40°06.5′N, 73°33.3′W	4/18/72	67	Grab
K47 – 1 – 11	40°01.8′N, 73°33.3′W	2/5/75	38	Grab
K47 – 1 – 12	40°04.1′N, 73°29.0′W	2/5/75	72	Grab
K47 – 1 – 8	40°10.0′N, 73°42.7′W	2/5/75	60	Grab
K19 – 5 – 16	49°17.0′N, 73°47.5′W	4/18/71	54	Grab
New York Bight				
K19 – 5 – 18	40°24.0′N, 73°47.5′W	4/18/71	39	Grab
K47 – 1 – 6	40°25.7′N, 73°48.1′W	2/5/75	28	Grab
G.B.S.	40°26.3′N, 73°48.1′W	4/17/72	27	Grab
K19 – 5 – 20	40°22.6′N, 73°37.1′W	4/18/71	23	Grab
Continental Shelf				
K19 – 5 – 21	40°28′N, 73°28′W	4/18/71	23	Grab
K19 – 5 – 22	41°18.3′N, 70°50.3′W	4/19/71	37	Grab

exposed to sources of PAH or that if exposed and PAH taken up, the fish metabolized the PAH. Several species from the study area possess the requisite enzyme systems to metabolize PAH (Stegeman, 1981).

Boehm and Hirtzer (1982) have reported on several measurements of hydrocarbons in biota on the inshore boundary of the study region, the continental shelf. These samples came from the Northeast Monitoring Program of NOAA-NMFS. The majority of the fish came from the Georges Bank — Mid-Atlantic Bight area mostly from water depths shallower than 200 m. Concentrations of petroleum hydrocarbons estimated from the gas chromatographic analyses and measurements by GC/MS were 1-327 μg/g dry weight (ppm) depending on species and location. Individual aromatic hydrocarbons or groups of aromatic hydrocarbons contained 0.2 to 85 μg/g dry weight (ppb).

Similar concentrations for petroleum hydrocarbons were reported in species from the southeast and Gulf portions of the U.S. continental shelf (Boehm and Hirtzer, 1982). The concentrations reported by these authors are similar to the survey measurement concentrations of petroleum hydrocarbons for two fish liver samples in the George's Bank region (Farrington and Teal, 1972).

Teal (1976) reported on survey measurements of hydrocarbons in benthic animals from the Nares Abyssal Plain (25°N, 62°W; 5,500-5,800 m) well to the east and south of our study region. Two of four animals analyzed contained indications of petroleum contamination at the 8 to 15 mg/g wet weight. If we use a reasonable ratio of wet weight to dry weight of 8 to 10, then concentrations would be in the range of 64 to 150 mg/g dry weight.

Burns and Teal (1973) and Morris et al. (1976) report on concentrations of hydrocarbons in biota of the Sargasso Sea. The samples contained the expected distribution of biogenic hydrocarbons. They noted the frequent presence of a complex mixture of hydrocarbons resembling weathered and biodegraded petroleum at concentrations ranging from 0.2 to 26 g/g wet weight.

Admittedly these data are sparse, but it is interesting to note that the limited number of samples analyzed both inshore and offshore of the ACSAR region have similar ranges of concentrations of petroleum hydrocarbons. Despite the survey nature of this data, this indicates that at least some species on the slope and rise region are probably contaminated with petroleum hydrocarbons.

Special Considerations

Many other contaminant organic compounds, such as DDT, PCBs, hexachlorobenzenes and freons have physical-chemi-

Table 5.9 Lipid and Hydrocarbon Concentrations in bottom sediments from ACSAR. (Farrington and Tripp, 1977)

Station	Extractable Lipid Conc. μg/g dry weight	Hydrocarbon μg/g dry weight f1	f2	f3	Total	(f1 f2 f3 / Lipid) x 100	(f1 / Lipid) x 100
Abyssal Plain							
K19-4-9	46.2	1.2	--	--	--	--	2.6
K19-5-3	21.0	1.0	0.2	0.1	1.3	6.2	4.8
K33-2-6	94.4	2.0	N.D.	4.0	--	--	2.1
K19-5-4	63.5	5.0	1.9	0.6	7.5	12	7.9
K19-5-5	149	4.7	--	5.1	--	--	3.2
K19-5-6	99.6	2.4	2.1	--	--	--	2.4
Continental Slope							
K19-5-7	96.4	5.9	4.6	4.5	15	16	6.1
K19-5-9	93.6	3.6	1.7	5.2	11	12	3.9
K19-5-13	159	5.6	2.4	8.1	16	10	3.5
Hudson Canyon							
K33-2-8	140	5.2	4.3	9.7	19	14	3.7
K33-2-9	226	4.7	1.7	7.1	14	6.0	2.1
K33-2-10	1440	55	6	7	68	4.7	3.8
K33-2-11	138	4.4	1.9	9.0	15	11	3.2
Hudson Channel							
K19-5-15	153	8.6	4.1	6.0	19	12	5.6
G187DG	991	58	15	40	113	11	5.9
G187MG	1380	93	32	66	191	14	6.7
K47-1-11	41.9	3	10	6	19	45	7.2
K47-1-12	672	25	7	28	60	8.9	3.7
K47-1-8	351	48	11	31	90	26	13.7
K19-5-16	2751	399	66	95	560	20	14.5
New York Bight							
K19-5-18	9070	1800	620	480	2900	32.0	20
K47-1-0*	2364	440	132	132	704	29.8	18.6
G.B.S.	3911	1200	250	293	1743	44.6	30.7
K19-5-20	166	25	4.5	5.8	35	21.3	15
Continental Shelf							
K19-5-21	159	6.6	2.3	7.0	1.6	10.1	4.2
K19-5-22	108	3.5	1.2	.29	5.0	4.6	3.3

*Losses of about 30% prior to weighing.

Table 5.10 Summary of PHN Hydrocarbons in Marine Tissues (ppb, wet weight). Brown and Pancirou (1979)

	Location Lat., Long.	Depth (m)	Benz(a)-anthrocene	Benzo(a)-pyrene	Pyrene	Methyl-pyrene	Other PNA's*	Dry wt./Wet wt.
Summer flounder (*Paralichthyl dentatus*)	37°53.5'N, 74°10.6'W	102	1	1	2	1	1.1-1.5	0.21
Scup (*Stenotomus chrysops*)	38°11.4'N, 74°09.1'W	66	1	1	2	1	1-3	0.26
Black Sea Bass (*Centropritus stricta*)	38°33.2'N, 74°05.0'W	57	1	1	1	1	0.1-1	0.23
Butterfish (*Paprilus triacanthus*)	38°37.7'N, 74°18.5'W	39	20	5	1	2	2-13	0.29
Sea Scallops (*Placopeskin magellianicus*)	38°47.0'N, 73°22.8'W	78	1.1	1	4.1	2.7	5-8.7	
Red Hake (*Arophysate chuss*)	38°47.0'N, 73°22.8'W	78	0.3	11	5	6	2-9	

*Includes fluoranthene, chrysene, triphenylene, benzo(e)pyrene, and perylene.

cal properties of volatility, solubility, air-sea exchange, and solid solution interactions similar to petroleum compounds. Although these compounds are not specifically within the stated purview of this review, we would be remiss in not pointing out that there are measurements for these compounds in the atmosphere, waters, sediments and biota in the ACSAR area which clearly document the incursion of these man-made compounds throughout the marine ecosystem of the area (NAS, 1979; Harvey et al., 1972, 1974a,b; Harvey and Steinhauer, 1976; Jonas and Pfaender, 1976; Bidleman and Olney, 1975; Bidleman et al., 1981; Stout, 1980; Sims et al., 1977; Boehm and Hirtzler, 1982; among others).

The presence of these compounds with an overwhelmingly predominant terrestial origin is clear proof that compounds can move from land via atmospheric or water advective processes (or a combination of processes) to the slope and rise. These data coupled with the aforementioned data for pyrogenic PAH and land plant n-alkanes offer substantial support that petroleum hydrocarbons and other fossil fuel compounds can be transported offshore. The reported detection of petroleum type hydrocarbons in sediments and biota of the region is not surprising given the possibility of long range inputs from land and the ubiquitous stock of pelagic tar in the region.

The biogeochemical cycle for organic pollutant inputs is derived from a number of studies and knowledge of oceanic processes. Despite the variety of potential and real sources of input for petroleum to the study region, it should be possible with careful high resolution analytical techniques to measure inputs from exploration and production activities if care is taken in pre- and post-operation sampling and analyses. The revolution in more extensive application of high resolution glass capillary gas chromatography mass spectrometry measurements which has been applied to recent Department of Interior (DOI) OCS environmental studies programs on the continental shelf areas so far has not been applied to more than a few samples in the slope and rise.

6 Biological Oceanography

P.H. Wiebe, E.H. Backus, R.H. Backus, D.A. Caron, P.M. Glibert, J.F. Grassle, K. Powers and J.B. Waterbury

Biogeographic Context

The ACSAR study area crosses the western extremities of two biogeographic regions — the North Atlantic Temperate Region and the North Atlantic Subtropical Region. The Gulf Stream divides the study area into northeastern and southwestern pieces as it turns east from Cape Hatteras; it is the northern edge of this current that forms the boundary between temperate and subtropical regions in this part of the North Atlantic. Nowhere else in the Atlantic is there a stronger biotic contrast than at this boundary, thus preventing description of the ACSAR area as a homogenous region. The northeastern part of the study area lies in the temperate province called "Slope Water", the southwestern part in the subtropical province called "Northern Sargasso Sea" (Backus et al., 1977). Although the Gulf Stream is considered to be a part of the northern Sargasso Sea, it is biotically distinct (Jahn and Backus, 1976).

The northern edge of the Gulf Stream forms a sharp southern limit to the normal range of many cold-water animals. The transgression of this boundary by such animals can be explained readily in terms of transport by cold-core rings. The reverse effect is not so clear; that is, the northern edge of the Gulf Stream does not form so sharp a northern limit to the range of warm-water animals inhabiting the northern Sargasso Sea. For example, it is far more common to find midwater fishes of various warm-water distribution patterns in the Slope Water than it is to find fishes of cold-water distribution patterns in the northern Sargasso Sea. This can probably be related (1) to differences in the effects that cold-core rings and warm-core rings have on the environments that they invade and (2) to differences between the pressures exerted by the two environments on foreign organisms.

According to the Ring Group (1981), the area of the northern Sargasso Sea affected by cold-core rings is about 3×10^{12} m^2. On the other hand, the area of the Slope Water, into which warm-core rings intrude, is only about 0.5×10^{12} m^2 (Jahn, 1976). If mass is preserved by the formation of a warm- core ring for each cold-core ring formed, then the gross effect of warm-core rings on the Slope Water is about six times that of cold-core rings on the northern Sargasso Sea. Thus, even if the midwater fish biomass of the Slope Water were three times that of the northern Sargasso Sea, expatriates from the Sargasso Sea in the Slope Water would be twice as available as Slope Water expatriates in the Sargasso Sea, assuming equal survival and uniform dispersal.

But a warm-core ring cannot be viewed simply as a vehicle by which plants and animals are carried into a foreign environment. If a warm-core ring has a volume of 3×10^{13} m^3 (Ring Group 1981) and eight rings per year enter the Slope Water (*ibid.*), whose volume is 0.5×10^{12} m^2 x 1000 m or 0.5×10^{15} m^3 (Jahn, 1976), then about half of the Slope Water will be replaced each year by the rings if the rings were to mix into the Slope Water completely. The common fate of a warm-core ring is to drift westward and coalesce with the Gulf Stream in the vicinity of Cape Hatteras after a lifetime of about seven months. There seem to be no estimates of how much of the average ring is mixed away in the Slope Water and how much of it is reabsorbed by the Gulf Stream. Thus, there is a large, virtually continual, input of Western North Atlantic Water from the northern Sargasso Sea into the Slope Water, as well as the input of some water of tropical origin from the Gulf Stream itself, that significantly contributes to the Slope Water's character as a habitat for plants and animals. It is mainly this contribution that makes the Slope Water, particularly that part of it which is of concern in the present report, biologically distinct from the other provinces of the North Atlantic Temperate Region (Backus et al., 1977) and possibly a place where certain organisms of otherwise warm-water distribution can reproduce as noted above.

Distribution of pelagic organisms depends critically on the differences in water mass while the comparative effects of the change in water depth from 200 to 4000 m across the study area are trivial. For benthic animals, however, almost the reverse is true: water depth is more important than the nature of the overlying water. Although benthic animals in the transition zone (250 to 300 m to about 1000 m) off New England may be affected by Gulf Stream rings (Grassle et al., 1979), most changes are independent of water masses. Changes with depth below the transition zone are more likely to result from changes in the amount of food reaching the bottom (Sanders and Hessler, 1969b; Rex, 1981; Rowe et al., 1982).

Microbiology

J.B. Waterbury and D.A. Caron

Microbiology in its broadest sense encompasses the biology of all those organisms too small to be seen by the naked eye, including the microalgae, the cyanobacteria (blue-green algae), the fungi, the protozoans, the bacteria and the viruses. The present report will be limited to a survey of our current knowledge of marine bacteria (including the cyanobacteria) and protozoans in the ACSAR area.

Bacteria

The bacteria possess the properties of procaryotic cells. Their enormous diversity is expressed physiologically. The principal forms of energy-yielding metabolism, aerobic respiration, photosynthesis and fermentation all exist in bacteria as well as in various groups of eucaryotic microorganisms. In addition, there are types of energy-yielding metabolism which are unique to bacteria, including variations of photosynthesis, anaerobic respiration, a large number of unique fermentations and the oxidation of reduced inorganic compounds. It is the existence of this physiological diversity among bacteria that makes assessing the ecological role of these organisms as a whole an extremely difficult, if not impossible, task.

The vast majority of the recent literature in marine bacteriology concerns studies that have been conducted on inshore waters. Bacteriological studies were conducted by Waksman and his collaborators off the coast of New England during the 1930s (See Zobell (1946) for a summary of this early work.) Their work was largely confined to the continental shelf and dealt with the enumeration of bacteria and correlating the presence and activity of bacteria with decomposition and the availability of organic matter, both in the water column and in sediments. The number of bacteriological studies that have specifically examined processes in the Slope Water off the eastern coast of the United States are limited.

Water Column

Hobbie and co-workers (1972) studied the distribution and activity of microorganisms at two stations within the ACSAR area. Station 1 (36°25′N, 74°43′W) was in Slope Water and station 2 (35°00′N, 73°00′W) was in the extreme western Sargasso Sea. The major conclusion from their data set was that the bacteria were not present in sufficient number to contribute a significant fraction of the living biomass or to contribute significantly to the respiratory processes measured. They also concluded that bacteria and not phytoplankton were responsible for heterotrophic uptake of the dissolved organic substrates tested. Their results indicated that the activity measured by the uptake of labeled organic substrates represented a small fraction of the total respiration as measured by respirometry and ATP and ETS activity. This led them to conclude that most of the observed respiration was non-bacterial or the bacteria were utilizing substrates they did not test. The organism responsible for the observed rates of respiration could not be identified. The actual bacterial cell counts at the two stations were low as a result of counting by phase contrast microscopy. If the acridine orange technique using epifluorescence microscopy had been available, the major conclusion drawn by the authors might have been different.

Ferguson and Palumbo (1979), in a study of the distribution of suspended bacteria in neritic waters south of Long Island, included one Slope Water station (39°40′N, 71°55′W). The bacterial profile showed concentrations approaching 10^6 cells ml^{-1} in the mixed layer, dropping to about 10^5 cells ml^{-1} below the thermocline and remaining relatively constant throughout the water column.

Burney et al. (1979) examined the effect of small scale nannoplankton and bacterioplankton distributions on concentrations of dissolved carbohydrates in the western Sargasso Sea. Water samples were collected from two isotherms at drogue buoy stations, one of which was in the study area (Station 1, 32°41′N, 74°31.6′W). The writers concluded that the combined activities of the plankton smaller than 20 μm regulated the dissolved carbohydrate concentration in the Sargasso Sea.

Packard and Williams (1981) compared the rates of respiratory oxygen consumption and electron transport activity (ETS) in surface water of the North Atlantic. Two of their stations (40°14′N, 67° 13′W and 40°12′N, 67°12′W) were in the ACSAR area. Respiration rates in Slope Water, 5.6 g O_2 day^{-1} m^{-2}, were slower than those measured in the Gulf of Maine. They concluded that oxygen consumption and ETS activity are related and that water column respiration exceeded primary production in July.

Cuhel et al. (1983) studied microbial growth and macromolecular synthesis at three stations in the northwestern Atlantic Ocean. Results indicate the utility of using inorganic nutrient uptake and subcellular incorporation patterns to measure growth and metabolism in natural microbial populations. The authors also stress the necessity of making time-course rather than end point incubations as evidenced by the marked deviation from linearity in many of their incubations.

Bacteria Associated with Particles ("Marine Snow"). The relative roles of free-living bacteria and bacteria associated with particles have been debated for some time. "Marine snow" seems to be present in the marine water column almost everywhere. The number and source of particles are variable, with the result that the degree of colonization by bacteria also varies. Field observations indicate that some flocculent aggregates are produced by zooplankton (Silver and Alldredge, 1981). They include feeding structures formed by larvaceans, pteropods, salps, veligers and polychaetes.

Jannasch (1973) reported that bacteria were absent from particles collected from surface waters near the ACSAR area (29°33′N, 67°35′W). He suggested that the absence of bacteria might be due to a recent origin or labile state of the particulate material. Hodson (1981), in a study including five stations within the ACSAR study area, observed that the per cell uptake of dissolved ATP by attached bacteria was one to two orders of magnitude faster than uptake by free-living bacteria. They suggested that the increased uptake of the former could be accounted for by their larger cell volumes.

Wiebe and Pomeroy (1972) studied the association of microorganisms with aggregates and detritus. In the open ocean they observed two types of particles: flat, plate-like flakes and flocculent particles. The flakes seldom contained recognizable bacteria or other microorganisms, whereas the flocs contained low numbers of bacteria. They concluded that the notion that bacteria coat particles in the ocean is generally not valid.

Deep-Sea Populations

During the last decade there has been a renewed interest in the microbiology of the deep sea. Three approaches have been used to examine deep sea microorganisms and their activities: 1) study of decompressed samples; 2) *in situ* studies; 3) laboratory studies of undecompressed samples. The first approach, initiated by ZoBell and co-workers in the 1940's, entails studies enumerating, isolating, and studying the physiology of bacteria present in deep sea water and sediments and in the gut flora of deep sea animals (e.g., Zobell, 1946). These studies use traditional microbiological techniques on decompressed samples to examine the effects of such properties as temperature and pressure. Results have been summarized in reviews by ZoBell (1970) and Morita (1976).

Two reports by Colwell and co-workers include stations in the ACSAR area (Tabor et al., 1981; Ohwada et al., 1980). Tabor et al. isolated bacteria from deep-ocean bottom water and described isolates that passed through 0.45 μm filters, showing that a significant relationship existed between decreased cell size and increased survival of bacteria isolated from the deep sea. In the study of Ohwada et al. (1980), the gut bacterial flora of animals collected from 570 to 2446 m depth were enumerated and characterized. The following conclusions were drawn from this study: 1) The number of culturable aerobic, heterotrophic bacteria was low in the animals that were collected from the greatest depths. 2) *Vibrio* spp. were the predominant isolates in ten of fifteen samples with *Photobacterium* and yeasts being predominant in the remainder. 3) *Pseudomonas, Achromobacter* and *Flavobacterium* comprised minor components of the gut flora of deep sea fish. 4) Strains of bacteria isolated from fish intestines were more barotolerant than those isolated from stomach.

Much of the interest in the study of deep-sea bacteria stemmed from the accidental sinking of the research submersible ALVIN in 1968. A packaged lunch was found to be well preserved after ten months at 1540 m (Jannasch et al., 1971). The food stuffs rapidly decomposed when brought to the surface and incubated at 4 °C, indicating that pressure had been responsible for slowing microbial degradation. See Jannasch and Wirsen (1977) and Jannasch (1979) for overviews of this work.

In Situ Studies. Jannasch's first approach (Jannasch and Wirsen, 1973; Wirsen and Jannasch, 1976) was to use ALVIN-deployed pressure containers containing serum-stoppered bottles. The bottles, containing a variety of substrates, were deployed at Deep Ocean Stations (DOS 1-3) in the North Atlantic at depths between 1830 and 3640 m. The pressure chamber was opened and the bottles inoculated and incubated for periods up to one year. Solid organic substrates (agar, starch, gelatin) placed on the sea floor in open containers showed almost no sign of disintegration (except for animal feeding marks) after exposure of one year. The microorganisms collected and incubated on the deep sea floor exhibited extremely slow metabolic rates, confirming the conclusion that the deep sea is extremely inefficient at recycling organic wastes.

In a study designed to examine the role of chemoautotrophic bacteria, Tuttle and Jannasch (1976) examined the utilization of thiosulfate on a mooring at 5300 m located at 34°02′N, 69°59′W, just outside the ACSAR area. These experiments demonstrated the potential for microbial thiosulfate utilization at elevated pressure and low temperature in seawater by both natural populations as well as by previously isolated pure cultures of thiosulfate-oxidizing bacteria.

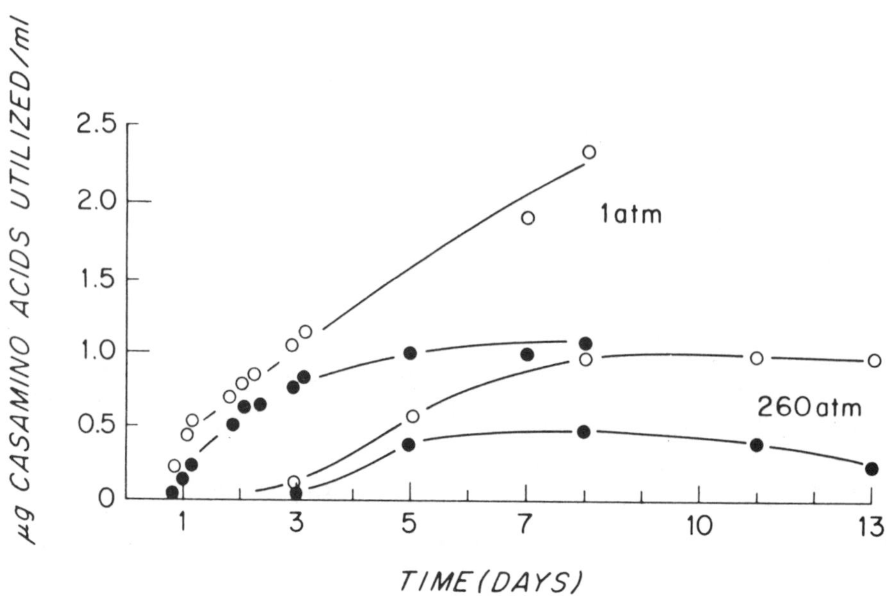

Figure 6.1 Incorporation (• •) and respiration (-o-) of ^{14}C from labeled casamino acids in a seawater sample retrieved from 2700 m before and after decompression. From Jannasch (1979).

Jannasch's second approach to *in situ* studies was to develop a "free vehicle" that would not require submersible deployment (Jannasch and Wirsen, 1980). The experiment described by Jannasch and Wirsen (station DOS-2, 38°18'N, 69°36'W, at 3580 m) showed the amount of acetate metabolites after decompression at 1 atmosphere and 3°C was 2½ times that metabolized *in situ*, confirming earlier observations that the combination of low temperature and high pressure greatly reduced rates of mineralization.

Laboratory Studies with Undecompressed Samples and Cultures. The *in situ* experimental methods described above have drawbacks. Most important, rate measurements are not possible, only end-point determinations can be made. To circumvent these problems Jannasch and Wirsen (1973) designed and built samplers and incubation chambers that would retain both *in situ* pressures and temperature. Colwell's group (Tabor and Colwell, 1978; Tabor et al., 1981) developed similar devices. One of Jannasch's and Wirsen's early time-course measurements is shown in Figure 6.1 (Jannasch, 1979). The incorporation and turnover of casamino acids by a mixed microbial population showed an increase upon decompression. They observed that the degree of pressure sensitivity depended upon the type of substrate used.

In a recent, more comprehensive study, Jannasch and Wirsen (1982) compared microbial activities in undecompressed and decompressed deep-seawater samples mostly taken within the ACSAR study area. Results showed that rates of incorporation and CO_2 production as well as total substrate utilization are generally lower at pressure than at 1 atmosphere control. Rates also were different for each of the four substrates used. With one exception (a water sample collected at 4500 m and incubated with glucose), the transformation of all the substrates showed an increased lag period at pressure when compared to the 1 atmosphere control.

Early in the study of microbial transformations in the deep sea, Jannasch and co-workers (Jannasch and Wirsen, 1973; Wirsen and Jannasch, 1976) noted that solid substrates incubated *in situ*, that were not screened to exclude larger organisms, showed feeding marks from small invertebrates. From these and other experiments, it became clear that a considerable part of the turnover of organic matter in the deep sea takes place in the guts of animals. The guts of animals such as arthropods, mollusks, echinoderms and fishes provide environments which are high in nutrients and favorable to bacterial growth.

In summary, advances in the study of deep sea microbiology in the last decade have shown that microbial activity is strongly influenced by the conditions characteristic of the deep sea. They are, in descending order of importance, low nutrient levels, except in localized areas such as invertebrate guts and where large pieces of organic input from surface waters (e.g., fish carcasses) are decomposing; low temperatures (typically between 2° and 3°C); and hydrostatic pressure. With respect to pressure, Jannasch and co-workers (Jannasch et al., 1982) believe that most of the free-living bacteria in the deep sea are barotolerant and may represent bacteria that are introduced to the deep sea as the result of particle flux from the surface. Barophilic bacteria can also be isolated from the deep sea and may represent a class of bacteria that are indigenous to the area.

Cyanobacteria

The cyanobacteria (blue-green algae) are a morphologically diverse group of bacteria that make their living by performing oxygenic photosynthesis in a wide variety of habitats. They are often important components of freshwater planktonic communities, where as many as 24 genera and over 100 species are known to be capable of forming extensive water blooms (Whitton, 1973; Fogg et al., 1973). In marked contrast to their freshwater counterparts, marine planktonic cyanobacteria are restricted to two principal general, *Trichodesmium* and *Synechococcus*. The organisms in the genus *Trichodesmium* are gas-vacuolated oscillatorian forms which are common in tropical oceans where they form extensive blooms (Fogg et al., 1973). Their filaments aggregate into bundles which float to the surface and are seen easily with the naked eye.

The second genus, *Synechococcus*, while well documented in freshwater planktonic communities, was not known to be an important marine phytoplankter until members of this genus were observed in large numbers by Waterbury et al. (1979) using epifluorescence microscopy and by Johnson and Sieburth (1979) using transmission electron microscopy.

In addition to these two principal genera a number of others are known from the planktonic marine environment (Marshall, 1981). Two that have been studied recently are *Richelia* (Sournia, 1970; Mague et al., 1974) and *Dichothrix* (Carpenter, 1972).

Trichodesmium

Field studies indicate that maximum rates of photosynthesis are achieved by *Trichodesmium* between 20° and 30°C (Aruga et al., 1975). Field observations indicate that blooms do not begin until the water temperature reaches approximately 20°C (Marumo and Nagasawa, 1976; Carpenter, 1983). Temperature appears to be extremely important in predicting where and when *Trichodesmium* will occur. This cyanobacterium has a narrow temperature range at which active growth can occur, with 20°C the apparent minimum and 30°C the maximum.

Distribution and Concentration. Data on the distribution and concentration of *Trichodesmium* are taken from a recent review by Carpenter (1983).

1. *Summer*: Hulburt (1962) reported 1.4×10^5 trichomes m^{-3} in August between Cape Cod and Bermuda. Carpenter and McCarthy (1975) observed a mean population of 1.0×10^3 trichomes m^{-3} in the western Sargasso Sea in late summer. Carpenter (1983) concludes that concentrations of 10^3 trichomes would be typical of the open North Atlantic during the summer. Higher concentrations have been reported outside of the Sargasso Sea during the summer. Dunstan and Hosford (1977)

observed 2 x 10^6 trichomes m^{-3} near the coast of Georgia. Blooms may also occur in the Slope Waters north of the Gulf Stream during the summer months in periods of calm weather.

2. *Autumn*: Carpenter (1983) concludes "the northern limit of the active population in the autumn is 45 °N, with about 10^3 trichomes m^{-3} occurring south of this, to 40 °N where the population averages 10^4 trichomes m^{-3}." During the fall inshore concentrations of 10^5 trichomes m^{-3} have been reported (Dunstan and Hosford, 1977; Marshall, 1971).
3. *Winter*: Dunstan and Hosford (1977) calculated a mean concentration of 1.5 x 10^5 trichomes m^{-3} off the Georgia coast, while Dugdale et al. (1961) found none in the Sargasso Sea when the surface temperature was 18 °C or lower. Carpenter (1983a) concludes that the northern limit of active populations in the winter is 30 °N for the open Atlantic Ocean. In the western Atlantic the northern limit for *Trichodesmium* is 30 °N (Carpenter, 1983a), but extends to 35 °N in the Gulf Stream in the winter (Dunstan and Hosford, 1977).

Primary Production. Estimates of primary production by *Trichodesmium* have been complicated by its fragility, with the result that cellular carbon doubling times vary widely. The study of Carpenter and Price (1977) contains the most extensive data set on *Trichodesmium* within the ACSAR study area and is summarized below. A number of their stations were in or very near the study area, but rates of primary production were not measured at any of the stations.

Synechococcus

Recent reports have shown that small unicellular cyanobacteria are widely distributed and present in large numbers within the euphotic zone of the world's oceans (Waterbury et al., 1979; Johnson and Sieburth, 1979). Preliminary studies have also indicated that these cyanobacteria contribute significantly to primary productivity (Li et al., 1983 and Waterbury et al., 1980).

Synechococcus is widely distributed in the world's oceans in surface waters between 5° and 30 °C. It is present in the tropical oceans throughout the year at concentrations varying from 10^3 to 10^4 cells ml^{-1}, and seasonally distributed in the temperate oceans ranging from a few cells ml^{-1} in the winter months when the water temperature falls below 5 °C to near 10^5 cells ml^{-1} during the summer months. *Synechococcus* appears to be excluded from waters below 5 °C. Figure 6.2 shows a typical vertical distribution of *Synechococcus* in the study area just north of the Gulf Stream. Within the ACSAR study area, *Synechococcus* concentrations vary between 10^3 and 10^5 cells ml^{-3} during the annual cycle.

Preliminary data indicate that *Synechococcus* is an important primary producer. In the Sargasso Sea it is responsible for 15 to 25 percent of the total primary productivity. In Slope Water, *Synechococcus* is responsible for 5 to 15 percent of the primary productivity and in inshore waters these unicellular cyanobacteria contribute about 5 to 7 percent of the total primary productivity.

Protozoa

Protozoa — singled-celled, eucaryotic organisms — constitute a highly diverse group of species. Taxonomically these organisms have received much attention but little agreement in recent years (Levine et al., 1980; Laval-Peuto, 1982). The groups pertinent to this discussion fall into two phyla by the most recent classification (Levine et al., 1980):

Phylum: Sarcomastigophora
- Subphylum: Mastigophora
  - Class: Phytomastigophorea (heterotrophic flagellates)
  - Class: Zoomastigophorea (heterotrophic flagellates)
- Subphylum: Sarcodina
  - Superclass: Rhizopoda (naked amoebae and foraminifera)
  - Superclass: Actinopoda (radiolaria and acantharia)

Phylum: Ciliophora (ciliates)

There is a great deal of variability in the abundance of these protozoan groups. Heterotrophic microflagellates are present in the most oligotrophic waters at densities of hundreds to thousands per cm^3, while shell-bearing sarcodines may occur at densities less than 1 per m^3 — a difference of 9 orders of magnitude. These differences in standing stocks reflect trophic and morphological differences between the groups. For example, bacterivorous (bacteria-eating) flagellates as small as 2 μm are common in the plankton, whereas omnivorous or carnivorous planktonic foraminifera can attain a few cm in diameter and colonial radiolaria can form colonies reaching more than 3 m in length (Swanberg, 1979) — a difference of six orders of magnitude. However, most protozoan species occur in the size range from 2 to 200 μm.

Recent studies have indicated that most of the respiration of the plankton is performed by organisms smaller than 30 μm, (Williams et al., 1981a) and protozoa are the dominant heterotrophs in this size fraction (Beers and Stewart, 1969). Furthermore, the weight-specific filtration and ingestion rates for protozoa are at least as great as those for larger zooplankton (Conover, 1982). As a result several investigators have incorporated these formerly overlooked organisms into models describing oceanic food webs (Pomeroy, 1974; Sieburth et al., 1978; Williams, 1981b; Sorokin, 1981; Azam et al., 1983).

Protozoan distributions and abundances are less well characterized for the oceanic realm than for littoral and neritic systems. This is due in part to the accessibility of the coastal environment, and in part to open ocean sampling techniques which are inappropriate for the collection of many protozoan taxa. The protozoology of the deep-sea benthos is especially poorly known because of problems with sampling, fixation and examination. Work in the Pacific

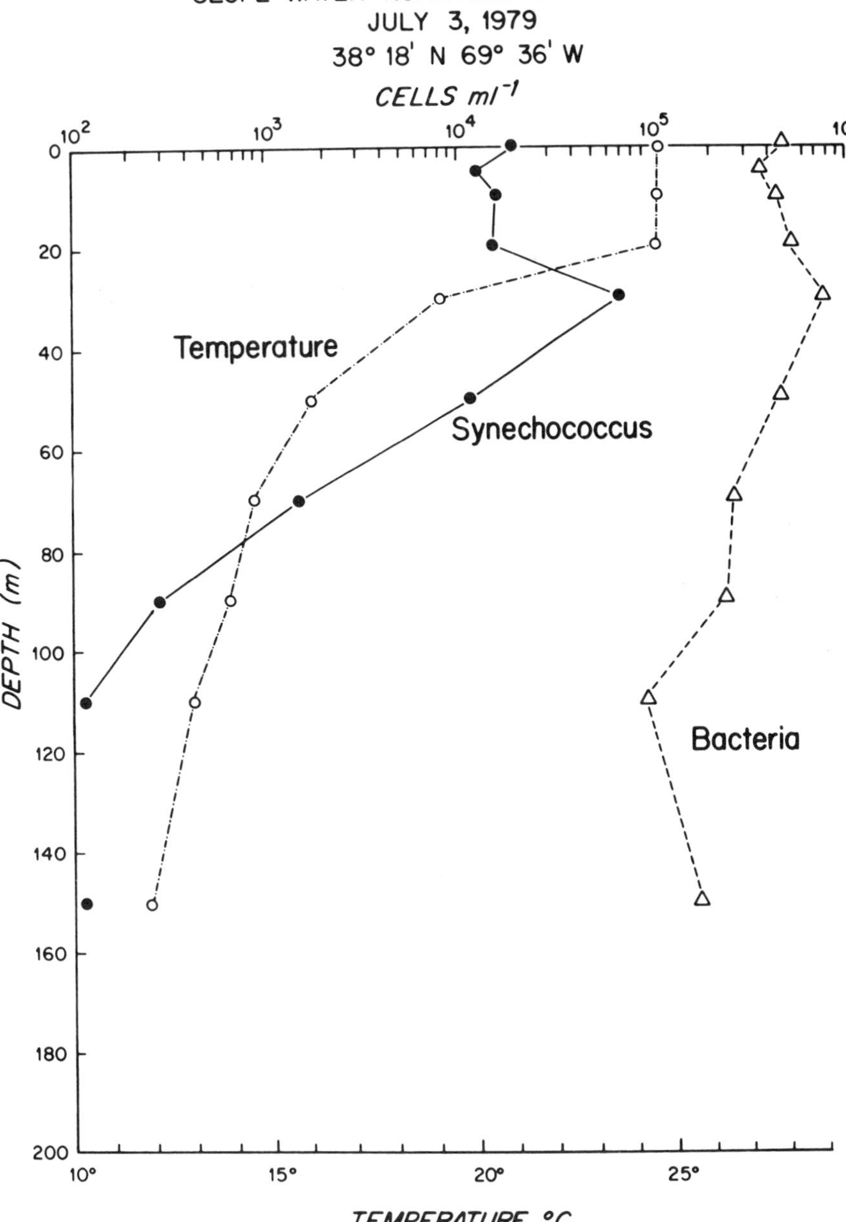

Figure 6.2 Vertical distribution of temperature, *Synechococcus*, and bacteria at a Slope Water station. (Waterbury, unpublished data).

(Burnett, 1977, 1979, 1981) has documented large populations of microbiota (16,500-26,900 cells/cm²), but similar studies have not been carried out in the Atlantic. For this reason discussion of benthic protozoa will be omitted from this review.

Flagellates

Heterotrophic flagellates are a ubiquitous component of plankton communities (Davis, 1982; Fenchel, 1982c; Davis et al., 1985). No distinction was made in most older literature between autotrophic and heterotrophic flagellates, and heterotrophic cells were generally included in the phytoplankton counts. In addition to true zooflagellate taxa, phytoplankton taxa with non-photosynthetic species which appear often in the plankton include the chrysomonads, cryptomonads, euglenids and dinoflagellates. Many of the known species of dinoflagellates are non-photosynthetic (Kofoid and Swezy, 1921), and most oceanic species are thought to be non-photosynthetic.

Few studies have been conducted on the distribution of heterotrophic micro-flagellates and still fewer studies within the ACSAR area. Three studies (Davis, 1982; Caron, 1984; Davis et al., 1985) have enumerated heterotrophic nanoplankton (2-20 mm cells, presumably heterotrophic flagellates) by epifluorescence microscopy at stations throughout the North Atlantic, including several stations within the study area. Those studies indicate average population densities for the Slope Water, Gulf Stream, and Sargasso Sea of $0.7x10^3$, $0.7x10^3$ and $0.8x10^3$ cells/ml, respectively, for subsurface samples in the mixed layer. The density of cells in the surface microlayer of the Slope Water is approximately twice as large. Nearshore waters contain densities up to an order of magnitude greater than oceanic densities. Flagellates are by far the most numerous protozoa in the plankton, with densities often rivaling or even dominating the number of phytoflagellates (Davis, 1982; Caron, 1984; Davis et al., 1985).

Taxonomic characterization of the species from stations within the study area were not performed, but Davis (1982) has characterized the culturable species of bacterivorous flagellates occurring in Narragansett Bay and for a transect across the North Atlantic at 24°30′N. In addition to heterotrophic species within several phytoplankton taxa, a number of true zooflagellate species were cultured. These isolates were dominated numerically by kinetoplastids, with the genus *Bodo* contributing the most species.

Vertical distributions of heterotrophic nanoplankton have been performed in the study area by Caron (1984) (Fig. 6.3). Population densities tend to decrease with depth, and populations in the mixed layer tend to decrease with distance from shore. Where subsurface peaks in bacteria and phytoplankton occur, heterotrophic flagellates often mirror these profiles. Davis et al. (1985) have noted a positive correlation between direct counts of bacteria and heterotrophic nanoplankton in the North Atlantic.

While Figure 6.3 depicts the gross distribution of heterotrophic flagellates in the study site, the microscale distributions of these organisms are much less predictable. Many species attach to surfaces (Fenchel, 1982a), and thus detrital material and other organisms represent microenvironments of elevated population densities. In addition, population numbers appear to undergo significant diel fluctuations (Burney et al., 1979; Davis et al., 1985) due to rapid growth rates of the flagellates (Fenchel, 1982b) and grazing by flagellate predators. Burney et al. (1981) noted a change in the heterotrophic nanoplankton concentration in the western Sargasso Sea from 420 to 1200 cells/ml over a diel cycle. These results suggest that flagellates are highly dynamic populations.

Recent work has shown that heterotrophic microflagellates (less than 20 μm) are significant bacterioplankton consumers (Haas and Webb, 1979; Fenchel, 1982b; Davis and Sieburth, 1984; Caron, 1984). This literature has been reviewed by Sieburth (1984). Measured feeding rates of flagellates vary from approximately 25 to 250 bacteria/flagellate/hr. Davis (1982) and Caron (1984) showed that significant feeding rates were still observable at bacterial concen-

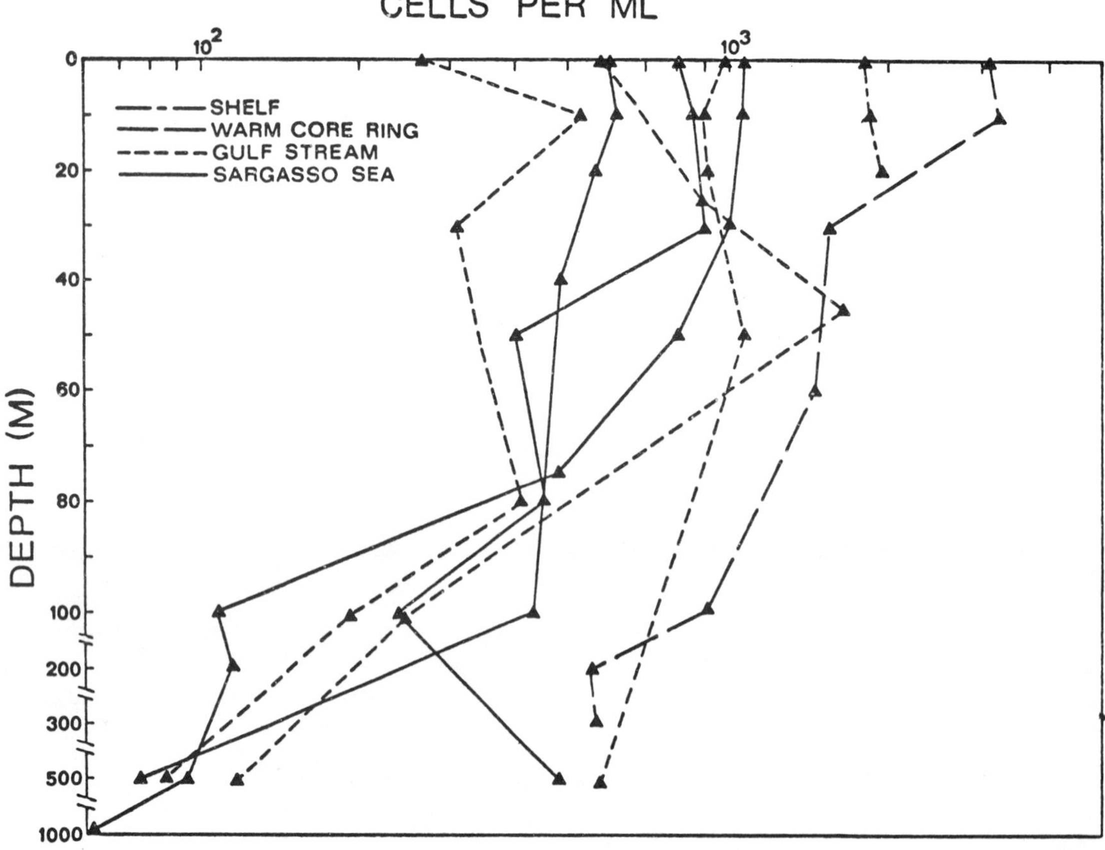

Figure 6.3 Vertical distributions of heterotrophic nanoplankton in the North Atlantic as shown by epifluorescence microscopy. Representative profiles from Caron, (1984).

trations approaching *in situ* concentrations. Correlations between the number of bacteria and the number of heterotrophic nanoplankton tend to strengthen this hypothesis and suggest an important role for bacterivorous flagellates as a mechanism whereby bacterial biomass becomes available to larger zooplankton (Azam et al., 1983). In addition to bacteria, microflagellates are also capable of consuming chroococcoid cyanobacteria (Johnson et al., 1982) and eucaryotic algae (Haas, 1982; Goldman and Caron, in press). However little more than anecdotal information exists on the distribution of microflagellates which are capable of ingesting algae, and more work is required to determine the magnitude of this predation.

Larger flagellates (e.g. dinoflagellates) have a well-documented ability to feed and grow on microalgae. Gold (1970) cultured a heterotrophic dinoflagellate using phytoflagellates as food. Kimor (1981) gave evidence for phagocytosis in a number of marine dinoflagellate species. However, as is the case with microflagellates, the importance of this predation is unknown due to the scant information that is presently available.

Finally, symbioses have been described between heterotrophic dinoflagellates and various photosynthetic organisms (see Taylor, 1982, for review). In the case of chroococcoid cyanobacteria invaginations of the cell wall may form special chambers which contain the cyanobacteria. Phagocytosis has not been observed, and the physiological relationship is not clear. For other phototrophs, the degree of integration with the host varies.

Sarcodines

The planktonic sarcodines are composed of four major groups; foraminifera, radiolaria, acantharia, and naked amoebae. The first three possess a rigid skeleton or test which can withstand plankton net collection. For this reason distributional information for these groups is most complete. The foraminifera possess a multi-chambered calcium carbonate test. Some radiolaria construct silica skeletons, but some species form no skeleton at all. The acantharia produce a skeleton composed of strontium sulfate. The non-testate or naked amoebae are traditionally thought of as benthic organisms, but their presence in the plankton has been recently substantiated.

Little pertinent information is available concerning the diet of acantharia, but radiolaria and foraminifera are known to accept a variety of organisms as food. Their prey includes algae, as well as a wide variety of zooplankters (Anderson and Bé, 1976; Anderson et al., 1979; Bé et al., 1977; Caron and Bé, 1984; Swanberg, 1979). Radiolaria also consume a significant amount of animal tissue. Swanberg (1979) found copepods, appendicularians, mollusc larvae, hydromedusae and tintinnids as prey in colonial radiolaria.

Many shell-bearing sarcodines possess symbiotic algae which contribute to their nutrition. These symbioses have been reviewed by Anderson (1980) for radiolaria, and by Bé et al. (1977), Lee (1980), and Taylor (1982) for planktonic foraminifera. Swanberg (1979) concluded that primary production by the symbiotic algae of colonial radiolaria was an insignificant fraction of the total primary productivity of the water. However, symbiont-derived nutrition may have a profound effect upon survival and growth of the sarcodine host (Bé et al., 1982; Caron et al., 1982a).

Foraminifera. — The foraminifera are perhaps the most intensively studied group of protozooplankton due to their importance in paleoecological work. While lacking fine-scale resolution, available data show a transition from low abundance to moderate abundance corresponding with changes in the water masses (Sargasso Sea to Slope Water) (Fig. 6.4). Boreal species extend into cold Slope Water (Fig. 6.5), while tropical and subtropical species occur in the Gulf Stream and Western Sargasso Sea (Fig. 6.6). Eighteen of the 20 species collected by Bé and Tolderlund (1971) have distribution ranges which overlapped the study site.

Cifelli (1962, 1965) found the highest concentrations of foraminifera in the Slope Water, and these samples were generally dominated by *Globigerina* species. Species diversity increased towards the Sargasso Sea, but density of foraminifera decreased. Species diversity and density decreased dramatically on the shelf. Maximum densities in the Slope Water occurred during spring and fall. Fairbanks et al. (1980) showed that the vertical distributions of 13 species of foraminifera were not uniform within the mixed layer for 3 stations in Slope Water. Vertical distribution of planktonic foraminifera appears to follow the deep chlorophyll maximum (DCM) (Fairbanks and Wiebe, 1980).

Diel changes in the vertical distribution of planktonic foraminifera have been suggested by Bé (1960). However, Boltovskoy (1973) was unable to observe a significant difference between day and night tows, and the question of vertical migration by these organisms remains unresolved. Short term changes (days) can be expected on a regular or irregular basis due to the reproductive cycle inherent in some species of foraminifera. *Hastigerina pelagica* undergoes gametogenesis on a lunar cycle, resulting in a monthly cycle in its abundance in surface waters (Spindler et al., 1978). Another species, *Globigerinoides sacculifer*, initiates gametogenesis at a specific time of day, but shows no synchrony as to which day gametogenesis occurs in laboratory cultures (Bé et al., 1983). These aspects of behavior may cause large fluctuations in the standing crop of planktonic foraminifera whose meaning cannot be explained on the basis of hydrographic data alone.

Radiolaria. — The radiolaria are another paleoecologically important group of planktonic protozoa. Life histories for these organisms are largely unknown, and species are much more numerous than the foraminifera. The radiolaria contain the largest protozoan structures alive, with some colonial radiolaria forming a gelatinous matrix up to 3 m long (Swanberg, 1979). Like foraminifera, densities of radiolaria generally range from less than 1 m^{-3} to more than 100 m^{-3}. Cifelli and Sachs (1966) compared numerical abundances of foraminifera and radiolaria along a transect which included stations within the ACSAR study area. Foraminifera generally outnumbered radiolaria, but radiolaria were dominant at warm-water stations. In most cases, radiolaria and foraminifera had coincident peaks in abundance.

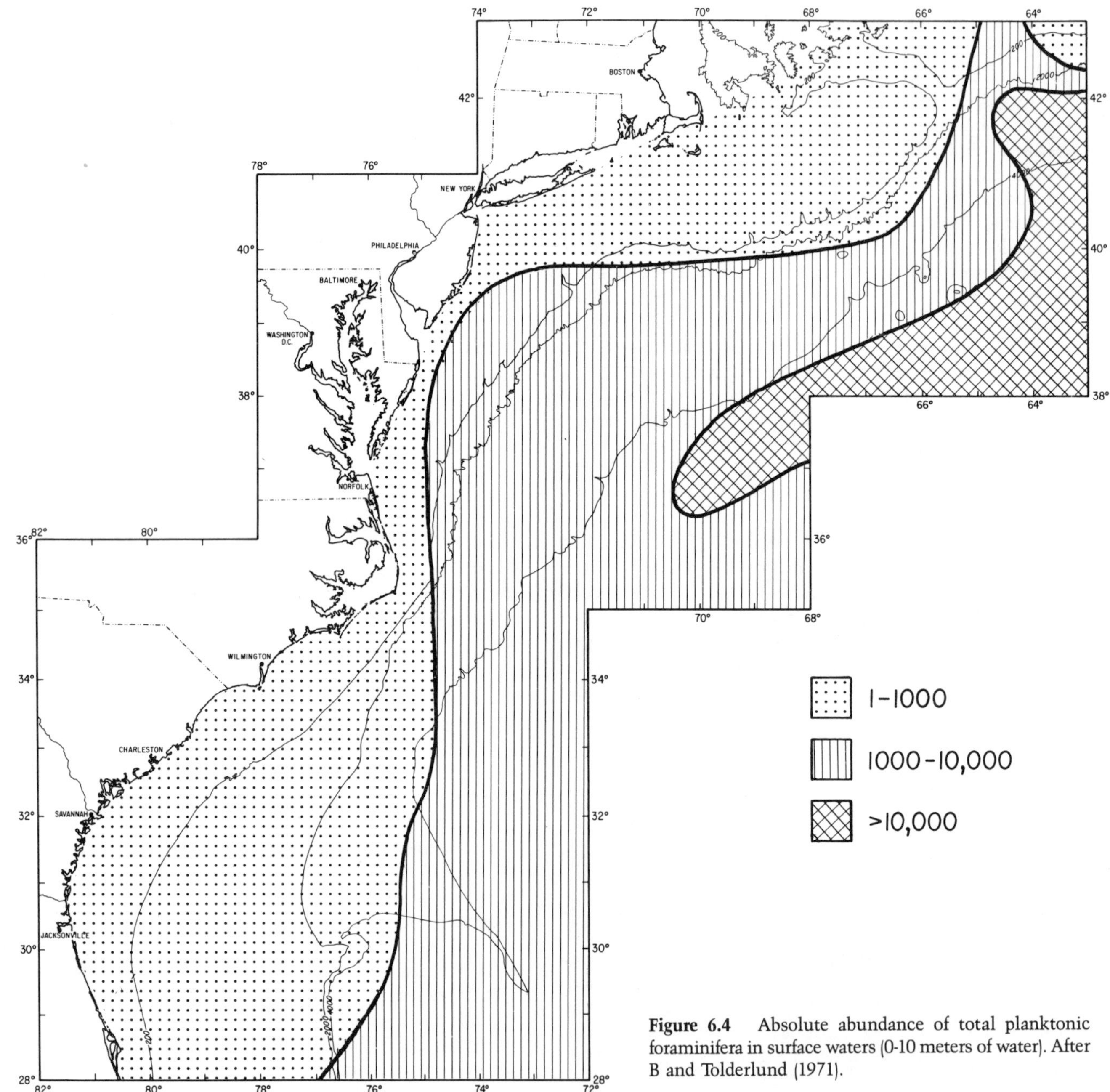

Figure 6.4 Absolute abundance of total planktonic foraminifera in surface waters (0-10 meters of water). After B and Tolderlund (1971).

Thus, one group does not appear to prosper at the expense of the other.

Extensive work has been performed on the colonial radiolaria collected by Swanberg (1979) throughout the North Atlantic. Colonial radiolaria were observed at 89% of the stations over a 4-yr period, although negative data are not conclusive because of the nature of the collecting method. Densities of radiolaria ranged from 0.04 to 540 colonies m^{-3}.

Acantharia. — The acantharia have been less intensively studied than the other shell-bearing sarcodines, probably owing to dissolution problems in preserved samples. Still, the abundance of this group is at least as great as other sarcodines; average concentrations were approximately 25, 24 and 39 m^{-3} for the Slope Water, Gulf Stream, and Northern Sargasso Sea, respectively (Bottazzi et al., 1971). In general, species diversity and abundance of acantharia are greater in warmer waters, as with radiolaria. Nonetheless, species in the Slope and Sargasso Sea samples are not observed in the southerly stations, indicating an endemic fauna for this region.

Amoebae. — Distributional information concerning the naked amoebae in the open ocean is rare. The study by Davis et al. (1978) is the only study from oceanic waters, including stations bordering the ACSAR area. Highest densities were observed in the surface microlayer (up to 100

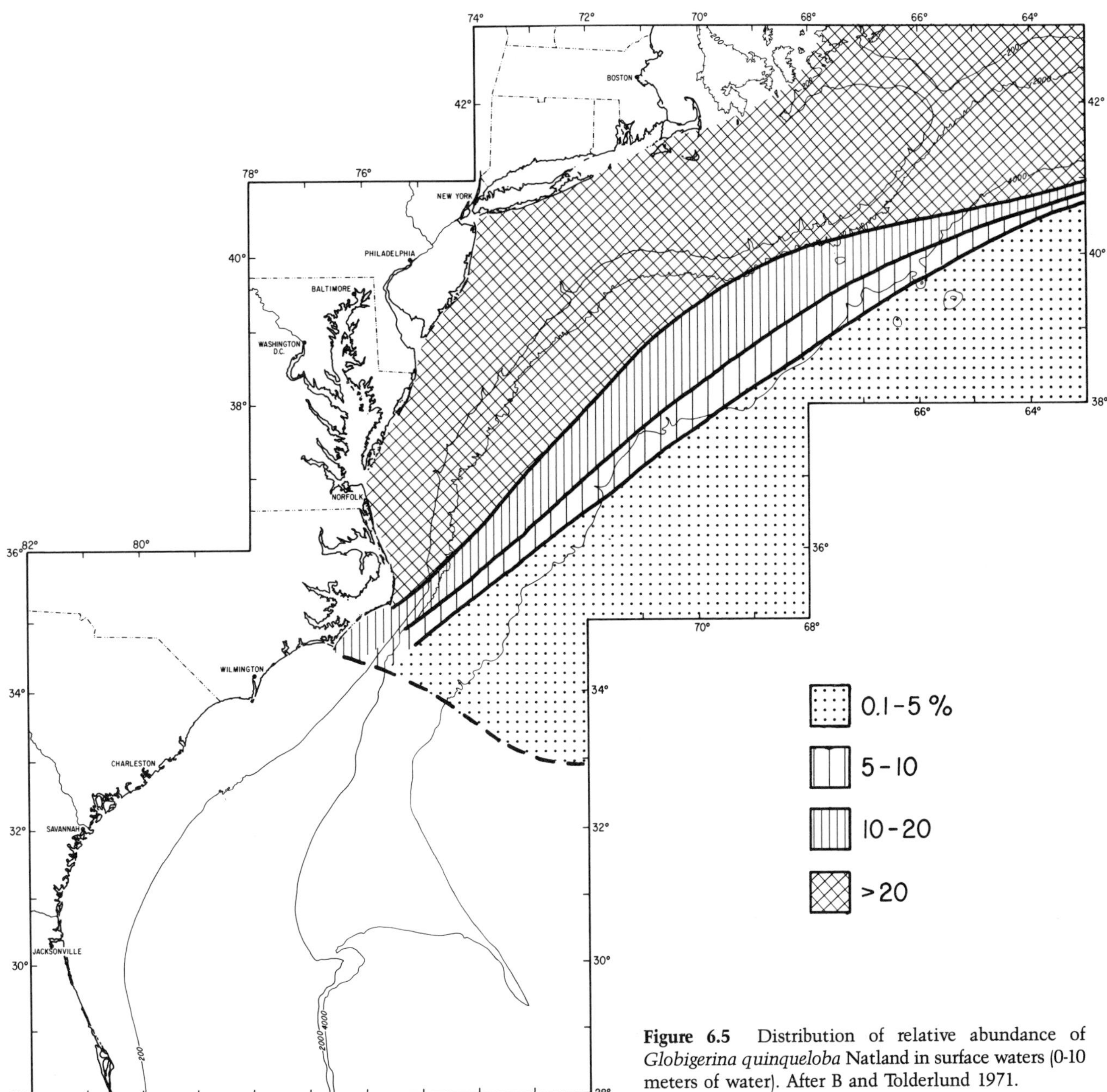

Figure 6.5 Distribution of relative abundance of *Globigerina quinqueloba* Natland in surface waters (0-10 meters of water). After B and Tolderlund 1971.

l^{-1}), while subsurface samples averaged about 1 amoeba l^{-1}. These densities are generally greater than for other sarcodines. When corrected for sample dilution, surface microlayer samples reached densities exceeding 10^3 amoebae l^{-1}. Seven families and 11 genera of amoebae were isolated during that study; three genera (*Acanthamoeba*, *Clydonella*, and *Platyamoeba*) accounted for more than half of these isolates and only one genus (*Clydonella*) was isolated from all three study sites.

The high density of amoebae associated with the surface microlayer is probably an indication of the particle-associated nature of these sarcodines. Amoebae are usually associated with particles in cultures (Davis et al., 1978), and high densities have also been observed on marine snow samples from the open ocean (Caron et al., 1982c).

The naked amoebae are probably the primary bacterivorous sarcodines in the plankton. Due to their low abundance in the water column relative to microflagellates, and to their manner of feeding (particle-associated), these protozoa are probably not important consumers of bacterioplankton. However, amoebae may play a role in the grazing of bacteria at interfaces. Davis et al. (1978) showed that the highest concentrations of amoebae in the North Atlantic occurred at the air/water interface, and densities of these protozoa in the neuston appeared to be positively correlated with densities of bacterioneuston.

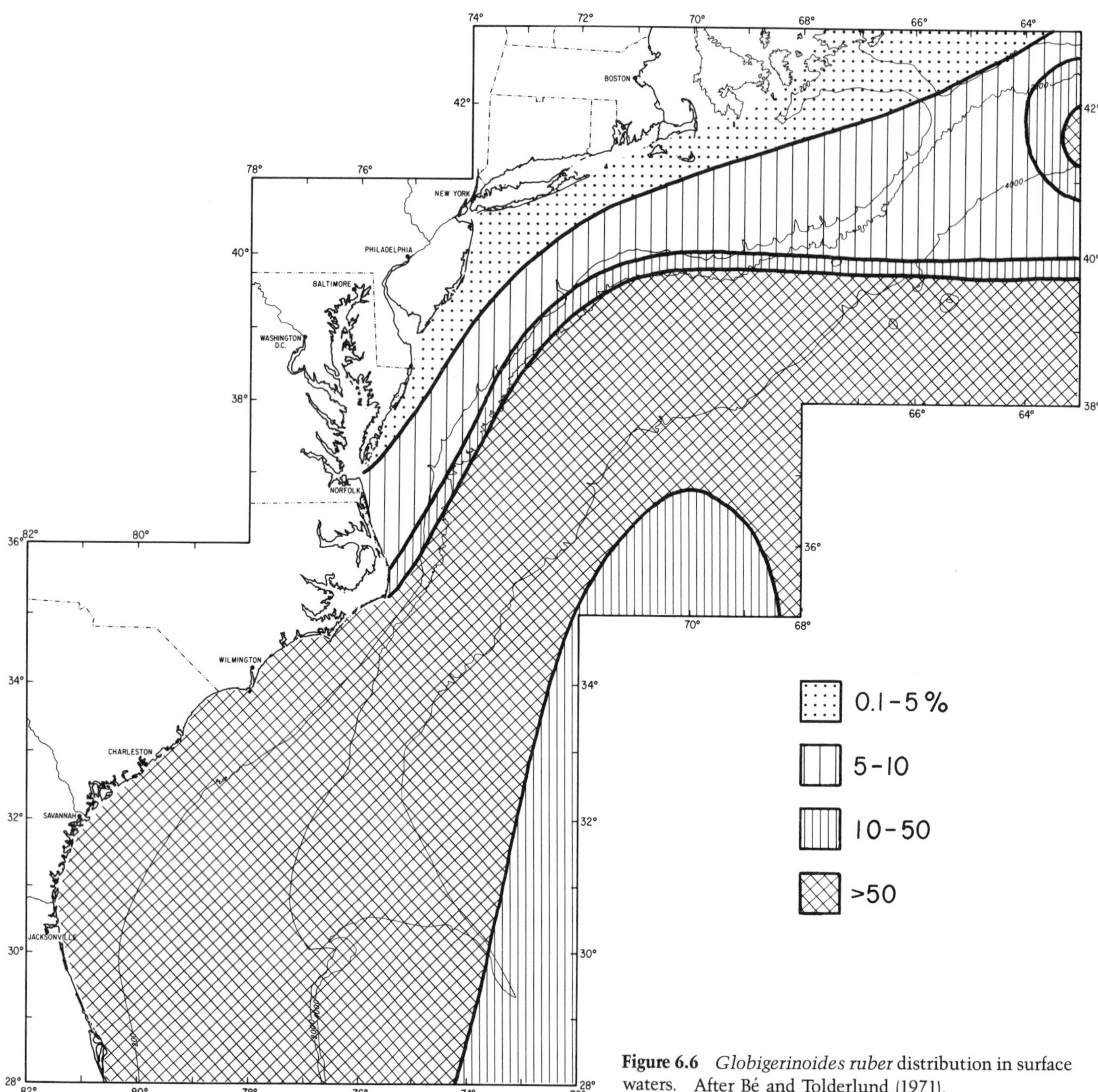

Figure 6.6 *Globigerinoides ruber* distribution in surface waters. After Bé and Tolderlund (1971).

Ciliates

The phylum Ciliophora contains a highly diverse fauna. However, only a relatively few species are known to be truly planktonic. Most pelagic marine ciliates are from the order Oligotrichida, which contains the familiar tintinnids, and the less well-known non-loricate oligotrichs.

Distribution patterns of oceanic ciliate populations in the ACSAR area are poorly known. Most investigations have been conducted in near-shore environments and have concentrated on tintinnids, which possess a preservable lorica used in identifying species. For example, Gold and Morales (1975) reported 34 species of tintinnids from New York Bight over a 1-yr period. Microplankton studies conducted in the Pacific have shown that ciliates represent a large percentage of the total number of organisms in the size class less than 103 mm (Beers and Stewart, 1969), and generally constitute larger populations and greater biovolume than the sarcodines. In general, densities of ciliates tend to be intermediate between flagellate and sarcodine densities. Non-tintinnid oligotrichs may be a large portion of the total ciliate numbers. These ciliates can constitute 71% of the microplankton biomass, while tintinnids comprise 5.5% (Laval-Peuto, 1982). At present, however, our knowledge concerning the distribution of non-loricate pelagic ciliates is very poor (Borror, 1980), and the taxonomy is outdated and insufficient (Laval-Peuto, 1982).

Of much less importance than the oligotrichs are a few taxonomically diverse ciliate species which occur primarily in coastal waters but have been found in the oceanic environment. These include the free-living species *Uronema* sp. (order Scuticociliatida) (Hamilton and Preslan, 1969) and *Mesodimium rubrum* (order Haptorida) (see review by Taylor et al., 1971), and the ectocommensal species *Ephelota gemmipara* (order Suctorida) (Sieburth et al., 1976), and *Myoschiston centropagidarum* (Hirche, 1974) and *Zoothamnium* sp. (Herman and Mihursky, 1964; Sieburth et al., 1976), both from the order Peritrichida. One peritrich species, *Zoothamnium pelagicum*, has been described as a truly planktonic species (Laval, 1968).

The major role for ciliates in oceanic plankton communities appears to be as consumers of phytoplankton and other small protozoa. A considerable amount of work has been performed to investigate the magnitude of ciliates grazing on coastal phytoplankton both in-situ and in cultures (Capriulo, 1982; Heinbokel, 1978; Heinbokel and Beers, 1979). Work performed with tintinnid ciliates indicates that ciliates can have a significant effect on phytoplankton communities in coastal waters. Extrapolation of these results to oceanic waters is difficult, and the quantitative effect of ciliates grazing on phytoplankton in the open ocean remains largely unknown.

Protozoa as Food

Protozoa are eaten by larger zooplankton, and in some cases may contribute significantly to the diets of zooplankton. The latter case is particularly true where the number of primary producers is low, such as in the deep-ocean (Hardy, 1974) or during the heterotrophic phase of plankton succession (Sorokin, 1977).

Bacterivorous microflagellates play a key a role in plankton communities by serving as a trophic link between bacteria and larger zooplankton. Heterotrophic flagellates are directly available to fine filter-feeding zooplankton, and attachment of flagellates to particles may make them available to coarse filter feeders (Kopylov et al., 1981). Ciliates are large enough to be directly available as a food source to many species of zooplankton. They may be sufficiently dense to constitute an important food source for copepods in neritic waters, and probably serve to supplement the diet of copepods in open water (Berk et al., 1977; Robertson, 1983).

Sarcodines appear to be less palatable to zooplankton than flagellates and ciliates. Predators for planktonic foraminifera and acantharians are not well known, although foraminiferan tests have been found in the stomachs of salps. Swanberg (1979) has observed that colonial radiolaria appear to be distasteful to fish and are avoided after an initial mouthing. Primary consumers of colonial radiolaria appear to be planktonic amphipods, although a few copepod species and turbellarians may also eat them.

Protozoa and Nutrient Regeneration

Although some disagreement exists, evidence continues to increase that small (less than 20 μm) protozoa have a significant role in regenerating major nutrients in the planktonic ecosystem (Stout, 1980; Taylor, 1982). Glibert (1982) has shown that organisms passing a 10 μm filter were responsible for most of the NH_4^+ remineralization in and around the study site. Burney et al. (1979) noted significant correlations over a diel cycle between carbohydrate concentrations and the microbial plankton less than 20 μm, indicating an active role by these organisms in controlling the concentrations of these materials. Protozoa also have been shown to increase the rate of decomposition of detrital material (Fenchel, 1977; Sherr et al., 1982) and the breakdown of zooplankton fecal material by acting as bacterial consumers and by physical disruption of feces (Gowing and Silver, 1982; Honjo and Roman, 1978; Pomeroy and Diebel, 1980).

Microenvironments and Protozoa

Recent work has established the importance of large aggregations of microorganisms as sites of intense protozoan activity in the plankton. This work has been primarily concerned with fragile, macroscopic detrital aggregates (marine snow). Marine snow has been shown to contain large populations of microorganisms (Caron et al, 1982c), and many protozoan species which are absent from the surrounding water occur on these particles (Silver et al., 1982; Caron et al., 1982c). Densities of culturable bacterivorous protozoa are up to 10^4 times as large on marine snow as in the surrounding water (Table 6.1). This raises questions as to the true habitat of some bacterivorous protozoa in the open ocean. In particular, the distribution of bacterivorous ciliates in the open ocean might be explained by the occurrence of these highly enriched aggregates in environments where the surrounding water contains too few bacteria to support ciliate growth (Caron et al., 1982b). Small particles may also have protozoa attached to them (Pomeroy and Johannes, 1968), and Goldman (1984) has proposed that microaggregates, like macroaggregates, may be important sites of high microbial activity and rapid nutrient cycling.

In addition to marine snow, a number of other aggregations exist which constitute important microenvironments for protozoan growth in plankton communities and whose distribution encompasses the study site. These include large algal aggregations such as *Rhizosolenia* mats (Carpenter et al., 1977), *Thalassiosira partheneia* colonies (Elbrachter and Boje, 1978; Caron, unpubl.) and *Oscillatoria* (*Trichodesmium*) bundles. *Rhizosolenia* mats contain large populations of protozoa throughout the matrix of the aggregate (Caron et al., 1982b), while *T. partheneia* colonies form hollow cylindrical colonies with protozoa living in the hollow center. High ciliate densities also occur sporadically on colonial radiolaria where they appear to cause breakup of the colonies (Swanberg, 1979). The presence of microenvironments of intense microbial activity in the plankton provide oases for the growth of protozoa which otherwise would not be able to survive on the dilute concentration of prey organisms in the surrounding water,

Table 6.1 MPN of Protozoa (No. m^{-1}) on marine snow. From Caron (1984).

Sample Location and Collection Dates	Number of Samples	Population	Parameter: Range	Parameter: Average	Concentration Factor (Snow: Control)
Sargasso (8/21/81-8/31/81)	16	Flagellates	3-2400	743	3,229
		Ciliates	UN-23	2.6	>3.919
		Amoebae	UN-23	2.9	260
Sargasso (2/19/82-2/24/82)	41	Flagellates	85.9-859	329	701
		Ciliates	UN	—	—
		Amoebae	UN-3.72	0.53	177
Gulf Stream (2/28/82-3/2/82)	23	Flagellates	390.3-1800	1000	1,970
		Ciliates	UN-57.0	19.59	10,590
		Amoebae	8.59-85.9	44.84	11,200
Gulf Stream (5/27/82-6/1/82)	29	Flagellates	174-23,000	1693	4,460
		Ciliates	0.34-94.2	8.02	10,690
		Amoebae	0.34-174	81.3	5,910
Warm Core Ring (5/19/82-5/21/82)	8	Flagellates	92.2-350	188.9	205
		Ciliates	UN-3.4	1.28	1,700
		Amoebae	UN-3.4	1.28	196
Georgia Shelf (2/27/82)	1	Flagellates	—	173	100
		Ciliates	—	—	—
		Amoebae	—	8.59	954

and thus may be important in explaining the distribution of some protozoan species in oceanic plankton communities.

Phytoplankton

P.M. Glibert

Standing Stocks

Regional and Seasonal Distributional Patterns

General seasonal phytoplankton dynamics were described for temperate waters as early as 1946 by G. Riley. Briefly, low-standing stocks of phytoplankton are usually observed during late fall and winter, followed by sporadic increases during early spring. As water column stratification becomes established in the spring, phytoplankton growth proceeds rapidly, and more or less continuously, until nutrients in the upper waters become depleted. Phytoplankton stocks then decline to a more modest level, maintained by nutrient regeneration processes and occasional mixing events, and remain at this level through the summer. Early fall may bring an autumn bloom, fueled with nutrients from below as an increase in vertical mixing is brought on by storm events or cooling.

Within this very general framework, many differences exist with regard to the large-scale phytoplankton distribution patterns observed in Slope Water and those in the northern Sargasso Sea or ring waters. These differences include the maximum phytoplankton standing stock reached on an annual basis and the seasonal rate of change in standing stocks both in the surface and with depth. For example, data from a variety of sources indicate the formation of a late spring to fall deep chlorophyll maximum in both regions, but a seasonal difference in the depth at which it occurs (Cox et al., 1982).

In Figures 6.7 and 6.8 vertical distribution profiles of chlorophyll are shown for the northern Sargasso Sea and Slope Waters. These data, compiled by Cox et al. (1982), are based on published and unpublished data from Wiebe et al. (1976), Ortner (1977), Ketchum and Ryther (1965), as well as data collected by them on R/V *Knorr* cruises 62, 65, and 75, and R/V *Endeavor* cruise 11. The most notable features of these profiles are that first, in Slope Water, phytoplankton standing stocks attain higher levels (>3.0 mg m^3 chlorophyll) in surface waters (<100 m) than in the northern Sargasso Sea, where maximum levels rarely exceed 0.5 mg m^3 chlorophyll. In addition, total standing stock in Slope Water remains fairly high after the spring bloom period, in contrast to standing stocks in the Sargasso Sea, which decline rapidly after the spring bloom and formation of the deep chlorophyll maximum. Finally, the intensity of the deep chlorophyll maximum is greater although longevity is shorter in Slope Water relative to the northern Sargasso Sea (Cox et al., 1982). The maximum depth of the deep chlorophyll maximum in Slope Water is approximately 75 m, occurring in September, while in the Sargasso Sea the maximum depth is approximately 85 m, occurring during July (Fairbanks and Wiebe, 1980).

There is some evidence that the maximum chlorophyll concentration within the deep chlorophyll maximum is strongly influenced by grazing pressure (Jamart et al., 1977; Ortner et al., 1980); therefore, regional and seasonal differences in zooplankton biomass and composition between

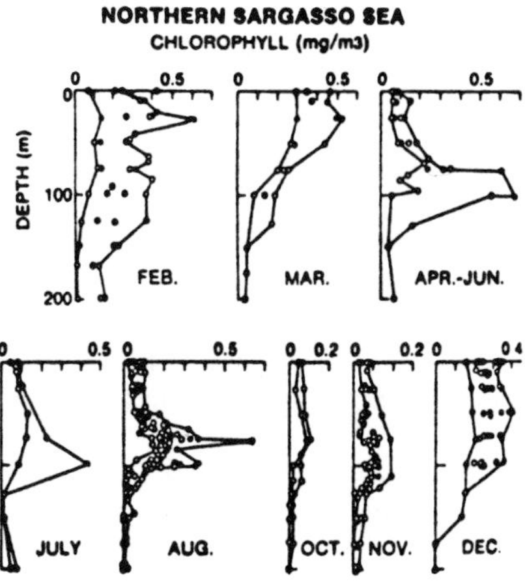

Figure 6.7 Composite vertical profiles of chlorophyll *a* from the northern Sargasso Sea. Solid lines represent arbitrary limits for high and low values from all the station data plotted and are not reflective of the values from any individual profile. Solid dots represent the maximum value for individual profiles, so the number of such dots in each composite plot is equal to the number of profiles used in its construction. In some cases, lower depth points are only represented by a single sttion, in which case the points are connected by a single line. Sources of data are listed in the text. From Cox et al. (1982).

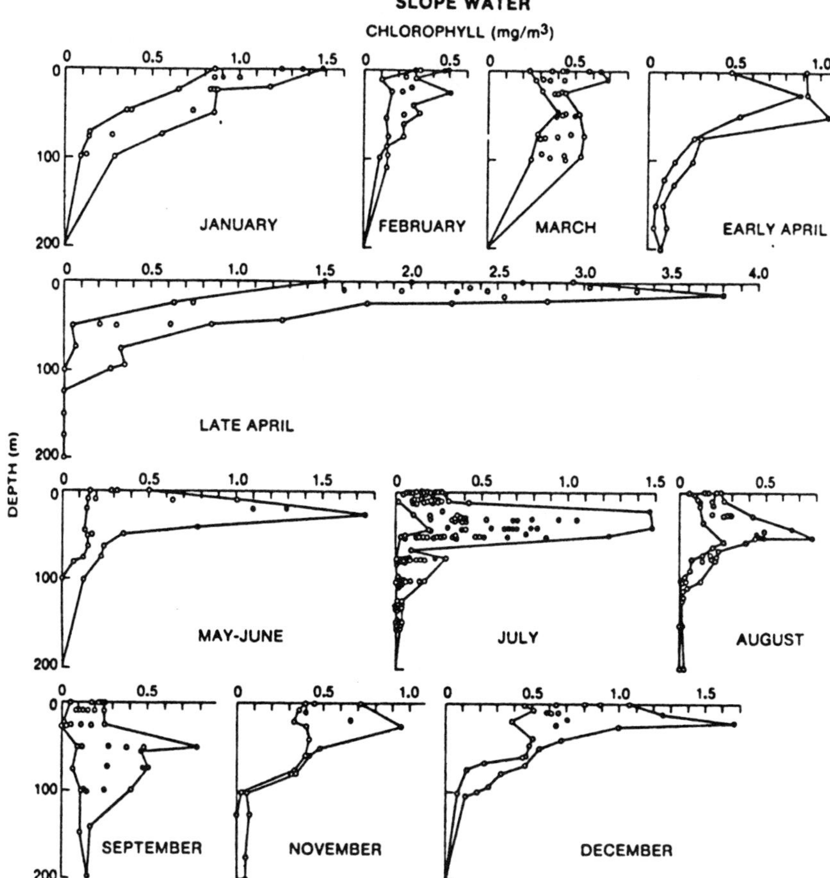

Figure 6.8 Composite vertical profiles of chlorophyll *a* from the slope water. Sources of data are listed in the text. From Cox et al. (1982).

Slope Water and the Sargasso Sea could help to explain the observed regional differences in depth profiles of chlorophyll (Figs. 6.7 and 6.8; Cox et al., 1982; see also the zooplankton section). For example, differences in the degree to which zooplankton could aggregate in the chlorophyll maximum layer in one region or the other would clearly impact the observed chlorophyll concentrations. Such aggregations of macro- and microzooplankton are more common at the deep chlorophyll maximum in the Sargasso Sea than in Slope Water.

Variability in Distribution of Phytoplankton

From the discussion above, as well as the data shown in Figures 6.7 and 6.8, there are certain general seasonal patterns as well as regional differences in phytoplankton biomass in Slope Water and in the northern Sargasso Sea. Yet a tremendous variability in phytoplankton biomass in both regions is observed on both large and small scales. One source of large-scale variability in phytoplankton biomass within the ACSAR region is the sharp boundary zone separating coastal and shelf water from Slope Water (Fournier et al., 1977; 1979). Such fronts may become regions of localized aggregations of phytoplankton biomass in surface waters, particularly during spring bloom periods, and may be sites of enhanced production as well (Uda, 1959; Fournier et al., 1979). Typically high concentrations of phytoplankton biomass develop in the surface waters on the stratified side of the frontal boundary (Slope Water) during the spring bloom, but nutrients may subsequently become depleted. In nearby shelf water, waters remain unstratified due to strong tidal currents. Although these currents may provide high nutrient levels by rapid vertical mixing, they also may limit the time during which phytoplankton are at light levels of sufficient intensity to cause high production (Pingree et al., 1975; Herman and Denman, 1979). In the front itself there are intermittent periods of thermal stability and nutrient renewal from tidal or wind-mixing (Fournier et al., 1979), resulting in enhanced production. Such appears to have been the case along a frontal boundary 100-200 miles south of Nova Scotia during spring 1977, where Herman and Denman (1979) measured chlorophyll concentrations of 1-2 mg m^{-3} in the surface coastal waters, 2-3 mg m^{-3} in Slope Water, and 4-7 mg m^{-3} in the front itself. In a winter set of observations off Nova Scotia, chlorophyll concentrations averaged 1.3 mg m^{-3} in shelf water, increased sharply to 4.5 mg m^{-3} along a 6.5° temperature increase, and dropped to ~1.0 mg m^{-3} in Slope Water (Fournier et al., 1979). No peak in biomass has been observed near the shelf break in late fall (Fournier et al., 1977). The available data thus serve to demonstrate that the shelf-break front may be a well-defined feature at least several months of the year (Fournier et al., 1979; Herman and Denman, 1979), which may significantly impact production as well.

The formation, development, and/or presence of Gulf Stream rings may also have a significant influence on the biomass distribution within Slope Water and the Sargasso Sea. Based on preliminary data, it appears that highest chlorophyll *a* values were observed near ring center, compared with the ring periphery, although considerable variability in concentrations occurred during any one month (G. Hitchcock, pers. comm.). Streamers or intrusions could have contributed to this variability. Given the fact that warm-core rings can occupy approximately 40% of surface Slope Water at various times, their impact on overall Slope Water biomass budget can be very significant. Likewise, cold-core rings, occupying 10-15% of the northern Sargasso Sea at any time, may also have a strong influence on mesoscale biomass structure.

Patchiness, independent of ring structure, also deserves comment. Phytoplankton patches on the order of 10-100 km may develop partly as a result of turbulent diffusion (Steele, 1976), but the environmental influences mediating patchiness development are not fully understood. Therriault and Platt (1981) have provided evidence, for one near-shore ecosystem off Nova Scotia, that during periods of low turbulent mixing patchiness was induced by local differences in phytoplankton production efficiency, whereas during periods of high turbulent mixing, spatial variations in phytoplankton distribution could become more homogeneous. This supports the notion that small-scale structure develops according to physiological dynamics until overriden by physical processes (Platt and Denman, 1980). Thus when wind stress is low, spatial variation in production and biomass may be ascribed to differences in physiological state of the phytoplankton, but when wind stress is high enough for surface layer mixing, it dominates other sources of variability (Therriault and Platt, 1981).

Species Composition and Distribution

Regional and Seasonal Phytoplankton Composition

The most extensive studies on the composition of phytoplankton within the ACSAR region have been conducted by Marshall (1971, 1976, 1984), who has summarized data from a variety of literature sources as well as from collections made during 42 cruises from 1964 to 1981 covering sampling areas from the Gulf of Maine and southeast of Nova Scotia to the Florida Straits. A total of 609 species of phytoplankton were identified (but later updated to over 900), which included 277 diatoms, 247 pyrrhophyceans, 54 coccolithophores, 9 silicoflagellates, 6 cyanophyceans, and 16 representatives of the Chlorophyta, Euglenophyta, Crytophyceae, and Xanthophyceae (Marshall, 1976). Of the species identified in the Marshall (1971) study, 76% occurred in only one of the three regions studied, the continental shelf, the Gulf Stream, or the Sargasso Sea.

In terms of relative abundance, the concentration of diatoms decreases seaward, whereas the concentration of coccolithophores increases significantly in pelagic waters, and waters above 23°C (Marshall, 1976). Dinoflagellate concentrations do not reach levels observed for diatoms, although isolated blooms may occur. Two other studies of phytoplankton composition in waters of the continental shelf (Fawley et al., 1980; Kalenak and Marshall, 1981) have also listed diatoms, dinoflagellates, and coccolithophores in decreasing order of abundance. In these latter two studies representatives of the blue-green algae, silicoflagellates, and an unidentified ultraplankton component were also noted. In Figures 6.9-6.11 are shown the seasonal distribution of total diatoms, dinoflagellates, and coccolithophores, as recently compiled by Marshall (1984). Highest concentrations of total phytoplankton were frequently observed adjacent to the lower New York Bay, the Delaware Bay, and the Chesapeake Bay, as well as in the Gulf of Maine, Georges Bank, and along the shelf margin.

Recently, another class of phytoplankton, the cyanobacteria, have been found to have a cosmopolitan distribution, and contribute significantly to primary productivity (Waterbury et al., 1979; Li et al., 1983). Because of their prokaryotic nature, however, a discussion of their distribution and abundance is included with the microbiology section of this volume.

Temporal and Spatial Variability

Several other studies on phytoplankton species composition and distribution have been conducted in the ACSAR region, but were more localized in scope. Hulburt and MacKenzie (1971) assessed the distribution of phytoplankton of the continental shelf of the southern United States during the winter of 1968. Marshall (1982) assessed the phytoplankton abundance and distribution in southeastern shelf waters of the United States. In contrast to several of the earlier studies, this study observed a large unidentified ultraplankton component in almost all collections, as well as a fair amount of patchiness in species dominance. Species identifi-

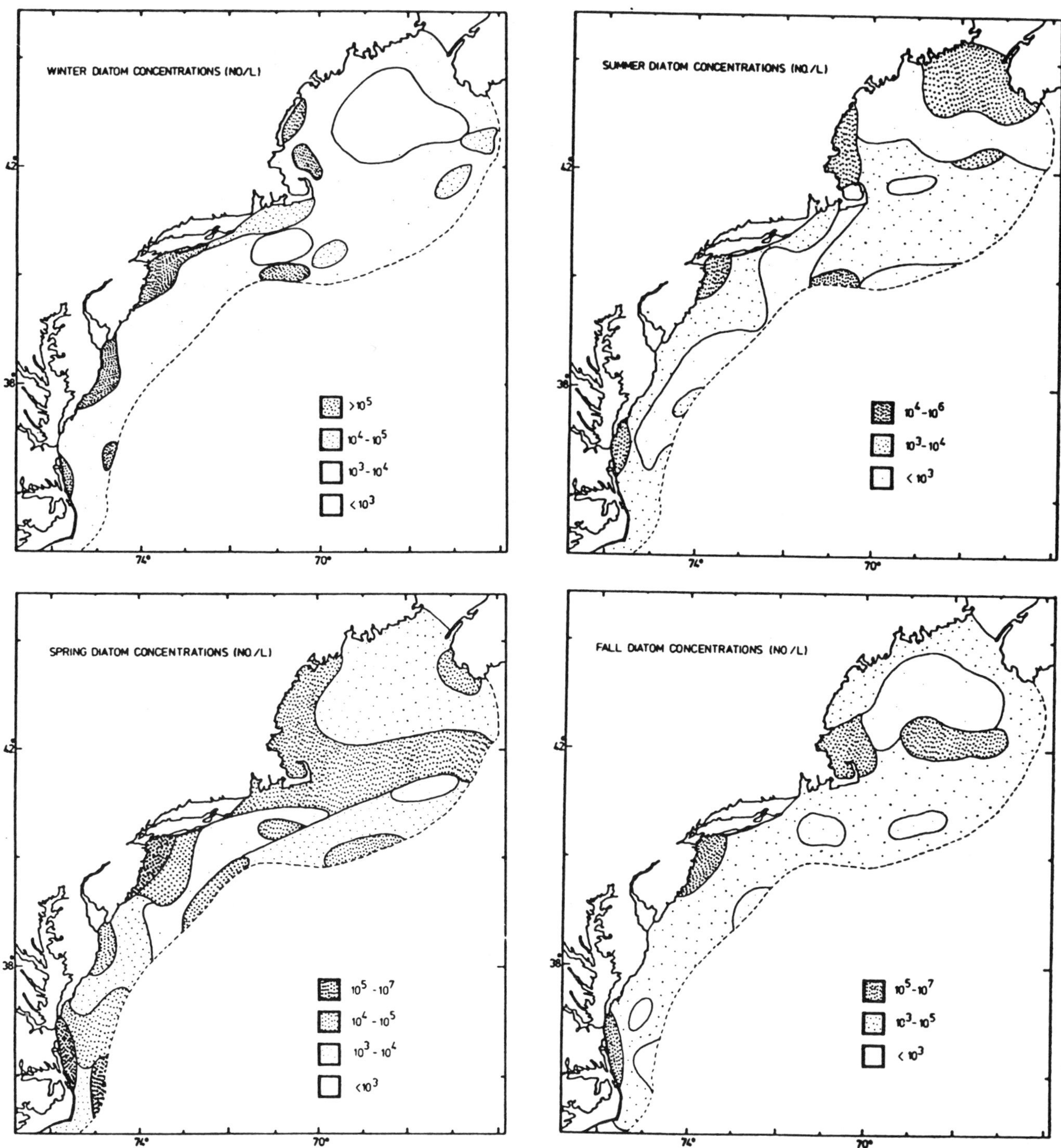

Figure 6.9 Seasonal distribution of diatoms in surface waters off the northeastern U.S. From Marshall, 1984.

cations of phytoplankton in and around Gulf Stream warm-core rings have also been made, but as yet are unpublished.

Phytoplankton composition in the northern Sargasso Sea, Slope Water, and in Gulf Stream cold-core rings was also described by Ortner et al. (1979) in relation to the physical and chemical properties of the water masses. They observed that although the physical and chemical properties of the cold-core rings frequently appeared intermediate between Slope Water and northern Sargasso Sea conditions, at no time did species composition appear intermediate. Rather, phytoplankton composition in northern Sargasso Sea samples was more similar to that in Slope Water samples than that in the rings. Diatoms appeared to be greatly reduced in numbers in the cold-core ring relative to the

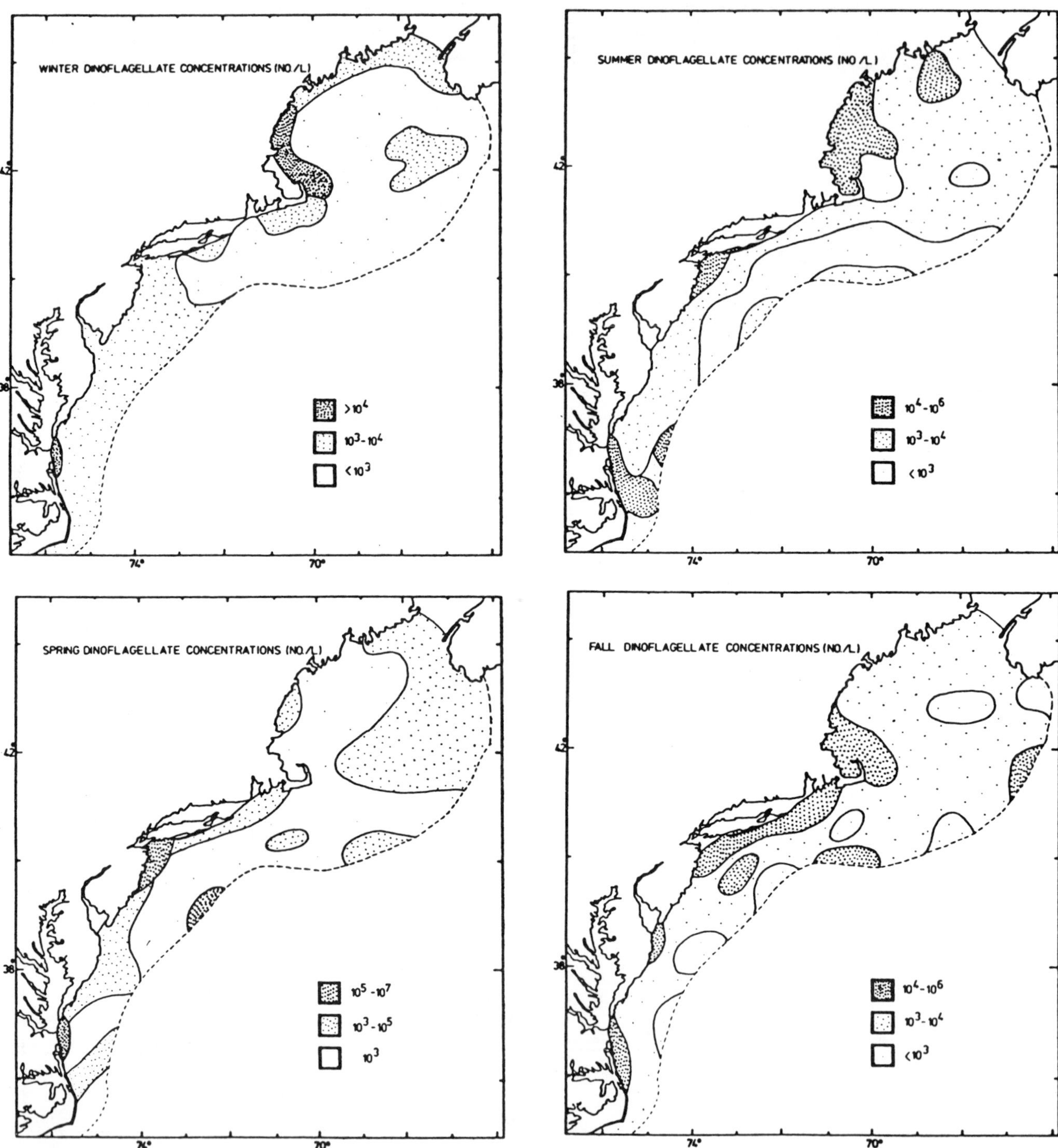

Figure 6.10 Seasonal distribution of dinoflagellates in surface waters off the northeastern U.S. From Marshall, 1984.

Slope Water or northern Sargasso Sea, while several dinoflagellate and coccolithophore species had higher abundances in the rings. Factors such as differential grazing pressure, nutrient flux, and physical mixing processes were hypothesized to be important in controlling these distributional patterns (Ortner et al., 1979).

In summary, the studies cited above have all indicated that a characteristic pattern of abundance as well as species composition exist for shelf waters, the Gulf Stream, and the Sargasso Sea, with considerable localized distributional patterns or patchiness. Clearly, the boundaries between regions are variable and mixing will occur, particularly from the meanderings of the Gulf Stream. The capacity of species to change or to tolerate new environmental conditions will determine their success in a new water mass (Marshall, 1971; Hulburt, 1983).

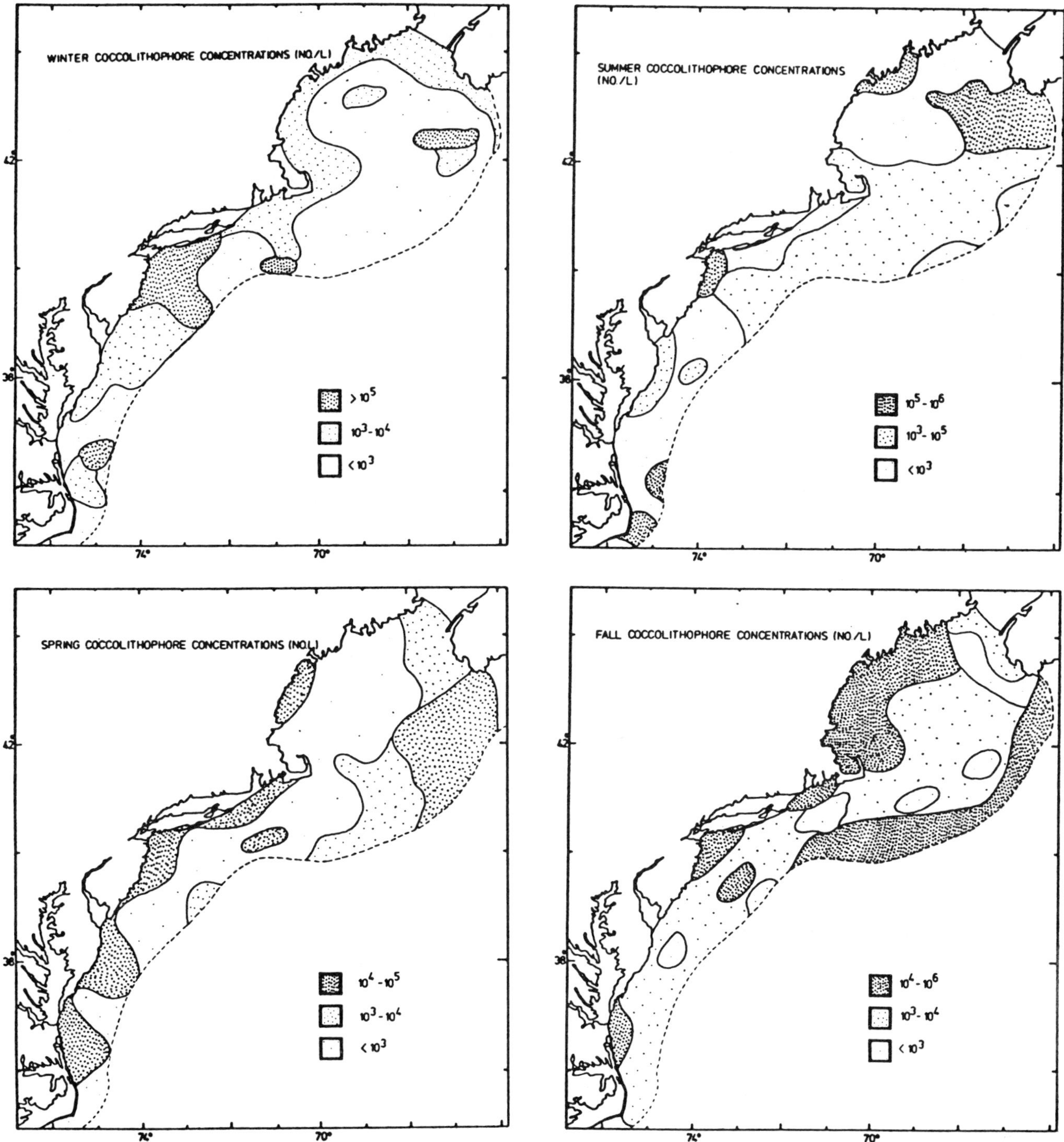

Figure 6.11 Seasonal distribution of coccolithophores in surface waters off the northeastern U.S. From Marshall, 1984.

Rates of Primary Productivity

Regional and Seasonal Primary Productivity Patterns

There have been numerous reviews over the past several decades of the extensive literature on seasonal cycles of production in the major oceanic regions (cf. Ryther, 1963; Cushing, 1975; Eppley, 1981). In general, seasonal patterns of production can be explained in terms of stratification of the water column and the rate of nutrient input. The spring bloom typically occurs when seasonal stratification begins to set in, thereby reducing the depth to which surface phytoplankton are mixed. Net phytoplankton growth occurs then because the mixing depth is shallower than the 'critical depth', the depth at which integrated water column photosynthesis equals respiration. When nutrients become depleted in the surface waters, growth rates and standing stocks decline. Development of a subsurface chlorophyll maximum may be noted following spring blooms (Figs. 6.7 and 6.8). A fall bloom also may be seen, if there is renewed nutrient input to the surface, or if grazing pressure becomes relaxed (Eppley, 1981).

Unfortunately few data sets exist for offshore waters where seasonal cycles of production have been assessed over the course of several years. Rather, most productivity measurements have been made as part of oceanographic cruises to specific areas at irregular times. In the Sargasso Sea, however, there has been one study (Menzel and Ryther, 1960; 1961) in which productivity rates were determined on a biweekly basis for two and a half years. This data set, shown in Figure 6.12, shows not only a clear seasonal pattern in productivity rates, with the maxima between January and April, but also major yearly differences in the amplitudes of the maximum and minimum productivity. These year-to-year differences were at least partially attributed to climatic differences, in that the winter of 1957-1958 was severe enough to destroy the thermocline by March, but the subsequent winters were milder and slight thermal gradients persisted all year (Ryther, 1963).

Malone et al. (1983; see also references cited therein) have recently reviewed seasonal as well as small-scale variations in the distribution and growth of phytoplankton on the continental shelf and adjacent Slope Waters. They observed that whereas phytoplankton biomass (as chlorophyll) is at a maximum during March-April and a minimum during July, production per unit biomass increases from a November-January minimum to a July maximum. Approximately 39% of the annual production for the region as a whole occurs during the spring diatom bloom period (February-April), and 54% during the period of water column stratification (May-October).

Within this broad perspective, biomass availability and productivity can be influenced by physical factors such as wind events and the shelf-break front (Malone et al., 1983). High phytoplankton growth rates and significant biomass accumulation occur during both spring and summer on the shelf side of the front. However, during the diatom bloom period, most of the biomass accumulates in surface waters due to stratification which sets in between storm events. In contrast, most of the biomass in the summer accumulates below the pycnocline (Malone et al., 1983).

Factors Regulating the Rate of Primary Productivity

Light alone rarely limits the rate of primary productivity in oceanic waters; it has been suggested that if light alone were limiting, then our estimates of primary productivity would be five to ten times those typically reported (Ryther, 1959; Vishniac, 1971). During winter however, inshore productivity may be limited by light availability (Fournier et al., 1979). Additionally, phytoplankton growth rates during the spring diatom bloom on the continental shelf may be light-limited, at least at times, as accumulation of biomass is high (Malone et al., 1983). The strategies by which phytoplankton adapt to different light levels have been the subject of many investigations, and have been reviewed by Falkowski (1980) and Harris (1978, 1980).

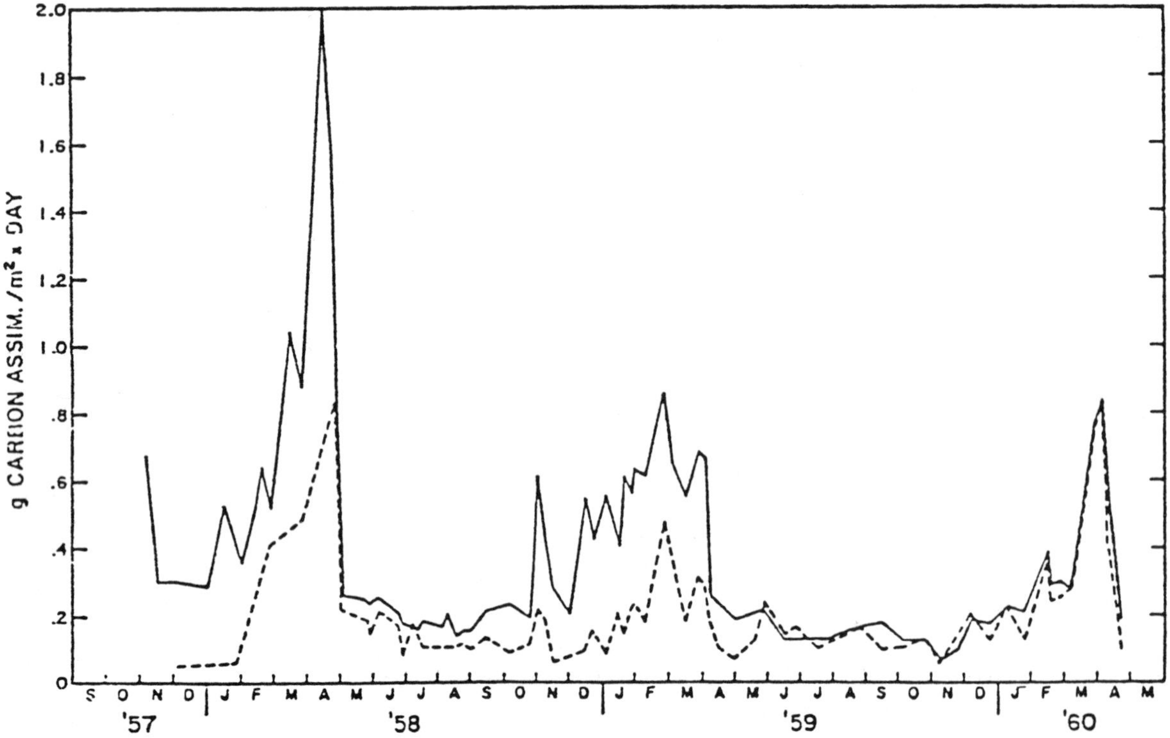

Figure 6.12 Gross (solid line) and net (broken line) primary production at station 'S' 1957-1960. (From Menzel and Ryther, 1961).

Temperature may be extremely important in controlling both the rate of primary productivity as well as seasonal succession of species. Eppley (1972), in reviewing the wealth of published data on algal growth rates, established that there was an upper limit to phytoplankton growth which could be described by the equation:

$$\mu_{max} = 0.851(1.066)^t$$

where μ_{max} is the specific growth rate in doublings/day, and t is the temperature of growth in degrees Celsius (<40°C).

This is an extremely useful expression, but it should be born in mind that other factors may prohibit certain species from attaining their maximal growth rate at a given temperature, and not all species will necessarily reach the same growth rate even when their optimal temperatures are similar (Goldman and Ryther, 1976). Malone (1976), for example, observed that for natural phytoplankton assemblages from the New York Bight, maximal photosynthesis (normalized to unit biomass) was similar at 4° and 24°, but significantly less at intermediate temperatures. This phenomenon was attributed to time periods of relative temperature stability: at the warmest and coolest temperatures, cells have the opportunity to adapt to the temperature regime and flourish, whereas during rapid warming or cooling, adaptation is more difficult. Glibert et al. (1985) also noted that maximal photosynthetic rates are not significantly different for summer and winter phytoplankton assemblages off Woods Hole. Hitchcock and Smayda (1977) have also observed large winter blooms at <2°C.

The nutritional requirements of phytoplankton for macro-nutrients (nitrogen and phosphorus) and micro-nutrients (vitamins and trace metals) have been well reviewed in the past several years (McCarthy, 1980; 1981; Nalewajko and Lean, 1980; Swift, 1980; Huntsman and Sunda, 1980; Bonin and Maestrini, 1981; Maestrini and Bonin, 1981). Vitamins and vitamin-like substances have been identified as essential for phytoplankton growth, but there is little evidence that their availability limits marine primary productivity. There is more evidence that trace metals may limit productivity, and may also be toxic to phytoplankton in their excess. It is now clear from several studies that it is the activity of the free metal ion which determines the availability or toxicity of the metal (Sunda and Guillard, 1976; Anderson and Morel, 1978). The free ion activity is, in turn, a function of the organic chelator concentration and pH.

Of the major elements required by phytoplankton, nitrogen is usually considered to be the nutrient limiting primary productivity. The role of NH_4^+ and NO_3^- in the physiological ecology of phytoplankton has received considerable attention during the past decade. Nevertheless, the degree to which nitrogen may be limiting phytoplankton growth in oligotrophic oceanic waters is still not well understood, in part because current analytical techniques restrict our ability to collect data on the temporal and spatial scales necessary to make meaningful interpretations of *in situ* rates of nitrogen utilization (McCarthy, 1980; Goldman and Glibert, 1983). Goldman et al. (1979), using chemical composition data, concluded that growth rates of natural oligotrophic phytoplankton were near maximal and not nutrient limited, and Glibert and McCarthy (1984), using a variety of indices of nutritional status, have demonstrated only mild, if any, nitrogen limitation for phytoplankton assemblages in the Sargasso Sea during summer. Resolution of the uncertainty regarding the degree of nitrogen limitation typical for oceanic phytoplankton would further our understanding of the role nitrogen plays in regulating marine primary productivity. Clearly one of the major recent advances in our understanding of the marine nitrogen cycle is the recognition that a small nutrient pool turning over rapidly (NH_4^+) can be as important, if not more important, to phytoplankton nutrition as a large nutrient pool with a longer renewal time (NO_3^-).

On a more global scale, we clearly lack sufficient data to state whether nitrogen fluxes to and from the oceanic region are balanced (McCarthy and Carpenter, 1983). Deficiencies are most pronounced in estimating the vertical flux of NO_3^-. This problem is confounded by the fact that a variety of physical processes may contribute to vertical nutrient flux on variable time and space scales. Additionally, given current analytical methodology, we are poorly equipped to measure small concentration changes with necessary spatial resolution. In this regard, collaborations between phytoplankton ecologists and geochemists are providing both new approaches (Jenkins, 1982; Altabet and Deuser, 1985; Altabet and McCarthy, 1985a,b) and new perspectives on global nitrogen cycling.

Macrozooplankton

P.H. Wiebe

Zooplankton are aquatic animals ranging from the smallest protozoans to the largest shrimps and jellyfish. Although many are able swim sizeable distances at moderate speeds and thus can perform diel vertical migrations of hundreds of meters, their large-scale horizontal distributions are determined by the ocean currents and the suitability of the physical, chemical, and biological components of the hydrographic regimes they encounter. Thus, they are distinguishable from the larger nektonic marine animals, which have control over their horizontal as well as their vertical distributions.

Standing Crop

Horizontal Distribution

Compared to the middle Atlantic continental shelf region, the zooplankton of the Slope Water, Gulf Stream, and northern Sargasso Sea have been poorly sampled until recently. Previous to the current work described below, the only long-term studies of zooplankton of the region were by Clarke (1940) and Grice and Hart (1962). Both studies were based on samples taken at stations on the continental shelf, in the Slope Water, and in the Sargasso Sea on a line from Montauk Point, Long Island, N.Y., to Bermuda. Clarke's samples were principally collected with ring nets equipped with scrim netting (about 800 μm mesh and with stramin netting (about 1500 μm mesh); they were taken in the upper 50 to 275 m on 10 cruises spaced irregularly throughout the years 1937 to 1939. Grice and Hart used ring nets equipped with 230 μm mesh and towed obliquely to between 100 and 200 m, on five quarterly cruises taken over a 15-month period (1959-1960). Such shallow sampling provided a biased estimate of the seasonal pattern of upper water column zooplankton.

Bé, et al. (1971) provided the most recent comprehensive summary of the zooplankton biomass in the upper 300 m for the entire North Atlantic Ocean. A sizeable number of samples were collected within the Slope Water/Sargasso Sea portion of the ACSAR area, but they were not separated out for detailed discussion.

Since 1972, studies of the zooplankton populations in Gulf Stream rings (see chapter 2 for description of the physical structure of these mesoscale eddies), the Sargasso Sea, and the Slope Water have been carried out by P. Wiebe and associates at Woods Hole Oceanographic Institution. For the first three years 1-m diameter ring nets and opening/closing Bongo Nets (McGowan and Brown, 1966) equipped with flowmeters were used to sample to depths of 800 m. With the development of the Multiple Opening/Closing Net and Environmental Sensing System (MOCNESS) in 1974 (Wiebe et al., 1976b), higher resolution stratified sampling of the region commenced. In 1981 and 1982, the zooplankton group at Woods Hole conducted two concurrent programs in the northern portion of the study area. The first was a 19-month time series study of the life history and population dynamics of Slope Water populations. During this investigation, great effort was made to sample unadulterated Slope Water, i.e. water not directly influenced by warm-core rings or the Gulf Stream. The second was a time-series study of the warm-core rings themselves and the zooplankton work was only a small part of total multidisciplinary effort (Warm-Core Rings Executive Committee, 1982).

The picture of the seasonal cycle of biomass in the northern Sargasso Sea outside the influence of cold-core rings is incomplete; because sampling has been spread out over a number of years, it must be presented as a composite picture rather than as a time-series for any one year. Values integrated from 800 to 1000 m appear to show a pattern considered typical of subtropical regions of a spring high and a late fall and winter low (Fig. 6.13). A similar pattern emerges if data from depth-specific tows integrated from 200 m to the surface are used. However, only the April data are substantially higher than other months and there are no data for the key months of May, June, and July. Thus, in the northern Sargasso Sea evidence for a coherent seasonal cycle of zooplankton biomass in the upper 1000 m is skimpy at best. Similar conclusions are reached for the Slope Water where extreme variability obscures any underlying seasonal pattern and ontogenetic vertical migrations (migrations associated with the reproductive cycle of the animal; i.e., eggs, larvae, and young live at or near the surface and adults live at depth) appear responsible for the apparent seasonal pattern in surface waters.

Using Slope Water time-series data, there appears to be a seasonal enhancement of biomass during the spring within

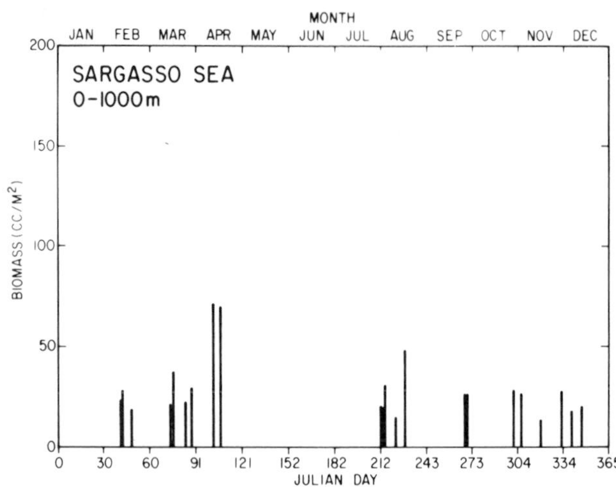

Figure 6.13 Seasonal cycle of zooplankton biomass (displacement volume) in the upper 1000 m of the Sargasso Sea based on tows taken between 1972 and 1982.

the upper 200 m, but very little evidence for this trend within the upper 1000 m as a whole (Fig. 6.14). Biomass of zooplankton in the upper 200 m of the water column, however, shows fairly pronounced changes although not neccessarily on a seasonal basis (Fig. 6.15). While data are not available for a definitive explanation for these differences, ontogenetic shifts in the vertical distribution of dominant copepod species such as *Calanus finmarchicus*, which come to the surface to spawn in the spring and whose progeny subsequently return to subsurface depths a month or two later, are likely to be responsible in part for the variations in the upper 200 m.

Within the upper 200 m in this data set values vary by a factor of about eight whereas Clarke (1940) and Grice and Hart (1962) reported variations up to 40. Much of the differ-

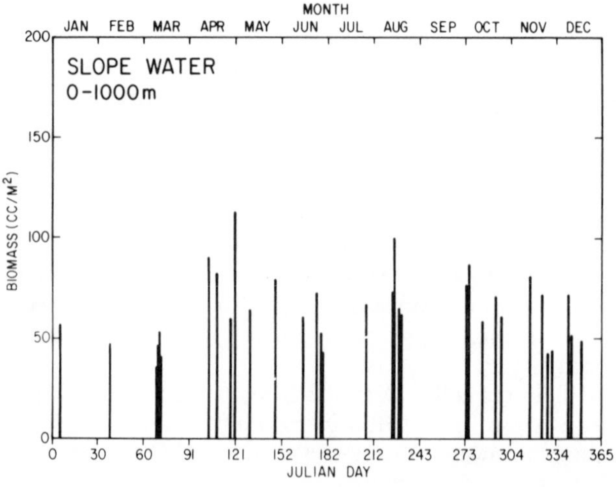

Figure 6.14 Seasonal cycle of zooplankton biomass (displacement volume) in the upper 1000 m from "hydrographically pure" Slope Water based on tows taken between 1972 and 1982.

ence can be explained by differences in mesh size and tow depths of the nets, but both investigations, without doubt unknowingly, sampled portions of warm-core rings with lower biomass rather than Slope Water.

Clark (1940) and Grice and Hart (1962) found about four times more zooplankton biomass in the Slope Water than in the Sargasso Sea within the depths they sampled. If the data of Wiebe and coworkers for the upper 1000 m are compared, the difference is only a factor of 2 or 3. Clark's physical data indicate that some of his samples were taken in cold-core rings rather than in Sargasso Sea water. His high values may have come from these collections.

Effect Of Interaction With Shelf Water, Including Seasonal Dependence. In addition to the bias introduced into the seasonal cycle of Slope Water zooplankton by the inadvertent sampling of warm-core rings or Gulf Stream meanders, bias can be caused by the sampling of Slope Water overridden by shelf water. Nutrient rich shelf water can extend out over the Slope Water either as a result of cold-water bulge formation or as part of the entrainment field frequently associated with the presence of warm-core rings. In the former case shelf water can penetrate into the Slope Water as much as 80 km (Wright, 1976b; Halliwell and Mooers, 1979; Mooers et al., 1979) whereas in the latter, shelf water can be drawn out and then entrained along the northern edge of the Gulf Stream. Because shelf water is relatively fresh (34‰, it is difficult to mix vertically with the underlying Slope Water; high nutrients and shallow mixed layers appear to give rise to enhanced production throughout the year. It is not certain how shelf water intrusions affect seasonal estimates of zooplankton standing crop.

Effect Of Cold- And Warm-Core Rings. The above discussion of seasonal cycles in biomass structure of the Slope Water and Sargasso Sea ignores the significant exchanges of water and biota between these two hydrographic regions as a result of the formation and presence of Gulf Stream rings. Cold-core rings occupy between 10 and 15 percent of the surface area of the northern Sargasso Sea at any given time and warm-core rings can cover as much as 40 percent of the Slope Water. They have, for a significant portion of their existence, a biomass structure and species composition distinctly different from the adjacent Sargasso Sea and Slope Water (Wiebe et al. 1976a; Jahn, 1976; Wiebe, 1976; Ortner, 1978; Wiebe and Boyd, 1978; Cox and Wiebe, 1979; The Ring Group, 1981; Wiebe, 1981b; Wiebe et al., in press). By virtue of the distribution of rings, there is in the northwestern Atlantic Ocean a mosaic pattern of expatriated communities interspersed throughout home-range communities. This pattern is continuously changing because of the horizontal movement of rings and because of hydrographic changes resulting from air-sea interactions and physical exchange processes with adjacent waters which foster change in ring biotic structure towards that of the surrounding water (Wiebe and Flierl, 1983). Thus, rings are responsible for major perturbations in the horizontal and vertical distribution of zooplankton biomass in the uppr r 800 to 1000 m of the water column.

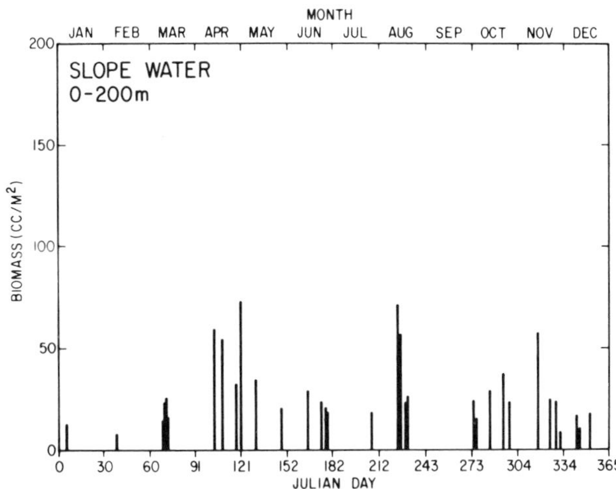

Figure 6.15 Seasonal cycle of zooplankton biomass (displacement volume) in the upper 200 m of the Slope water based on tows taken between 1972 and 1982.

In the case of cold-core rings, zooplankton biomass in the upper 1000 m is generally higher in the ring core than in the surrounding Sargasso Sea for at least a year after formation. For the five different rings reported by Ortner et al. (1978) and Wiebe et al. (1976a), standing stock of zooplankton was from 1.3 to 1.8 times larger. Data from four additional rings support these findings (The Ring Group, 1981).

As much of the work on the biological structure of warm-core rings has occurred recently (Warm-Core Rings Executive Committee, 1982), most of the data have not been published. However, some preliminary information is available about the zooplankton biomass structure of warm-core rings (Wiebe, 1982b, Wiebe et al., in press). The data were collected on six cruises between September 1981 and October 1982, the four middle ones being to ring 82-B.

Ring 82-B was first sampled in March 1982 about 3 weeks after it was formed. The waters of the ring center were isothermal at 17.7 °C from the surface to 330 m and the salinity was 36.6‰, indicating winter-mixed northern Sargasso Sea water. By April, winter mixing had cooled the core waters to 15.6 °C and had extended the depth of the mixed layer to 440 m. During both of these sampling periods, the total zooplankton biomass per m² in the ring center was significantly lower than in adjacent hydrographic regimes. Between March and April, biomass increased in ring 82-B by about 50 percent and in the Slope Water by about a factor of three (the result of a major spring bloom of phytoplankton). In the ring core, production and biomass were also high but because the mixed layer was several hundred meters deep (i.e. seasonal stratification had not yet occurred in the ring) concentrations of phytoplankton were much lower than in the Slope Water.

Data for June indicate a major change. Between April and June, the ring surface waters had warmed sufficiently to form a shallow mixed layer and seasonal thermocline. Ship of opportunity observations in May showed sharp increases in phytoplankton biomass concentration in the ring.

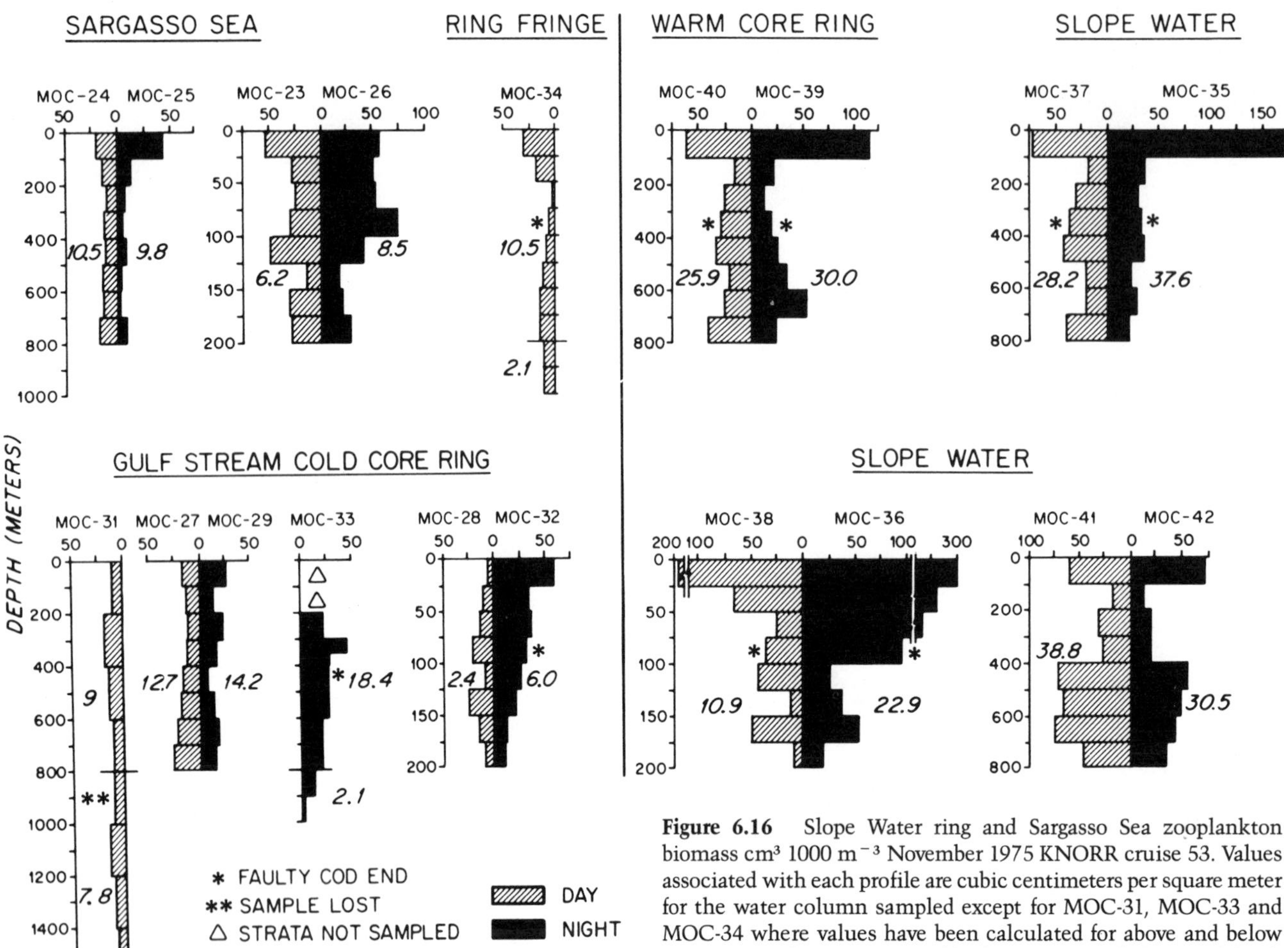

Figure 6.16 Slope Water ring and Sargasso Sea zooplankton biomass cm³ 1000 m⁻³ November 1975 KNORR cruise 53. Values associated with each profile are cubic centimeters per square meter for the water column sampled except for MOC-31, MOC-33 and MOC-34 where values have been calculated for above and below 800 m. Note that 0-200 m profiles have an expanded vertical depth scale. From Ortner et al., 1978.

Levels were greater in the ring than in the Slope Water, where levels had decreased.

During July, the ring underwent at least one interaction with the Gulf Stream during which it lost a considerable portion of its mass (Joyce et al., 1984). At the beginning of the sampling period in August, a Gulf Stream meander swept over part of the ring to depths of at least 75 to 100 m. The biomass in ring 82-B, while higher in August than in June, was not significantly lower than in the Slope Water. However, variability in the Slope Water was extreme with the jellyfish, *Pelagia pelagia*, dominating one Slope Water station and strongly affecting the estimates of the Slope Water zooplankton mean state. Ring 82-B biomass was substantially higher than in either the Sargasso Sea or the Gulf Stream.

In a comparison of the evolution of zooplankton standing crop in cold- and warm-core rings, one fact stands out. Elevation of biomass in warm-core rings can take place much faster than the decline in biomass in cold-core rings to Sargasso Sea levels. This is probably a result of the higher frequency of interactions that warm-core rings undergo with the Gulf Stream and continental shelf waters.

Vertical Distribution

The literature contains only a few quantitative studies of the vertical distribution of zooplankton biomass in the study area of the northwestern Atlantic Ocean. Sampling below 300 m was rare until the mid-1970's when a fairly extensive sampling program in the upper 1000 m of this area was begun by Wiebe and co-workers using MOCNESS. There are currently five different MOCNESS systems designed for capture of different size ranges of zooplankton and micro-nekton (Wiebe et al., in press). Each system is designated according to the size of the net mouth opening and in one case the number of nets it carries.

The distribution of biomass vertically throughout the upper 1000 m of the study area is not uniform with depth (Fig. 6.16). Because of the extreme variability in the Slope Water, it is difficult to describe quantitatively a typical biomass profile. However, zooplankton biomass is generally highest at night in the upper 100 to 200 m in both the Slope Water and northern Sargasso Sea. Values typically range between 50 and 500 cc/1000 m³ in the Slope Water and between 15 and 400 cc/1000 m³ in the Sargasso Sea (Ortner et al., 1978; Wiebe, 1981a; Wiebe et al., in press). In the Slope Water, there is often a substantial subsurface peak in biomass between 300 and 600 m (up to 200 cc/1000 m^3) which is not evident in the Sargasso Sea. Extreme deviations from the typical pattern in the Slope Water result from blooms of salps or jellyfish. For example, the salp, *Salpa aspera*, dominated the zooplankton biomass in August 1975 and was still present in substantial numbers in November 1975 (Wiebe et al., 1979).

There is very little published information that relates directly to the biomass structure of shelf water overriding the Slope Water. Samples taken in 1981 and 1982 as part of the Warm-Core Rings Program should provide the data necessary to evaluate the effect of such intrusions.

Effects Of Cold- And Warm-Core Rings. Highest biomass occurs near the center of a cold-core ring and usually declines on the flanks. However, in ring 'Bob', within and below the region of intense circular currents, a zone of low biomass was evident at intermediate depths on both sampling occasions (Fig. 6.17). Because the Gulf Stream supports a lower standing crop of zooplankton than either the northern Sargasso Sea or the Slope Water for a portion of the year, this unusual feature may have been introduced at the time of ring formation, or it may have resulted from the strong interaction with the Gulf Stream at age 2 months (The Ring Group, 1981).

Another feature in most cold-core rings is that, compared to the Sargasso Sea, an unusually large proportion of the biomass occurs between depths of 200 and 1000 m. There are times when older rings have significantly less surface layer (0-200 m) biomass than the surrounding waters and the generally higher ring standing crop is due entirely to the larger subsurface biomass. This was the case of cold-core ring D which was sampled at ages 6 and 9 months (Ortner et al., 1978).

The relationship between the deep chlorophyll maximum (DCM) and epizooplankton biomass and abundance distributions in the Sargasso Sea, cold-core ring D and the Slope Water was examined by Ortner et al. (1980) during summer and fall 1975. Total zooplankton biomass was significantly enhanced in or adjacent to the seasonal thermocline in all three hydrographic regimes. The DCM in these regions was also predictably associated with the seasonal thermocline, thus giving rise to a significant correlation between zooplankton biomass and the DCM.

In addition to measuring total biomass, Ortner et al. (1980) counted a number of categories of functional zooplankton groups in the Clarke-Bumpus samples. In August, the distribution patterns of particular zooplankton functional groups or taxa were consistent with those of zooplankton biomass in the three areas. In general, for the northern Sargasso Sea, medium copepods, larvaceans, various copepod developmental stages, molluscs, chaetognaths, and tintinnids were concentrated in the 75-100 m depth interval, i.e., DCM depths. Other groups were concentrated near the surface or were too variable to be categorized. Variations to this pattern were evident in the Slope Water and ring D. Fall mixing substantially changed the vertical temperature and salinity structure of the upper water column, largely erasing the DCM in the Slope Water and Ring D, but not in the Sargasso Sea. In the latter region, there was still a subsurface peak in biomass around the base of the seasonal thermocline and nearly all functional groups had distributions centered on the DCM which was still present. In ring D, although the DCM was not evident and subsurface peak in zooplankton biomass was very small, many of the functional groups had nocturnal peak abundances at the sharpened thermocline. In the Slope Water, zooplankton biomass was concentrated at or near the surface, yet some groups (for example, tintinnids, nauplii, and medium copepods) still showed enhanced numbers at the permanent thermocline.

For warm-core rings, the vertical biomass structure has been examined in more detail. In April, when ring 82-B was 2 months old, the median (50%) depth of biomass in the ring center was below 200 m whereas in the Slope Water it was about 50 m. The transition between the distribution in the ring core and the Slope Water was abrupt and coincided with the changes in the vertical temperature and salinity structure. A similar pattern is evident in the daytime data except that cumulative percentage biomass depths were 50 to 100 m deeper.

In June (ring age 4 months), the median biomass depth shoaled to between 100 and 200 m in the ring core and deepened to between 200 and 300 m in the Slope Water. The night data are more complete, but the trend is evident in the day data as well. The August data (ring age 6 months) show similar vertical distribution of biomass at night in both Slope Water and ring, and a fairly deep (<200 m) median depth of biomass. Similar distributions were found in the Sargasso Sea and Gulf Stream during this cruise.

The evolution of biomass as a warm-core ring ages can be compared with that in cold-core rings and in the Sargasso Sea. As described above, the median depth of biomass in ring 82-B at night started out deep (400 m) in March and progressively shoaled through June (Fig. 6.18). In August there was a downward shift due in part to interactions with the Gulf Stream. Day biomass was distributed generally less than 100 m deeper than at night. Cold-core rings show the opposite trend, with a deepening of median biomass with increasing age (Fig. 6.18) that is larger than in warm-core rings; the delta averages approximately 200 m.

Vertical distribution of biomass in Slope Water from the cold-core ring cruises is similar to that obtained on the warm-core ring cruises (Fig. 6.18), but the dramatic upward shift in biomass concentration occurred earlier in the Slope Water. In the Slope Water, the shoaling and subsequent submergence is believed associated in part with ontogenetic migrations of species such as *Calanus finmarchicus*. Ontogenetic migrations did not seem a factor in the biomass shift in ring 82-B. Diel shifts in Slope Water biomass also are similar and generally 100 m or less in magnitude. The distribution of biomass at night in the Sargasso Sea is no deeper and often shallower than in either the warm- or cold-core rings or the Slope Water (Fig. 6.18). Furthermore, there is a stronger diel migration pattern in the Sargasso Sea, with daytime median biomass 150 to 250 m below the nighttime level.

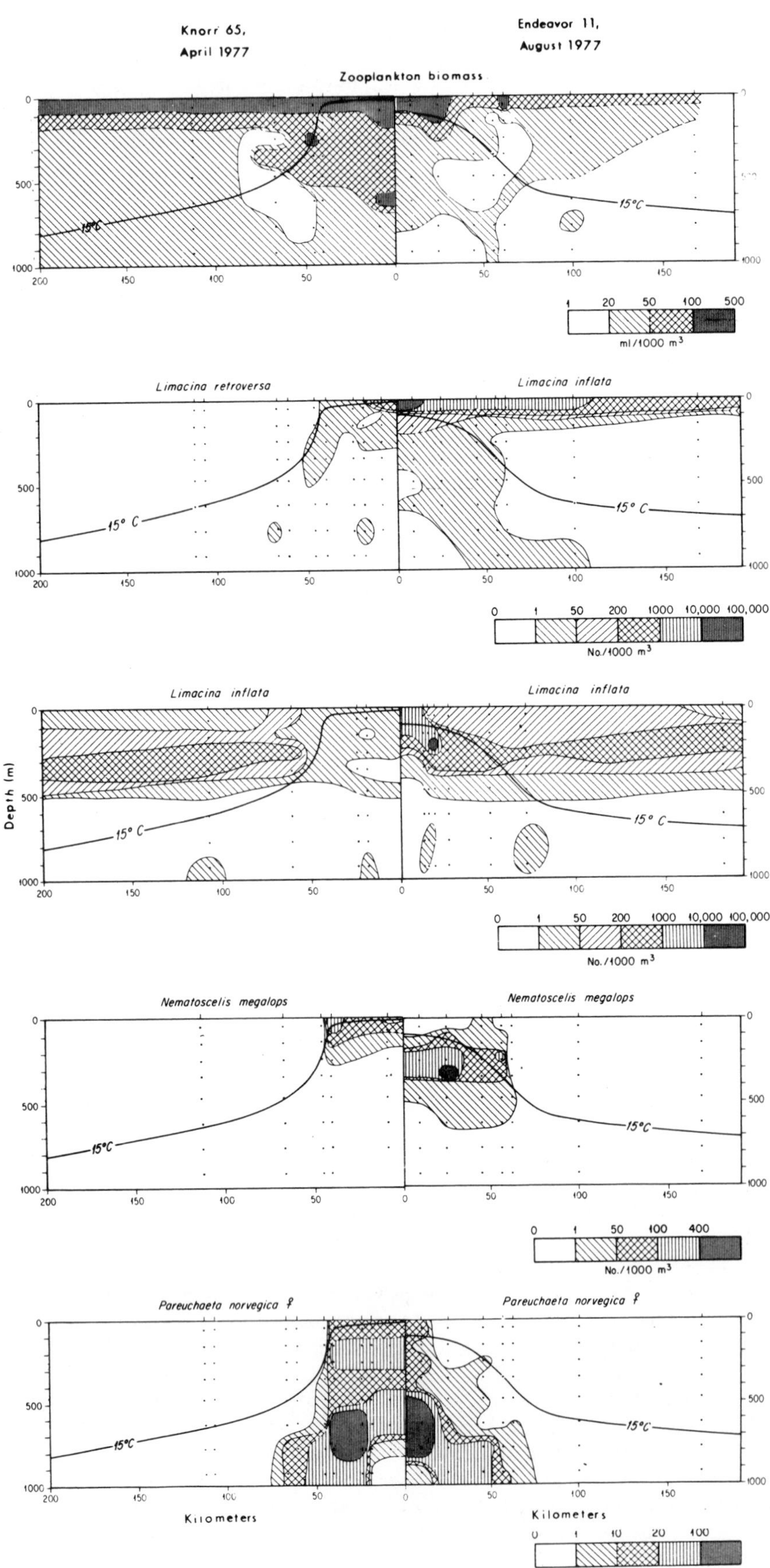

Figure 6.17. Vertical sections of zooplankton biomass [measured as displacement volume (55)] and the abundance of warm-water *(Limacina inflata)* and cold-water *(Nematoscelis megalops*, and *Pareuchaeta norvegica)* zooplankton indicator species from the center of ring Bob out to 150+ km for April 1977 (left) and August 1977 (right). Collections were made with a MOCNESS; the solid dots denote the center of the oblique portion of the tow taken with one of eight nets. The ring extended out to about the 80-km mark as indicated by the depth of the 15 °C isotherm; beyond was the Sargasso Sea. The cold-water pteropod *Limacina retroversa* already had disappeared from ring Bob by August. The left-right pair for *L. inflata* show daytime distributions. The vertical pair allows a day (above)-night (below) comparison for August and shows diel vertical migration. By August *Nematoscelis* and *Pareuchaeta* lay deeper in the water column and *Pareuchaeta* had become less abundant. From the Ring Group (1981).

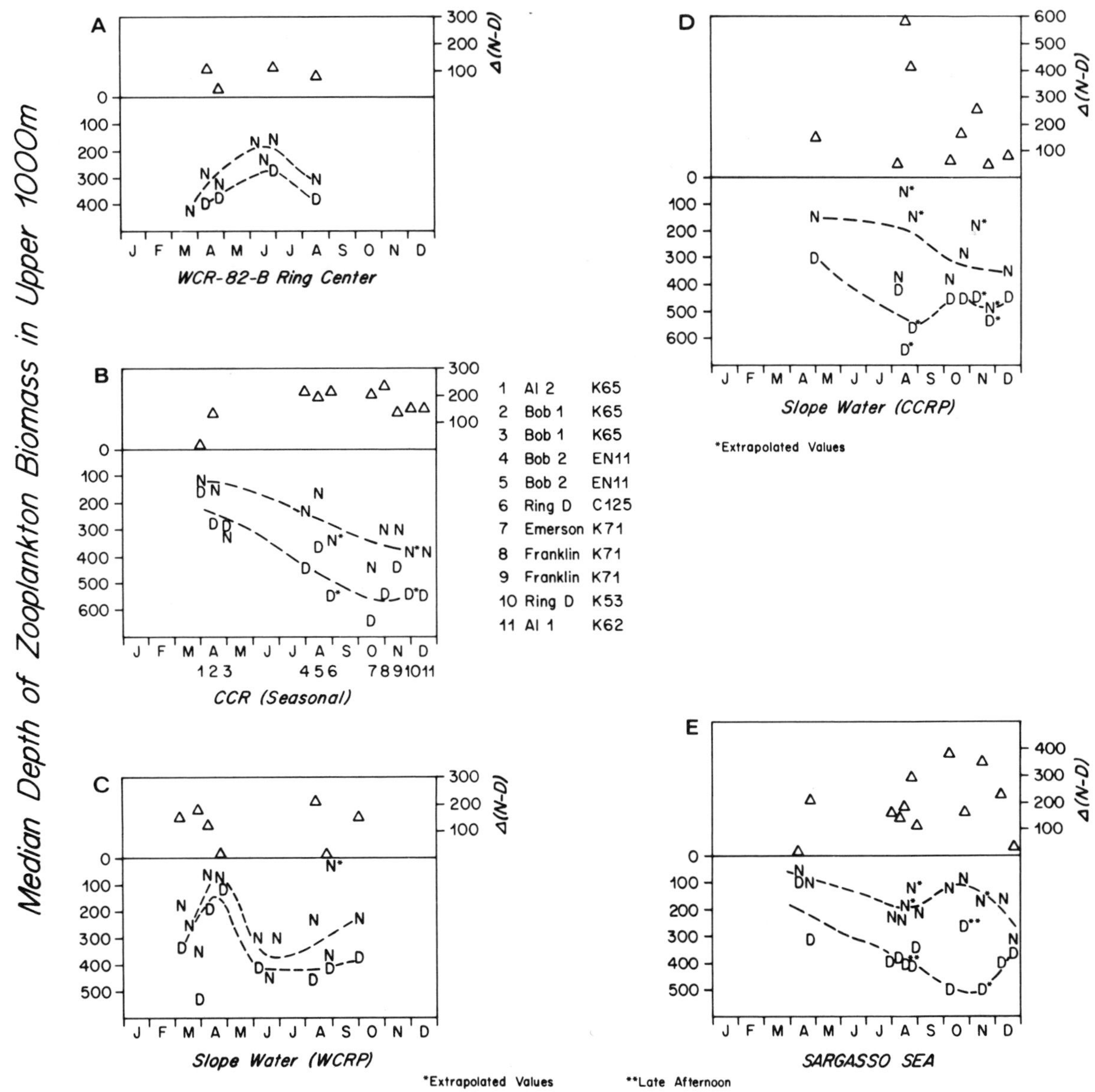

Figure 6.18 *(upper left).* Ring 82-B center day and night median depth of zooplankton biomass in the upper 1000 m and the difference between day and night medians plotted for the four time-series cruises (from Wiebe et al., in press). N = night, D = day. *(Upper right).*Cold-core ring Day and night median depth of zooplankton biomass in the upper 1000 m and the difference between day and night medians plotted for rings sampled in 1976 and 1977 (from Wiebe et al., in press). N = night, D = day. *(Lower left).*Slope Water day and night median depth of zooplankton biomass lower left in the upper 1000 m and the difference between day and night medians plotted for the four time-series cruises of 1982 (from Wiebe et al., in press). N = night, D = day. *(Lower right).*Sargasso Sea day and night median depth of zooplankton biomass in the upper 1000 m and the difference between day and night median plotted for the cruises of 1976 and 1977 (from Wiebe, Barber, and Boyd, in preparation). N = night, D = day.

Species Composition

Principal Groups and Relation to Biogeographic Provinces

A general discussion of the biogeographic patterns in the ACSAR area has been given above. Only the previous work of Grice and Hart (1962) includes counts of the vast majority of species which occurred in their samples thereby enabling a comparison of the relative importance of major taxonomic groups of the zooplankton. Dominant taxa include copepods, amphipods, chaetognaths, euphausiids, thecosomes, ctenophores, cnidaria, and thaliacia. Species of gelatinous zooplankton are generally under-represented because many species are damaged beyond recognition in nets or disentegrate in the preserving fluid (Harbison et al., 1978). Chenoweth (1978) has summarized what is known about the zooplankton species composition in most of the study area up to the mid-1970s. This review will dwell mainly on work published recently.

Horizontal Distributions

Typical Seasonal Distributions. There are few good seasonal data on the abundance, reproductive capacity or timing, size frequency distribution, and the rate of individual growth for most species in the ACSAR area. Point source sampling off Miami by Stepien (1980) provides an update of studies of plankton of the Florida Current by Pierce (1951), Moore (1953), Moore et al. (1953), Miller et al. (1953), Lewis (1954), Moore and Corwin (1956), Moore and O'Berry (1957), Pierce and Wass (1962), and Wormelle (1962). But Stepien's sampling was designed to look at reversals in the deep flow of the Florida Current; seasonal fluctuations in abundance can not be readily determined from her data.

Wormuth (1981) and Cheney (1982) examined some of the samples taken with the 9-net MOCNESS-1 by Wiebe and coworkers (described above), which provide seasonal patterns in abundance. Wormuth studied the vertical distribution and seasonal variation of nine of the most abundant or most frequently occurring pteropods in the northwest Sargasso Sea using 14 tows to 1000 m and 5 tows to 200 m. Variability in abundance was so great that he found that none of the seasonal patterns was significant although five of the nine species had highest values in the spring and one had highest values in the summer.

Cheney (1982) identified 21 species of chaetognaths from 52 9-net MOCNESS-1 tows in the Slope Water (18 tows), northern Sargasso Sea (18 tows), Gulf Stream (2 tows), and Gulf Stream rings (14 tows). For the dominant 18 species, mean abundances in the above hydrographic regimes (Table 6.2) showed major differences between the Slope Water and northern Sargasso Sea for all but 2 of these species (*S. enflata* and *K. pacifica*). Seven species were most abundant in the Slope Water, while nine were most abundant in the northern Sargasso Sea. In the Gulf Stream collections, abundances of chaetognaths were between those observed in the Slope Water and northern Sargasso Sea, although species composition was more similar to the northern Sargasso Sea fauna. The Gulf Stream is, however, a gradient region and similarity with other hydrographic regimes depends to a large extent on the placement of the tows within the gradient.

Cheney (1982) also examined seasonal pattern in integrated numbers (0- 1000 m) of individuals of each species. Cheney concluded that there was no evidence for seasonal change in chaetognath numbers in the Slope Water. On the other hand, 7 of the 17 species had significant shifts in seasonal abundance in the northern Sargasso Sea. Highest values occurred in April for 12 of 17 species; most species not showing such a peak were bathypelagic species. While Wormuth (1981) was forced by high variability in his pteropod data to conclude that seasonal fluctuations were not significant in the northern Sargasso Sea, the fact that highs for five out of nine of his species came at the same time as the chaetognath highs suggests a spring enhancement in abundance for some pteropod species.

Table 6.2 Mean chaetognath species abundance (No. m^{-2}) in the Slope Water (SW), Northern Sargasso Sea (NSS), Gulf Stream (GS) and cold core rings (CCR). The number of tows on which the average is based is given in parentheses. R represents the ratio of SW to NSS or NSS to SW abundance, whichever is larger; P is the probability of equal SW and NSS abundances as tested by the Mann-Whitney U test (*** = $p<0.01$, ** = $p<0.1$, * = $p<0.05$, NS = not significant). (From Cheney, 1982).

Species	SW		NSS		GS		CCR		R	P
				Slope Water Species						
E. bathypelagica	1.8	(9)	0.05	(14)	0.74	(2)	0.02	(9)	36	***
E. fowleri	5.6	(9)	2.8	(14)	3.4	(2)	5.7	(9)	2.0	**
E. hamata	43.5	(14)	0.87	(14)	4.5	(2)	30.6	(14)	50	***
S. helenae	7.6	(17)	0.08	(18)	5.9	(2)	0	(14)	95	*
S. macrocephala	39.5	(8)	7.6	(14)	29.6	(2)	44.9	(9)	5.2	***
S. maxima	21.1	(8)	2.0	(14)	3.5	(2)	45.5	(9)	11	***
S. tasmanica	339	(17)	0.30	(18)	60.5	(2)	69.3	(14)	1130	***
				Northern Sargasso Sea Species						
K. subtilis	3.2	(15)	103	(16)	23.9	(2)	31.0	(14)	32	***
P. draco	4.9	(17)	160	(18)	75.1	(2)	29.6	(14)	33	***
S. bipunctata	3.7	(17)	42.9	(18)	14.6	(2)	33.6	(14)	12	***
S. decipiens	14.7	(16)	150	(18)	90.3	(2)	144	(14)	10	***
S. hexaptera	3.2	(17)	101	(18)	38.5	(2)	34.9	(14)	32	***
S. lyra	8.9	(15)	179	(18)	51.6	(2)	56.5	(14)	20	***
S. minima	66.0	(18)	110	(18)	21.7	(2)	46.5	(14)	1.7	***
S. planctonis	2.0	(8)	7.6	(14)	3.1	(2)	3.4	(9)	3.8	**
S serratodentata	36.7	(17)	342	(18)	110	(2)	33.0	(14)	9.3	***
				"?" Species						
K. pacifica	13.5	(17)	24.8	(18)	25.1	(2)	1.9	(14)	1.8	NS
S. enflata	252	(17)	118	(18)	106	(2)	143	(14)	2.1	NS

Using the same samples as Cheney and Wormuth plus earlier collections with meter nets and Bongo nets, Wiebe (in manuscript) found that the total number of adolescent and adult euphausiids in the upper 1000 m of the Slope Water and northern Sargasso Sea also reflects the differences in hydrography. Numbers of individuals are substantially higher and considerably more variable in the Slope Water than in the northern Sargasso Sea. Although the seasonal picture is incomplete, there appears to be a spring high and a late fall and winter low in the northern Sargasso Sea. The extreme variability in the Slope Water obscures any underlying seasonal pattern that may exist, but minima occur in late fall and winter. The few samples from the southern Sargasso Sea only permit the observation that euphausiid numbers overlap the northern Sargasso Sea at the low end of the scale. This is consistent with observations of biomass, numbers of individuals, and species composition at other trophic levels. For example, Backus et al. (1969) found midwater fish biomass and numbers of individuals were substantially lower in the southern than in the northern Sargasso Sea. Hulburt et al. (1960) found a similar change in the numbers of phytoplankton cells and species composition from north to south. Lower primary production was measured in the southern Sargasso Sea by Ryther and Menzel (1960). There are too few tows to examine seasonal cycles in the Gulf Stream.

Thirty-three species of euphausiids have been recognized in the zooplankton collections taken by Wiebe and co-workers described above. Six species are temperate or arctic boreal species (Mauchline and Fisher, 1968) and are generally restricted to the Slope Water and to cold-core rings (Wiebe et al., 1976a, b; Wiebe, 1976; Wiebe and Boyd, 1978; Cox and Wiebe, 1979; Ring Group, 1981; Wiebe and Flierl, 1983). These species do not, however, have identical distributions. *Meganyctiphanes norvegica* occurs in maximum numbers near the continental slope, sporadically in the more open waters of the Slope Water, and only incidentally in cold-core rings. *Thysanopoda acutifrons* is usually in low numbers in the Slope Water and in cold-core rings, with strong evidence of submergence in the Slope Water as compared to its vertical distribution further to the north (Einnersson, 1948); this trend is accentuated in rings. Submergence is also a factor in the distribution of *Thysanoessa Longicaudata*, but this species occurs more regularly and in high numbers than the previous two species. Only *Euphausia Krohnii*, *Nematoscelis megalops*, and *Thysanoessa gregaria* typically occur in the upper 300 m and are the usual numerical dominants in the Slope Water and in young rings.

All of the other species have tropical or sub-tropical affinities. Species typical of the northern Sargasso Sea are *Bentheuphausia amblyops*, *Euphausia americana*, *E. mutica*, *E. tenera**, *E. hemigibba**, *E. gibboides*, *Thysanoessa parva*, *N. atlantica*, *Nematobranchion boopis*, *N. flexipes*, *Stylocheiron abbreviatum**, *S. carinatum**, *S. longicorne*, *Thysanopoda pectinata*, *T. obtusifrons*, *T. monocantha*, and *T. tricuspidata*. The four asterisked species are frequently dominant numerically and account for a majority of the variability at northern Sargasso Sea stations. These four species, as well as *E. americana*, *E. mutica*, *T. parva*, and *N. atlantica*, regularly occur in low to moderate abundance in the Slope Water region under the influence of warm-core rings or meanders of the Gulf Stream, but rarely in the Slope Water itself (Cox and Wiebe, 1979). Some species, *E. hemigibba*, *S. abbreviatum*, *S. carinatum*, and *T. obtusifrons*, also occur in the southern Sargasso Sea at more than half the stations, but in proportionately lower numbers.

The remaining species, *Euphausia brevis**, *Nematoscelis microps**, *N. tenella**, *Nematobranchion sexspinosus*, *Stylocheiron affine**, *S. elongatum**, *S. maximum*, *S. suhmii,**, *Thysanopoda aequalis**, and *T. orientalis*, are characteristic of southern Sargasso Sea stations. The asterisked species account for a sizeable proportion of the variability at these stations. Some of these species (e.g., *E. brevis*, *N. microps*, *S. affine*, and *S. elongatum*) can, however, also dominate in the northern Sargasso Sea as well, and do occur in the Slope Water under the influence of rings or the Gulf Stream.

Of the six euphausiid species characteristic of the Slope Water, only two, *E. krohnii* and *N. megalops*, show strong evidence of seasonal variation in their numbers. *E. krohnii* has an abundance maximum between May and July, while *N. megalops* peaks between August and October. The abundances of the other four species are so variable that a seasonal periodicity is not readily visible. Fluctuations of tropical and sub-tropical euphausiids found in the Slope Water are also without a distinct seasonal influence.

Seasonal peaks in abundance are evident in nine species in the northern Sargasso Sea; *E. brevis*, *E. hemigibba*, *E. tenera*, *N. microps*, *N. tenella*, *S. abbreviatum*, *S. carinatum*, *T. parva*, and *T. aequalis*. Except for *T. parva*, the peaks all occur in the spring. *T. parva*, a bathypelagic form, peaks in the fall and strangely disappears from the zooplankton samples taken throughout all of the northwestern Atlantic between February and May (Wiebe and Flierl, 1983).

Effects Of Shelf Water Interactions. The effect of the entrainment of shelf water into the Slope Water on the species composition of the Slope Water zooplankton has not been studied in detail, although programs now funded to study warm-core rings and seasonal cycles in the Slope Water will certainly contribute towards filling this void. Cheney (1982) noted that the dominance of the boreal coastal chaetognath species, *Sagitta elegans*, in the Slope Water reported by Grice and Hart (1962) reflected entrainment of shelf water.

Effects Of Cold- And Warm-core Rings. Rings are also sites of strongly contrasting species composition compared to surrounding waters. Because rings begin life containing water from the opposite side of the Gulf Stream, their patterns of zooplankton species distribution are strongly dependent upon their age or state of decay, the affinity of a species for a particular hydrographic regime, the vertical distribution of the species, and the particular composition of the Slope Water or Sargasso Sea Water population at the time of ring formation. This latter point is especially important both because the evolution of the species composition within a ring in terms of absolute abundance is strongly dependent on the starting composition and because absolute abundances vary strongly due to seasonal cycles which may be proceeding differently within or outside the ring or due

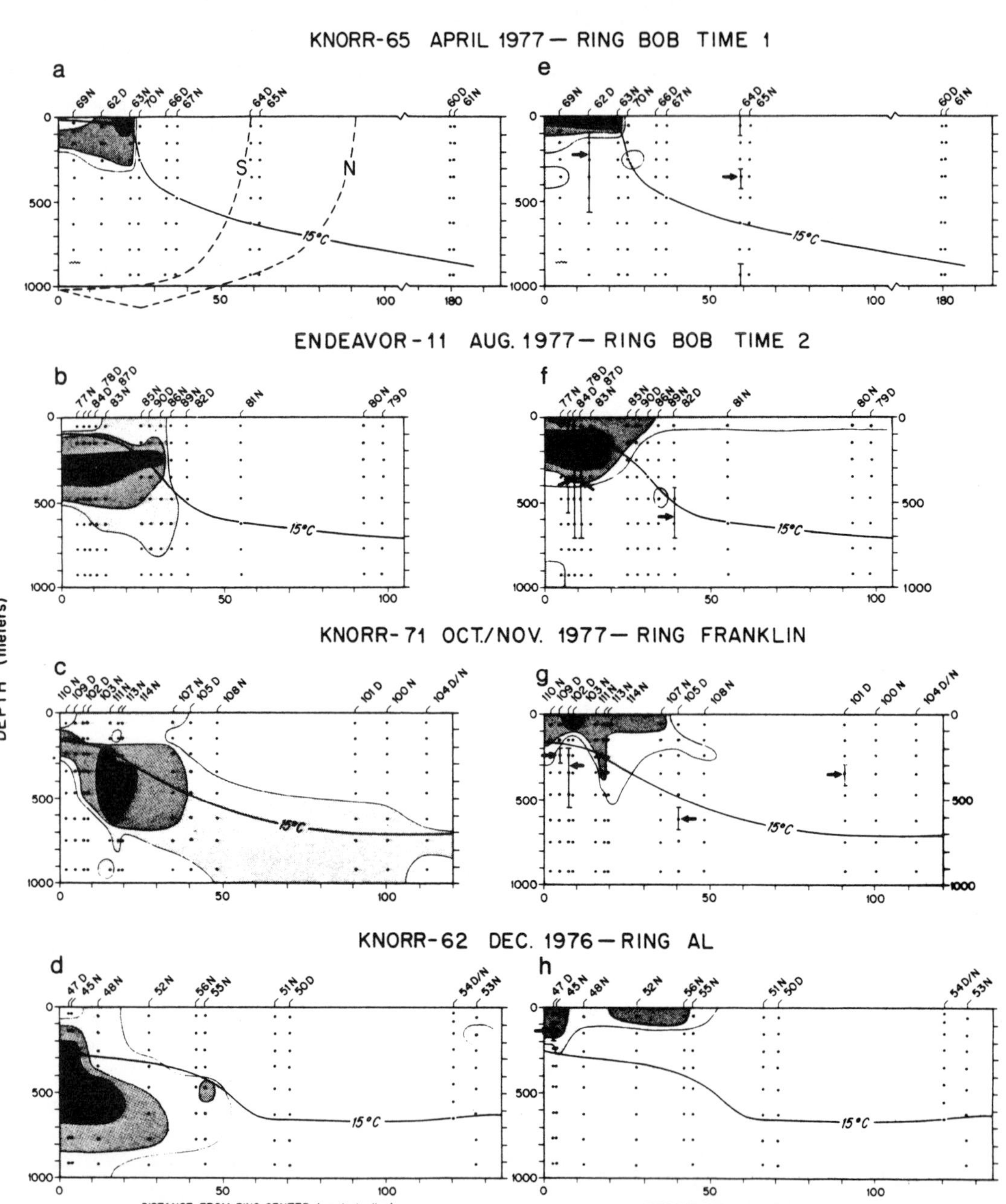

Figure 6.19 Cold-core ring/Sargasso Sea vertical sections of abundance of the Slope Water species *Nematoscelis megalops, Euphausia krohnii, Stylocheiron carinatum,* and *E. tenera.* Four cruises are illustrated for each species, two to ring "Bob", KNORR 65 and ENDEAVOR 11, one to ring "Franklin", KNORR 71, and one to ring "Al", KNORR 62. The solid line is the depth of the 15°C isotherm. For species which show strong diel vertical migration, night data are contoured, and day data are given as the range (I) with an arrow indicating the center of the distribution. The number/letter combinations along the top of each section are

to patchiness of species in the parent water mass (Cox and Wiebe, 1979; Ortner et al., 1979; Wiebe, 1976; Wiebe et al., 1976a; Wiebe and Boyd, 1978; Wiebe and Flierl, 1983). This discussion focusses first on cold-core rings because they are presently best known.

Wiebe et al. (1976a) showed that in cold-core rings 3-11 months of age there was a gradual transformation in euphausiid species composition from one dominated by species characteristic of the Slope Water to one more similar to the adjacent Sargasso Sea waters. It also appeared that the decay rate of the Slope Water species assemblage was much more rapid than that of the physical properties characterizing a ring, especially below 200 m.

More extensive details of the changes in distribution and abundance of euphausiids in aging cold-core rings are presented by Wiebe and Flierl (1983). In a young ring such

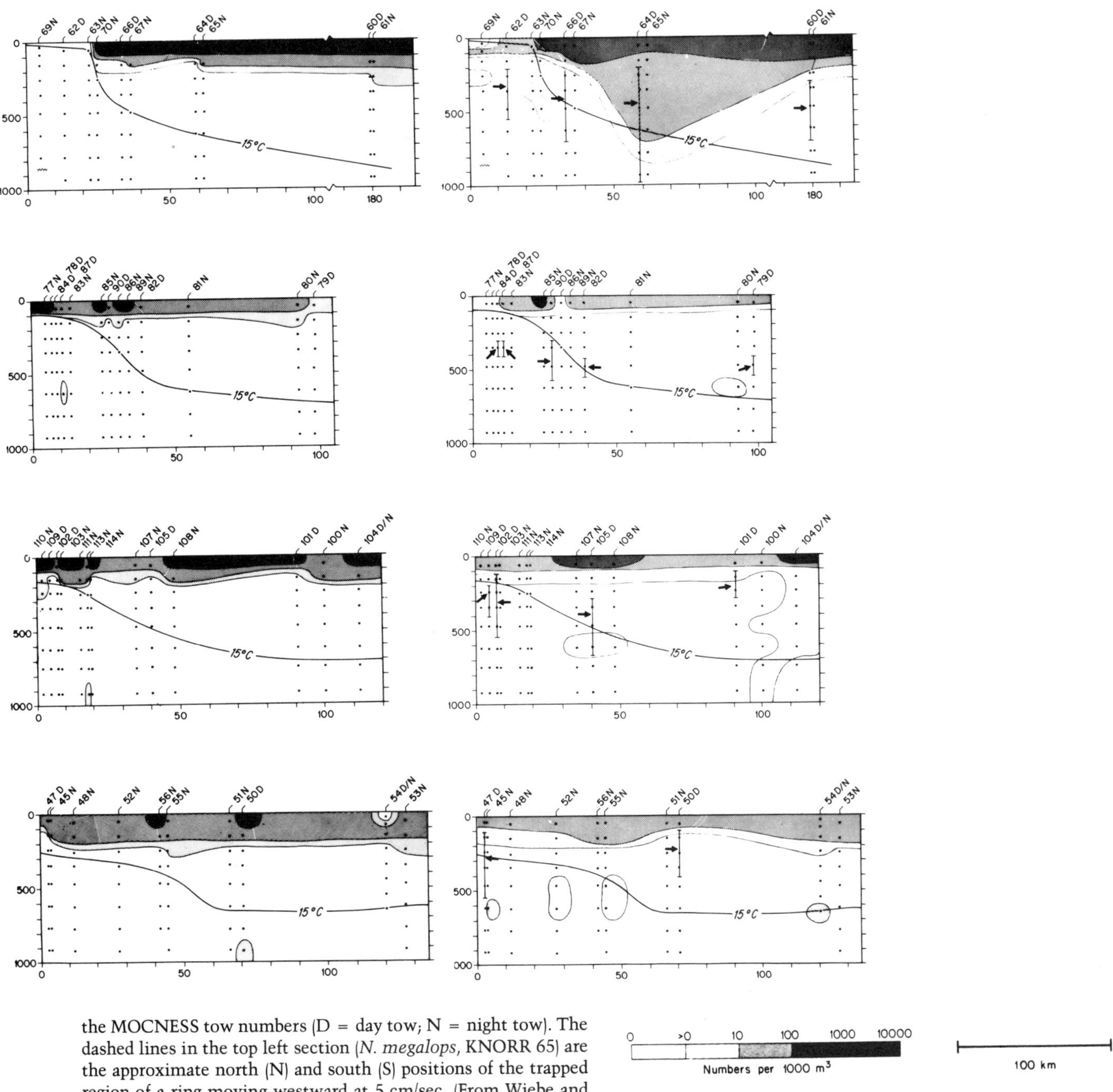

the MOCNESS tow numbers (D = day tow; N = night tow). The dashed lines in the top left section (*N. megalops*, KNORR 65) are the approximate north (N) and south (S) positions of the trapped region of a ring moving westward at 5 cm/sec. (From Wiebe and Flierl, 1983).

as Bob at age 2 months, species endemic to the Slope Water (for example *Nematoscelis megalops* (Fig. 6.19), *Euphausia krohnii*, and *Thysa noessa longicaudata*) were mostly or wholly restricted to the ring center. Species in other taxonomic groups, such as the copepod *Pareuchaeta norvegica* and the pteropod *Limacina retroversa*, showed very similar patterns.

A number of euphausiids found in the Sargasso Sea during all seasons show the opposite pattern. Virtually none of the species of the *Stylocheiron* was present in the center of Bob (Fig. 6.19). Similarly, of the two abundant species of *Nematoscelis* with subtropical/tropical distributions, only a few individuals of *N. microps* were found in the ring (Wiebe and Flierl, 1983). In contrast, two of the three species of *Euphausia* which were present in the Sargasso Sea in reasonably large numbers were also present in ring Bob in moderate numbers (Fig. 6.19). The pteropod, *Limacina inflata*, showed a similar distribution pattern. Only *E. brevis* was totally absent from the ring center area. Another species, *Thysanopoda aequalis*, was present in low numbers within the ring, but not at the centermost staton.

For the euphausiids, the pattern of cold-core ring evolution that emerges is the following: (1) Warm water species living permanently at or near the surface and those which perform diel migrations invade a ring more quickly than do species which live at subsurface depths of 150 to 600 m. However, even for these rapid invaders, there is often a tendency for population numbers to be lower within the ring compared to adjacent seas for half a year or more. (2) Vertical migrators migrate to shallower depths in young rings and non-migrators show a strong tendency to shoal. (3) Cold-water species persist within the ring core for extended periods. In some species, population numbers in middle-aged rings exceed levels at the time of formation (e.g., *D. krohnii* and *N. megalops*). Other species such as *T. longicaudata* can show rather drastic declines in numbers during this same period in rings like Bob. (4) Cold-water species, such as *N. megalops* and *T. longicaudata*, which show submergence as a ring ages, appear to be dispersed out of a ring at depths of 400 to 1000 m. For the shallower dwelling species like *E. krohnii*, which can survive surface water modification, dispersal appears to take place near the surface. (5) The species compositional structure of the ring core remains distinctly different from the surrounding Sargasso Sea for 6 to 8 months after formation in spite of apparent exchanges of species into and out of the ring.

The changing pattern of species abundance as cold-core rings age has been studied for three other groups of zooplankton: chaetognaths by Cheney (1982); amphipods by Hart and Wormuth (1982); copepods by Cowles (1982). Cheney (1982) presented abundance patterns for 16 species of chaetognaths from the center of ring Bob to the Sargasso Sea during the second period of sampling of this ring. Species classified as northern Sargasso Sea species (see above) showed low abundance in the ring core and a monotonically increasing abundance to the Sargasso Sea except for *S. bipunctata*, which had a peak abundance in the ring fringe. Five of the seven species characteristic of the Slope Water showed the reverse pattern of higher abundance in the ring core and either lower abundance or absence in the Sargasso Sea. One Slope Water species, *S. helenae*, was absent from the ring entirely and another, *S. bathypelagica*, was present in small numbers in only one tow from the ring core.

Cold-core ring D (Wiebe and Boyd, 1978; Boyd et al., 1978) was also used by Cheney to examine temporal changes in chaetognath species composition in an older ring. This ring was sampled at 6 and 9 months of age. Except for *S. macrocephala* which maintained abundances in ring D equivalent to its Slope Water levels, Slope Water species either declined in abundance or disappeared during the period. The expected opposite pattern was also observed for most of the Sargasso Sea species although none save *S. decipiens* reached abundance levels in ring D as high as generally observed in the Sargasso Sea. *S. decipiens* actually attained numbers considerably above the level normally found in the Sargasso Sea and Cheney suggested that it was opportunistically exploiting the hybrid ecological conditions present in the ring. Similar observations were made by Wiebe and Flierl (1983) with regard to *S. carinatum* and the Ring Group (1981) with regard to *Limacina inflata*. There are also midwater fish species which appear to exploit conditions in middle-aged rings (Backus and Craddock, 1982 and below).

The data summarized by Cowles (1982) for the copepods *Calanus finmarchicus*, *Rhincalanus nasutus*, *Pleuromamma robusta*, and *P. borealis*, which are characteristic Slope Water species and *P. gracilis* and *P. abdominalis*, which are important in the Sargasso Sea, show essentially the pattern described above. Ring Bob had a larger total copepod biomass and a larger proportion of cold-water species than ring D. Nevertheless, significant numbers of warm-water species were present in the near-surface portion (upper 200 to 400 m) of the core waters of both rings. Evolution of the pelagic amphipod species composition in these same rings was parallel to the above groups (Hart and Wormuth, 1982).

Although substantial information about the species composition of warm-core rings will soon be available, the only currently published data come from Cox and Wiebe (1979) who examined the potential role of warm-core rings as a source for the expatriated oceanic zooplankton species that occur on the Middle Atlantic Bight shelf. Abundance data for 31 euphausiid species at three Slope Water stations, two Gulf Stream stations, two warm-core ring stations, and a composite shelf station made up of data from Grice and Hart (1962) were presented. The species composition of warm-core rings was most similar to the Gulf Stream and to a station on the Slope Water under the influence of a warm-core ring (Cox and Wiebe, 1979).

Vertical Distributions

Typical Seasonal Distributions. Very little published information relates specifically to the seasonal pattern of vertical distribution of zooplankton from the Slope Water or the northern Sargasso Sea, although this will change for the Slope Water when data from the extensive samples being analyzed by Wiebe and co-workers are reported. The works of Deevey (1971) and Deevey and Brooks (1971, 1977) provide seasonal data for a number of zooplankton taxa, but are not reviewed here because the sampling had limited vertical resolution and the collection site was near Bermuda. The works of Wormuth (1981), Cheney (1982), and Wiebe and Flierl (1983) provide some insight into the changes that are to be expected for the pteropods, chaetognaths, and euphausiids living in the northern Sargasso Sea, although the seasonal coverage is sparse.

The most exhaustive analyses of vertical structure for both the Slope Water and the northern Sargasso Sea have been done by Cheney (1982) for the chaetognaths from 52 9-net MOCNESS-1 tows. Based on average vertical distribution in these two hydrographic regimes, 9 species were considered epipelagic (0- 200 m) — *K. pacifica*, *P. draco*, *S. bipunctata*, *S. enflata*, *S. helenae*, *S. hexaptera*, *S. minima*, *S. serratodentata*, and *S. tasmanica*; 4 species were mesopelagic (200-1000 m) — *K. subtilis*, *S. decipiens*, *S. lyra*, and

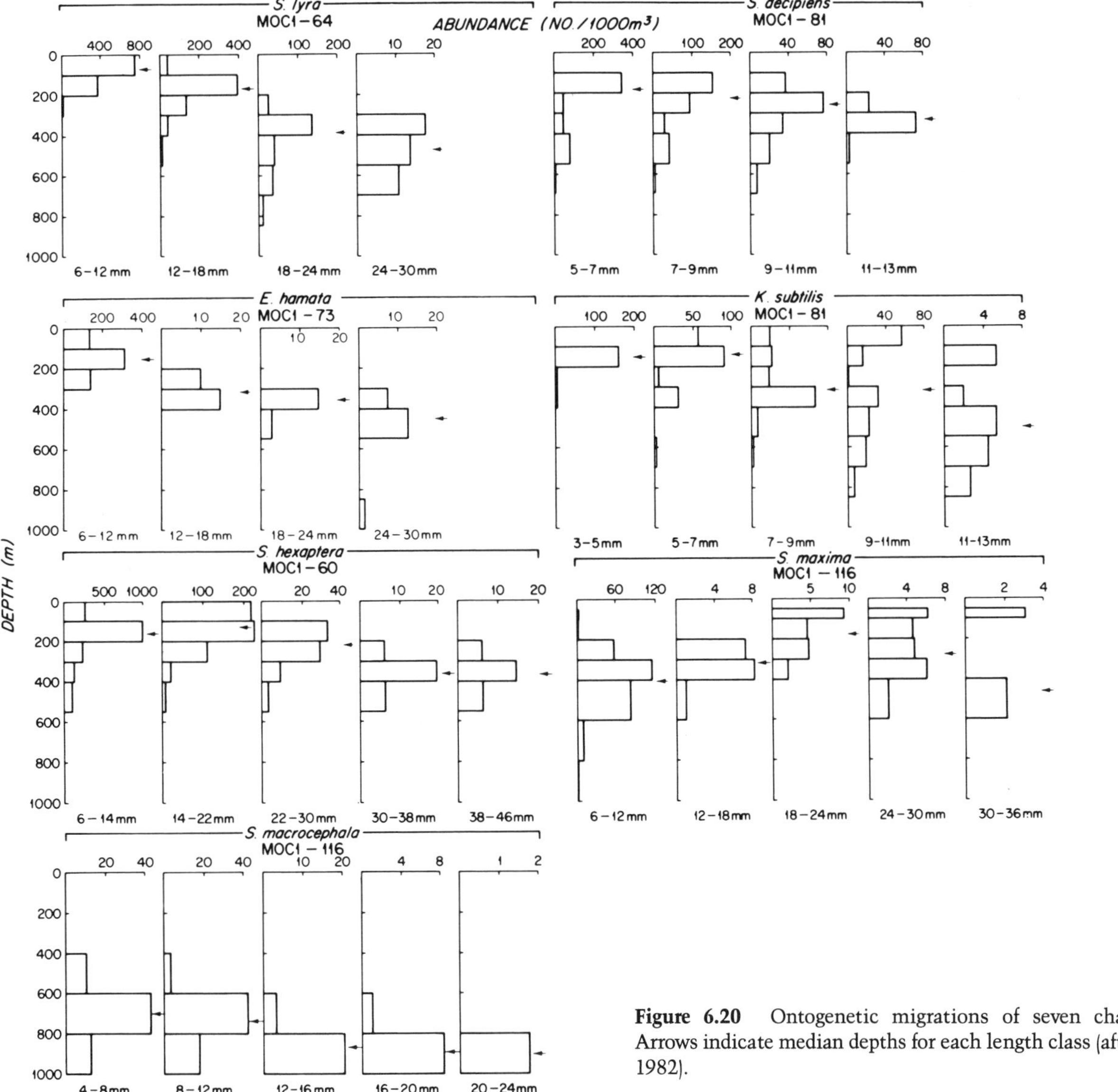

Figure 6.20 Ontogenetic migrations of seven chaetognaths. Arrows indicate median depths for each length class (after Cheney, 1982).

S. planctonis; and 5 were bathypelagic (>1000 m) — *E. bathypelagica*, *E. fowleri*, *E. hamata*, *S. macrocephala*, and *S. maxima*. For all epipelagic species, the fraction of the population below 100 m was larger in the northern Sargasso Sea than in the Slope Water. Most mesopelagic species showed no shift in vertical distribution between these two regimes. In contrast, of the bathypelagic species only the population center of abundance of *E. bathypelagica* did not shift upward 100 to 500 m in the Slope Water.

None of these species showed evidence of significant diel vertical migration, although it would not have been discernable in the upper 100 m because of the coarseness of the sampling. Ontogenetic migrations were, however, quite evident in the seven mesopelagic and bathypelagic species for which size frequency data were obtained (Fig. 6.20). The typical migration resulted in small individuals predominating near-surface and large individuals predominating at the bottom (e.g., *S. lyra*, *E. hamata*, *S. decipiens*, and *S. macrocephala*). Cheney (1982) pointed out that spatial or temporal variation in ontogenetic shifts in vertical distribution coupled with the decreases in abundance with increasing size could give rise to shifts in the overall vertical distribution of a species which would be unrelated to changes in the physical environment, although they could appear to be. He did not, however, present data to show seasonal shifts in vertical distribution nor did he relate ontogenetic migrations to the seasons.

The northern Sargasso Sea vertical distributions of the nine pteropods described by Wormuth (1981) show three distinct patterns. Three species, *Creseis acicula* (Fig. 6.21), *C. Virgula concia*, and *Limacina trochiformis* were epipelagic non-migrators; their centers of distribution (day, night, and year-round) were within the upper 100 m. A mesopelagic non-migrator, *Clio cuspidata*, lived below 300 m; its cen-

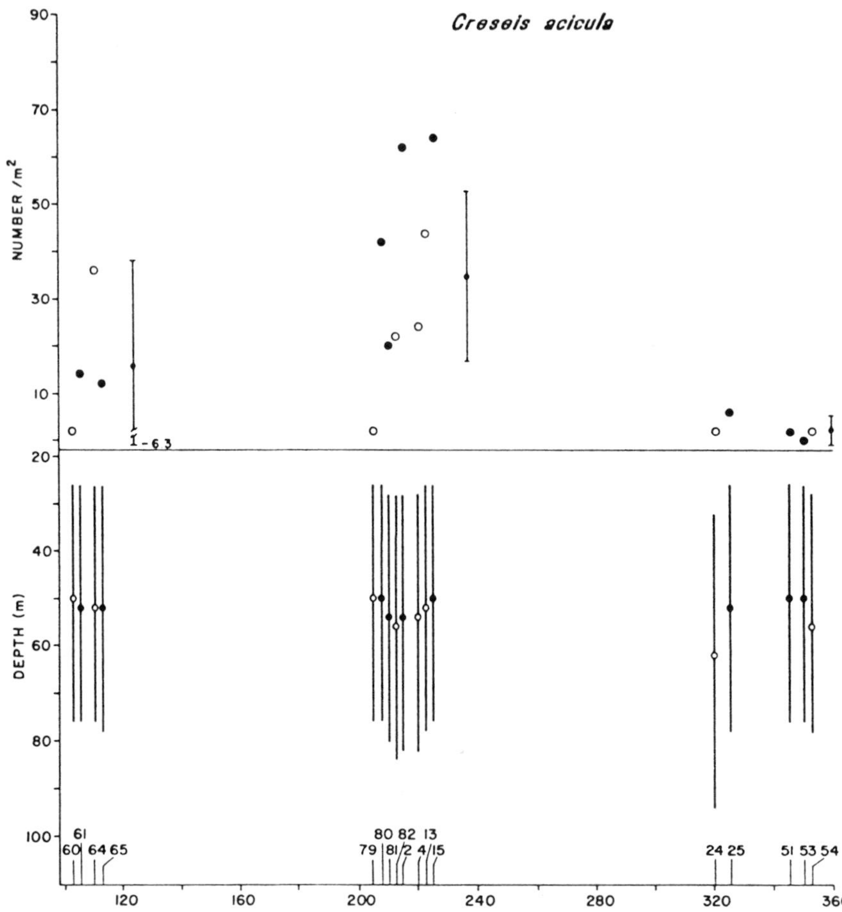

Figure 6.21 Numbers per square meter and vertical distributions at different Julian dates for *C. acicula*. Open circles represent day tows, closed circles night tows. Vertical lines show means (dots) and 95% confidence intervals (ends of lines) for each time period. (Bottom) The top of each line represents the depth of the 25th the dot the 50th, and the bottom of the line the 75th percentile of the population. The numbers above the x-axis are the tow numbers (offsets <5 days have been made for clarity). From Wormuth (1981).

ter of distribution was quite variable, ranging typically between 450 and 800 m (Fig. 6.22). There were four strong vertical migrators, *C. pyramidata* (Fig. 6.23), *L. inflata, L. lesueuri*, and *Styliola subula*, which typically were at depths of 200 to 500 m during the day in the upper 50 to 100 m at night. *L. bulimoides* was a much weaker diel migrator moving from 120 to 160 m during the day to above 50 m at night.

Effects Of Shelf Water Interactions. The effects of shelf water overflow on the vertical distribution of Slope Water zooplankton has not been studied in detail. Cheney (1982) showed that the shelf water chaetognath, *Sagitta elegans*, was most abundant in the upper 50 to 100 m of two tows taken in the entrainment field east of warm-core ring 'Q', but he did not indicate what effect the presence of shelf water had on the other species.

Effects Of Warm- And Cold-core Rings. The effects of cold-core rings on the vertical distribution of zooplankton species in the regions have been studied only for the euphausiids (Wiebe and Boyd 1978; Wiebe et al., 1982; Wiebe and Flierl, 1983) and chaetognaths (Cheney, 1982). The warm-water euphausiids which penetrate cold-core rings exhibit a common reaction of shoaling. In the vicinity of a ring, it often involves truncation of the lower portion of the depth distribution by 100 to 300 m and for deep-dwelling species, elevation of the upper limit by about 100 m. The pattern is most pronounced in young rings and less evident in the older rings. For non-migrators living near the surface (*Stylocheiron carinatum, S. suhmii*) the shoaling is subtle (Fig. 6.19). These species typically range the upper 200 m in the Sargasso Sea, but become restricted to the upper 100 m in young rings. For deeper-living non-migrators (*S. affine, S. elongatum*), shoaling is more dramatic. The day and night vertical distributions of migrators of *Euphausia, Nematoscelis*, and *Thysanopoda* (Fig. 6.19) show pronounced shoaling relative to the patterns of vertical distribution in the Sargasso Sea.

The Slope Water species, *Nematoscelis megalops*, only occurs south of the Gulf Stream in association with cold-core rings. Data presented by Wiebe and Boyd (1978) and Wiebe et al. (1982) show that *N. megalops* typically lives in the upper 600 m, with most individuals above 300 m both day and night. A similar pattern was observed in ring D at 6 months of age (August 1975), except a larger fraction of the population was present below 300 m and individuals occurred down to 800 m. On the second cruise to ring D

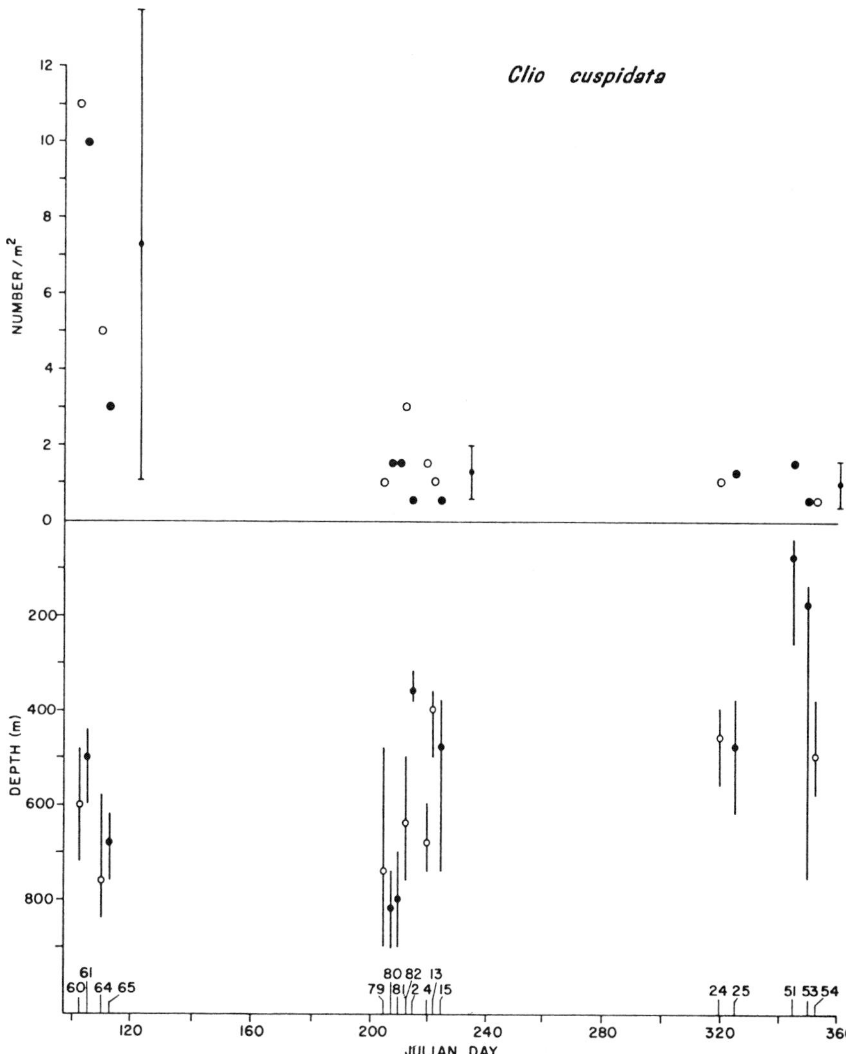

Figure 6.22 Numbers per square meter and vertical distributions at different Julain dates for *C. cuspidata.* Symbols are in Figure 6.21. Y axes are scaled separately for each species (from Wormuth, 1981).

(November 1975), the distribution of this species had shifted significantly downward with the major portion occurring below 300 m. On the third cruise to ring D (June 1976), there were no *N. megalops* in the single night sample taken in the ring core. A similar pattern was evident in vertical distributions in rings Al, Bob, Emerson, and Franklin (Fig. 6.19). The older the ring, the deeper the distribution of this species. A similar observation can be made for the cold-water species, *Thysanopoda acutifrons,* but not for *Euphausia krohnii* or *Thysanoessa longicaudata,* although their abundance declined with ring age.

Changes in chaetognath vertical distribution in cold-core rings is inextricably related to the pattern of ontogenetic migration and changes in the size frequency distribution. Thus, in ring D, Cheney (1982) found that the five significant shifts in vertical distribution between August and November 1975 were for populations which became shallower. All were species which Cheney had shown to have strong ontogenetic migrations; if individuals were smaller on average in November, then their vertical distribution should have been shallower if they were ontogenetically migrating. In support of this hypothesis, individual size for the one species, *Sagitta lyra,* that was measured in all samples he examined showed that the average size of the population in November (11.8 mm) was about half that in August (22.0 mm).

Relation To Chemical And Physical Factors

The spatial and temporal patterns of plankton distributions summarized above are the result of physical and biological processes acting alone and in concert. Haury et al. (1978) reviewed major causative factors affecting plankton populations and the spatial and temporal scales upon which they might be expected to affect planktonic biotas. Although it seems self-evident that biological factors such as competition, predation, social interaction and reproduction should have a profound effect on plankton patterns, for the ACSAR area there are few directly relevant data. Nor have there

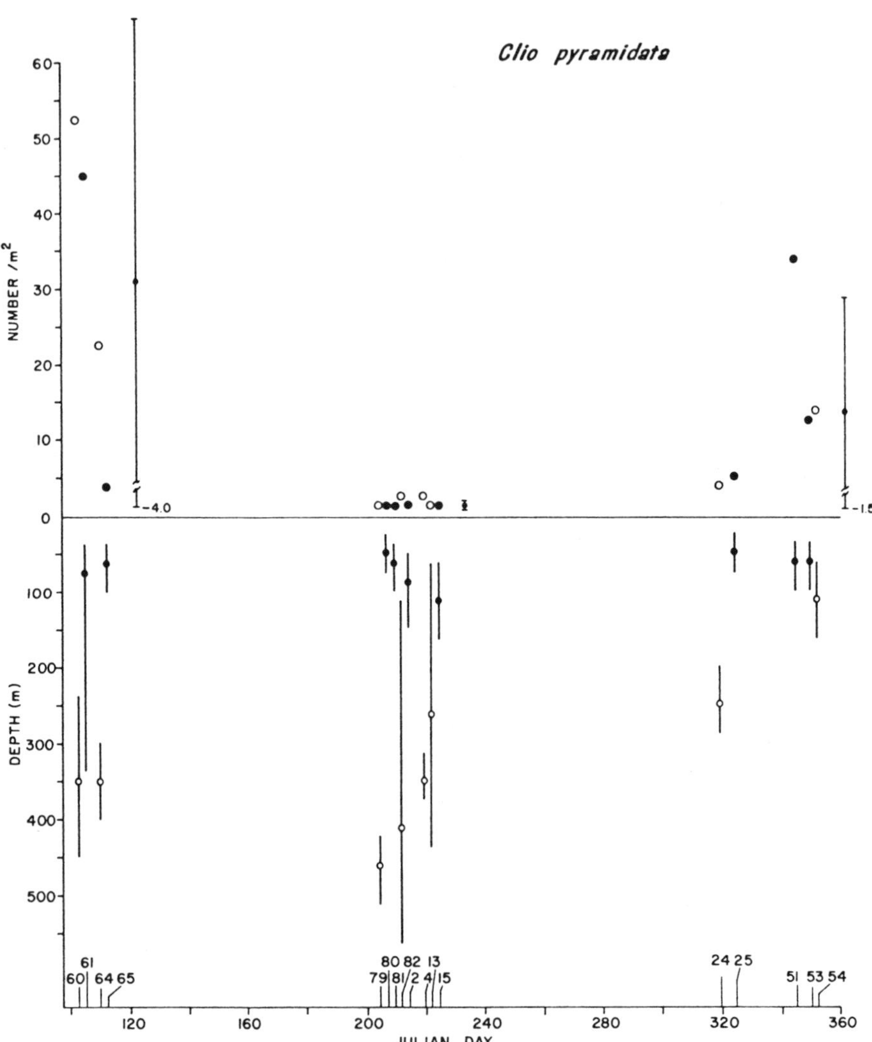

Figure 6.23 Numbers per square meter and vertical distributions at different Julain dates for *C. pyramidata*. Symbols are in Figure 6.21. Y axes are scaled separately for each species (from Wormuth, 1981).

been many attempts to rigorously define the physical-chemical structure of the habitat of oceanic plankton in the area. The early work of Moore and co-workers (cited above) may have been severely compromised by fluctuations in vertical and horizontal distributions caused by (then unknown) flow reversals of the Florida Current (Stepien, 1980). This discussion will focus on the few recent reports which attempt to link changes in spatial pattern to the physical-chemical environment.

In the case of epipelagic and mesopelagic diel migrators, light is clearly one of the most important abiotic factors directly influencing vertical distributions. Vertical temperature structure also seems to play a crucial role in determining the vertical distribution of some zooplankton but not others.

For example, vertical migration of warm water euphausiids of the genera *Euphausia*, *Nematoscelis*, and some *Thysanopoda* takes place regardless of season or vertical temperature structure (Wiebe, unpublished data). While temperature structure at or near the surface does not seem to affect the migration pattern of warm-water euphausiid species, abnormally colder deep water does inhibit the daytime depth of migration. Thus, in the Slope Water and cold-core rings, shoaling (100 to 300 m) occurs for both migrating and non-migrating species (Wiebe et al., 1976a; Wiebe and Flierl, 1983).

Although vertical temperature structure would appear to be a major factor causing the shift in the vertical distribution of most if not all of these species, Wiebe and Flierl (1983) suggest it is not the only factor regulating these patterns. For one thing, when warm-water species are dispersed into a colder regime, temperature compensation is not perfect; i.e., species are living at colder temperatures than is typical in their home range. Light penetration, at least for some species, appears to set an upper limit for shoaling. *Stylocheiron elongatum* individuals, for instance, do not occur at light levels higher than 10 $\mu W/cm^2$; optimal level appears to be between 10^{-2} and 10^{-1} $\mu W/cm^2$. Optimal temperature appears to be 16-18°C, a range which frequently occurs in rings and in the Slope Water above the apparent upper limit for light. Thus individuals in these regions seem to be unable, because of light, to adjust their vertical distribution to find optimal temperatures. It remains to be determined whether temperature structure in combination with light levels sets limits on the day-time vertical distribution of warm-water diel migrating species of euphausiids listed above.

For some cold-water species vertical temperature structure appears to be a major determinant of their vertical distribution. For the non-migrator *N. megalops*, Wiebe and Boyd (1978), Wiebe et al. (1982), and Wiebe and Flierl (1983) found that the central 50 percent of the adult portion of the population generally stayed within about ±2°C of the 10°C iosotherm in both the Slope Water and cold-core rings. In the Slope Water, this species is generally distributed above 300 m both day and night, while in aging rings, the center of the vertical distribution deepens coincidently with the sinking of isotherms and isohalines to below 300 m and deeper in an apparent attempt to stay in an 'optimal' temperature and salinity regime. Wiebe and Boyd argued that vertical temperature structure was the main factor to which *N. megalops* was responding in withdrawing from near surface layers. This change in vertical distribution as a ring ages appears to bring about changes in the physiology and biochemistry of this species which ultimately leads to local extinction. Data from Boyd et al. (1978) provide a picture of the ring population of *N. megalops* being physiologically stressed by the environmental changes associated with ring decay. In rings aged between 6 and 9 months, respiration rates of individuals declined from a fifth to a twentieth of the rates determined for the Slope Water populations. Furthermore, in older rings, adult males disappeared, production of eggs and larvae appeared to cease, and growth rates of individuals were markedly reduced relative to Slope Water individuals (Boyd et al., 1978). The conclusion was that as a ring decays, *N. megalops* tends to live deeper in the water column, away from the relatively food-rich surface layers. Food levels are reduced to a point inadequate for growth and reproduction. In spite of a drastic lowering of the metabolic rate, body energy stores are used. Thus in older rings, individuals of this species appear to be in a state of starvation, and this probably is a major factor for demise of the ring population.

Cheney (1982) found that mixed layer depth played a major role in determining the vertical distribution of a majority of the epipelagic chaetognath species. *Krohnitta pacifica, Sagitta bipunctata, S. enflata, S. lyra, S. minima*, and *S. serratodentata* all appear limited primarily to the surface mixed layer. Cheney presumed that colder temperatures in the pycnocline limited depth, but he emphasized the need for higher resolution data. The median depth of only one epiplelagic species, the cold-water form *S. tasmanica*, was strongly correlated with temperature. As surface waters warmed and deepened, the center of vertical distribution was also found to be deeper.

For selected mesopelagic and bathypelagic species and one epipelagic species, *S. hexaptera*, Cheney observed that differences in vertical distribution were largely a function of differences in population size structure and population abundance as noted above. The fact that there were significant changes in size structure from one hydrographic regime to another gave rise to spurious correlations between median depth and environmental variables such as temperature and salinity for some species. This appeared to be the case for *S. lyra* and *K. subtilis*. It was also possibly true for the bathypelagic species *S. macrocephala* and *E. hamata*, although Cheney concluded that there were insufficient data to distinguish between control of vertical distribution by vertical temperature structure or population size structure for these species. Two species, *S. decipiens* and *S. hexaptera*, both ontogenetic migrators, showed no significant correlations to measured environmental variables nor were there significant shifts in vertical distribution between the Slope Water and the Sargasso Sea.

Light penetration did not appear to be a significant factor controlling the vertical distribution of any chaetognath species; there were no significant correlations between median depth of abundance and light at 100 m for any species.

While Cheney did not find salinity to be a major factor shaping the vertical distribution of the chaetognaths studied, he made one interesting observation. Epipelagic species which typically inhabited the surface Slope Water were severely reduced in numbers when low salinity water of shelf water origin was present in the Slope Water. This may reflect an intolerance to lower salinity water, but it is more likely due to physical displacement of the species and their habitat.

Wormuth (1981) examined the relationship of temperature, salinity, and total zooplankton biomass to the abundance of each of the nine most abundant pteropods in his samples from the Sargasso Sea. Separate analyses were done for day and night tows to avoid complications caused by diel migration of some of the species. For the migrating species (*Limacina inflata, L. bulimoides, L. lesueuri*, and *S. subula*), none of the variables accounted for much of the population abundance variation during the day. In contrast, temperature was the dominant factor correlating with abundance changes for the migrators at night and for the non-migrators both day and night. Total zooplankton biomass was significant for only three species, *Creseis acicula, C. virgula concia*, and *Clio pyramidata*, and salinity was even less significant. Wormuth concluded that the depth distribution of these species showed no significant seasonal response to different thermal structures.

Neuston

E.H. Backus

The term "neuston" is generally defined as the plant and animal community that inhabits the narrow, uppermost layer (perhaps 10 or 20 cm) of the ocean. Some authors have produced a complex of descriptive categories within the neuston. Hempel and Weikert (1972) group neuston components into three major ecological categories: (1) "euneuston" — organisms with maximum abundance in the immediate vicinity of the surface where they stay day and night; (2) "facultative neuston" — organisms which concentrate at the surface only at certain hours, mostly during darkness, and (3) "pseudoneuston" — organisms whose maximum concentrations lie at deeper layers, but whose range of vertical distribution reaches the surface layer during certain hours. Zaitsev (1970) provides the following terms: (a) "pleuston" — hydrobionts whose bodies are situated partly in the water and partly in the air (e.g., Portugese Man-of-War, *Physalia*) and (b) "epineuston" — the aerial surface film organisms (e.g., the marine water strider, *Halobates*).

The importance of the surface layer in the economy of the sea was first stressed by Zaitsev (1970), based largely on studies in the Black Sea. Neuston fauna are dominated by Crustacea and Cnidaria (Hempel and Weikert, 1972; Morris, 1975) and within these groups copepods and siphonophores are most abundant. The surface layer is also important to a wide variety of fishes (David, 1965; Craddock, 1968). Grant (1977, 1979) found evidence that this zone in the Middle Atlantic Bight serves as an "incubator" for the reproductive stages of numerous fishes and crustaceans (see also Zaitsev, 1970), and also noted important differences between the neuston and subsurface zooplankton. In contrast, deep-water studies in the northwest Atlantic (Morris, 1975) have found the neuston layer to be impoverished compared with the subsurface.

Biomass and Faunal Composition

Distribution and Diel Variations. No study has specifically examined the neuston of the ACSAR region. Since several biogeographic regions occur within the area, it is assumed that the neuston community shows complexity associated with these different regions and their interactions. Insight into parts of this community can be gained from neuston studies from adjacent waters.

Neuston biomass in the North Atlantic reflects the general level of productivity in the underlying waters (Morris, 1975). Higher neuston biomasses occur in the productive temperate and boreal coastal waters. Thus the northeastern part of ACSAR likely has a higher neuston biomass than the southwestern areas (Gulf Stream and Sargasso Sea). Morris (1975) found that in autumn the waters of the Scotian Shelf and Slope Water have about four times the standing stock of neuston of the Gulf Stream or Sargasso Sea. Daytime neuston biomass falls within two ranges: (a) eutrophic waters such as the temperate-boreal seas (northern part of the report area) where biomass ranges between 50-100 mg m^3; and (b) oligotrophic waters, such as the southeastern part of the report area (Sargasso Sea), where neuston wet weights generally average less than 25 mg m^{-3}.

The general faunal composition of the neuston is affected by the diel vertical migration of animals (Craddock, 1968; Hempel and Weikert, 1972; Morris, 1975; Grant et al., 1979). There are very few euneustonts and the community is dominated by facultative animals from subsurface depths. Thus the majority of neuston animals have a maximum abundance during hours of darkness, therefore the faunal structure is temporally quite variable. Hempel and Weikert (1972) concluded that the magnitude of the nocturnal increase depended on several factors such as the abundance and composition of the subsurface zooplankton, hydrographic features of the water column, and depth to bottom. The deep ocean (>200 m) neuston at night are less enhanced by benthic migrants than in shelf regions (Morris, 1975).

Seasonal Changes

Grant's (1977, 1979) studies in the Middle Atlantic Bight include three stations beyond the 200-m isobath, some of which were seasonally in Slope Water. Neuston collections showed a progressive change from a highly structured and predictable faunal pattern in coastal waters to a relatively unpredictable one at the shelf edge. The latter is dependent on incursions of offshore water and the presence or recent passage of Gulf Stream warm-core rings. Copepods dominated the neuston fauna at these deep stations in all seasons except in spring when the salp *Thalia democratica* was prominant. Other important neustonts were amphipods (*Parathemisto*), euphausiids (*Thysanoessa*), and hake larvae (*Urophycis*). A list of seasonal numerical dominants is given (Table 6.3).

As part of the Middle Atlantic Bight study, Smyth (1980) found crab larvae (*Callinectes*) in significantly greater abundance in neuston tows compared to subsurface collections. Megalopae of *Callinectes* were present at outer stations in winter and spring together with other decapod forms of southern origin.

The National Marine Fisheries Service (NMFS) of the National Oceanic and Atmospheric Administration conducts seasonal ichthyoplankton and hydrographic (MARMAP) surveys which regularly tow neuston nets on

Table 6.3 Seasonally Dominant Zooplankters in the Deep Water Neuston of the Middle Atlantic Bight (after Grant, 1977; Grant et al., 1979).

Season		
Fall:	*Parathemisto gaudichaudii*	
	Pleuromamma gracilis	*Idotea metallica*
	Centropages typicus	*Paracalanus* spp.
	Nannocalanus minor	
	Unidentified Fish eggs	
	Lestrigonus bengalensis	
	Temora stylifera	
	Thysanoessa spp.	
Winter:	*C. typicus*	*Calanus finmarchicus*
	Anomolocera ornata	*P. gracilis*
	P. gaudichaudii	*N. minor*
	Metridia lucens	*Clausocalanus arcuicornis*
	Urophycis spp.	*Euphausia* spp.
Spring:	*Thalia democratica*	
	C. typicus	
	P. gaudichaudii	
	Sapphirina ovatolanceolata	
	P. gracilis	
Summer:	*Labidocera* spp.	
	T. stylifera	
	L. bengalensis	
	Undiaula vulgaris	
	Penilia avirostris	
	N. minor	

the continental shelf with some stations in water deeper than 200 m. Although most of these data (1972 to present) have not been examined, 16 spring and summer transects with deep shelf-edge stations were analyzed in the course of a seabird feeding ecology study (Powers and Backus, in review). Twelve deep stations were examined in a gross taxonomic fashion (e.g., copepods, fish eggs, etc.) to determine dominant groups and particle size classes. All stations were dominated numerically by copepods, except one which was dominated by crab larvae. Other important neuston were amphipods, fish eggs, and euphausiids. Neuston tows were made in early 1973 at 50 stations seaward of the 200-m isobath between Cape Canaveral and Cape Fear (28° to 34°N) as part of two MARMAP cruises (Mathews and Pashuk, 1977). Ichthyoplankton catches in the offshore neuston were bigger and more widespread in May than in February-March (Powles and Stender, 1976). In winter, larvae of Mullidae, Carangidae and Mugilidae were most numerous; a few larvae of other families (Sciaenidae, Clupeidae, Gadidae, Bothidae and Scombridae) were found, mostly north of 32°N. In May the most numerous catches were of Carangidae, Mugilidae, and Scombridae, with some larval Pomatomidae. Significant differences between day and night hauls were found only with Serranidae, which were caught only at night on both cruises.

The Slope Water surveys of the Soviet Fishery Research Vessel *Stvor* in the fall of 1981 included 142 neuston tows, which have not been analyzed. In February, 1983, 50 neuston tows were made from the Canadian R/V *Alfred Needler* along the northern edge of the Gulf Stream between 31°N and Cape Hatteras; Chief Scientist was T.W. Rowell of the Department of Fish and Oceans, Halifax. Neuston sampling along the shelf break south of New England was done by the U.S. Coast Guard in conjunction with MARMAP during all seasons between 1975 and 1982. The data are being prepared by Tossi and Benway.

Backus et al. (1977) made extensive neuston collections in the Slope Water south of Cape Cod and Georges Bank (300-2000 m). They found that many mesopelagic fishes occur regularly in the neuston of this region. *Gonichthys cocco*, by far the most common (Table 6.4), has a strong affinity for the neuston layer (on one cruise 3,000 were caught neuston fishing while only 14 were caught mid-water trawling; Craddock, 1968). Subpolar/temperate, temperate, subtropical/tropical, and tropical species were represented. While exploring primarily the mesopelagic fish fauna, neuston tows also caught other fish of this zone. Dolphin (*Coryphaena hippurus*), butterfish (*Peprilus triacanthus*), flying fishes (*Exocetidae*), and filefishes (*Alutera* and *Monocanthus*) were most common (Craddock, unpubl. data). A species list based on 270 summer tows is given in Table 6.4. It is interesting to note the regular presence of the commercially important species hake (*Urophysis* spp.), mackerel (*Scomber* sp.), and unidentified flatfish.

Three other well-known neustonts are present in the study area. *Halobates micans*, the marine water strider is an epineustont (Scheltema, 1968). Portugese Man-of-War (*Physalia physalis*) and *Velella* (the "by the wind sailor") are both siphonophore pleustonic animals (David, 1965). All three occur regularly in the subtropical waters south of Cape Hatteras and in the Gulf Stream but the seasonality of their occurrence is not known. These animals appear in the temperate northern part of the region mostly in summer or perhaps with Gulf Stream warm-core rings.

Table 6.4 Epipelagic Fishes Found in the Neuston of the ACSAR region.

Epipelagic	Mesopelagic
Bonito - *Sarda* sp.	*Gonichthys cocco* *
Butterfish - *Peprilus triacanthus*	*Centrobranchus nigroocellatus**
Dolphin - *Coryphaena hippurus*	*Gempylus serpens*
Flatfishes (larval)	*Diaphus dumerilii*
Filefishes - *Alutera* and *Monocanthus*	*Hygophum hygomii*
Flying Fishes - *Exocetidae*	*Myctophum punctatum*
Hakes - *Urophycis* spp.	*M. nitidulum*
Jacks - *Caranx* spp.	*M. obtusirostre*
Lizardfishes - *Synodus* spp.	*M. asperum*
Mackerel - *Scomber* sp.	*M. affine*
Mullet - *Mugil* sp.	*Symbolophorus veranyi*
Puffer - *Sphaeroides* sp.	
Triggerfish - *Bolistes* spp.	
Pilotfishes - *Naucrates* spp.	

* common

Plastics and Petroleum Wastes

Plastics and petroleum wastes are well-known components of the neuston (Backus, 1968; Morris, 1971). Colton et al. (1974) analyzed neuston tows taken during the first NMFS MARMAP survey in summer 1972. Sixty-nine percent of neuston samples collected in coastal, slope, and Gulf Stream waters between Florida and Cape Cod contained various types of plastic particles. None, however, were found in coastal and Gulf Stream waters south of Cape Lookout, N.C. While the greatest concentrations were in the coastal waters of southern New England and Long Island, plastics occurred regularly in lesser amounts in the off-shelf waters of the report area. Plastics are currently not known to have serious deleterious effects on environments, but they are non-biodegradable and their concentrations in the study region presumably are increasing.

No study has examined the quality and distribution of neuston petroleum wastes but they are known to occur regularly in the report area at various depths mostly as a result of shipping traffic. A further discussion is found in Chapter 5 of this book.

Mesopelagic Fishes

R.H. Backus

Mesopelagic fishes are a conspicuous element in the marine fauna everywhere seaward of the edges of the continental shelves. Although the mature stages of many species escape small midwater trawls, a number of species mature at lengths as short as 25 or 30 mm, and it can be said that as a whole the mesopelagic fish fauna comprises small species. Distributions of some of the larger species are discussed in the next chapter.

These fishes inhabit the water column from the surface to a little beyond the limit of penetration of daylight — about 1000 m. Many species, if not most, make a pronounced light-controlled diel vertical migration, spending the night somewhere in the upper 100 m and the day 500 m or more deeper. Most of the species are large-eyed, large-mouthed, sharp-toothed carnivores, eating such things as copepods, euphausiids, and other fishes smaller than themselves. Most are bioluminescent. Many have gas-filled swimbladders and thus are effective sound-scatterers, responsible for the so-called "deep scattering layers" so often conspicuous on echo-sounder records.

Zoogeographic Background

The North Atlantic mesopelagial can be divided into six faunal regions, four ocean-spanning ones — the Atlantic Tropical, the North Atlantic Subtropical, the North Atlantic Temperate, and the Atlantic Subarctic Regions — plus two small, marginal regions, the Mauritanian Upwelling Region and the Gulf of Mexico. As explained earlier, the ACSAR area spans the western extremities of two of these regions. The northeastern part of the study area lies in the Slope Water, a province of the North Atlantic Temperate Region, while the southwestern part of the study area lies in the northern Sargasso Sea, a province of the North Atlantic Subtropical Region. From a zoogeographical standpoint, the Florida Current and its continuation, the Gulf Stream, are considered special parts of the northern Sargasso Sea. The first bounds the southwestern part of the study area at its western extremity, the second divides the study area into its two main parts — a northeastern temperate one and a southwestern subtropical one.

Mesopelagic fishes and other pelagic species of *temperate* and *subpolar-temperate* distribution find the southern limit of their western North Atlantic range at the Gulf Stream-Slope Water boundary, i.e., the boundary between the northeastern and southwestern parts of ACSAR. The same boundary sets the northern limit in the west for species of *subtropical* and *tropical-subtropical* distribution. *Tropical* and *tropical-semisubtropical* species originating in the Caribbean Sea, the westernmost province of the Atlantic Tropical Region, are swept north by the Florida Current and Gulf Stream. (Backus *et al.*, 1977).

Not all species are limited by the Gulf Stream-Slope Water boundary, of course. *Temperate-semisubtropical* ones, for instance, normally live in northern Sargasso Sea and Slope Water alike. However, species living on both sides of this important boundary may have different vertical distributions in the two domains. The Gulf Stream rings, as has been demonstrated throughout this chapter, also play a major role in mesopelagic fish distribution.

Mesopelagic Fish Research In the Study Area

Studies of mesopelagic fish are patchier than those of the mesozooplankton and therefore any discussion is also fragmented. The principal studies of mesopelagic fishes in the ACSAR area are (chronologically) as follows:

Backus et al. (1969) contrasted the mesopelagic fish faunas of the northern and southern Sargasso Seas, while Backus et al. (1970) described the distribution of mesopelagic fishes in the equatorial and western North Atlantic and gave zoogeographic information for both the Slope Water and northern Sargasso Sea.

Jahn and Backus (1976) compared the mesopelagic fish faunas of Slope Water, Gulf Stream, and Northern Sargasso Sea. Jahn (1976) studied mesopelagic fishes in cold-core Gulf Stream rings and necessarily compares Slope Water and the northern Sargasso Sea.

Krueger et al. (1977) summarized what was learned in an investigation of mesopelagic fishes at Deepwater Dumpsite 106, located in the ACSAR area at 38° 50'N, 72° 15'W. During part of this study the dumpsite was under the influence of warm-core rings, and part of the time was occupied by more typical Slope Water.

Backus and Craddock (1977) and Backus et al. (1977) described faunal regions and provinces for the Atlantic Ocean and provided information about mesopelagic fishes in the Slope Water and northern Sargasso Sea.

Backus and Craddock (1982) studied mesopelagic fishes in cold-core Gulf Stream rings and made comparisons of the fauna of the rings with those of northern Sargasso Sea and Slope Water.

The most abundant and ecologically important families are Myctophidae, the lantern-fishes, and Gonostomatidae, sometimes called "pearl-sides". (We use Gonostomatidae in a broad sense to include Gonostomatidae, Photichthyidae, and Sternoptychidae.) For instance, in an elaborate study of the upper 1000 m off the Canary Islands, Badcock (1970) found that myctophids and gonostomatids together made up about 80% of the midwater fish fauna as a whole, and this may be typical of mesopelagic communities in general.

The Mesopelagic Fish Fauna of the Slope Water

The best description of the mesopelagic fish fauna of the Slope Water is found in Backus and Craddock (1982). The basically temperate character of the Slope Water fauna is well shown in Table 6.5. The four most abundant species (accounting for 76 percent of the specimens) are of *subpolar-temperate* (*Bentho-sema glaciale*, 51% of the total number of specimens), *temperate* (*Ceratoscopelus maderensis*, 13%), and *temperate-semisubtropical* (*Hygophum hygomii* and *Lobianchia dofleini*, 13%) affinities. Most of the remaining species in Table 6.5 are broadly distributed in the tropical and subtropical Atlantic and probably reflect both the more or less continuous input of alien species into the Slope Water by warm-core rings and modification of the Slope Water as a habitat by the same mechanism. Indeed, four species — *Diaphus dumerilii*, *Lampanyctus alatus*, *Lepidophanes guentheri*, and *Myctophum affine* — having otherwise *tropical* ranges may actually reproduce in the Slope Water.

Table 6.5 Slope Water (temperate) mesopelagic fishes, 20 most abundant, and comparison with the Sargasso Sea. (from Backus and Craddock, 1982)

Rank*	Species	Number caught	Catch/ 10000m³	Sargasso Sea rank	Sargasso Sea rate
1.	*Benthosema glaciale*	1550	10.14	-	.03
2.	*Ceratoscopelus maderensis*	381	2.49	3	1.00
3.	*Hygophum hygomii*	209	1.37	5	.95
4.	*Lobianchia dofleini*	176	1.15	19	.24
5.	*Lampanyctus alatus*	92	.60	-	.14
6.	*Sternoptyx diaphana*	77	.50	12	.42
7.	*Lepidophanes guentheri*	64	.42	-	.20
8.	*Benthosema suborbitale*	55	.36	-	.14
9.	*Hygophum benoiti*	53	.35	14	.34
10.	*Notolychnus valdiviae*	47	.31	6	.73
11.	*Lampanyctus crocodilus*	41	.27	16	.32
12.	*Ceratoscopelus warmingii*	39	.26	1	2.38
13.	*Myctophum affine*	37	.24	-	.02
14.	*Gonostoma elongatum*	36	.24	11	.44
15.	*Notoscopelus resplendens*	34	.22	20	.24
16	*Bolinichthys indicus*	31	.20	4	.97
17	*Diaphus dumerilii*	31	.20	-	.02
18.	*Diogenichthys atlanticus*	31	.20	8	.55
19.	*Hygophum taaningi*	29	.19	-	.15
20.	*Lampanyctus cuprarius*	28	.18	10	.45

*With samples of the size taken, ranks after 5 or 6 should not be regarded seriously according to Miller and Wiebe (McGowan 1971).

Jahn and Backus (1976) described the Slope Water mesopelagic fish fauna from two dozen collections in which the Slope Water criterion (in addition to a simple geographical one) was that the depth to 15 °C was less than 200 m. The three most abundant species in their set — *Lobianchia dofleini, Benthosema glaciale,* and *Ceratoscopelus maderensis* — are among the top four in the Slope Water set of Backus and Craddock (1982). Most samples taken in the Slope Water (excluding warm-core rings) will show principal species much like Table 6.5, which can be taken as descriptive of the mesopelagic fish fauna in the northeastern, temperate part of the study area. The actual catch-rates shown can be expected to vary widely depending upon seasonal and other factors.

The Mesopelagic Fish Fauna of the Northern Sargasso Sea

The 20 most abundant species in the northern Sargasso Sea (Table 6.6) show a more equitable distribution than does the Slope Water set. The four most abundant species account for only 45 percent of the total, while 10 to 11 species are required to account for the 76 percent that the four most abundant species in the Slope Water set comprised.

The distribution patterns for the first ten fish on the list are, with one exception, those to be expected for a subtropical faunal province: *tropical-subtropical* (*Ceratoscopelus warmingii* and *Bonapartia pedaliota*, 23% of total specimens), *tropical-subtropical-temperate* (*Argyropelecus hemi-*

Table 6.6 Northern Sargasso Sea mesopelagic fishes, 20 most abundant, and comparison with the Slope Water (from Backus and Craddock, 1982)

Rank	Species	Number caught	Catch/ 10000m³	Slope Water rank	Slope Water rate
1.	*Ceratoscopelus warmingii*	311	2.38	12	.26
2.	*Argyropelecus hemigymnus*	171	1.31	-	.16
3.	*Ceratoscopelus maderensis*	131	1.00	2	2.49
4.	*Bolinichthys indicus*	127	.97	16	.20
5.	*Hygophum hygomii*	125	.95	3	1.37
6.	*Notolychnus valdiviae*	96	.73	10	.31
7.	*Lampanyctus pusillus*	78	.60	-	.16
8.	*Diogenichthys atlanticus*	72	.55	18	.20
9.	*Bonapartia pedaliota*	66	.50		
10.	*Lampanyctus cuprarius*	59	.45	20	.18
11.	*Gonostoma elongatum*	58	.44	14	.24
12.	*Sternoptyx diaphana*	55	.42	6	.50
13.	*Pollichthys mauli*	52	.40		
14.	*Hygophum benoiti*	45	.34	9	.35
15.	*Lepidophanes gaussi*	43	.33	-	.01
16.	*Lampanyctus crocodilus*	42	.32	11	.27
17.	*Vinciguerria attenuata*	42	.32	-	.10
18.	*Argyropelecus aculeatus*	36	.28	-	.07
19.	*Valenciennellus tripunctulatus*	35	.27	-	.18
20.	*Lobianchia dofleini*	32	.24	-	1.15

gymnus and *Notolychnus valdiviae*, 16%), *temperate-semi-subtropical* (*Hygophum hygomii* and *Lampanyctus pusillus*, 13%), *subtropical* (*Bolinichthys indicus* and *Lampanyctus cuprarius*, 12%), and *tropical-subtropical-Slope Water* (*Diogenichthys atlanticus*, 4%). A *temperate* species, *Ceratoscopelus maderensis*, 8%, is the exception. It occurs in the Sargasso Sea set by virtue of cold-core ring transport (Backus and Craddock, 1982).

Jahn and Backus (1976) described the midwater fish fauna of the northern Sargasso Sea from 20 collections. The most abundant species in their set, *Ceratoscopelus warmingii*, is the same as that in the set of Backus and Craddock (1982) and their top four species fall within the top eight.

The Mesopelagic Fish Fauna of the Gulf Stream and the Slope Water and Sargasso Sea Further Compared

The mesopelagic fish fauna of the Gulf Stream itself has not been thoroughly studied, although it is known to carry some vestiges of a tropical fauna. Jahn and Backus (1976) studied sets of mesopelagic fishes from Slope Water (200-m temp. <15°C), Gulf Stream (200-m temp. 15-17.5°C), and northern Sargasso Sea (200-m temp. >17.5°C). A similarity measure showed the Slope Water set to be about equally distinct from the Gulf Stream and northern Sargasso Sea sets (percentage of similarity 39 and 36, respectively), while the last two were somewhat similar (PS 57). A cluster analysis showed the Gulf Stream set to be intermediate between the distinct Slope Water and northern Sargasso Sea sets, but suggested that the Gulf Stream fauna was not simply a mixture of the other two faunas.

The Northern Sargasso Sea Fauna as Modified by Cold-Core Gulf Stream Rings

The southwestern, subtropical part of the study area can be expected to have a fauna normal for the Northern Sargasso Sea (Table 6.6), except in those parts temporarily occupied by cold-core Gulf Stream rings. Because the cold-core rings entering this part of the study area are aged, the fauna that they carry is not much different from the normal one of the northern Sargasso Sea. Table 6.5 — Slope Water (temperate) mesopelagic fishes, 20 most abundant, and comparison with the Sargasso Sea.

Table 6.7 shows the 20 most abundant species at 15 stations at various places in several cold-core rings, the depth to 15°C at these stations varying from 100-450 m. A cold-core ring in the southwestern, subtropical part of the study area would have a mesopelagic fish fauna intermediate between the faunas shown in Tables 6.6 and 6.7.

Table 6.7 Mesopelagic fishes in cold-core Gulf Stream rings, and comparison with those of Slope Water and northern Sargasso Sea. See text.

Rank	Species	Number caught	Catch/ 10000m³	Slope Water rank	Slope Water rate	Sargasso Sea rank	Sargasso Sea rate
1.	*Ceratoscopelus maderensis*	1223	7.29	2	2.49	3	1.00
2.	*Benthosema glaciale*	933	5.56	1	10.14	-	.03
3.	*Lampanyctus pusillus*	272	1.62	-	.16	7	.60
4.	*Argyropelecus hemigymnus*	256	1.53	-	.16	2	1.31
5.	*Ceratoscopelus warmingii*	169	1.01	12	.26	1	2.38
6.	*Bolinichthys indicus*	164	.98	16	.20	4	.97
7.	*Hygophum benoiti*	161	.96	9	.35	14	.34
8.	*Lampanyctus crocodilus*	117	.70	11	.27	16	.32
9.	*Notolychnus valdiviae*	84	.50	10	.31	6	.73
10.	*Vinciguerria attenuata*	82	.49	-	.10	17	.32
11.	*Lampanyctus photonotus*	70	.42	-	.07	-	.23
12.	*Lobianchia dofleini*	62	.37	4	1.15	19	.24
13.	*Lepidophanes guentheri*	57	.34	7	.42	-	.20
14.	*Argyropelecus aculeatus*	52	.31	-	.07	17	.28
15.	*Valenciennellus tripunctulatus*	51	.30	-	.18	18	.27
16.	*Diogenichthys atlanticus*	47	.28	18	.20	8	.55
17.	*Diaphus mollis*	43	.26	-	.04	-	.16
18.	*Gonostoma elongatum*	36	.21	14	.24	11	.44
19.	*Lampanyctus alatus*	36	.21	5	.60	-	.14
20.	*Benthosema suborbitale*	30	.18	8	.18	-	.14

The Fauna of the Westernmost Slope Water as Modified by Warm-Core Gulf Stream Rings

The way in which the Slope Water mesopelagic fish fauna is modified by the presence of warm-core rings is being studied by R. H. Backus and colleagues at the Woods Hole Oceanographic Institution. Presumably a newly formed warm-core ring has a mesopelagic fish fauna at its center very much like the one of the northern Sargasso Sea (Table 6.6). This fauna presumably becomes more and more like the fauna of the Slope Water (Table 6.5) with ring age. It is possible, if not probable, that the westernmost part of the Slope Water has a mesopelagic fish fauna somewhat different from the remainder of the Slope Water because of a continual replenishment of warm-water species and quasi-permanent modification of the habitat by warm-core rings.

The Displacement Volume of Mesopelagic Fishes in the Study Area

There are relatively few good data on the displacement volume of mesopelagic fishes in the study area. Generally speaking the northeastern, temperate part supports a larger standing crop than the southwestern, subtropical part. Gulf Stream cold-core rings are intermediate in this regard, but there are no data as yet for Gulf Stream warm-core rings.

The general inverse relation between biomass of myctophids plus "gonostomatids" (excluding *Gonostoma elongatum* and *Cyclothone* spp.) and depth to 15°C (Fig. 6.24 illustrates the reduction in standing crop by several-fold that others have described in going from Slope Water to northern Sargasso Sea (Grice and Hart, 1962, for epizooplankton, about 4:1; Jahn and Backus, 1976, for mesopelagic fishes with about 60 percent of the fishing effort above 200 m, about 4.5:1; Jahn, 1976, for mesopelagic fishes 0-1000 m, about 2:1; and Ortner et al., 1978, for zooplankton 0-750 m, 3.5:1). Biomass data for the deep-living *Cyclothone* spp. suggest a similar relationship to that for myctophids-"gonostomatids," but the Slope Water/northern Sargasso Sea ratio appears to be less — 2 or 3:1.

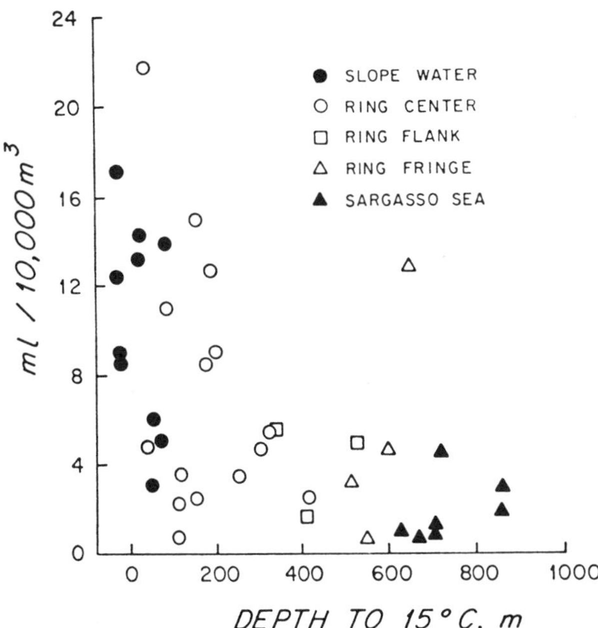

Figure 6.24 Myctophid-"gonostomatid" biomass as a function of depth to 15°C. The four points to the left of depth zero represent stations at which the surface temperature was less than 15°. From Backus and Craddock (1982).

Cetacea — Whales and Porpoises

R.H. Backus

Introduction

There are no books or papers devoted precisely to the relative abundance of the cetaceans of the ACSAR area. However, there are systematic reports of sightings of cetaceans for parts of this area and for immediately adjacent ones. From these and from the more general literature for the western North Atlantic it is possible to state what cetacean species probably occur in the study area and to assign them, although less certainly, to one of four general abundance categories — *abundant, common, uncommon,* or *rare.* We refer to the temperate part of the study area from Cape Hatteras north and east to off Georges Bank as the "northeastern part" and to the subtropical part from Cape Hatteras south to off central Florida as the "southwestern part".

Specific sightings of cetaceans are provided for the shoreward or western part of the northeastern temperate half of the present study area by CETAP (1981, undated, and 1982) and by Powers et al. (1982). The area of coverage of the CETAP reports extends seaward to a depth of about 2000 m from Cape Hatteras to the northeast beyond the limit of ACSAR. The study by Powers et al. included a small amount of observing just seaward of the 200-m isobath and had latitudinal limits similar to the CETAP study. There has been less systematic observing in the southwestern, subtropical part of the present study area, i.e., south of Cape Hatteras. Some observations, mainly for the shoreward part of the area, are to be found in Schmidly (1981). Recent useful general works for the western North Atlantic include Katona et al. (1975) and Leatherwood et al. (1976). General remarks are also based to some extent upon observations by the writer.

No values in terms of numbers of animals are attached to the four abundance categories used; they are merely relative within the area of concern. For some species, parts of the study area are probably among those parts of the North Atlantic where the species is most abundant. Thus, the apparently anomalous designation of "common" might be given for an endangered species, as it has been for the sperm whale, *Physeter catodon*. The systematic classification followed below is from Leatherwood et al. (1976).

The species fall into both of the main groups of cetaceans — the Mysticeti or baleen whales, mostly large, and the Odontoceti or toothed whales, which except for the large sperm whale are medium-sized to small. Several species of toothed whales in the 3- to 6-m class are sometimes called "blackfish". Small odontocetes are generally called "dolphin" or "porpoise". We prefer the latter and use it here, because it is the term in general use with mariners, the people who most often see these animals, and avoids confusion with the fish called "dolphin".

The baleen whales are conspicuous migrants, moving through shoreward parts of the study area to its northern parts and beyond in summer to feed, then to its southernmost parts and beyond in winter to reproduce (Fig. 6.25). Such a seasonal movement is also well known in the sperm whale, and probably is present in most cetaceans on some scale or another. These movements result in a seasonal change in the species present, particularly if the temperate and subtropical halves of the study area are considered separately.

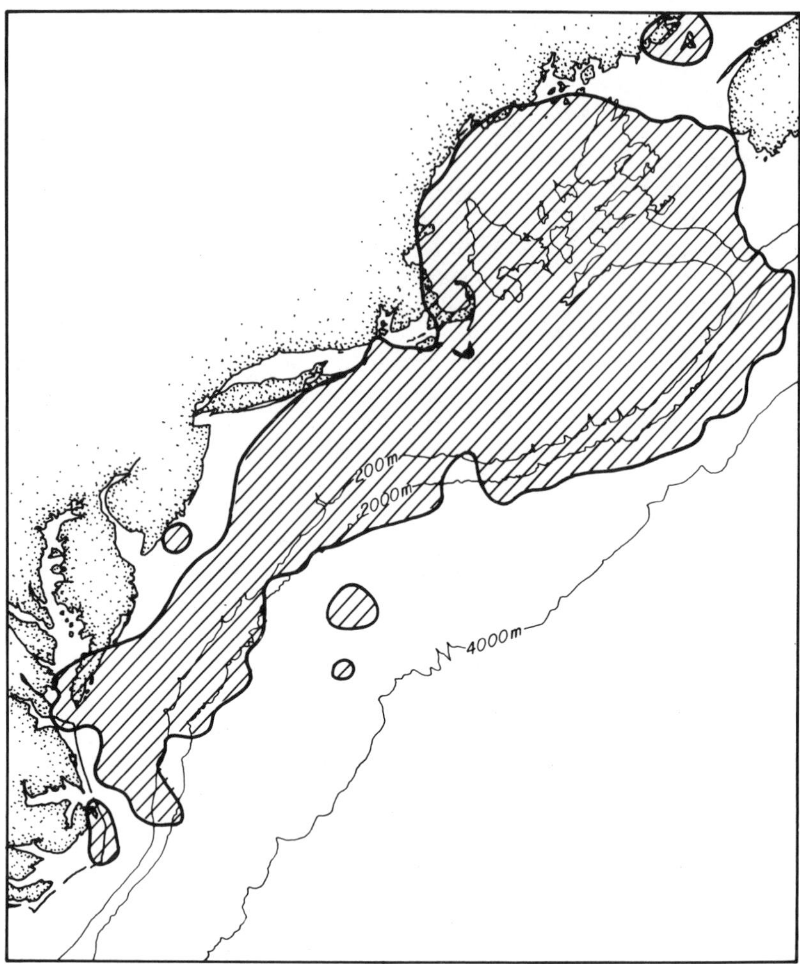

Figure 6.25 The distribution of baleen whales (after CETAP 1981).

Because the seasonal movements of cetaceans are only known in a general way and because the region of the report is so narrow in the onshore-offshore dimension, it is impossible to know with any precision the extent to which some cetaceans actually use the area. The baleen whales, which in the North Atlantic are mostly inhabitants of the continental shelves, are no doubt mostly confined to the shoreward, western side of the area (compare Figs. 6.25 and 6.26).

A rank order of abundance of cetaceans in the ACSAR area for three size-classes (Table 6.8) tries to take into account numbers of animals and seasonal usage; that is, a few animals passing rapidly through the area during migration would be low on the list; many animals resident throughout the year, high on the list. The boundary between size classes is not sharp. It should be emphasized that this is educated guesswork.

Mysticeti — Baleen Whales

Balaenopteridae — Rorquals

Balaenoptera acutorostrata — minke whale. This small whale occurs in the temperate, northeastern part of the study area, principally in spring and summer and especially north of about 40°N. Here it is probably *uncommon*, for this is an inshore animal as a rule. The species withdraws to the south in fall, probably mainly but not completely beyond the southern limits of the study area. CETAP (1982) has provided some sightings in the north near the 2000-m isobath.

Balaenoptera physalus. — fin whale (endangered). The fin whale is an *uncommon* to *common* species in the study area, both as a migrant and as a resident. Spring and summer numbers are greater than fall and winter ones, although the seasonal contrast is probably not so great in the study area as it is further shoreward over the continental shelf where the species is much more abundant. The fin whale occurs in small numbers more or less throughout the area in winter, but in summer is found mostly or wholly in the northeastern temperate part of the area. Two stocks may be involved, northern and southern ones with seasonally overlapping ranges. A number of sightings have been made on the shoreward edge of the present study area by CETAP (1982) and by Powers et al. (1982), but like most baleen whales this is an animal of the continental shelf, and its numbers diminish rapidly with distance from the shelf edge.

Figure 6.26 The distribution of Odontocete whales and dolphins (after CETAP 1981).

Table 6.8 Rank order of Cetacean Abundance in ACSAR area. (* denotes endangered species).

Large cetaceans	*Medium-size cetaceans*	*Small cetaceans*
*Physeter catodon**	*Globicephala melaena*	*Tursiops truncatus*
*Balaenoptera physalus**	*Globicephala macrorhynchus*	*Grampus griseus*
Balaenoptera acutorostrata	*Pseudorca crassidens*	*Delphinus delphis*
*Balaenoptera borealis**	*Orcinus orca*	*Lagenorhnychus acutus*
*Megaptera novaeangliae**	*Ziphius cavirostris*	*Stenella coeruleoalba*
*Eubalaena glacialis**	*Hyperoodon ampullatus*	*Stenella plagiodon and frontalis*
Balaenoptera edeni	*Mesoplodon spp.*	*Stenella longirostris*
*Balaenoptera musculus**		*Steno bredanensis*
		Lagenorhynchus albirostris
		Stenella clymene
		Kogia breviceps
		Kogia simus
		Feresa attenuata
		Lagenodelphis hosei
		Peponocephala electra

*endangered species

Balaenoptera musculus. — blue whale (endangered). This biggest of animals occurs in the northeasternmost part of the study area in unknown numbers, but undoubtedly should be classed as *rare*. It is possible that the species moves through the area as a migrant — south in fall, north in spring — but its wintering ground and migration paths are not known. There appear to be no sightings of record for the present study area, although CETAP (1982) reported two sightings just to the north.

Balaenoptera borealis. — sei whale (endangered). This poorly known rorqual is probably *rare* to *uncommon* in the study area. Like other large whales it moves north and south with the seasons, but in our area it may live offshore more than the other baleen whales. Judging from CETAP (1982), it is commonest in the study area in spring and in its northeasternmost part. It may be resident in the southwesternmost part of the study area in winter.

Balaenoptera edeni. — Bryde's whale. This whale is hard to identify and has often been confused with other rorquals; thus, its distribution is poorly known. It is mainly a nearshore species. If it occurs in the study area at all, it will be on the shoreward side of the southwestern, subtropical part. There appear to be no sightings of record here, but Schmidly (1981) reports a number of strandings on the shore just to the west. The species should be classified as *rare* in the study area.

Megaptera novaeangliae — humpback whale (endangered). This whale is probably *rare* to *uncommon* in the study area, for as a resident it mostly occurs in the shallower water to the west of our area. CETAP (1982) reported a few sightings near the 2000-m isobath at 40-41°N. Some humpbacks probably move through the area on their annual north-south migrations. Schmidly (1981) summarizes a number of observations based on captures made years ago in the Straits of Florida just to the south of the southern limit of the study area. These captures seem to have been of individuals making their northward migration in early spring, but some may have been of resident animals; therefore, there may be some wintering individuals on the southwest fringes of the ACSAR.

Balaenidae — Right Whales

Eubalaena glacialis. — right whale (endangered). This *rare* species moves north in summer, south in winter, like the other baleen whales. It mainly occurs shoreward of the northeasternmost part of the study area in spring. Its migration paths are very poorly known, but they too probably lie mainly shoreward of the study area. CETAP (1982) made a few sightings near the 2000-m isobath in the northeasternmost part of the study area, but the CETAP data show that numbers diminish rapidly seaward of the shelf edge. There may be some wintering animals on the shoreward edge of the southwestern, subtropical part of the study area. Schmidly (1981) reports sightings and strandings all along this coast from southern Florida to Cape Hatteras.

Odontoceti — Toothed Whales

Ziphiidae — Beaked Whales

Mesoplodon bidens, M. densirostris, M. europaeus, and *M. mirus.* — These four species are everywhere *rare*, but might be encountered in the study area, judging from the distribution of their strandings. According to CETAP (1982) sightings, these whales are more often to be seen over the slope than over the continental shelf. The first species probably is found only in the northeasternmost part of the study area. Schmidly (1981) reports strandings of the last three species along the coast from Florida to Cape Hatteras.

Ziphius cavirostris — goose-beaked whale. This *rare* species might be encountered anywhere in the study area. The few CETAP sightings (CETAP, 1982) indicate that this is principally an offshore animal, more to be seen in the study area than over the continental shelf to the west. Schmidly (1981) reports a number of strandings in the Bahamas and along the coast from Florida to Cape Hatteras.

Hyperoodon ampullatus — bottlenosed whale. This little known whale is probably *rare* and confined to the northeastern part of the study area, where its numbers may increase in winter due to the southward movement of animals that summer to the north. CETAP (1982) reported two sightings that come from the study area.

Physeteridae — Sperm Whales

Physeter catodon — sperm whale (endangered). This largest of toothed whales is *common* in the study area, perhaps being most numerous on the traditional sperm-whaling grounds off Cape Hatteras in the westernmost Slope Water and Gulf Stream (Townsend, 1935). There is a withdrawal to the south in winter, and some animals may use the southwesternmost part of the study area as a wintering ground. CETAP (1982) reported numerous sightings that fall within the study area. Schmidly (1981) summarizes numerous captures from years ago. These cover the southwestern subtropical part of the study area and come at all times of year.

Kogia breviceps — pygmy sperm whale and *Kogia simus* — dwarf sperm whale. These are widely distributed, hard to distinguish, little seen offshore species. They may be *uncommon* to *rare* in the southwestern part of the study area, *rare* in the northeast. CETAP (1982) reports a single sighting in the study area. Schmidly (1981) reports numerous strandings for the coast west of the southwestern part of the study area. In these *K. breviceps* is commoner than *K. simus*.

Stenidae

Steno bredanensis — rough-toothed porpoise. This is a poorly known warm-water species not easy to identify, and it might be *common* in the southwestern subtropical part of the study area. Like other warm-water porpoises, it may come north of the Gulf Stream into the Slope Water in summer or with Gulf Stream warm-core rings. CETAP (1982) reported one such occurrence at about 39°N (latitude of Cape Henlopen). Schmidly (1981) reports two strandings to the west of the southwestern part of the study area.

Delphinidae — Porpoises

Peponocephala electra — melon-headed blackfish. It is possible that this poorly known and apparently *rare* animal is to be found in the southernmost part of the study area. There are no records of its occurrence there, but it is known from the Lesser Antilles (Schmidly, 1981).

Feresa attenuata — pygmy killer whale. This warm-water species, which is everywhere rare to uncommon, has been reported for the northeastern part of the study area (CETAP, 1982, one sighting). There appear to be no sightings of record in the southwestern part of the study area, but it is possible that the species is more abundant there than in the northeast. Schmidly (1981) records a few strandings in Florida just southwest of the southern limit of the study area. The species should be considered *rare* in our area.

Pseudorca crassidens — false killer whale. This blackfish, though poorly known, is probably *common* in the southwestern subtropical part of the study area. Schmidly (1981) reports some strandings just to the west. It may occur in the western part of the temperate Slope Water in summer or as a part of warm-core Gulf Stream rings. CETAP (1982) reported one sighting on the 2000-m isobath off Cape Hatteras.

Globicephala melaena — pothead whale. This blackfish is *common* to *abundant* in the temperate part of the study area, where its numbers may be higher in the shallower parts for most of the year. It is possible, however, that there is an offshore movement in winter. CETAP observations (made for *Globicephala* spp., but probably pertaining mainly to *G. melaena*) suggest that numbers are highest right on the shelf edge (CETAP, 1982).

Globicephala macrorhynchus — short-finned pilot whale. This blackfish replaces the preceding species in the southwestern subtropical part of the study area, where it is *common*. Schmidly (1981) reports sightings here and strandings to the west all along the coast from Cape Hatteras to the Straits of Florida.

Lagenodelphis hosei — Fraser's dolphin. This little known porpoise occurs in the Caribbean Sea (Caldwell et al., 1976), which makes it likely that it will be found sometime in the southernmost part of the study area. It should be classified as *rare*.

Tursiops truncatus — bottle-nosed porpoise. This species is *common* in the northeastern part of the study area, but *uncommon* in the southwestern subtropical part, where it is more of an inshore animal. There appears to be a pronounced north-south, summer-winter movement. CETAP (1982) and Powers et al. (1982) report numerous sightings of this animal for the temperate part of the study area, mainly along the inner slope and shelf edge. Schmidly (1981) reports some sightings and numerous strandings in the latitude of the southwestern subtropical part of the study area, but these are wholly within the 200-m isobath except at Cape Hatteras.

Grampus griseus — grampus. This small toothed whale or big porpoise is probably *common* to *abundant* throughout the study area. There appears to be a northward movement in spring and summer into the temperate part of the study area, the converse in fall and winter into the subtropical part, although the species is found in some numbers in both places in all seasons. CETAP (1982) data suggest that the species is most common on the shelf edge.

Stenella longirostris — spinner. This poorly known porpoise is probably *uncommon* in the southwesternmost part of the study area and rare or absent elsewhere. It may be one of those warm-water species that come north into the Slope Water in summer or with warm-core Gulf Stream rings. CETAP (1982) reported a few sightings in deep water, the northernmost of which was between 39° and 40°N.

Stenella clymene — short-snouted spinner. This little-known *Stenella* probably occurs in the southwestern part of the study area. There are a few stranding records outside the study area to the west (Schmidly, 1981). It probably should be classed as *rare* to *uncommon*.

Stenella plagiodon and *S. frontalis* — spotted porpoises. These porpoises, which are difficult to distinguish, are *common* to *abundant* in the subtropical part of the study area, and are often found north of the Gulf Stream in the westernmost part of the Slope Water, perhaps in summer or associated with warm-core rings. CETAP (1982) reported a number of sightings for the temperate northeastern half of the study area, particularly from its southwesternmost part. Schmidly (1981) gives some records for the shoreward edge of the southwestern part of the study area.

Stenella coeruleoalba — striped porpoise. This porpoise is *common* to *abundant* in the study area, but its occurrence north of the Gulf Stream may be associated with summer or warm-core rings. CETAP (1982) reports numerous sightings (in the northeastern, temperate part of the study area).

Orcinus orca — killer whale. This whale is probably *rare* to *uncommon* in both parts of the study area. There were a few sightings by CETAP (1982) in the northeast, both inshore and near the 2000-m isobath, and a few by Schmidly (1981) to the west and south of the southwestern part.

Lagenorhynchus acutus — white-sided porpoise. This species occurs in the northern reaches of the temperate part of the study area, from about Hudson Canyon north. It is mainly a species of the continental shelf, but there appear to be onshore-offshore movements with the seasons such that it may be *uncommon* in the study area in winter although *rare* at other times. CETAP (1982) reported some sightings for the shoreward part of the study area. It is not known where the main part of the population of this species, which is common inshore, spends the winter.

Lagenorhynchus albirostris — white-snouted porpoise. This *Lagenorhynchus* is a colder-water species than the preceding one; its range is mostly north of the study area and so it is classed as *rare* to *uncommon* there. CETAP (1982) reports a few sightings for the study area.

Delphinus delphis — saddleback porpoise. This species is *common* to *abundant* shoreward of the study area, but, being an animal of the outer shelf, probably should be classed as *uncommon* to *common* in the study area itself. The numerous sightings reported by CETAP (1982) suggest that many individuals summer north of the study area, moving back into it in winter. Judging from Schmidly (1981) the species may be less abundant in the southwestern part of the study area than it is in the northeast.

Benthos

J.F. Grassle

The Fauna of Soft Sediments

Standing Crop

Deep-sea fauna are generally divided into three classes (not including microorganisms) on the basis of size and taxonomic position. The largest size group, the *megafauna*, are defined as animals visible in photographs. The *macrofauna* are the animals retained by screens ranging from 250 to 500 mm. Individuals belonging to *meiofaunal* groups such as foraminifera, copepods, nematodes, and podocopid ostracods are generally excluded from the macrofauna. Biomass of macrofauna and megafauna in the ACSAR region has been estimated from wet weight (Haedrich et al., 1980; Haedrich and Rowe, 1977; Rowe et al., 1982) of formalin-preserved specimens with shells. These numbers have large errors because of the carbonate in large molluscs and echinoderms and may not be very useful (Khripounoff et al., 1980; Mills et al., 1982). Smith (1978) and Smith et al. (1978) used estimates corrected for shell weight and 0.297 mm screens and found the same order of magnitude of biomass that Rowe (1983) found using weights uncorrected for shells and 0.420 mm screens. These data only show trends and do not provide statistically reliable estimates of biomass from any particular site.

The most comprehensive sampling of deep-sea macrofauna was done in the 1960s by Sanders and Hessler (summarized in Sanders and Hessler, 1969) on a transect from Gay Head on Martha's Vineyard, Massachusetts, to Bermuda. The macrofauna below 200 m depth are generally reduced in size (Sanders et al., 1965). Haedrich and Rowe (1977), Haedrich and Polloni (1976), Wenner and Musick (1977), and Wenner (1979a) indicate an increase in average weight of individual demersal fish going from 500 m to 2500 m, however, Wigley et al. (1975) and Wenner (1978) found an inverse relationship between depth and size in decapod invertebrates.

Table 6.9 Depth, latitude, longitude, number of animals collected, and number of animals per square meter for (the transect stations included in the present study). (from Sanders et al., 1965).

Station	Depth (m)	Latitude	Longitude	No. animals in sample	No. animals/ m^2
55	75	40°27.2′N	70°47.5′W	3791	13073
C 1	97	40°20.5′N	70°47′W	3082	5314
Sl. 2	200	40°01.8′	70°42′	6455	12910
Sl. 3	300	39°58.4′	70°40.3′	11907	21263
Sl. 4	400	39°56.5′	70°39.9′	4439	6081
D 1	487	39°54.5′	70°35′	5115	8669
E 3	832	39°50.5′	70°35′	3008	2979
F 1	1500	39°47′	70°45′	997	1719
G 1	2086	39°42′	70°39′	1120	2154
GH 1	2500	39°25.5′	70°35′	365	521
GH 4	2469	39°29′	70°34′	299	467
HH 3	2870	38°47′	70°08′	636	748
II 1	3742	37°59′	69°32′	*	*
II 2	3752	38°05′	69°36′	391	1003

*Sample excluded from quantitative analysis because of small size.

Data from the Gay Head-Bermuda transect show megafauna densities at 300 m depth on the continental slope up to 21,000/m^2 and densities at about 5000 m depth from 33 to 92 per m^2 (Table 6.9). Changes with depth in numbers of individuals and biomass from box core samples are given in Fig. 6.27. The biomass reduction with depth is similar for macrofauna and megafauna (Haedrich and Rowe, 1977). Smith and Hinga (1983) also found a reduction in biomass with depth. Off North Carolina meiofaunal densities per 10 cm^2 were 442.4 ± 196.7 at 400 m, 891.9 ± 350.1 at 900 m and 73.5 ± 46.0 at 4000 m depth. Zonation of fish and megafauna are summarized in Grassle et al. (1975) and Haedrich et al. (1975). Biomass and density of fish decrease with depth but the average size of individuals may increase.

Horizontal Variation in Density

Little is known about variation in population density along isobaths since the same methods have never been used over the latitudinal range of eastern North America. Most of the work on macrofauna has been done off the northeastern U.S. and the meiofauna studies have been off North Carolina. Regressions of density of meiofauna indicate a decline with depth irrespective of horizontal location (Thiel, 1979). An unpublished study on fish from the slope and rise of the U.S. middle Atlantic U.S. coast found similar results to those of Haedrich et al. (1980). Musick (1976) found the most rapid change in species composition of demersal fish

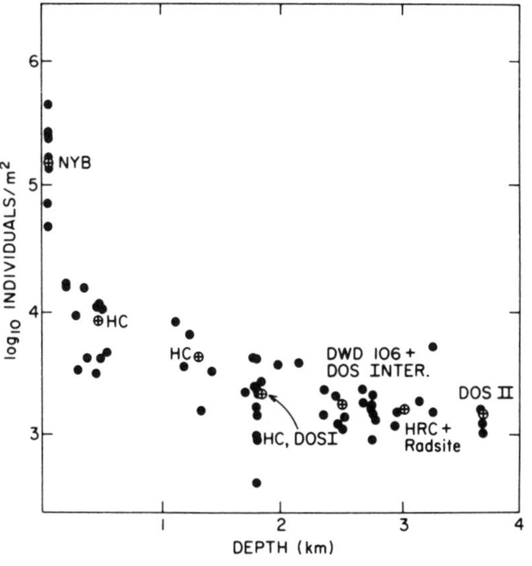

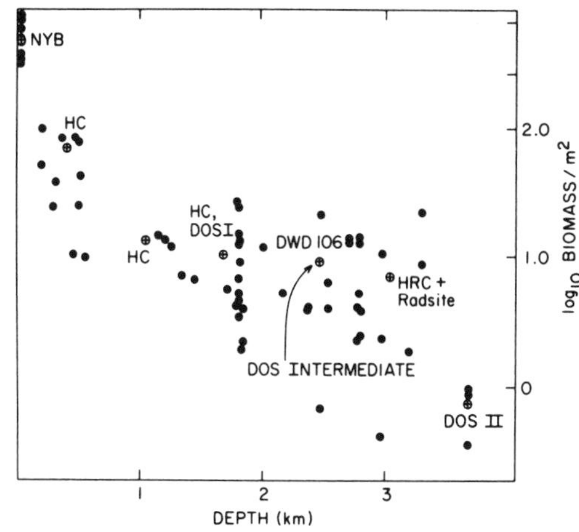

Figure 6.27 Numbers of animals and wet-weight biomass per square meter vs depth. (Rowe et al., 1982). Solid circles are individual samples and open circles are averages for areas defined in Table 6.30.

at 150-200 m, 400-600 m, 950 m, 1350-1525 m, 1930-2130 m and 2700 m. Boundaries between zones proposed by Haedrich et al. (1980) south of New England are at 270 m, 650 m, 1300 m, 2000 m, 2500 m, 3200 m, and 3800 m.

Species Distributions

Species in the Region Over the last decade there have been a number of monographs describing the deep-sea fauna in the ACSAR area. The most important deep-sea group in terms of numbers of individuals and species is the polychaetes. Hartman (1965) and Hartman and Fauchald (1971) summarized results from anchor dredge and epibenthic sled samples mainly from the Gay Head-Bermuda transect. The number of species recognized increased and several families were revised in the second monograph (Hartman and Fauchald, 1971). The second volume (1971) has a list of 46 of the most abundant species from all samples studied. These polychaete species have a broad depth distribution, however, only three species are very abundant at depths below 1500 m. *Myriochele* nr *heeri*, *Melinnata americana*, and *Paraonis uncinatus* are the only species on this list most common at depths below 1500 m. In an environment where single species seldom make up more than 5 percent of the fauna, it is very difficult to single out individual species as more or less important. When results from a greater number of quantitative samples (box cores) and more taxonomic studies of individual families are available from the Atlantic (such as the doctoral dissertation of Maciolek 1983), it should be possible to identify numerically dominant species within particular depth ranges. For example, *Aurospio dibranchiata* is clearly one of the more common polychaete species at depths from 1700 to 3600 m (Grassle, 1977; Rowe et al., 1982; Maciolek, 1981). According to Maciolek (1981), this species was identified as both *Prionospio cirrifera* and *Prionospio steenstrupi* in Hartman (1965) and *Laonice antarcticae* in Hartman and Fauchald (1971). This same species was called spionid (undescribed) by Grassle (1977) and spionid A by Rowe et al. (1982). A species of *Glycera* called *Glycera mimica* by Hartman (1965) and Grassle (1977) and *Glycera capitata* by Rowe et al. (1982) is also common at these depths. Grassle (1977) and Rowe et al. (1982) also found *Poecilochaetus fulgoris* and *Pholoe minuta anoculata* to be very abundant at 1700-1800 m.

A summary of bivalve distributions by Sanders and Allen is not yet complete. The bivalve systematic studies published thus far include Allen and Sanders (1966, 1969, 1973), Allen and Turner (1974), Allen and Morgan (1981), Sanders and Allen (1973, 1977). Some of the work is summarized in Allen (1979) but abundance of named species is not discussed. Grassle (1977) and Rowe et al. (1982) found that *Nucula cancellata* was among the most abundant species at 1700-1800 m depth south of New England. The same two studies also showed that the aplacophoran mollusc, *Prochaetoderma* sp. A, is very common at 1700-1800 m south of New England. Scaphopods are also common but taxonomic studies of this group are still in progress.

Gastropods are not very abundant in the deep sea but Rex (1972, 1973, 1977, 1981, 1983) has summarized the distribution and ecology of this group. Below shelf depths quantitative samples do not provide enough individuals for community and population studies. From anchor dredge samples, Sanders et al. (1965) found less than one gastropod per m^2 from depths below 2000 m and no gastropods below 4500 m depth. At depths below 1500 m, 2442 individuals were collected from large nonquantitative epibenthic sled samples. Of these 2442 animals, 904 are *Cithna tenella* from depths below 3800 m. The next most common species was *Mangilia bandella* with 185, found from 2500 m to 3800 m depth. At depths between 478 m and 1102 m, there were 970 *Alvania americana* out of a total of 3704 gastropods collected.

Ophiuroids are among the most abundant deep-sea benthic groups however, they are represented by only a few species (Sanders, 1977). The main study in the western Atlantic is the doctoral dissertation and subsequent publications by Schoener (1967, 1968, 1969, 1972). Of the nearly 30,000 individuals considered by Schoener, 10,098 were of a single species, *Ophiura ljungmani*, from a single sample. There is some uncertainty about identification of juveniles of this species but approximately 90 percent of the juveniles are likely to have been correctly identified as *O. ljungmani*. Although a number of species including *O. ljungmani* have been shown to reproduce seasonally in the eastern Atlantic (Gage and Tyler, 1981), *O. ljungmani* is the only deep-sea species in the western Atlantic where there is good evidence for seasonal reproduction.

The taxonomic groups with highest species diversity in the deep sea are the peracarid crustacean orders Amphipoda, Cumacea, Isopoda, and Tanaidacea. The systematics have yet to be worked out for most of these groups despite extensive work on amphipods by Mills (1967, 1971, 1972a, b); Cumacea by Jones (1973, 1974) and Reyss (1974, 1978); Isopoda by Chardy (1974, 1976), Hessler (1967, 1968, 1970a,b), Kensley (1982), Siebenaller and Hessler (1977, 1981), Thistle and Hessler (1976, 1977), Wilson (1976, 1980a, b, 1981), Wilson and Hessler (1974, 1980, 1981); and tanaids by Gardiner (1975).

Mills reviewed the number of species and distribution of amphipods on the Gay Head-Bermuda transect. Ampeliscid amphipods gradually drop out on the upper slope and phoxocephalids and lysianassids are important at all deep-sea depths. In four epibenthic sled samples from 1300 to 2900 m depth, the number of species ranged from 20 to 43 and a single sample from about 4700 m had 17 species.

The number of species of Cumacea per epibenthic sled trawl at depths from 1300 to 2900 m were 15 to 32 (Jones and Sanders, 1972). At depths around 4700 m, two samples had 10 and 11 species. Of 100 species found on the Gay Head-Bermuda transet, 37 percent are new or undescribed. A single sample may have as many as 8 congeneric species of *Campylaspis* or 7 congeneric species of *Leucon*.

Samples from the same series of epibenthic sled hauls yielded 51 isopod species from 2900 m and 39 isopod species from 3800 m on the Gay Head-Bermuda transect (Hessler et al., 1979). Comparison with transects elsewhere in the Atlantic suggest that the North American Basin along the east coast of the United States has a somewhat lower diversity of species. In the tanaids, only the Neotanaidae have been studied (Gardiner, 1975). This difficult and highly diverse peracarid order is particularly in need of further study.

Other groups that have been studied from the Gay Head-Bermuda transect include the tunicates (Monniot and Monniot, 1968, 1970, 1975, 1976a, b, c, 1978; Monniot, 1971, 1979); sipunculids (Cutler, 1973; Cutler and Duffy, 1972; Cutler and Doble, 1979); oligochaetes (Erseus, 1979a, b, 1982); echiurans (Datta-Gupta, 1981); pycnogonids (Child, 1982); crinoids (Clark, 1977); and Pogonophora (Southward,

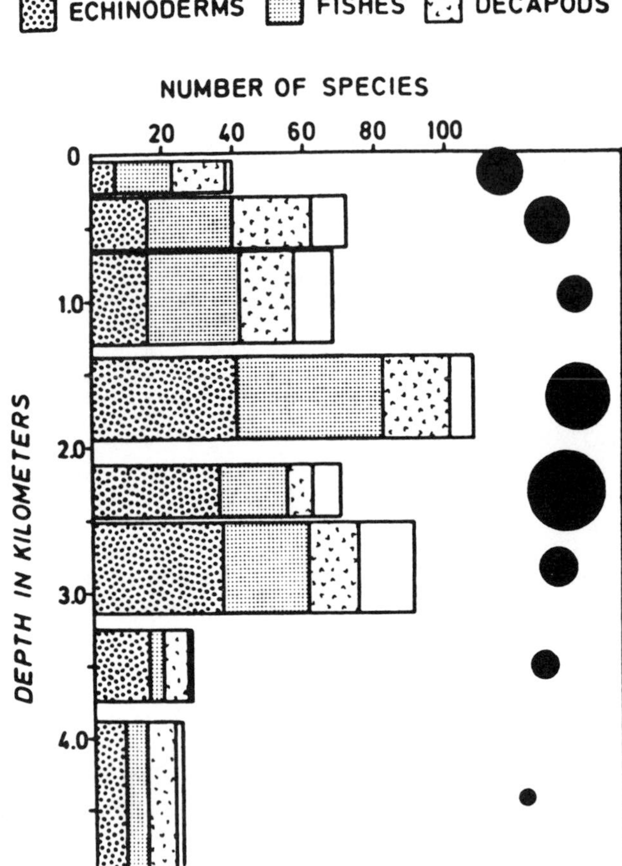

Figure 6.28 Vertical extent of each of the 8 zones (regions of relatively little faunal change) summarized along depth axis; discontinuities are the faunal boundaries (regions of relatively rapid faunal change). Length of each bar indicates total number of megafaunal species subdivided into number of echinoderms, fishes, decapod crustaceans, and other groups (clear area). Filled circles to right represent megafaunal biomass. From Haedrich et al. (1980).

1968, 1971; Southward and Brattegard, 1968). From these studies the fauna of the North American Basin off the United States is much better known than for other regions of the ocean, but much more taxonomic work is needed.

A few taxonomic studies of western Atlantic meiofauna have been completed (Benson, 1975; Hope and Murphy, 1969a, b, 1970; Hope 1977; Humes, 1974; Bartsch, 1980, 1982) but ostracodes, nematodes, harpacticoid copepods, and mites are still poorly known. The Foraminifera have been comparatively well-studied by Buzas and Gibson (1969), Buzas and Culver (1970), Culver and Buzas (1982), and Sen Gupta and Strickert (1982). Diversity of foraminiferan species increases at depths below 200 m.

Some deep-sea species are widely distributed and others are restricted or endemic to particular ocean basins. Endemism has been discussed by Jones and Sanders (1972) and Sanders (1977) in terms of the proportion of species known only from the North American Basin. Of the species of Cumacea collected on the Gay Head-Bermuda transect, 49 percent are not found elsewhere. High endemism is likely to characterize most of the peracarid groups such as isopods, amphipods, and tanaids because they brood their young and lack a planktonic stage of development. The ophiuroids and bivalves tend to have the broadest distributions(Sanders 1977). Many of the megafaunal species, such as *Hyalinoecia tubicola* and *Ophiomosium lymani*, have much more restricted distributions than had previously been thought (Grassle et al., 1975). For most groups too few samples have been studied with uniform criteria to determine broad-scale zoogeographic distributions. In a well-studied group, the tunicates, endemism has been diminishing as more samples become available from additional areas (Monniot and Monniot, 1978).

Megafaunal distribution with depth is given in Haedrich et al. (1980) and Grassle et al. (1975). Figure 6.28 illustrates the changes with depth. It is possible to identify depths of more rapid changes in one group or another but the boundary between zones is somewhat arbitrary. The most rapid transitions occur at depths shallower than 2000 m.

The methods of study and incompleteness of most of the taxonomy make comparisons along depth zones difficult. The fish on a transect off New England differ from those off Greenland at similar depths (Haedrich et al., 1980). Markle and Musick (1974) studied changes in fish species composition along the 900 m contour from approximately 36° to 40° N and found a shift from an association dominated by *Glyptocephalus cynoglossus* and *Phycis chesteri* to one dominated by *Synaphobranchus kaupi*. Several species were found only in the north and others were found only in the south. Although Haedrich et al. (1975) found a zonation similar to that of Rowe and Menzies (1969) and Musick (1976), a direct quantitative comparison of species distributions cannot be made from the published data.

Horizontal Variation in Macrofaunal Distribution. A portion of the deep-sea fauna is surprisingly homogeneous along depth contours. A station at 1400 m off New England shares 48 percent of its bivalves with a station off the west coast of Africa. About 30 percent of the bivalves are common to both sides of the Atlantic (Allen, 1979). Other groups such as peracarid Crustacea are more dissimilar with horizontal distance (Grassle et al., 1979). Tunicates have been studied from all of the Atlantic transects sampled from Woods Hole ships. The western Atlantic tunicates have closest affinities with species from off Labrador and the eastern Atlantic. As with other groups, a few species are cosmopolitan and others have restricted distributions (Monniot and Monniot, 1978). Along the east coast of the United States, Cape Hatteras may be a geographic boundary for upper slope fauna (Cutler and Doble, 1979).

Along the Gay Head-Bermuda transect five samples from a depth range of 4800-4862 m and three samples from the depth range of 2862-2891 m showed high similarity of polychaete species even though the stations were 100 miles or more apart. There are not enough samples taken with similar methods to define zoogeographic boundaries along depth zones. The zonation of macrofauna is described by Sanders and Hessler (1969), Sanders (1977), Grassle et al. (1975), and Rex (1981); some examples are shown in Figures 6.29 and 6.30. Names of common species are given in Rowe et al. (1982) (Table 6.10). Zonation within each major faunal group is a little different, suggesting that zones should not be thought of as discrete ecological entities.

The vertical and horizontal distributions of meiofauna are unknown since, except for Foraminfera, few taxa are identified to species. The distribution of major groups off North Carolina at 400 and 4000 m are shown in Table 6.11.

Fauna of Hard Surfaces

Aside from a few isolated submersible observations (Grassle et al., 1975) the main information on hard surfaces in the ASCAR region comes from the work of Hecker at Lamont-Doherty Geological Observatory (Hecker et al., 1980). The hard-surface epifauna is not as dense as in shallow water, although canyons may develop a lush epifauna (Hecker, 1982).

Hecker et al. (1983) found that canyons differ from the slope environments in that increased environmental heterogeneity results in faunal differences. In Baltimore Canyon there are three faunal zones: 800-1400 m, 1300-1600 m, and 1600-2050 m. *Ophiomusium lymani* dominates the deeper zone but is relatively rare at depths less than 1600 m allowing the suspension/filter feeding species to dominate the middle depths. Densities are generally low in depths of 800-1400 m. The fauna at depths less than 800 m is extremely variable and relationships to substratum, depth, and geography are less obvious. Results from studies in Lydonia Canyon show similar results (Hecker et al., 1983).

In contrast to the findings of Haedrich et al. (1980), Hecker et al. (1983) found that additional faunal groups (mainly associated with hard substrata) are located in canyons. The canyons sampled by Haedrich et al. (1980), Alvin and Hudson Canyons, have limited exposure of hard substrata, and trawls are not an effective means of sampling the high topographic relief that occurs. Continous observations from manned submersibles or towed photographic vehicles are needed to describe the highly patchy faunal assemblages that occur in canyons.

Extensive banks of coral are known from the Blake Plateau at depths between 650 and 850 m (Stetson et al., 1962; Milliman et al., 1967; Stetson et al., 1969). The banks are made up principally of *Dendrophyllia profunda*. *Lophelia prolifira* is abundant on the crest of the banks and *Bathypsammia*, *Caryophyllia*, and *Balanophyllia* are dominant between the banks. Deep-water coral mounds have been reported from the southern Blake Plateau and the slopes of Little Bahama Bank, just south of the ASCAR study area (e.g., Neumann et al., 1977; Mullins et al., 1981). Dominant coral genera are *Bathypsammia* and *Solenosmilia*, but numerous other species also are present. These corals trap finer-grained sediment, ultimately resulting in growth of deep-water coral mounds which can host numerous other invertebrate fauna, including many molluscs, echinoderms, and crustaceans. To date there has been no detailed biological study of these mounds, although they compose one of the dominant communities on the Blake Plateau.

Table 6.10 Ten most abundant 'species' in each sample set (percentages in parenthesis).

32 m	203 to 570 m	1141 to 1437 m	1707 to 1815 m	2351 to 2673 m	2749 to 3264 m	3659 m
Tharyx acutus (29.8)	Oligochaete "A" (29.9)	*Cossura longocirrata* (16.1)	*Nucula cancellata* (10.3)	*Glycera capitata* (5.8)	*Sipuncula* spp. (8.2)	Scaphopod "Spp." (7.6)
Prionospio steenstrupi (23.5)	*Cassura longocirrata* (5.3)	*Heteromastus filiformis* (10.8)	*Poccilochaetus fulgoris* (8.9)	Oligochaeta "Spp." (5.6)	Spionid "A" (6.0)	*Ophelina abranchiata* (7.6)
Nucula proxima (13.5)	*Tharyx acutus* (4.6)	*Axinulus ferruginosus* (7.7)	*Glycera capitata* (5.7)	*Prochaetoderma* sp. (A) (5.4)	Oweniid "Spp." (5.5)	*Ampharete* "A" (6.5)
Cassura longocirrata (8.1)	*Minuspio cirrifera* (4.0)	*Tharyx acutus* (7.1)	Nemertean "Spp." (2.8)	*Leptognathia* C (4.2)	*Glycera capitata* (4.2)	Oweniid "Spp." (3.3)
Mediomastus ambiseta (8.0)	*Terebellides stroemi* (3.2)	*Axinulus* sp. (6.5)	*Harpiniopsis* sp. (2.8)	*Notomastus latericeus* (4.0)	*Ophiura* "A" (3.6)	*Sigambra tentanculata* (3.3)
Asabellides oculata (6.4)	*Paraonis geacilis* (3.1)	*Nucula granulosa* (4.7)	*Polycarpa delta* (2.6)	*Tharyx* "B" (3.3)	*Ehlersia anoculata* (3.3)	Sipuncula "Spp." (3.3)
Euchone incolor (1.7)	*Siphonodentalium* sp. (2.7)	*Ceratocephale loveni* (4.7)	*Aricidea neosuccia* (2.6)	*Ophiura ljungmani* (3.3)	*Typhlotanais* "G" (2.4)	
Dorvillea caeca (1.2)	*Chaetozone setosa* (2.6)	*Glycera capitata* (4.5)	*Prochaetoderma* sp. (A) (2.6)	Spionid "A" (2.4)	Oligochaeta "Spp." (2.2)	
Thracia myopsis (1.1)	Nemertean "Spp." (2.3)	*Paramphinome jeffreysi* (2.8)	Spionid "A" (3.0)	*Sipuncula* "Spp." (2.6)	Cirratulid "Spp." (1.9)	
Aricidea jeffreysi 0.9)	*Falcidens caudatus* (2.9)	*Falcidens caudatus* (2.4)	*Pholoe minuta* (2.4)	*Malletia estheriopsis* (2.3)	Scaphopod "Spp." (1.7)	

There were 48 taxa among the 10 most abundant species in the seven sample sets.
*Next 14 "ranking" species at 2.2 percent with two specimens each.
"Spp." indicates more than one species is included. (From Rowe et al., 1982)

Table 6.11 Numbers (10 cm^{-2}) of meiofauna at 400 m. The numbers 24301-24305 represent individual boxcores and the single digits refer to the subsample from each boxcore (see Table 1 for how each subsample was processed). SD: Standard Deviation.

Taxon	24301					24302						24303				24305						x	of all meio-fauna
	6	7	1	9	3	8	4	6	1	7	8	1	3	6	7	2	4	4	7	1	8		
Foraminifera	101	152	39	126	136	136	39	110	51	136	136	87	140	75	89	136	136	238	523	136	136	136.1	30.8
Nematoda	264	145	353	520	184	231	238	94	352	45	51	79	59	173	334	227	257	175	76	196	137	199.5	45.1
Copepoda	103	30	46	234	27	66	39	17	55	21	9	24	9	37	138	20	43	43	16	11	7	47.4	10.7
Unidentified	23	10	52	60	35	85	4	12	31	8	10	3	47	24	26	0	42	4	0	1	0	22.7	5.1
Polychaeta	13	15	15	44	10	22	10	7	15	5	11	6	8	7	36	6	6	17	5	1	1	12.3	2.8
Turbellaria	17	8	0	2	0	0	8	9	2	0	0	5	7	3	5	3	10	17	4	1	1	4.9	1.1
Gastrotricha	19	9	0	23	0	0	13	5	4	6	10	0	0	2	7	8	22	2	11	3	1	6.9	1.6
Oligochaeta	3	3	0	0	1	3	3	1	0	3	1	1	1	0	0	0	1	5	2	0	0	1.3	0.3
Ostracoda	7	3	1	15	3	2	4	2	2	0	0	4	0	1	1	2	1	1	1	0	0	2.4	0.5
Tardigrada	1	1	1	18	0	13	2	2	2	0	2	3	0	4	9	1	3	1	0	1	1	3.1	0.7
Konorhyncha	4	0	0	6	0	2	0	1	1	2	1	0	0	0	2	0	2	0	0	1	0	1.0	0.2
Other groups	3	8	3	9	4	16	4	8	6	1	2	5	3	2	6	4	10	4	0	2	0	4.8	1.1
Total	558	384	510	1057	400	576	364	267	521	227	233	217	274	328	653	407	533	507	638	353	284		
x ± SD											442.4		± 196.7										

[A]Forams not counted. Numbers given are estimates based on mean of all other samples.

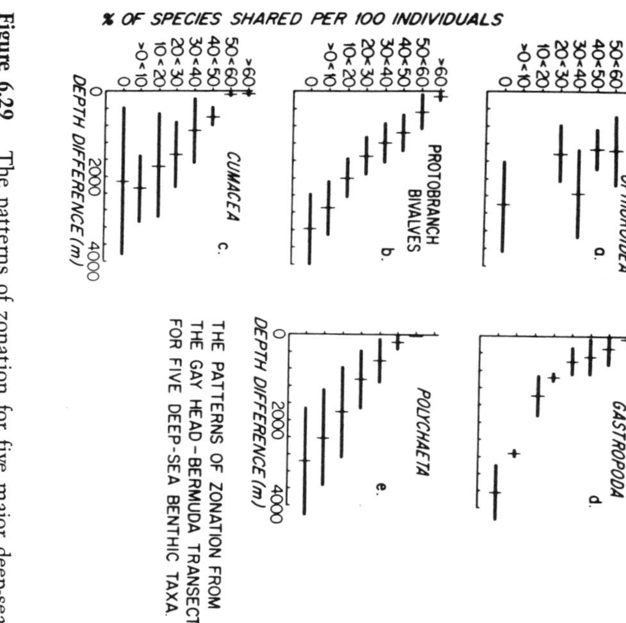

Figure 6.29 The patterns of zonation for five major deep-sea benthic taxa from the Gay Head-Bermuda transect. Mean depth differences and standard deviations are compared for station pair groupings on the basis of the percentage of species shared (from Sanders, 1977).

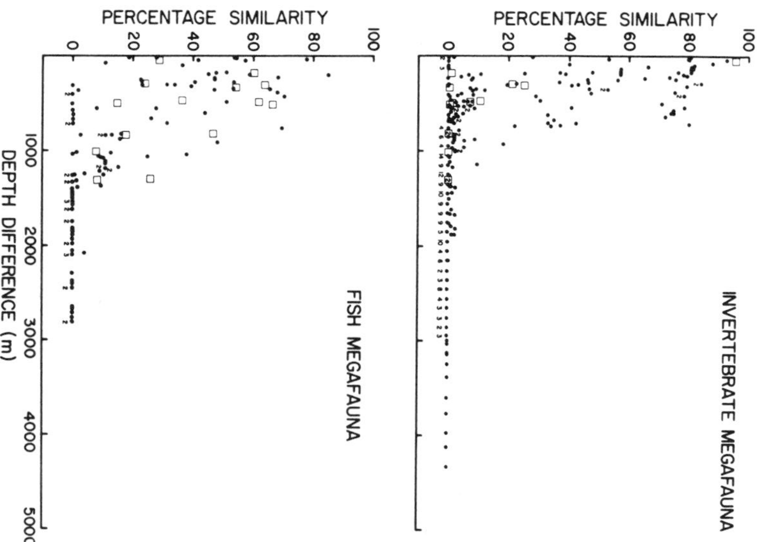

Figure 6.30 (top) Percentage faunal similarity values for samples of invertebrate megafauna plotted against difference in depth to selected reference stations. (bottom) Same comparisons for the fish megafauna. From Rex (1981).

Diversity

Using several methods of expressing diversity including number of species per sample, Sanders and Hessler (1969), Sanders (1968), and Hessler and Sanders (1967) described the relatively high diversity of species of the polychaete and bivalve fraction of the deep-water samples from the Gay Head-Bermuda transect. Previous studies had indicated that deep-sea species diversity was low. An increase in the diversity of the entire fauna was observed on the continental slope compared with the fauna of the continental shelf off North Carolina (Grassle, 1967, 1972). The diversity of most groups is now seen to increase to intermediate depths, and then to decline in deeper water (Rex, 1981) (Fig. 6.31). From trawl data, megafaunal diversity also appears greatest at intermediate depths (Haedrich et al., 1975, 1980).

Several explanations have been made for the increase in diversity in the deep sea and the changes in diversity within the deep-sea environment (Rex, 1983). An intermediate disturbance argument (Connell, 1978) is increasingly favored (Grassle and Sanders, 1973; Grassle, 1977; Rowe et al., 1982; Rex, 1981, 1983). In greatly disturbed environments few species are able to recover rapidly enough to maintain themselves. In very infrequently disturbed environments competitive exclusion is likely to result in fewer species. A balance between these processes has been called a dynamic equilibrium (Rex, 1983). The spatial scale of disturbance is not discussed in most theoretical papers and this is particularly important in the deep sea. Grassle and Sanders (1973) and Jumars (1976) argue that deep-sea spatial heterogeneity results from small-scale biogenic disturbance. In the deep sea the sources of disturbance are falls of organic material from the surface (such as wood, algal remains, or carcasses of animals), mud slumps, the activities of large animals (such as snails moving over the bottom), or sessile animals such as glass sponges projecting above the bottom resulting in an accumulation of fine particles and/or an increase in microbial activity. Predation by large animals, such as rattail fishes or various echinoderms, may be regarded as another form of disturbance. Disturbances not only allow reduction in competition, but also adaptations of individual species to the sequence of conditions following different kinds of disturbances. For example, the polychaete species increasing in the vicinity of wood sunk from the surface differ from those increasing as a result of deposition of particles around glass sponges, or as a result of defaunated spaces left by the feeding activities of predators and scavengers.

Another hypothesis relates the number of species to total area of an entire region. The theory stems from correlation between species numbers and island areas on land (MacArthur and Wilson, 1967). This hypothesis was rejected by Rex (1981) because the species-area correlation does not hold in the deep sea. The relationship between species diversity and sediment texture (Gray, 1974) has also been rejected by Rex (1983).

A high proportion of the increased diversity in the deep sea may also relate to commensal or mutual interactions. These relationships are still very poorly understood.

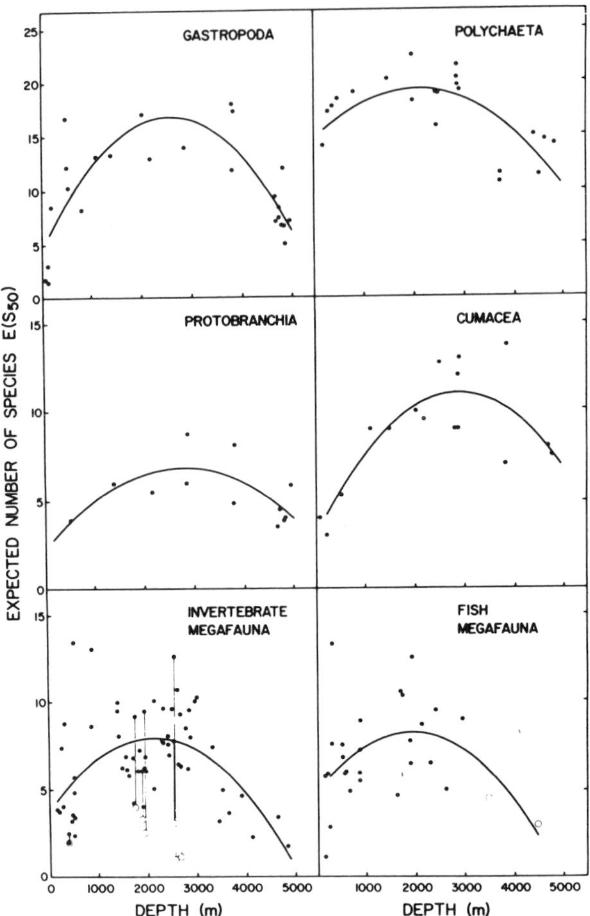

Figure 6.31 Depth gradients in species diversity for the Gastropoda (79,80), Protobranchia, Polychaeta, Cumacea, invertebrate megafauna, and fishes. All samples are from the western North Atlantic south of New England. From Rex (1981).

Proportion of Planktonic Larvae and Reproduction

The general life histories of major deep-sea taxa are summarized by Sanders (1977). The peracarid Crustacea brood their young and have no planktotrophic dispersal. Most bivalves have a lecithotrophic larva which may indicate near-bottom dispersal of larvae. The groups with planktotrophic development are the ophiuroids (Schoener, 1972) and gastropods (Rex and Waren, 1982). Shallow water studies indicate that planktonic feeding stages increase dispersal ability of a species; however, it is unknown whether this is true in the deep sea. Rex and Waren (1982) suggest that the predominantely predatory gastropods with planktotrophic development have greater dispersal ability than other deep-sea species.

In the few species of bivalves from the western Atlantic examined, there appears to be no seasonality of reproduction (Allen, 1979; Scheltema, 1972). Egg numbers in the protobranch bivalves range from two for the genus *Microgloma* to several hundred (Sanders and Allen, 1973), and within the genus *Nucula*, egg number declines with increasing depth (Scheltema, 1972). Some species of polychaetes are in reproductive condition at all times of the year (Scheltema, 1972).

The gastropod *Benthonella tenella* appears not to reproduce seasonally but some seasonal variations cannot be discounted (Rex et al., 1979). Recruitment appears to be variable and infrequent. Another gastropod, *Alvania pelagica*, shows a lower proportion of energy devoted to reproduction and greater longevity than continental slope populations.

Crangonid, glyphocrangonid and nematocarcinid shrimp show asynchronous, year-round reproduction (Wenner, 1978, 1979b). The only estimate of larval dispersal range is for the red crab, *Geryon quinquedens*, from upper slope depth. The larvae of this species are thought to disperse over distances of several hundred kilometers (Kelley et al., 1982).

Genetic Variation

Species of deep-sea megabenthic invertebrates show a higher genetic variability (measured using electrophoretic techniques to examine polymorphisms in soluble enzyme loci) than their shallow-water relatives (Murphy et al., 1976; Doyle, 1972; Schopf and Gooch, 1971). This result is supported by studies on genetic variation of deep-sea animals in the Pacific (Ayala and Valentine, 1978). Most of the work has been done with coarse mesh trawls collecting large mobile scavengers with good dispersal ability (Grassle and Grassle, 1978). Genetic variation in these species is likely to be maintained by selection for different genotypes within each local biotic environment in the range of each species.

Feeding Types and Proportion of Predators

The majority of deep-sea animals feed at the sediment-water interface by removing particles from the sediment surface. The proportion of motile vs. sessile polychaetes increases on the Hudson Canyon Rise and at 3600 m on the Gay Head-Bermuda transect, and Rowe et al. (1982) attributed this to increased current activity. Filter feeding animals are rare.

The proportions of predatory gastropods (Rex, 1973, 1976) and predatory tunicates (Monniot and Monniot, 1978) increase with depth in the deep sea. Predatory bivalves never make up more than five percent of the bivalve individuals in the deep sea (Allen and Turner, 1974). Most Aplacophora or solenogasters are predators on Foraminifera, hydroids, crustacean eggs and possibly small worms and crustaceans (A. Scheltema, 1981). The only other major macro-faunal invertebrate group with substantial numbers of predators is the Polychaeta. A relatively high proportion of predators is found in association with large concentrations of wood, presumably feeding on the primary consumers rather than the wood itself (Turner, 1977). Many of the species classified as predators may also be scavengers or omnivores, so that food web interpretations are difficult even though relatively few species are involved.

Tietjen (1971) has studied changes in the feeding type of nematodes with depth; deposit feeders increase and epigrowth feeders decrease in deeper water. Most of the invertebrates big enough to be visible on the sediment surface (megafauna) are nonselective predators and/or scavengers (Haedrich et al., 1980). Nematocarcinid shrimp feed on the bottom after a more pelagic existence early in life (Wenner, 1979b).

Fish spend varying amounts of time feeding on the bottom depending on jaw morphology and stage of development (McLellan, 1977; Haedrich et al., 1980; Sedberry and Musick, 1978). Some deep-ocean fish may show some preferred prey (Haedrich and Polloni, 1976); however, most are nonspecific in their feeding (Haedrich et al., 1980; Sedberry and Musick, 1978). Bottom-feeding fish are heavily parasitized because the benthic invertebrates they feed on act as intermediate hosts (Campbell, 1983, McLellan, 1977).

Growth and Colonization Rates

The only studies of deep-sea colonization rates of soft sediments in the western Atlantic deep sea are those of Grassle (1977). Rates of colonization of disturbed deep-sea sediments are very low in comparison to shallow water. After two years on the bottom at about 1800 m depth, the populations were an order of magnitude lower in the trays than in the surrounding community, and few of the individuals had reached maturity. In similar experiments with relatively high concentrations of organic material in the sediments, more rapid colonization may occur (Grassle, unpubl.; Desbruyeres et al., 1980; Hecker, 1982).

Rate of colonization of wood on the bottom is relatively high (Turner, 1973) while colonization of rock or artificial surfaces is very slow (Grassle, unpubl.). Where there is extra food, a few groups such as polychaetes in the genera *Ophryotrocha*, *Capitella* and *Prionospio* may become abundant. This occurs in the vicinity of wood islands where the feces of animals living on the wood are spread over the surrounding sediments (Grassle, unpubl.).

Respiration Rates

Smith (1978; Smith and Hinga, 1983) has measured whole benthic community respiration along the Gay Head-Bermuda transect and finds a decline with depth and distance from land (Table 6.12). It is unknown whether this result is primarily the result of a decrease in microbial respiration or the respiration of the fauna. An approximate energy budget has been drawn by Rowe and Gardner (1979) (Fig. 6.32). Deep-sea scavenging fish also have lower rates of respiration than fish in shallow water (Smith, 1982).

Table 6.12 Sediment community respiration from 10 stations in the Western North Atlantic with associated environmental parameters[A] (from Smith, 1982).

Station	Sediment Community Respiration ml O m h	Depth (m)	Distance from Shore (km)	Annual Primary Productivity $gC\,m^{-2}y^{-1}$	Bottom Water Temp. °C	Dissolved Oxygen Bottom ml/l	Benthic Abundance/ m^2	Benthic Biomass mg wet wt/m^2	Sediment Organic Carbon mg/g/dry wt	Sediment Organic Nitrogen mg/g dry wt	Particulate Organic Carbon Flux $g\,m^{-2}\,y^{-1}$
DOS-1	0.50	1,850	176	120	4	7.05	3,218	9,450	10.0	1.1	—
DWD	0.46	2,200	172	100	3	6.34	22,988	556	12.1	1.5	6.3
ADS	0.35	2,750	259	160	3	6.52	8,764	8,764	13.3	1.6	2.3
HH	0.20	3,000	291	160	3	6.15	2,146	653	9.1	1.1	2.3
DOS-2	0.21	3,650	352	100	3	6.54	1,632	771	13.0	0.9	4.2
JJ	0.09	4,670	497	68	3	6.43	753	220	0.8	0.1	—
KK	0.04	4,830	612	68	3	6.04	285	180	6.9	0.7	—
NN	0.07	5,080	880	72	3	6.25	117	78	6.4	0.9	0.7
MM	0.02	5,200	806	72	3	6.15	259	142	6.4	0.9	0.7
77DE	1.31	1,345	148	85	4	5.65	—	—	15.6	—	5.4

[A]Adapted from Hinga et al., 1979; and Smith, 1978a.

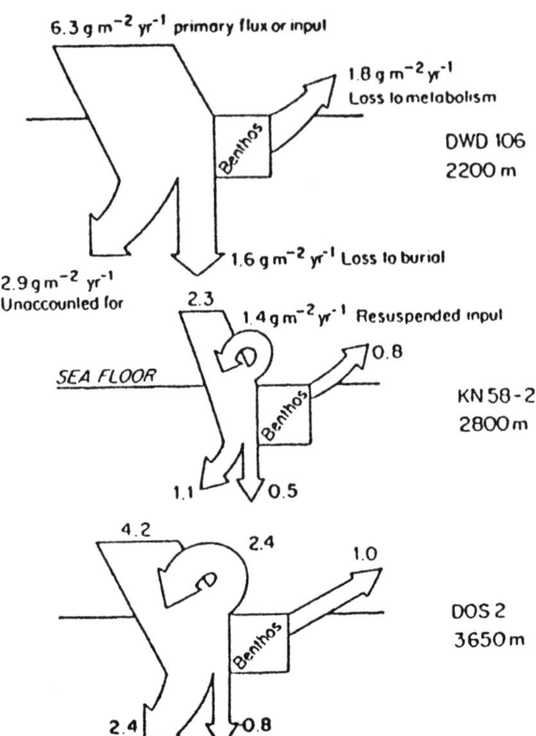

Figure 6.32 Organic carbon budget for the deep North Atlantic, in g cm^{-1} yr^{-1}. From Rowe and Gardner (1979).

Seabirds

K.D. Powers

The term "seabird" refers to pelagic birds from the groups: Procellariidae (including albatrosses, shearwaters, petrels and storm-petrels), Pelecaniformes (including gannets, tropicbirds, frigatebirds and pelicans), and the Lari-Limicolae (including phalaropes, jaegers, skuas, gulls, terns and alcids). Our knowledge of pelagic distributions of seabirds in the western North Atlantic has been rapidly increasing over the past decade, although a considerable amount of the data are scattered in unpublished reports, personal files of individuals, and current research. Jespersen (1924) began the quantitative description of the pelagic distribution of seabirds in the western North Atlantic. Following him, other important contributions were made by Wynne-Edwards (1935), Baker (1947), Rankin and Duffey (1948), Moore (1941, 1951), Palmer (1962) and Butcher et al. (1968). Surveys in the last ten years include: Brown et al. (1975), Brown (1977), Rowlett (1973, 1980), Lee and Booth (1979), Lee and Rowlett (1979), Powers (1983) and Powers and Brown (in press). Additional seabird distribution data from the area of interest lies with the Manomet Bird Observatory (MBO), David Lee (N.C. State Museum, Raleigh, NC), Richard Rowlett (Ocean City, MD), and Peter Stangel and Christopher Haney (Zoology Dept., University of Georgia, Athens, GA). The distribution of seabirds from the ACSAR study area makes particular use of data presented by Brown (1977), Powers (1983), and Powers and Brown (in press); citations to these manuscripts were not made but left as understood.

One problem with seabird observation data is that different investigators have used different methods of measuring abundance, so that quantitative relationships between different sources of data cannot be addressed. In comparing the two most popular abundance estimators used in the North Atlantic, density (birds km^{-2}) and abundance (birds seen per 10 min), Powers (1982) found that the former method minimizes inflationary effects caused by a number of bird behaviorial and observer counting biases which are not controlled in the latter method. Seasonal bird densities, as presented here, provide a means to compare variability in abundance within a given species (not among different species) between different oceanographic habitats of the ACSAR study area.

Species Accounts

Fulmars

Northern fulmars occur throughout boreal, subarctic and arctic waters of the North Atlantic. In the western North Atlantic they occur as far south as Virginia in late winter and early spring (December to April). The majority of fulmars south of the Grand Banks off Newfoundland concentrate on the north and east flanks of Georges Bank from December to April.

Shearwaters and Petrels

Five shearwaters occur seasonally in the western North Atlantic. Greater and sooty shearwaters breed in the southern hemisphere and spend the austral winter north of the equator. Greaters are principally found in boreal and subarctic waters of the western North Atlantic from May to December. The majority of the population resides on shelf waters from Georges Bank northeast to the Grand Banks, although spring (April-May) and fall (September-December) migrations occur through the ACSAR study area. The distribution of sooties is similar to that of greaters except that they have a clockwise migration in the North Atlantic; thus, most of the population is in the western part from April to July and in the eastern part by August, from which the birds return to the South Atlantic (Phillips, 1963). Cory's shearwaters breed on islands in the eastern North Atlantic and in the Mediterranean. Non-breeding birds are found throughout subtropical waters of the North Atlantic from June to November. In the western North Atlantic Cory's are most abundant on the shelf from Long Island to the Great South Channel. Of the smaller "black-and-white" shearwaters, many remain in boreal waters of the North Atlantic from April to November, although birds in the western part must pass through the ACSAR study area during their migrations to and from wintering areas off Brazil (Spencer, 1972). Audubon's are found principally in slope waters from June to at least October; they are probably the most common shearwater in the ACSAR study area during that time.

Black-capped petrels breed on islands and adjacent mainlands of the Caribbean (Wingate, 1964) and the closely allied and endangered Bermuda petrels breed only on Bermuda (Murphy and Mowbray, 1951). The pelagic distributions of these species are not well known. Bermuda petrels have not been recorded away from Bermuda (Clapp et al., 1982), but black-capped petrels are regularly seen from April to November in slope and Sargasso waters south of 35°N (Morzer-Bruyns, 1967; Lee and Rowlett, 1979).

Storm-Petrels

Wilson's storm-petrels breed in the southern hemisphere and a large population spends the austral winter in the western North Atlantic from April to November. Their seasonal migrations pass through the ACSAR study area (Roberts, 1940). From April to May they are found in greatest abundance along the shelf-edge of the Middle Atlantic Bight and from June to August in the southwestern Gulf of Maine and on northern Georges Bank. Leach's storm-petrels breed from Massachusetts to Labrador in eastern North America. Their pelagic distribution is mainly centered around the larger colonies off Newfoundland. Leach's occur in the ACSAR study area from April to November. White-faced storm-petrels breed on islands in the eastern North Atlantic and in the South Atlantic (Cramp et al., 1977). Sight records of this species are scattered throughout the subtropical parts of the western North Atlantic from August to October.

Gannets

Northern gannets breed in eastern Canada and migrate to shelf waters off the eastern United States and Gulf of Mexico during the period from October to May. The center of their winter distribution occurs in the Middle Atlantic Bight. Large numbers of gannets aggregate around the large fleets of trawlers fishing in the canyons of the Bight, particularly in the vicinity of Hudson Canyon off New York and New Jersey.

Phalaropes

In the western North Atlantic red and red-necked phalaropes are seasonal migrants principally in shelf waters. Reds are the most abundant species in spring (April to May); neither species is abundant in fall (September to November). In spring the majority of red phalaropes migrate in a "corridor" between the 60 and 200-m isobaths from Cape Hatteras to Georges Bank. Small numbers of both species spend the winter in shelf waters from Chesapeake Bay south to Florida (Weston, 1953; Lee and Booth, 1979; Rowlett, 1980).

Jaegers and Skuas

Three jaegers, all circumarctic breeders, occur in the western North Atlantic. Jaegers are migrants through the ACSAR study area in spring (April-May) and fall (September-November). Pomarine and parasitic jaegers are the more common species; long-tailed jaegers are rarely sighted in the western North Atlantic south of Newfoundland.

Two skuas occur in the western North Atlantic. Great skua breeds from Iceland east to the British Isles. It is found throughout the year off the northeastern United States, but is most common from November to March in shelf waters. South polar skuas migrate into the western North Atlantic from the southern hemisphere. They probably occur as migrants in the ACSAR study area from May to October.

Gulls

Eight gulls, all of which breed in eastern North America, occur in shelf waters off the northeastern United States. These include: glaucous, Iceland, great black-backed, herring, laughing, ring-billed, Bonaparte's and Sabine's gulls and black-legged kittiwakes. Great black-backed and herring gulls are the most common in the ASCAR study area. From November to April they concentrate around the commercial fishing fleets at the shelf-edge in the Middle Atlantic Bight and Georges Bank.

Terns

Seven terns potentially occur in the ASCAR study area. Common and arctic terns are spring (April-May) and fall (August-October) migrants from wintering areas in the southern hemisphere and breeding areas along the coast of the northeastern United States and Canada. Royal and sandwich terns are coastal inhabitants from Chesapeake Bay south to the Gulf of Mexico. Black terns breed in the interior parts of North America, but their fall migration is coastal south of Cape Hatteras where flocks may stray into the ACSAR study area. Sooty and bridled terns are pelagic during their nonbreeding season. Sooties breed off southern Florida, but autumn hurricanes will "blow" them into the ACSAR study area. Bridled terns, which breed on islands in the Caribbean, regularly occur from August to October in shelf-edge and slope waters of the ACSAR study area south of 36°N.

Alcids

Five alcids (razorbill, common and thick-billed murre, dovekie, and Atlantic puffin) have pelagic ranges in the western North Atlantic. None of these species regularly occurs away from the shelf. Georges Bank and the adjacent shoal waters to its west are the southern limit of any significant numbers of these species.

Seasonal Distribution

Since only Powers (1983) and Powers and Brown (in press) presented seabird densities (birds km^{-2}) for the ACSAR study area, quantitative evaluationis limited to those publications and to a compilation of unpublished MBO data from 1978-1982. The study area is divided into five regions: Georges Bank edge (GBe), Middle Atlantic Bight edge (MABe), Slope Water north of 35°N (SLn), over the Slope Water from 28°-35°N (SLs) and shelf water from 28°-38°N (SHs). The regions include depths from 100-500 m for GBe and MABe, 500-4000 m for SLn, <200 m for SHs, and >200 m for SLs. Data from 1978 were considered separately because of the extensive foreign fishing activity along the shelf break from Hudson to Lydonia canyons in that year.

The year was divided into winter (Dec-Feb), spring (Mar-May), summer (Jun-Aug) and autumn (Sep-Nov). Observation effort from the MBO seabird data files for these seasons from 1978 through 1982 is summarized in Table 6.13 for the three regions (GBe, MABe and SLn) north of 35°N latitude. Mean densities of seabirds by species from 1979-1982 are summarized by region in Table 6.14 (GBe), Table 6.15 (MABe), Table 6.16 (SLn), and Table 6.17 (SHs and SLs). Mean densities of seabirds by species from 1978 only are given for GBe (Table 6.18), MABe (Table 6.19) and SLn (Table 6.20).

Table 6.13 Seabird observation effort for each region north of 35°N by season and year from 1978 through 1982. Effort is measured by area (km^2) and number of transects (in parentheses).

Region	Year	Winter	Spring	Summer	Autumn
Slope	1978	6.0(7)	16.8(15)	180.2(187)	43.6(46)
(SLn)	1979	76.7(70)	116.3(106)	359.6(335)	175.0(184)
	1980	17.6(19)	4.1(4)	63.1(47)	38.7(26)
	1981			7.5(6)	
	1982	2.8(2)	129.3(109)	236.4(185)	56.4(43)
Georges	1978	10.4(11)	43.2(46)	19.1(20)	23.0(25)
Bank	1979	9.9(11)	3.4(4)	17.6(27)	48.7(61)
edge	1980	12.6(10)	2.0(2)	16.7(16)	12.4(9)
(GBe)	1981		98.4(82)	70.9(56)	3.4(6)
	1982		37.0(29)	5.0(4)	12.7(12)
Middle	1978	22.7(25)	22.7(21)	52.0(59)	16.8(20)
Atlantic	1979	16.0(18)	75.8(79)	85.9(54)	34.7(40)
Bight	1980		24.0(26)	55.1(40)	27.0(18)
edge	1981	12.5(10)	34.7(28)	42.3(34)	96.5(77)
(MABe)	1982	18.5(14)	103.1(85)	72.5(61)	95.8(91)

Winter

Total bird densities in winter (1979-1982) were lower than 10 birds km^{-2} for each region except MABe, where 132.7 birds km^{-2} were recorded (Table 6.15). North of 35°N, fulmars, gannets, great black-backed and herring gulls, black-legged kittiwakes, and dovekies were most common. South of 35°N, no species was common and species richness (number of species) was greater on shelf waters (SHs) (13 species) than over slope waters (SLs) (7 species). Excluding herring gulls, which are ship-followers, no species density exceeded 0.1 bird km^{-2} in SLs.

The most important distribution feature during winter was the concentrations of gannets and great black-backed and herring gulls at the shelf break in the Middle Atlantic Bight (Table 6.14). These concentrations involved thousands of each species and they were all associated with large foreign fishing fleets in the Hudson Canyon area approximately 160 km southeast of New York City. Concentrations of these three species were also found with fishing fleets in the MABe region in 1978 (Table 6.18).

Table 6.14 Mean bird densities (birds km $^{-2}$) by season on Georges Bank edge (GBe) water from 1979 through 1982.

Species*	Season			
	Winter	Spring	Summer	Autumn
NOFU	0.7(±1.2)	6.8(±0.2)	<0.1(±0.2)	0.4(±0.6)
COSH			0.2(±0.7)	0.2(±0.4)
GRSH	0.1(±0.2)	<0.1(±0.1)	1.1(±1.8)	3.7(±6.4)
SOSH		0.1(±0.3)	0.1(±0.3)	
AUSH			<0.1(±0.1)	
WISP		1.5(±2.7)	5.0(±7.1)	
LESP		<0.1(±0.2)	0.6(±1.5)	
NOGA	0.1(±0.2)	0.1(±0.2)		0.3(±1.4)
REPH		34.4(±31.0)		0.5(±1.4)
POJA	<0.1(±0.2)	0.1(±0.3)	<0.1(±0.1)	0.1(±0.4)
PAJA			<0.1(±0.1)	
UNSK	<0.1(±0.2)			<0.1(±0.1)
GLGU		<0.1(±0.1)		
GBBG	0.3(±0.5)	1.4(±3.5)		0.3-1.0)
HEGU	0.5(±0.7)	0.7(±1.3)		3.2(±8.1)
LAGU			<0.1(±0.1)	
BLKI	1.8(±3.1)	3.7(±7.1)		0.6(±1.8)
DOVE	0.2(±0.5)			0.2(±0.4)
ATPU		0.1(±0.3)		
Total	9.6	48.9	7.0	9.5

*Key to species codes is given in Apendix 6.1.

Table 6.15. Mean densities of birds (birds km $^{-2}$) by season on Middle Atlantic Bight edge (MABe) waters from 1979 through 1982.

Species	Season			
	Winter	Spring	Summer	Autumn
NOFU	1.0(±3.5)	1.1(±6.4)	<0.1(±0.1)	
COSH			0.3(±2.1)	0.6(±2.9)
GRSH			1.3(±7.7)	1.1(±4.6)
SOSH		<0.1(±0.2)	0.1(±0.3)	
MASH			<0.1(±0.1)	
AUSH			0.1(±0.2)	<0.1(±0.1)
WISP		0.6(±4.4)	7.0(±19.3)	0.4(±1.0)
LESP		<0.1(±0.1)	<0.1(±0.1)	<0.1(±0.1)
NOGA	28.9(±50.9)	19.0(±79.0)	<0.1(±0.1)	<0.1(±0.1)
REPH		34.4(±234.0)		0.2(±2.5)
RNPH	0.1(±0.1)	<0.1(±0.1)		
POJA				0.2(±0.7)
PAJA			<0.1(±0.1)	
UNSK	<0.1(±0.2)	<0.1(±0.1)		
GLGU		<0.1(±0.1)		
ICGU		<0.1(±0.1)		
GBBG	56.7(±115.5)	25.4(±92.9)	<0.1(±0.2)	0.1(±0.6)
HEGU	44.4(±79.3)	37.1(±103.3)	0.1(±0.1)	1.4(±4.2)
RBGU				<0.1(±0.1)
LAGU		<0.1(±0.1)		
COTE			0.1(±0.1)	0.1(±0.1)
BLKI	1.2(±3.5)	0.5(±2.7)		0.1(±0.4)
RAZO		<0.1(±0.1)		
DOVE	0.4(±1.3)	<0.1(±0.1)		
ATPU	0.1(±0.2)	<0.1(±0.1)		
Total	132.7	118.1	8.8	4.1

*Key to species codes is given in Apendix 6.1.

Table 6.16 Mean bird densities (birds km^{-2}) by season in slope water (SLn) north of 35°N from 1979 through 1982.

	Season			
Species*	Winter	Spring	Summer	Autumn
NOFU		0.2(±0.6)	<0.1(±0.1)	<0.1(±0.1)
COSH		<0.1(±0.1)	0.2(±1.6)	0.1(±0.1)
GRSH		<0.1(±0.1)	0.6(±3.8)	0.1(±0.2)
SOSH			<0.1(±0.8)	
MASH		<0.1(±0.1)	<0.1(±0.1)	<0.1(±0.1)
AUSH			0.5(±3.4)	0.1(±0.2)
BCPE			<0.1(±0.1)	0.5(±3.8)
WISP		0.3(±1.8)	4.4(±15.1)	0.6(±1.0)
LESP		0.1(±0.3)	0.2(±0.9)	<0.1(±0.2)
WFSP			<0.1(±0.1)	
NOGA	<0.1(±0.1)	0.7(±3.7)		<0.1(±0.2)
REPH		1.5(±4.3)		<0.1(±0.1)
RNPH		0.2(±0.4)		
POJA		<0.1(±0.1)	<0.1(±0.1)	<0.1(±0.1)
PAJA		<0.1(±0.1)		
UNSK	<0.1(±0.1)	<0.1(±0.1)		
ICGU	<0.1(±0.1)			
GBBG	0.2(±0.8)	0.4(±0.5)	<0.1(±0.1)	0.1(±0.1)
HEGU	2.4(±5.6)	4.0(±19.3)	<0.1(±0.1)	1.0(±3.1)
RBGU				<0.1(±0.1)
LAGU				<0.1(±0.2)
BLKI	0.4(±2.4)	0.2(±0.9)		<0.1(±0.2)
ARTE			<0.1(±0.1)	
COTE			<0.1(±0.1)	<0.1(±0.1)
BRTE	<0.1(±0.1)		<0.1(±0.1)	
SOTE				<0.1(±0.1)
DOVE	1.8(±2.5)			
Total	4.8	7.6	5.9	2.4

*Key to species codes is given in Appendix 6.1.

Spring

During 1978-82, the greatest seasonal densities of seabirds were typically found in spring. North of 35°N total seabird densities were 48.9-118.1 birds km^{-2} along the shelf break (Tables 6.14-6.15), but only 7.6 birds km^{-2} in slope water (Table 6.16). The spring migration of arctic-bound red phalaropes occured along the outer shelf off the northeastern United States between the 60 and 500-m isobaths. Local densities often exceeded 1000 birds km^{-2} on the shoreward edge of the shelf/slope front (Powers and Backus, 1981; Powers and Brown, in press). In the MABe region the concentrations of gannets and great black-backed and herring gulls with foreign fishing fleets continued until April when densities of these species declined substantially (Table 6.16). In 1978 foreign fleets were actively fishing the seaward edge of Georges Bank, which caused fulmars, gannets and great black-backed and herring gulls to concentrate there in that year (Table 6.18). These concentrations in 1978 even spilled over into slope water (SLn) within the 1000-m isobath (Table 6.19).

South of 35°N, the effort data were limited to March (Table 6.13), but again densities were relatively low (3 birds km^{-2}) (Table 6.17). Although no species was numerically dominant in March, the shearwater, storm-petrel and phalarope migrations north during spring must have passed through the SLs region during April and May. In addition, Audubon's shearwater and black-capped petrel are found in the SLs region during spring (Lee and Booth, 1979; P. Stangel, personal communication).

Summer

In summer north of 35°N total densities of seabirds were typically low (<10 bird km^{-2}). Wilson's storm-petrel was the most abundant species with densities of 4.4-7.0 birds km^{-2} (Tables 6.14-6.16). Audubon's, greater and Cory's shearwaters were commonly found in low numbers. Flocks of Audubon's shearwaters (100 birds) were sometimes found along the northern edge of the Gulf Stream in the SLn region. No data were available for the areas south of 35°N at this time, but in August black-capped petrels were found in Gulf Stream water north of 35°N. This suggests a greater abundance in the SLs region to the south (cf. Lee and Booth, 1979; Powers, 1983).

Foreign fishing activities along the shelf break are limited during summer to a few squid fleets. Their importance in attracting birds appeared to be limited to greater shearwaters. In 1978 the pattern of distribution and abundance of seabirds north of 35°N was similar to that observed in 1979-1982, except that densities of Wilson's storm-petrels were slightly greater in that year (Tables 6.17-6.19).

Table 6.17 Mean bird densities (birds km^{-2}) by season on shelf (SHs) and slope (SLs) waters from 28°-35°N off the southeastern United States, 1979-1983.

	Winter		Spring		Autumn	
Species*	SHs	SLs	SHs	SLs	SHs	SLs
NOFU	<0.1(±0.1)		<0.1(±0.1)			
COSH					1.2(±6.6)	0.8(±0.8)
MASH			0.1(±0.1)			
AUSH		0.1(±0.1)			0.4(±2.8)	0.6(±0.9)
BCPE		<0.1(±0.1)				1.7(±2.8)
WISP			<0.1(±0.1)		<0.1(±0.1)	0.1(±0.1)
NOGA	0.5(±0.4)	0.1(±0.1)	0.7(±0.7)	0.2(±0.1)		
BRPE	<0.1(±0.1)					
REPH	0.1(±0.1)					
RNPH	0.1(±0.1)		0.1(±0.1)		0.4(±4.3)	
UNPH	0.7(±0.6)		0.2(±0.3)			
POJA		<0.1(±0.1)				
PAJA					<0.1(±0.1)	
GLGU	<0.1(±0.1)					
ICGU			<0.1(±0.1)			<0.1(±0.1)
GBBG	<0.1(±0.1)		<0.1(±0.1)	<0.1(±0.1)	<0.1(±0.1)	
HEGU	3.6(±12.0)	1.8(±5.7)	0.7(±0.7)	0.4(±0.7)		
RBGU	<0.1(±0.1)					
LAGU	0.7(±0.9)	<0.1(±0.1)	<0.1(±0.1)		0.7(±1.1)	
BOGU	0.6(±0.9)	<0.1(±0.1)	0.2(±0.2)			
BLKI	<0.1(±0.1)			<0.1(±0.1)		
ROTE	0.1(±0.1)		<0.1(±0.1)			
COTE			0.1(±0.1)		<0.1(±0.1)	<0.1(±0.1)
BRTE			<0.1(±0.4)			0.2(±0.1)
SOTE			0.1(±0.4)			0.2(±0.1)
SATE			0.2(±0.4)			
BLTE			0.6(±3.4)			0.1(±0.1)
Total	7.7	2.0	2.9	0.6	2.7	3.6

*Key to species codes is given in Appendix 6.1.

Table 6.18 Mean bird densities (birds km^{-2}) by season on Georges Bank edge (GBe) waters in 1978. Foreign fishing activity was heavy in this area from March through June.

	Season			
Species*	Winter	Spring	Summer	Autumn
NOFU	6.9(±12.1)	75.7(±456.0)	0.2(±0.7)	1.1(±3.0)
COSH			0.1(±0.4)	0.1(±0.5)
GRSH	0.9(±2.1)		4.2(±10.4)	10.9(±26.1)
SOSH		0.1(±0.2)	0.2(±0.6)	
AUSH				0.1(±0.2)
WISP		0.9(±2.8)	14.8(±20.6)	0.3(±1.2)
LESP			0.1(±0.3)	
NOGA		16.3(±62.3)		0.1(±0.5)
REPH		0.1(±0.7)		0.3(±1.3)
POJA		0.1(±0.4)		0.1(±0.3)
UNJA				0.2(±0.4)
UNSK				<0.1(±0.2)
GBBG	0.9(±0.9)	53.6(±282.8)		0.3(±1.1)
HEGU	0.6(±1.3)	69.3(±356.1)		1.0(±3.7)
BLKI	3.6(±6.5)	0.6(±1.8)	0.1(±0.5)	
RAZO		<0.1(±0.4)		
TBMU		0.3(±1.5)		
UNMU	0.6(±1.2)	0.6(±2.4)		
DOVE	2.4(±5.9)	0.3(±1.7)		
ATPU		<0.1(±0.2)		
Total	16.2	217.9	19.7	14.2

*Key to species codes is given in Appendix 6.1.

Table 6.19 Mean bird densities (birds km $^{-2}$) by season on Middle Atlantic Bight edge (MABe) waters in 1978. Foreign fishing activity was heavy in this area from January through March and sporadic from May to July.

Species*	Winter	Spring	Summer	Autumn
	Season			
NOFU	0.2(±1.1)	0.5(±0.9)	<0.1(±0.2)	0.1(±0.5)
COSH			0.2(±0.5)	
GRSH		2.1(±6.4)	6.8(±30.9)	1.4(±2.9)
SOSH		0.8(±2.6)	0.2(±0.8)	
MASH			<0.1(±0.1)	
AUSH			<0.1(±0.1)	
WISP		52.2(±212.9)	18.1(±102.6)	
LESP			<0.1(±0.1)	
NOGA	3.8(±7.7)	2.4(±3.6)		0.1(±0.3)
REPH		5.8(±11.5)		
RNPH		<0.1(±0.1)	0.4(±3.5)	
POJA				0.6(±1.6)
PAJA				0.1(±0.4)
UNJA			<0.1(±0.1)	0.2(±1.0)
UNSK	0.1(±0.4)		<0.1(±0.1)	
GLGU	<0.1(±0.2)			
ICGU	<0.1(±0.2)			
GBBG	169.4(±753.8)	1.7(±3.6)		0.3(±0.8)
HEGU	55.3(±214.8)	1.3(±3.1)		3.1(±5.1)
LAGU			<0.1(±0.3)	
BLKI	24.0(±108.0)	<0.1(±0.2)		0.2(±0.7)
UNTE		0.5(±1.8)	<0.1(±0.1)	
UNMU		<0.1(±0.2)		
Total	252.6	67.4	25.9	6.1

*Key to species codes is given in Apendix 6.1.

Autumn

In autumn, as in summer, total densities of seabirds were typically low (10 birds km $^{-2}$) throughout the regions north of 35°N. Densities of greater shearwaters may have increased over summer estimates along the shelf-break regions (Tables 6.14-6.15), but not farther offshore in SLn (Table 6.16). The fall migration of red phalaropes is not comparable to that of spring along the shelf break. They apparently take a more direct route from the Canadian arctic (Orr et al., 1982) to suspected wintering areas off northwest Africa. Other than ship-following herring gulls, no other species is particularly common at this time.

In autumn south of 35°N total densities of seabirds were also low (2.7-3.6 birds km $^{-2}$), both on shelf (SHs) and slope (SLs) waters (Table 6.17). The most abundant species in slope water were Cory's and Audubon's shearwaters and black-capped petrels. This pattern was probably evident in summer as well. Audubon's shearwaters may be locally abundant along the edge of the Gulf Stream (cf. Lee and Booth, 1979). Bridled, sooty and black terns were also found in slope water from at least late August to probably October, but again they were not abundant (Table 6.17).

Ecological Considerations

Seasonal total densities of seabirds over the continental slope off the northeastern and southeastern United States are relatively low compared to the shelf and shelf break areas off New England. Seasonal densities throughout North Atlantic Slope Water were always less than 10 birds km $^{-2}$ (in summer) (Powers and Brown, in press). Foreign fishing fleets are an important factor in concentrating certain species of seabirds (usually fulmars, greater shearwaters, gannets, and great black-backed and herring gulls) along the shelf break, particularly in winter and spring from Hudson Canyon in the Middle Atlantic Bight east to Lydonia Canyon on Georges Bank. The large influx of red phalaropes at the shelf break in April and May is tied to oceanographic factors, not fishing activity (Powers and Backus, 1981).

A paucity of any significant concentrations of prey near the surface in Slope Water may explain the low densities of seabirds relative to shelf regimes further north. The isothermal layer above the thermocline is deep in shelf break areas during summer and fall and throughout the year in Slope Water. These nutrient-poor waters set this area apart from a boreal shelf system, which gets completely mixed in winter from storms and low air temperatures. Well-mixed waters are more productive because a supply of nutrients is maintained in near-surface waters. Thus an increased growth in phytoplankton stocks stimulates production in higher trophic levels because more energy is available for consumption at each successive link in the food chain. On a smaller scale, fronts are probably one of the most important factors controlling the distributions of seabirds in these deep-water areas. The shelf/slope front in spring and the

Table 6.20 Mean bird densities (birds km^{-2}) on slope water (SLn) north of 35 °N latitude during 1978. Foreign fishing activity was heavy over the continental shelf edge from January through June.

Species*	Season			
	Winter	Spring	Summer	Autumn
NOFU	5.1(±6.6)	16.2(±42.5)	<0.1(±0.1)	0.3(±0.9)
COSH			0.2(±1.6)	0.1(±0.3)
GRSH		1.1(±3.1)	0.6(±2.3)	3.1(±7.5)
SOSH		0.1(±0.2)	0.1(±0.4)	
AUSH			0.2(±0.9)	0.1(±0.2)
WISP		3.6(±5.3)	8.0(±23.2)	0.2(±0.8)
LESP			2.6(±5.3)	
NOGA	0.2(±0.6)	8.6(±29.6)		<0.1(±0.1)
REPH		1.6(±2.7)	0.1(±0.3)	0.1(±0.3)
POJA		0.1(±0.2)		0.1(±0.5)
UNJA				0.1(±0.3)
UNSK			<0.1(±0.1)	
GBBG	3.8(±4.1)	11.8(±36.5)		0.2(±0.8)
HEGU	14.4(±23.3)	3.1(±10.2)		0.7(±1.7)
BLKI	4.2(±10.3)	0.2(±0.5)		0.5(±1.7)
TBMU		0.2(±0.8)		
UNMU	0.4(±1.0)	0.4(±0.9)		
DOVE	4.5(±7.8)			
ATPU		0.1(±0.5)		
Total	32.6	47.1	11.9	5.6

*Key to species codes is given in Appendix 6.1.

northern edge of the Gulf Stream in summer and early fall may be mechanisms which provide local aggregations of prey at the surface for birds like phalaropes, shearwaters and storm-petrels.

Seabird distribution north of 35 °N is relatively well understood at present, but data are limited further south. Although seabird densities there are undoubtedly low, (1) the migration corridors of phalaropes, shearwaters and storm-petrels in spring and fall and (2) the relationship of seabird interactions with the shelf/slope front and the front associated with the edge of the Gulf Stream south of 35 °N need further examination. Also it is not clear whether or not the endangered petrel, (*Pterodroma cahow*), which breeds only in Bermuda, ranges over part of the area of interest (SLs). This species is not easily distinguished in the field from the black-capped petrel.

Appendix 6.1 Key to species codes used in Tables 6.46-6.52, including common and scientific name.

Species Code	*Common Name*	*Scientific Name*	*Species Code*	*Common Name*	*Scientific Name*
NOFU	Northern fulmar	*Fulmarus glacialis*	ICGU	Iceland gull	*L. glaucoides*
COSH	Cory's shearwater	*Calonectris diomedea*	GBBG	Great black-backed gull	*L. marinus*
GRSH	Greater shearwater	*Puffinus gravis*	HEGU	Herring gull	*L. argentatus*
SOSH	Sooty shearwater	*P. griseus*	RBGU	Ring-billed gull	*L. delawarensis*
MASH	Manx shearwater	*P. puffinus*	LAGU	Laughing gull	*L. atricilla*
AUSH	Audubon's shearwater	*P. lherminieri*	BOGU	Bonaparte's gull	*L. philadelphia*
BCPE	Black-capped petrel	*Pterodroma hasitata*	BLKI	Black-legged kittiwake	*Rissa tridactyla*
WISP	Wilson's storm-petrel	*Oceanites oceanicus*	ARTE	Arctic tern	*Sterna paradisaea*
LESP	Leach's storm-petrel	*Oceanodroma leucorhoa*	COTE	Common tern	*S. hirundo*
WFSP	White-faced storm-petrel	*Pelagodroma marina*	SATE	Sandwich tern	*S. sandvicensis*
BRPE	Brown pelican	*Pelecanus occidentalis*	ROTE	Royal tern	*S. maximus*
NOGA	Northern gannet	*Sula bassanus*	SOTE	Sooty tern	*S. fuscata*
BRBO	Brown booby	*S. leucogaster*	BRTE	Bridled tern	*S. anaethetus*
REPH	Red phalarope	*Phalaropus fulicaria*	UNTE	Unidentified tern	*S.* sp.
RNPH	Red-necked phalarope	*P. lobatus*	BLTE	Black tern	*Chlidonias niger*
POJA	Pomarine jaeger	*Stercorarius pomarinus*	RAZO	Razorbill	*Alca torda*
PAJA	Parasitic jaeger	*S. parasiticus*	TBMU	Thick-billed murre	*Uria lomvia*
UNJA	Unidentified jaeger	*S.* sp.	UNMU	Unidentified murre	*U.* sp.
UNSK	Unidentified skua	*Catharacta* sp.	DOVE	Dovekie	*Plautus alle*
GLGU	Glaucous gull	*Larus hyperboreus*	ATPU	Atlantic puffin	*Fratercula arctica*

7 Human Activities and Impacts

A.G. Gaines, M.E. Silva, and S.B. Peterson

Environmental, Regulatory and Political Considerations

The coastal waters of the eastern United States are among the most intensely used and managed of the world. Nevertheless, it is evident that the future will see even increased exploitation of resources in this area and an increasing population of people seeking their livelihood and recreation in coastal related activities.

The ACSAR region is demarked roughly by the 50-mile and 200-mile offshore limits (Fig. 7.1). The 200-mile fisheries zone established by the Magnuson Fisheries Conservation and Management Act (MFCMA) approximates the outer limit of the ACSAR within U.S. jurisdiction. Portions of the Blake Outer Ridge, the Blake Spur and a small area near the seaward terminus of the Hudson Canyon lie outside the 200-mile limit (Fig. 7.1).

Constraints on human activities in the OCS are imposed by natural features of the area as well as by political and regulatory factors. Furthermore, impacts or potential impacts of human activities in the OCS depend on the nature of the undisturbed environment, against which the impact is measured or perceived. The distribution of physical and living environments, and of certain species of organisms as well as of economic activities and political jurisdictions, is given in 125 maps prepared jointly by National Oceanic and Atmospheric Administration and the Council on Environmental Quality (NOAA/CEQ, 1980). This data atlas is a valuable resource for people interested in the ACSAR area and the adjacent continental shelf. An automated computer-based inventory, accessible by narrowly defined geographic grid, is presently being developed by NOAA (D. Basta, personal communication). The system contains updated information on topics covered in the atlas and should aid in manipulation of large quantities of data in support of decision-making.

National Trends in Environmental Regulations

Use of the OCS is directly affected by national trends in the focus, extent, and enforcement of regulations. Public concern over chemical contamination of the environment and destructive environmental effects surrounding practices of industry and government increased during the 1950s and 1960s and was embodied in federal legislation, policy, and infrastructure largely in the 1970s. The decade of the 1970s, for example, saw establishment of the Council on Environmental Quality and the Environmental Protection Agency; Coastal Zone Management legislation and plans; consolidation of diverse federal ocean functions into the National Oceanic and Atmospheric Administration; as well as numerous laws designed to protect one aspect or another of the nation's air, waters, biota, and special environments and habitats.

The problems resulting from this intensive effort to stave off environmental damage are illustrated by the issue of ocean dumping (see review of Farrington et al., 1982; Lahey, 1982). In essence, the five federal statutes affecting disposal of society's waste materials protect the air by prohibiting incineration, protect the ocean by prohibiting ocean dumping, and protect ground and surface water by prohibiting landfill or deep-well injection. The overall effect has been to shift the burden of receiving society's wastes to the medium least regulated at the moment, or in the case of sludge disposal, leaving a city with no viable options.

In 1981, the National Advisory Committee on Oceans and Atmosphere (NACOA) prepared a special report to the President (NACOA, 1981) drawing attention to this difficulty and indicating that regulations on waste disposal should avoid the single purpose approach of the 1960s and 1970s; instead, regulations should identify the medium on which a given waste material would have least impact. Although this recommendation may have been influenced by pragmatism and current value judgments, it also is supported by recent scientific concepts pertaining to the flexibility of ecosystems in accommodating changes in the supply of materials or energy, their so-called "assimilative capacity."

Environmental Impact, Pollution and Assimilative Capacity

The National Environmental Policy Act of 1969 established the requirement for an environmental impact statement to accompany any proposed federal actions "...significantly affecting the quality of the human environment". In one sense the terms "environmental impact", "pollution", and "assimilative capacity" are similar in that each calls for a subjective identification of an "acceptable" amount of environmental change resulting from human activities. The term "environmental impact" carries no connotation of whether environmental changes associated with a given human activity are regarded as beneficial or otherwise; in comparison, the term "pollution" implies a deleterious change or that any change is deleterious. The assimilative capacity is de-

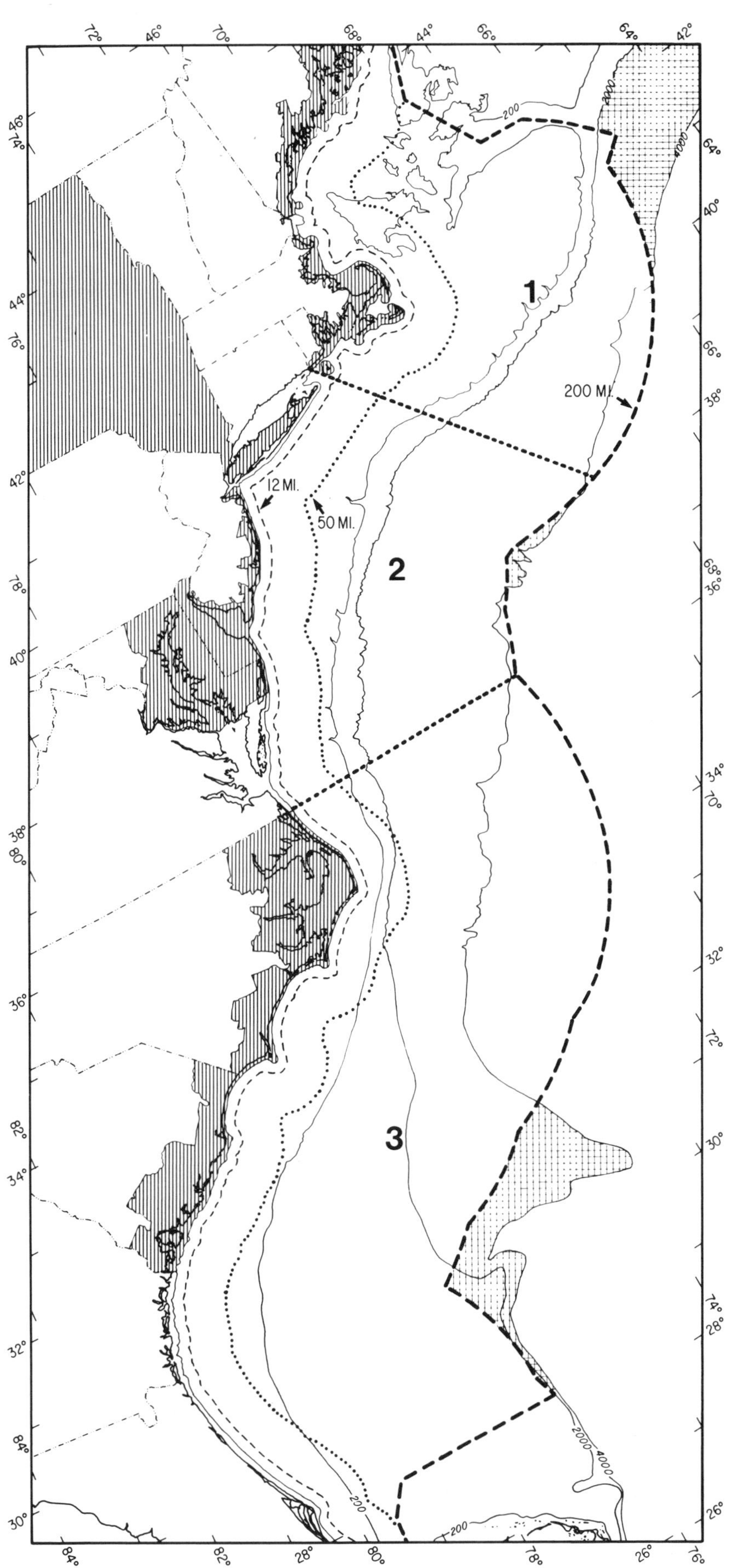

Figure 7.1 Bathymetry and certain jurisdictional delimitations of the ACSAR area. Numbers 1-3 indicate regional fisheries management council jurisdictions; heavy dashed line indicates boundary of the U.S. fishery conservation zone (modified from NOAA/CEQ, 1980). To date all coastal Atlantic states except Virginia and Georgia have approved CZM programs.

fined as the amount of material that could be contained within a body of seawater without producing an unacceptable biological impact (NOAA, 1979). The term carries the connotation that some level of environmental change is acceptable. It also contains a functional basis for quantifying assimilative capacity, although determination of what is "unacceptable" still leaves wide margins of variability.

Assimilative capacity can in principle be assessed by successive additions of a pollutant to a water body; the impact becomes evident at an "endpoint". Like other terms relating to the effect of human activity on the environment, value judgments are involved in the selection of the endpoint. For example, a committee assessing the impact of industrial wastes on plankton in the ocean selected death of all plankton in the wake of the discharging vessel as an endpoint. Another committee might have selected death of the most sensitive species as the endpoint, and very different conclusions regarding the assimilative capacity of the ocean site would have resulted. The concept of assimilative capacity accurately connotes the ability of organisms to endure a range of environmental conditions and of ecosystems to accept variations in the nature and rate of materials and energy flux.

Applicable Domestic and International Regulations and Treaties

All of the ACSAR area considered here lies outside the 3-mile territorial sea and all but a small area near Cape Hatteras lies outside 50 miles (Fig. 7.1). Most of the area is within the 200-mile fisheries zone established by the MFCMA and the 200-nautical mile Exclusive Economic Zone established by the 10 March 1983 Reagan Proclamation; a portion of the Blake Outer Ridge defined by the 4000 m isobath falls beyond 200 miles. The ACSAR area is free of international disputes, since the recent settlement of a U.S.-Canada border dispute on Georges Bank by the World Court. This ACSAR region could be affected by legislation concerning marine scientific research (H.R. 703), establishment of a U.S. exclusive economic zone (H.R. 2061), and the Presidential Proclamation regarding a U.S. 200-mile Exclusive Economic Zone. Domestic and international regulation of ocean dumping is reviewed by Park and O'Connor (1981); Pararas-Carayannis (1973) gives an historical synopsis of the River and Harbor Act of 1899 as well as other U.S. legislation up to 1972.

Domestic Laws and Regulations

At present, at least 35 domestic laws pertain to activities in the ACSAR region (Bureau of Land Management, 1981; Table 7.1). Coastal states with approved Coastal Zone Management programs (Fig. 7.1) can exercise the power of Federal Consistency review, stipulated by the Coastal Zone Management Act (as amended). This gives states with approved CZM programs the power to review federal decisions (such as OCS exploration or production plans) affecting their waters, and to legally challenge those which are inconsistent with the state's coastal policies.

The Magnuson Fishery Conservation and Management Act of 1976 (MFCMA; P.L. 94-265) regulates foreign and domestic fishing within 200 miles of the U.S. coast. This law established regional fishery management councils; the New England Fishery Management Council in Saugus, Mass.; the Mid-Atlantic Fishery Management Council in Dover, Del.; and the South Atlantic Fishery Management Council in Charleston, S. C. are important for ACSAR. These management councils develop fishery management plans which define an optimum yield for each species or species group and identify the tonnage needed to support the domestic fishery; any remainder is allocated to foreign fishing interests by the State Department following guidelines negotiated with the council(s).

The Magnuson Fishery Conservation and Management Act of 1976 provided that the management of highly migratory species (e.g., some species of tuna) should be left to international or regional organizations. However, there have been many efforts to have bluefin tuna included in the category of coastal fish because they are found predominantly in coastal waters. Regardless, all tunas remain under the management and conservation of agreements forged by international fisheries commissions. Although the Act exempts tunas from the expansion of U.S. jurisdiction over fisheries resources, it does define "fishing" to include any "activity which can reasonably be expected to result in the catching, taking or harvesting of fish." Since billfish and sharks fall within the definition of "fish," they are managed by the management councils. Rules for incidental catch of billfish and sharks taken in the catch of tuna by foreign vessels can be regulated within the Fishery Conservation Zone (FCZ) (NOAA, 1978). The Law of the Sea Convention also describes fishery management provisions, but since the U.S. has not signed the treaty, and the Reagan administration remains opposed to it, our fisheries are managed under domestic law. The U.S. is, however, party to the Geneva Convention negotiated in 1958, which, unless the U.S. ratifies the Law of the Sea, remains technically the "law of the sea" for U.S. citizens. The Geneva Convention defines a territorial sea without stating the breadth of that sea; in the U.S., a 3-mile territorial sea has been adopted. The Convention also provides for a contiguous fishery zone not to exceed 12 miles from the baseline.

International Agreements and Treaties

Activities in the OCS area are subject to several international treaties. The International Convention for the Prevention of Pollution of the Sea by Oil regulates and restricts the intentional discharge of oil and oily mixtures by ships. For tankers over 150 gross tons (GRT) the discharge of oil or oily mixtures is prohibited within 92.6 km of the nearest land. Ships other than tankers over 500 GRT can discharge only when as far as practical from the nearest land. The International Convention on the Prevention of Marine Pollution by Dumping of Wastes and Other Matter ("the London Dumping Convention") regulates ocean dumping and as of 1979 was ratified or acceded to by 43 governments including the United States (Park and O'Connor, 1981). This agreement prohibits ocean dumping of organohalogens, mer-

Table 7.1 Domestic laws pertaining to activities in ACSAR area.

Administrative Procedure, 5 USC 551-559, Including Provisions of the Freedom of Information Act, Privacy Act, and the Government in the Sunshine Act.
Clean Air Act
Crude Oil Windfall Profits Tax Act of 1980
Deepwater Port Act of 1974
Department of Energy Organization Act
Emergency Natural Gas Act of 1977
Emergency Petroleum Allocation Act of 1973
Endangered Species Act of 1973
Energy Policy and Conservation Act
Energy Reorganization Act of 1974
Energy Supply and Environmental Coordination Act of 1974
Environmental Quality Improvement Act of 1970
Federal Energy Administration Act of 1974
Federal Water Pollution Control Act (as amended; Clean Water Act)
Fish and Wildlife Act of 1956
Fish and Wildlife Coordination Act
Land and Water Conservation Fund Act of 1965
Marine Mammal Protection Act of 1972
Magnuson Fishery Conservation and Management Act of 1976
Marine Protection, Research and Sanctuaries Act of 1972 (Ocean Dumping Act)
Marine Resources and Engineering Development Act of 1966, including Coastal Zone Management Act of 1972
Mining and Minerals Policy Act of 1970
National Advisory Committee on Oceans and Atmosphere Act of 1977
National Environmental Policy Act of 1969
National Environmental Protection Act of 1972
National Historic Preservation Act
National Ocean Pollution Research and Development and Monitoring Planning Act of 1978
Natural Gas Act
Natural Gas Pipeline Safety Act of 1968
Natural Gas Policy Act of 1978
Occupational Safety and Health Act of 1970
Outer Continental Shelf Lands Act of 1953 and Amendments of 1978
Pipeline Safety Act of 1969
Ports and Waterways Safety Act
Submerged Lands Act
Withdrawal of Lands for Defense Purposes Act

cury, cadmium and their compounds, persistent plastics, high-level radioactive wastes, and biological and chemical warfare agents.

Several international fisheries agreements may need to be considered in OCS development. These include, for example, Northwest Atlantic Fisheries Organization (NAFO) (the successor to the International Commission for Northwest Atlantic Fisheries), the International Commission for Conservation of Atlantic Tunas (ICCAT) and the Convention for the Conservation of Salmon in the North Atlantic Ocean (see U.S. DOI Compilation of Laws related to Mineral Resource activities on the OCS, Vols. I and II, 1981, for more detail).

The Law of the Sea (LOS) Convention as negotiated over the last decade may have to be considered for any activity outside U.S. territorial waters. Under the Geneva Convention, the U.S. has access to the seabed and subsoil to a depth of 200 m or to where the depth of the superadjacent waters admit to exploitation of the natural resources. The Geneva Convention also defines the freedoms of navigation, overflight, fishing, and laying of submarine cables and pipelines.

The most recent law of the sea conference, called UNCLOS III, defines the continental shelf of a coastal state (Article 76, Paragraph 1) as "the sea-bed and subsoil of the submarine areas that extend beyond its territorial sea throughout the natural prolongation of its land territory to the outer edge of the continental margin, or to a distance of 200 nautical miles from the baselines from which the breadth of the territorial sea is measured where the outer edge of the continental margin does not extend up to that distance." If the continental margin extends beyond 200 nautical miles, various formulas, often confusing, can be applied to establish its outer limit. Furthermore, the coastal state can exercise sovereign rights over the continental shelf for the purpose of exploring and exploiting its natural resources; these rights are exclusive in the sense that no one may undertake these activities without the express consent of the coastal state. The term "natural resources" includes the mineral and other non-living resources of the seabed and subsoil together with sedentary species of living organisms — long ago defined as those organisms on or under the seabed, unable to move except in constant physical contact with the seabed or subsoil.

Fisheries

Data Sources

Most data on commercial fisheries in the ACSAR area are collected by the National Marine Fisheries Service (NMFS) in two ways: first, through spring and fall surveys; second, by interviews with fishermen in the major fishing ports along the east coast. This latter information was collected uniformly in New England beginning in the early 1960s, but only in the last four years have similar data been collected south of New York. Thus our ability to generalize about commercial fishing in the southern half of the area is constrained by lack of time-series information. Recreational fishing surveys have been done only twice in the last decade, and because of the expense, are not expected to be carried out more frequently than every five years.

Foreign fishing will continue to be monitored by NMFS and the Coast Guard within the 200-mile fishing limit. Any foreign fishery which develops outside that area will be of great concern to U.S. commercial fishing interests, but not subject to U.S. regulations. Currently NAFO, the Northwest Atlantic Fisheries Organization, would be the responsible international agency for sponsoring research in the northwest Atlantic outside the 200-mile limit. Since the U.S. is not a member of NAFO, our influence on their research programs is not great.

Current Commercial Fishing Activities

Active commercial fisheries in deep water on and over the edge of the continental shelf include tilefish, billfish, sharks and lobster. Other commercial fisheries for cod, haddock, silver hake, and redfish also exist, although none of these fisheries is done predominantly in the ACSAR area. There are no commercial fisheries of record on the rise, although various trawl surveys were done in that area in the late 1960s and early 1970s by the U.S.S.R. and the German Democratic Republic (G.D.R.)

Since passage of the MFCMA in 1976, the fisheries off the U.S. east coast have been dominated by U.S. and Canadian fishermen. Before that time, there were valuable commercial fisheries in deep water pursued by Bulgaria, Cuba, F.R.G., G.D.R., Japan, Poland, Romania, Spain and the U.S.S.R. Japan continues to fish in the slope area for squid, butterfish, mackerel and tuna, with a permit from the State Dept. for the first three fish. As mentioned before, no permit is needed to fish for highly migratory species such as tuna. Poland and the U.S.S.R. have had permits to fish off the Atlantic coast for some of these species. Those permits were rescinded when the U.S. government disagreed with non-fishing policies of those governments, but were reinstated after several years. It is likely that permits will be granted for foreign fishing on squid, herring, whiting, mackerel, and other species within the U.S. 200-mile fishery conservation zone throughout the 1980s.

Using data from the weighout files of NMFS for commercial landings from New England, and to a lesser extent, the Middle Atlantic fishing ports, Lange et al. (1981) analyzed species composition of otter trawl catches by statistical areas, depth zones, and months. They identified 9 major and 29 minor fisheries in the offshore waters. Total catch in the Mid-Atlantic winter-summer fishery has increased since 1974, with a significant peak in 1976. The major summer species caught are summer flounder and tilefish. However, lobster, silver hake and industrial catches (principally menhaden) have predominated this fishery at various times in the past. Total catch in the Middle Atlantic spring-autumn fishery has also increased since 1974, although a significant decline occurred in 1978. The major component of this fishery was scup, with catches of butterfish and several flounder species.

Lange et al. (1981) also characterized Georges Bank/Southern New England groundfish fishery during the late 1970s as having large catches of cod, yellowtail flounder, winter flounder, and haddock. However, the deep-water area of Georges Bank was not fished much by the U.S. commercial fleet in the years covered by their report, producing less than 0.1% of total catch. The deep-water fishery is potentially important to U.S. groundfish fishermen, but is not now fished because the costs of steaming to that area are higher and the traditional species less abundant. The deep-water fishery has a greater proportion of lobster and red crab than does the area fished predominantly by U.S. fishermen. Lobster was the primary catch between 1968 and 1978, while before that time cod, haddock, and silver hake were important. From 1977-79 43% of the catch from deep water was lobster; silver hake comprised 20% and redfish 19% (Lange et al., 1981).

No small vessels participate in the deep-water fishery, and large boats' effort has fluctuated from 2 to 55 days per year. Medium-ton-class vessels are predominant in this fishery, and their average time fishing in the area has ranged from 7 to 67 days per year. Total catch from the deep-water areas averaged 105 MT per year from 1971 to 1981. Catches between 1965 and 1969 were greater, averaging 380 MT for U.S. commercial vessels. Increases in operating costs — expecially fuel after 1973 — probably explain the decline, since the abundance of most species sought by U.S. fishermen has remained the same or increased since 1971 (Lange et al., 1981).

Fish distribution charts from the spring and fall surveys done by NMFS and foreign fishery scientists show potentially where the fish can be caught. These charts, updated annually, show silver hake and mackerel in water deeper than 200 m along the slope in the spring in the area from 36 to 42 °N. The same series of spring surveys show *Loligo* present in these depths and latitudes along with alewives, butterfish, lobster, red hake, redfish, scup, *Illex*, sea bass, and summer flounder. Fall surveys show butterfish, lobster, silver hake, red fish, herring, *Illex* and *Loligo* squid in slope waters (NMFS, 1960-80).

Landings by port are summarized for 1981 in Figure 7.2. Potential commercial fisheries depend upon the demand for the catch in U.S. and world markets, the cost of catching fish in the offshore waters, the advantage of catching the fish year-round rather than in the season most of them are available near shore, and the availability of species. For example, mackerel and whiting have the potential to be very valuable offshore fisheries when abundant and concentrated. Those same species are less valuable to the commercial fleet if dispersed in the offshore areas because the costs of catching them grow rapidly with distance from shore. Since foreign fleets are kept out of nearshore waters because of potential conflicts with coastal fishermen, foreign boats are most likely to fish in slope waters. But such a fishery would be under permit from the U.S. as long as the foreign fishermen were within 200 miles of shore. Species most likely to be sought for an intensive slope fishery by foreign boats include squid, mackerel, hake, and herring.

Current Recreational Fishing

The 1978-79 Marine Recreational Fishery Statistics survey reports estimates of participation, catch and effort by recreational fishermen from Nov. 1978 to Oct. 1979. Data of several types were recorded, including total number of fish caught and brought ashore in whole form from which length/weight samples were obtained, fish filleted, discarded dead, used for bait, and those caught and released alive. For the Mid-Atlantic offshore region, species include sea basses, bluefish, Atlantic Bonito, catfishes, Atlantic croaker, cunner, American eel, summer flounders, winter flounder, flounders, hakes, herrings, Atlantic mackerel, mackerels and tunas, white perch, yellow perch, porgies, scup, sea robins, spotted sea trout, sharks, dogfish, skates and rays, spot, striped bass, tautog, toadfishes, weakfish, and windowpane (Human Sciences Research Inc., 1979). The offshore fishery for many of these was curtailed by high fuel costs in the last decade. However, offshore fisheries do exist for cod and pollock because they can be taken year-round in the water deeper than 40 m in the area extending from Block Island, R.I. to Cape May, N.J. (Freeman and Walford, 1974).

Current estimates are that about 90% of commercial sportfishing for groundfish is conducted within 20 miles of shore (Nicholson and Ruais, 1979), although the division of catch between the territorial seas (from 0 to 3 miles) of the coastal states and the Fishery Conservation Zone (3-200 miles) is not known.

There are two possible explanations for this distribution of effort: first, charter and party boats are licensed by the U.S. Coast Guard, and for many of those boats, the license restricts travel to within 20 miles of a harbor of safe refuge; the second reason is cost, both for fuel and time. Boats for hire prefer short fishing trips to reduce operating costs and to enable them to make several trips each day. Privately owned boats can find many sportfishing opportunities within 30 or 40 miles of shore with a few exceptions, such as substantial recreational fisheries for tilefish, whiting, mackerel, billfish, tuna, and shark, which exist in offshore areas. Privately owned recreational fishing boats (i.e., those which do not take passengers for hire) are free to fish wherever and whenever they like.

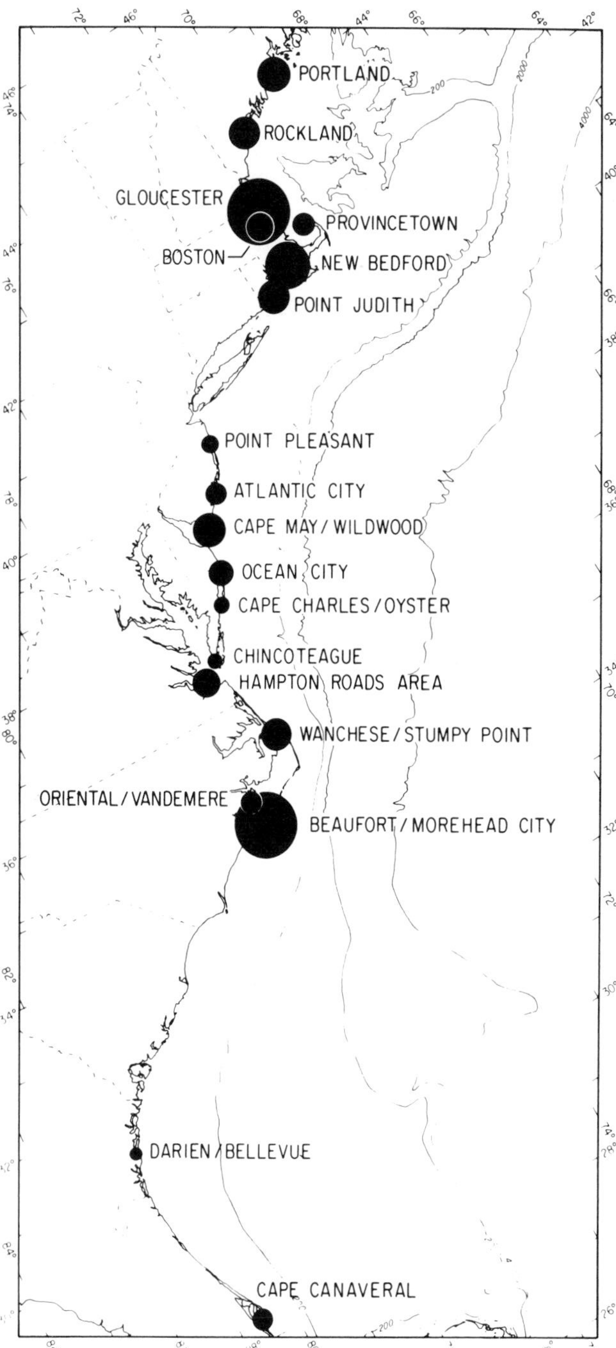

Figure 7.2 The relative commercial fish landing (weight) for 19 leading ports of the U.S. east coast (from data in National Marine Fisheries Service, 1982).

Status of Species

Billfish and Sharks

Fishing for billfishes and sharks in the Atlantic Ocean is mostly in waters varying in depth from 100 m to well over 1000 m (e.g., Fig. 7.3). The fishery is essentially a surface one, although canyons, seamounts, and other major morphological features are distinctive features where the fishery is most intense. Sport fish include tunas, billfishes and sharks, and associated species — dolphin fish, wahoo, king mackerel, and great barracuda. Except possibly for some mid-water fishes or some of the squids, no other resource presently of equal commercial importance exists in the

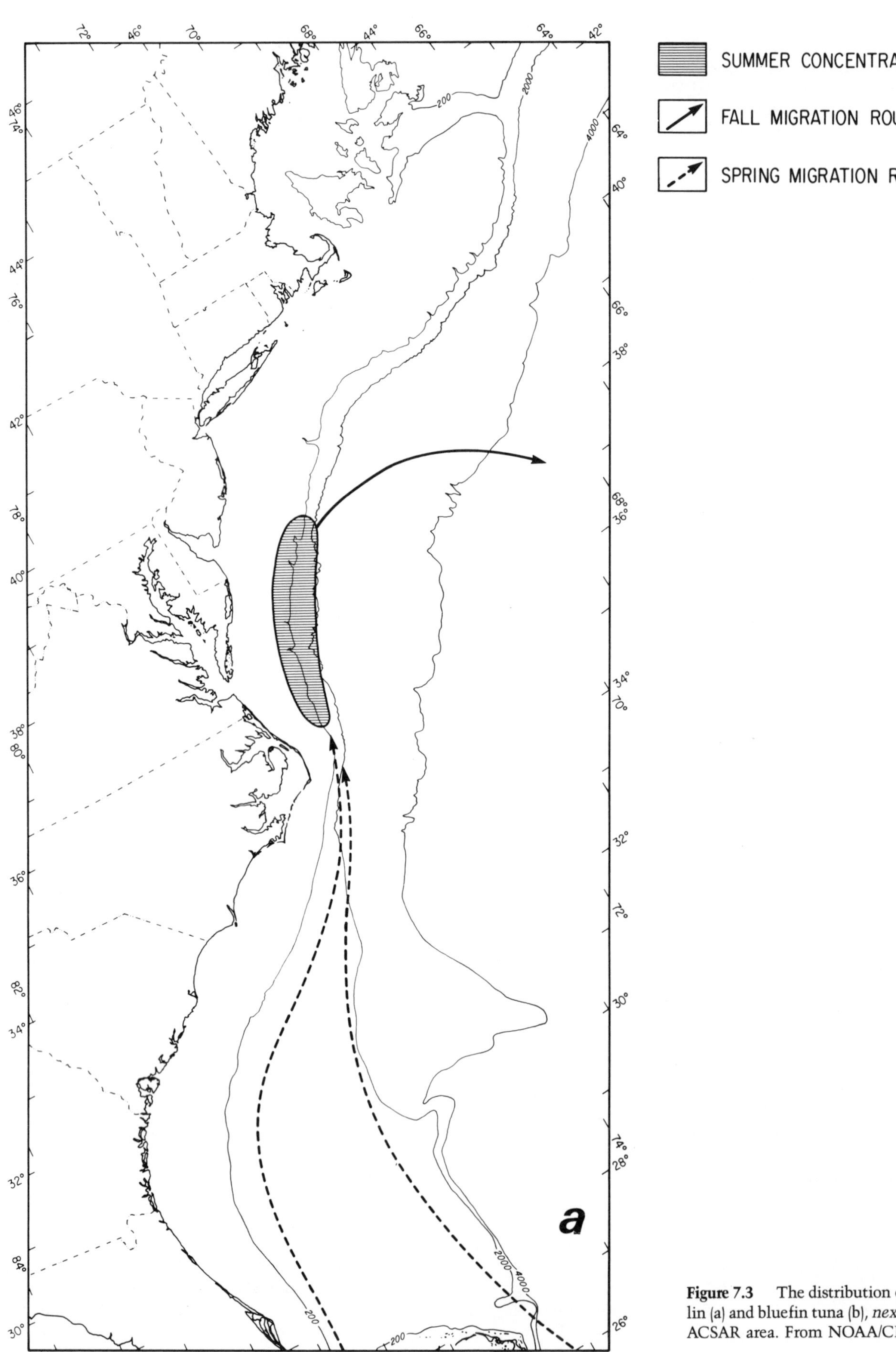

Figure 7.3 The distribution of white marlin (a) and bluefin tuna (b), *next page*, in the ACSAR area. From NOAA/CEQ (1980).

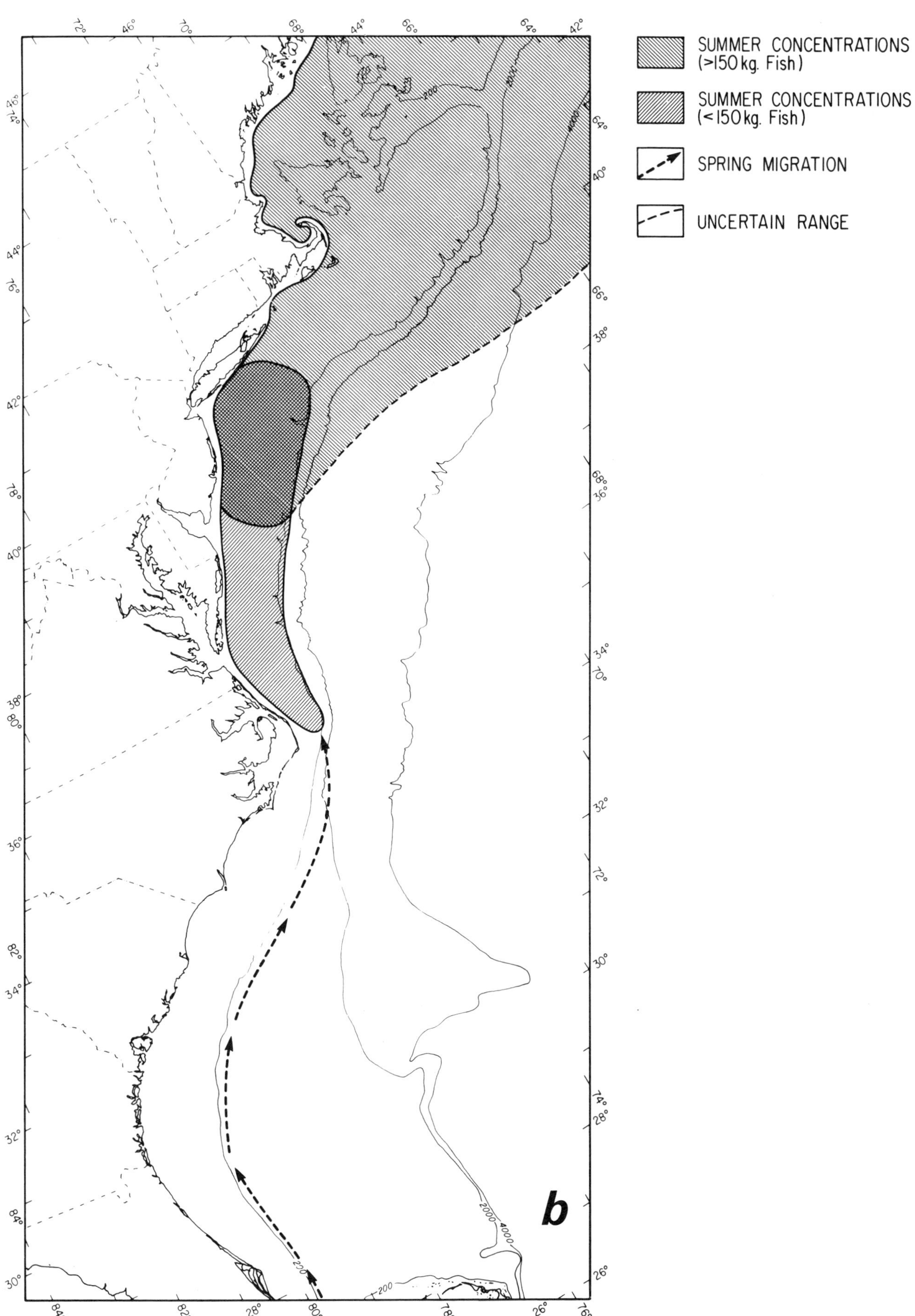
SUMMER CONCENTRATIONS (>150 kg. Fish)
SUMMER CONCENTRATIONS (<150 kg. Fish)
SPRING MIGRATION
UNCERTAIN RANGE
b
200
2000
4000

same area or at the same seasons as these pelagic fishes. Yellowfin tuna, albacore, bluefin tuna, bigeye tuna, skipjack tuna, blackfin tuna; sailfish, blue marlin, white marlin, longbill spearfish, swordfish, longfin mako, blue shark, shortfin mako shark, night shark, oceanic white tipshark, scalloped hammerhead shark, silky shark, thresher shark, tiger shark, bignose shark, porbeagle shark, spinner shark, bigeye thresher shark, great hammerhead shark, smooth hammerhead shark and Galapagos shark are all included in the commercial and recreational fisheries of the U.S. east coast (NOAA, 1978b; Fig. 7.2).

The sport fishery for billfishes is seasonal, particularly for the marlins. The fishery is most intense from April through October, but is heavily dependent on the weather rather than availability of billfishes since fishing is done a considerable distance from shore (South Atlantic Fishery Management Council, 1979). The only species for which maximum sustainable yield (MSY) estimates have been made are blue marlin, 4000 MT; white marlin, 1,900 MT; sailfish/spearfish, 960 MT.

Both commercial and recreational fisheries for pelagic sharks (i.e., sharks other than dogfish) are pursued with longlines, handlines, net trawls, and gill nets. The most intensive commercial fishery has been by foreign fleets; current commercial catches are low, but are likely to increase; in 1981 landings were valued at nearly $2 million. Sharks are popular recreational fish, especially with charter boats because their greater abundance almost ensures that the recreational fishermen will catch something, even while fishing for swordfish, billfishes and tunas. U.S. commercial catch of large pelagic sharks in the waters off North Carolina to East Florida varied between 3 and 598 MT over the last decade, with an average of about 55 MT; the by-catch of sharks in the commercial swordfish fishery was estimated at 1020 MT for 1978; MSY estimates were 41,000 MT. Estimated recreational catch for large pelagic sharks is approximately 1872 MT. There has also been a substantial by-catch of shark, estimated to be over 2000 MT, in the Japanese longline fishery for swordfish during the last decade (Mid-Atlantic Fishery Management Council, 1980a). Estimated catches by U.S. and Canadian fishermen of swordfish by longline have ranged as high as 5000 MT per year, although the average in the late 1970s has remained below 2000 MT; MSY in 1978 was estimated at 5800 MT. Swordfish are also caught commercially and recreationally by harpoon.

The Fishery Conservation and Management Act of 1976 left management of highly migratory fish up to international organizations; however, there have been many efforts to have bluefin tuna included in the category of coastal fish because they are found predominantly in coastal waters. As yet, tuna are managed under agreements forged by international fisheries commissions.

Bluefish

Commercial and estimated recreational catches of bluefish (*Pomatomus saltatrix*) have undergone approximately a fourfold increase during 1960-78. Estimates of MSY (Maximum Sustainable Yield) range from 85,800 to 92,100 MT, and 1975-78 catches averaged 88,200 MT (Anderson and Almeida, 1979). The value of commercial landings was $3.2 million and does not reflect the enormous recreational value of this fishery.

From Billingsgate Shoal off Cape Cod, Mass., to about Cape Lookout, N.C., angling for bluefish is particularly important (Fig. 7.3). Recreational fishing in this area accounts for 83% by weight of all bluefish caught. Young of the year and yearlings are caught in the bays and sounds while older bluefish, some weighing as much as 30 lbs, are caught offshore. Commercial fishing is done by trawling and seining, although in some states limitations on gear type are set to favor recreational fishing interests. The foreign catch of bluefish amounts to less than 2 percent. There appear to be two major areas and seasons of spawning along the U.S. east coast: one offshore near the inner edge of the Gulf Stream from southern Florida to North Carolina in the spring (chiefly in April and May), and the other in the Middle Atlantic Bight (i.e., Cape Hatteras to Cape Cod) over the continental shelf in the summer (chiefly June through August) (Mid-Atlantic Fishery Management Council, 1980b).

Tilefish

A domestic commercial longline fishery for tilefish (*Lopholatilus chamaeleonticeps*) has developed in the area between Cape Hatteras and Cape Cod (Fig. 7.4), with catches increasing from about 30 MT in 1968-69 to approximately 3800 MT in 1979. In 1981 the value of the catch was estimated at $7.5 million. The fish occur along the outer continental shelf from Nova Scotia to south of Florida at depths of 80 to 540 m. They are abundant in the southern New England-Middle Atlantic area where a commercial fishery has existed since 1915. In that area, the tilefish generally occur at depths of 80-440 m and at temperatures from 9 to 14.5° C. Fish have been observed from Norfolk to Lydonia Canyons, as well as in burrows and depressions near boulders and obstructions. The principal fishery is from ports in New York, New Jersey, Massachusetts, and Rhode Island. Both longlines and bottom trawls are used although the former dominate. Ireland, Japan, and Spain were the only foreign countries to report catches, and it is possible that some tilefish were caught by distant-water fleets during the 1960s and 1970s but reported as "other finfish". Since 1977 they have been reported as by-catch in the distant-water fleet fisheries for silver hake, red hake and squid. There is no allowable foreign catch for tilefish; it is required that they be discarded. The recreational fishery for tilefish developed in 1968 and party-, charter-, and private-boat activity was high during the early and mid-1970s. Since 1968, estimated annual catches have ranged from 5 to 340 MT. The recreational effort was greatly reduced by 1978 due to increased fuel costs and decreased size and availability of tilefish (Turner et al., 1981).

Squid

During late spring and summer, long-finned squid genera may be found in harbors and estuaries, particularly in south-

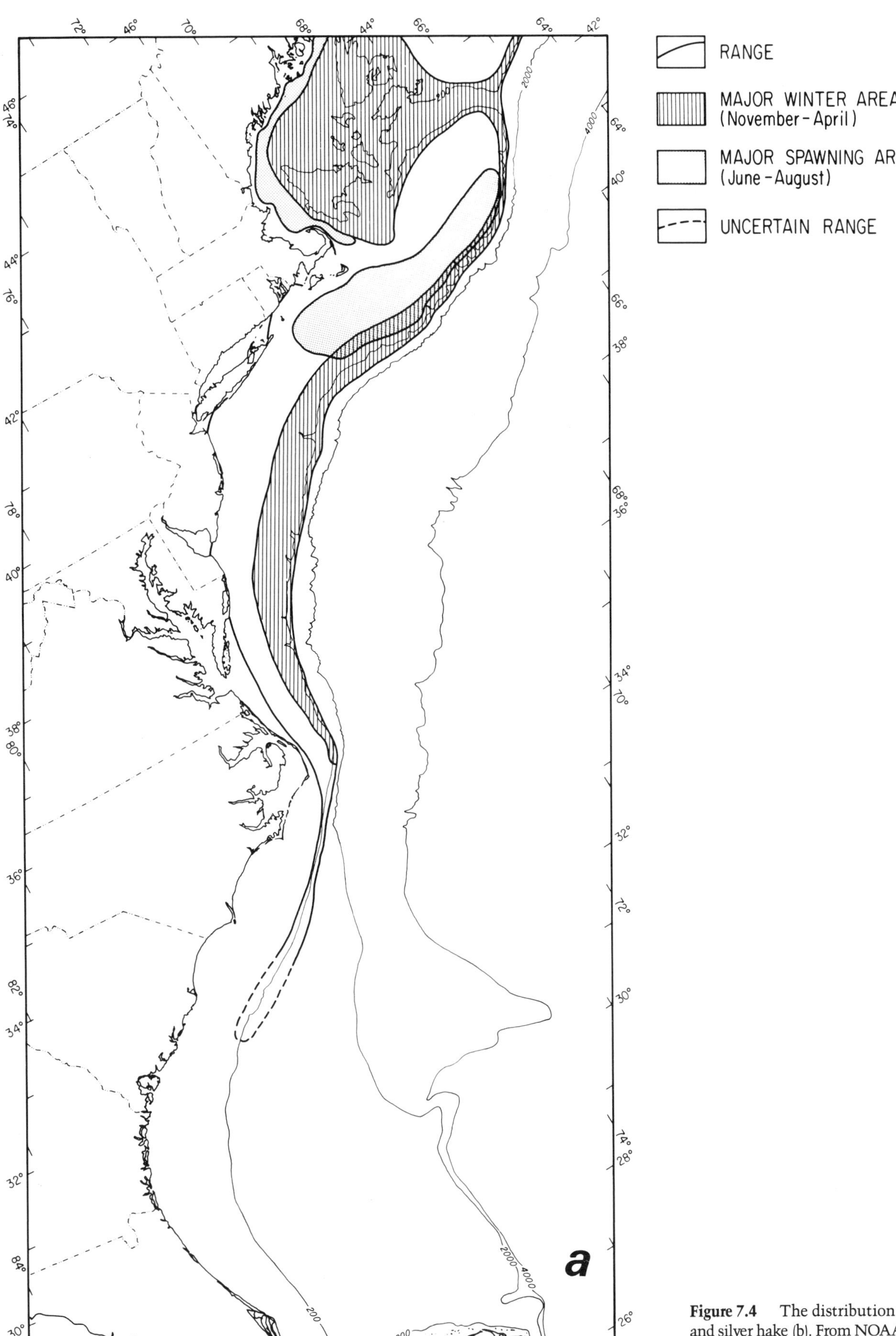

Figure 7.4 The distribution of tilefish (a) and silver hake (b). From NOAA/CEQ (1980).

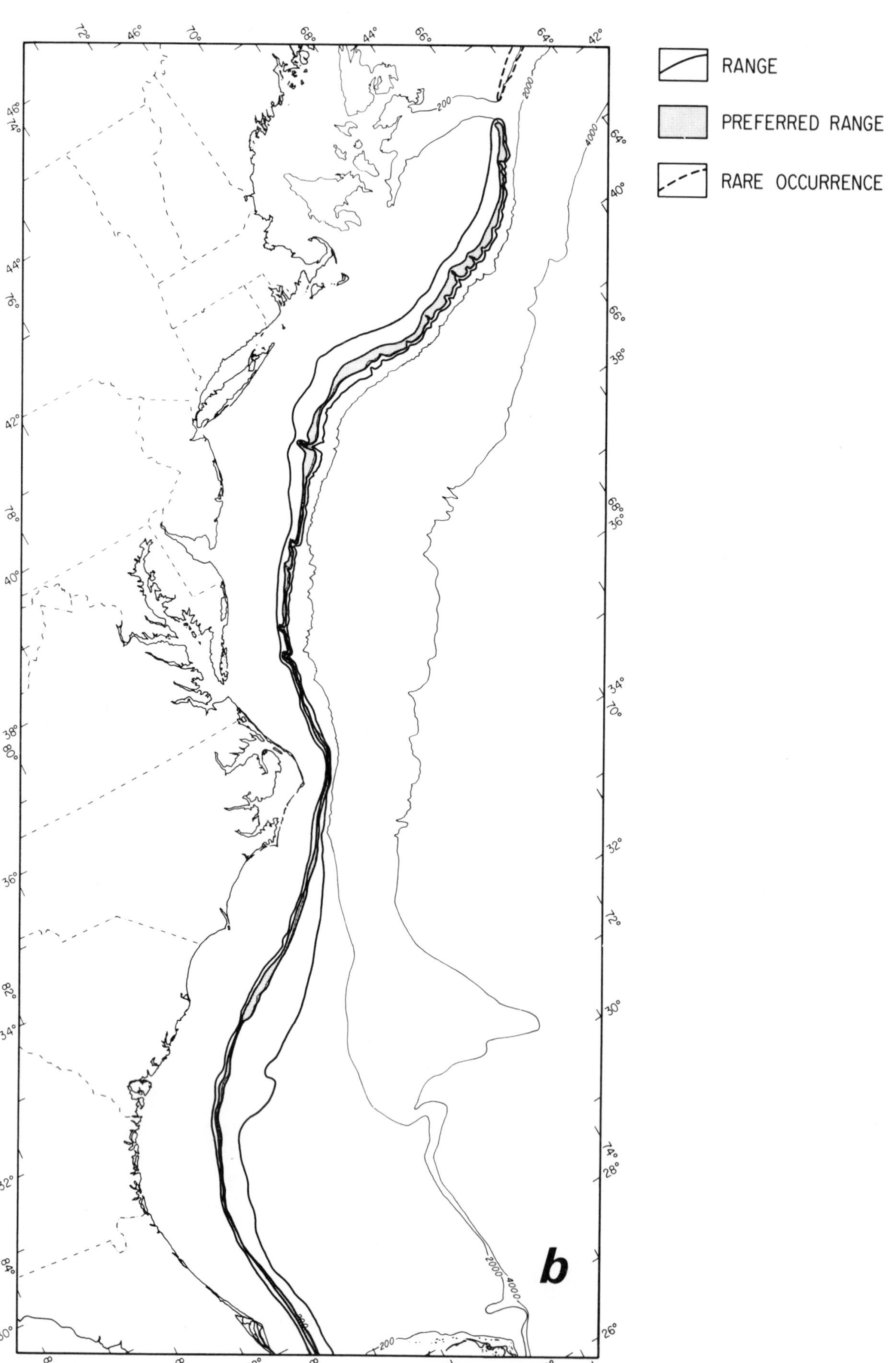
RANGE
PREFERRED RANGE
RARE OCCURRENCE
b
200
2000
4000

ern New England. In the fall, concentrations appear in the southern New England and Hudson Canyon area in water less than 110 m deep. NMFS spring bottom trawl surveys show maximum concentrations of *Loligo* in depths of 111-183 m. Size distribution correlates with depth in both spring and fall with the largest individuals usually taken at the greatest depths. MSY estimates are 40,000 MT for *Illex* and 44,000 MT for *Loligo*. The total allowable level of foreign fishing was 20,000 MT for *Illex* and 30,000 MT for *Loligo* (Mid-Atlantic Fishery Management Council, 1978). In previous years, foreign catches of squid have been two to three times the amount now allocated, and continued requests from foreign nations for squid allocations are expected if a U.S. commercial fishery does not expand. The 1981 value for squid caught in the Atlantic was $2.3 million. Although there is no "recreational" fishery for squid, squid are important as bait for many anglers.

Hake

Two representatives of the genus *Merluccius* are found off the Atlantic coast; the silver hake, *M. bilinearis*, which favors the continental shelf waters (Fig 7.4), and the American hake, *M. albidus*, a fish of the deeper continental slope. The ranges of both species overlap at the outer edge of the shelf and the commercial catch in that area may include both. The winter-spring distribution of silver hake (whiting) is along the outer continental shelf and slope and the summer-fall distribution is along the inner shelf and on the shoaler banks.

The U.S. fishing effort, valued at about $7.4 million, is conducted primarily in the summer and fall in depths less than 30 fathoms. Foreign fishing has traditionally been heaviest during the winter/spring along the outer edge of the shelf. The whiting fisheries were important to the foreign trawl fleets which caught approximately 80% of the 125,000 MT harvested per year through the mid-1970s. Foreign countries involved in that fishery include Bulgaria, Cuba, F.R.G., G.D.R., Japan, Poland, Romania, Spain and the U.S.S.R. In the 1973-75 period, 160,000 MT were taken from off New England and the middle Atlantic states. In 1976, 93,000 MT were taken and in 1977 115,000 MT were taken from those areas (Combs, 1977). The recreational fishery is particularly important along the New York and New Jersey coasts, but none is recorded or anticipated for New England (New England Fishery Management Council, 1978).

Total catches by New England and mid-Atlantic fishermen averaged about 16,800 MT during 1955-59, declined to 9,952 MT in 1960, and then increased steadily to 137,400 MT in 1966. Catches dropped sharply to 50,900 MT in 1967 and have since fluctuated between 19,200 and 67,000 MT. In 1978 the US commercial and recreational catches were estimated to be 11,405 and 4,000 MT respectively. The foreign catch that same year was 10,765 MT (Almeida and Anderson 1979).

The silver and American hake fishery in the mid-Atlantic area has a by-catch of mackerel, herring, squid, and red hake. There is potential bycatch of lobster for American fishermen in the offshore lobster pot fishery. During January-March, there are only a few lobster fishermen in the offshore area, but during April-June there is a substantial pot fishery along the edge of the continental shelf in waters greater than 150 m depth (NMFS, 1977).

Mackerel

There have been active commercial and recreational fisheries for mackerel (Fig. 7.5) in the U.S. starting in the early 1600s. Peak U.S. catch was in 1884; recent U.S. harvests have been below 5000 MT annually for commercially caught fish and value was low--approximately $800,000. There was a major foreign harvest in the early 1970s, approximately 400,000 MT per year. Catch fell to less than 100,000 MT in 1976. Geographical distribution of this fishery changes from year to year; thus it is difficult to predict the importance of the fishery in the ACSAR area.

Groundfish

Both cod and haddock are present in the deep waters off the edge of the continental shelf of northern states, although the fishery there is limited. U.S. commercial fishery from deep water accounts for 0.1% of the catch, and thus may be valued at about half a million dollars. Since cod is available in the deep water year round, it may be attractive to a limited number of recreational fishermen.

Herring

Herring is annually worth over $7 million to the U.S. commercial fishing fleet, but less than 5% of that is caught outside the 3-mile limit. There are stocks of herring on Georges Bank and in the deep water over the edge of the shelf, but these stocks have become depleted, catches falling from several hundred thousand metric tons to a few thousand MT in the last two decades (Sinderman, 1979). Although existing stocks are low (as is the demand for Atlantic herring) the potential exists for a substantial commercial fishery when the stocks recover. In the past, eastern European nations caught 95% of the offshore herring, but in the future, those stocks may be sought primarily by Canadian or U.S. fishermen.

Lobster

American lobster (*Homarus americanus*) is widely distributed off the northeastern coast of the U.S., from Maine to North Carolina and from the intertidal zone out to 700 m. In the U.S. there are two principal areas of harvest: the inshore waters from Maine to New Jersey out to a depth of 40 to 100 m; and the outer shelf and upper slope from Corsair Canyon to Cape Hatteras in depths of 100 to 600 m. The inshore areas account for the greater share of production (about 83% in 1978). There may be numerous local populations of lobster indigenous to offshore canyons, with maximum separation between populations in winter months. Landings reached a high in 1979 of 16,863 MT, valued at $72.3 million. Maine is the leading lobster-producing state with 55% of total landings, but all of that is from inshore waters. In Massachusetts, 42% of lobster

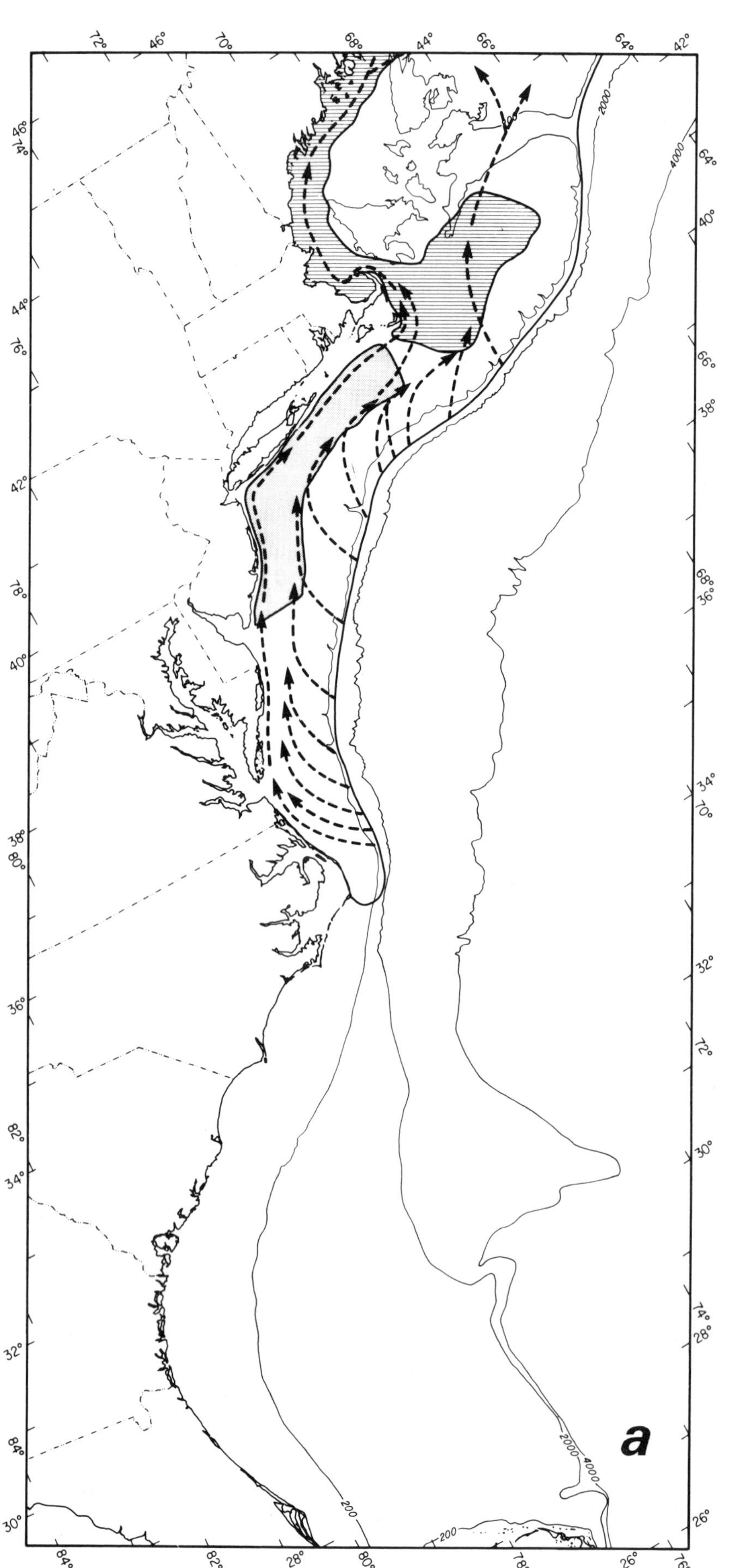

Figure 7.5 The distribution of Atlantic Mackeral spring/summer (a) and fall/ winter (b), *next page*, from NOAA/CEQ (1980).

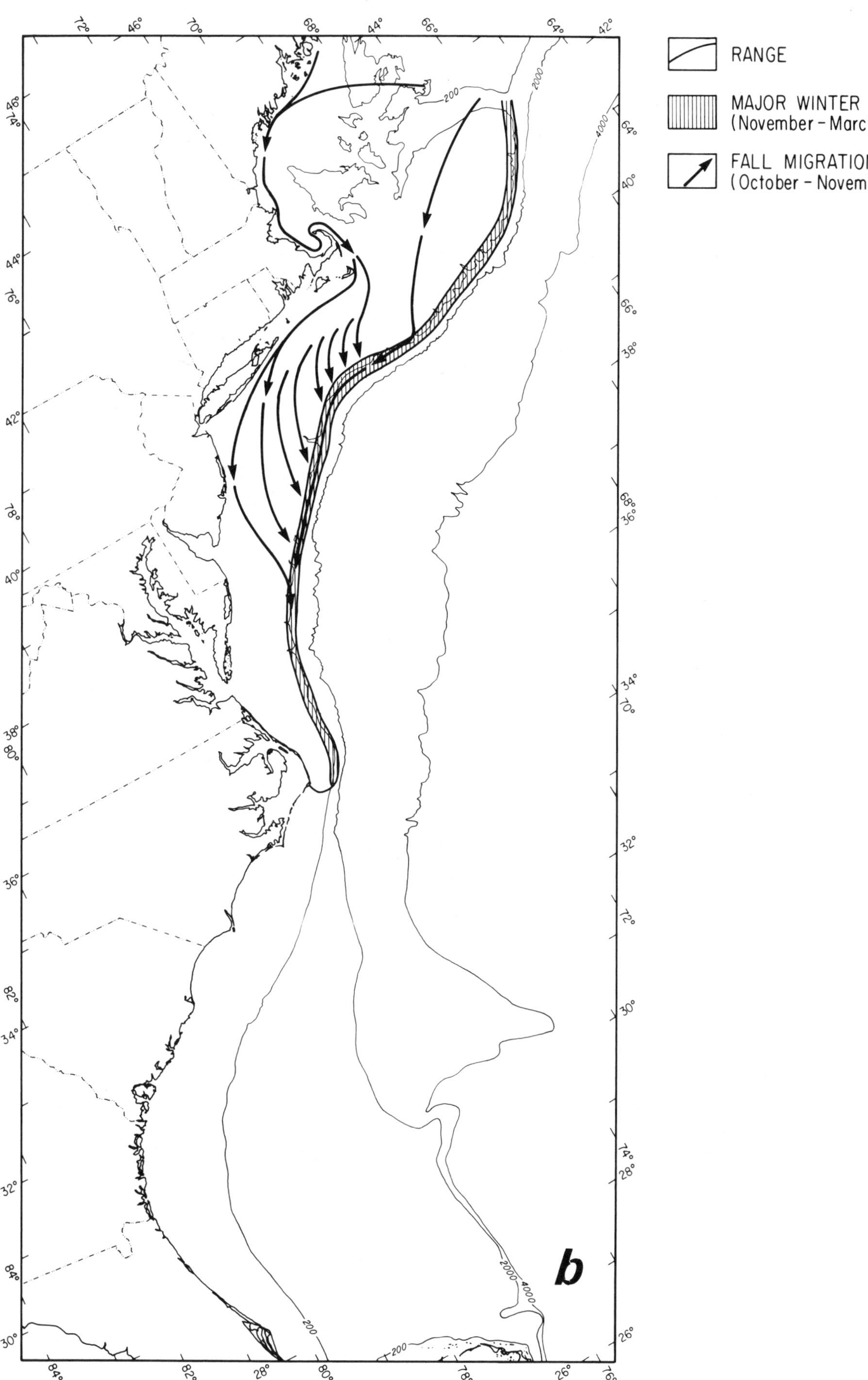

RANGE
MAJOR WINTER AREA
(November - March)
FALL MIGRATION ROUTE
(October - November)
b

landed was from the offshore trap and otter trawl fishery; 77% for R.I.; 48% for New York; 59% for New Jersey; 81% for Delaware; 91% for Maryland, and 100% for Virginia (New England Fishery Management Council [NEFMC], 1982).

Red Crab

The canyon areas along the edge of the continental shelf have populations of red crab — in particular the area between Veatch and Atlantis Canyons — in quantities attractive to commercial otter trawl fishermen. The commercial fishery began in 1973, but marketing problems made the fishery unattractive for many fishermen; by 1980 there were only two full-time red crab fishing boats in the northeast. Most of the commercial fishery is in the areas of Atlantis, Block and Hudson Canyons. In 1979, 1220 MT were landed worth $917,000. The red crab ranges from 300 to 1000 m (Gerrior, 1981), but the most profitable (largest catch and highest price per pot) fishery is from depths ranging from 535 to 620 m (Gerrior, 1981). Seasonal changes affect red crab distribution, with deeper water fishing in spring and winter.

Oil and Gas

Legal Structure Surrounding Exploitation

As with most human activities on the outer continental shelf and beyond, the extraction of oil and gas resources is heavily regulated (Table 7.1). A primary issue is which level of government has responsibility for the management of oil and gas resources in the ACSAR area. Currently this responsibility is vested in the federal government (Breeden, 1976; Ball, 1982). The Outer Continental Shelf Lands Act of 1953 (OCSLA) codified the right of the federal government, through the Secretary of the Interior, to grant oil and gas leases in offshore areas more than three nautical miles from shore. Though amended in 1978 (Krueger and Singer, 1979; Vild, 1979; Jones et al., 1979), the OCSLA as amended remains the primary statute for the development and regulation of oil and gas activities in the ACSAR area. Among other things, the OCSLA and its amendments (OCSLAA) prescribe the way in which leases are to be established and maintained, and the system of bids and royalties which will apply to leases. It requires that the Secretary of the Interior shall periodically prepare a five-year leasing program which will include a schedule of proposed lease offerings. The five-year plan (the current operable five-year plan is the 1982-1987; see Bureau of Land Management [BLM], 1982) is to indicate the size, timing, and location of leasing activities (U.S.C. 1344). In terms of safety and environmental regulations, enforcement responsibilities are mandated to, in addition to the Secretary of the Interior, the Secretary of the Army (pertaining to the Army Corps of Engineers) and the Secretary of the Department under which the Coast Guard is operating.

An important aspect of the OCSLA for the development of oil and gas is the provisions of Section 19, added by the 1978 Amendments. These require that state governments likely to be affected by leasing activities be kept informed of those activities by the Secretary of the Interior. Further, the Secretary must take into account any comments those states may have. Generally, it is felt that this will increase the ability of states to participate in OCS decisions (Vild, 1979). At the same time, however, Section 19 complicates the process of oil and gas leasing and development.

While the Federal government is responsible for leasing, the states have an important regulatory tool in the "consistency" provision of the Coastal Zone Management Act of 1972 (Section 307 (c) (2)). Under this provision, any state with a Department of Commerce approved coastal zone management plan may try to block a federal activity (such as the leasing of oil and gas tracts, see *California* v. *Watt*, 1982) if that state determines that the activity, or some part thereof, is inconsistent with its coastal zone management plan (Brewer, 1976; Deller, 1980; Behr, 1979; Best, 1979; California v. Watt, 1982). In the years to come, it is likely we will see increased federal-state conflict with regard to governance of offshore resources.

Current and Projected Activity

Until very recently, there has been very little OCS oil and gas development between 28°N and 42°S in water depths of 200 to 4000 meters in the northwest Atlantic, and not much is known regarding the area's potential. In fact there has been no development or production although there has been some exploration and some gas discoveries have been made. While uncertainty remains about future production levels, there is some optimism about the future (Sumpter, 1979). On the slope, there is an ancient submerged reef structure that some feel could provide economically recoverable resources (Edgar and Bayer, 1979). Under the DOI five-year plan for 1982-1987 (U.S. DOI, 1982), several lease offerings are scheduled. While earlier sales in the northwest Atlantic have focused on the shallower waters of the continental shelf, recent and pending lease offerings (numbers 52, 76, 78, and 82) are generally moving off the shelf and down the slope into waters of more than 200 meters (Fig. 7.6).

Since no commercial finds have occurred to date, any projections about future activity must be based on resource estimates. There are three commonly used methods for estimating undiscovered oil and gas resources: the volumetric approach which, based upon past geologic knowledge, estimates the total oil and gas that may exist in promising sedimentary rock formations; engineering projections, which suggest future production through use of mathematical formulae based on historical trends; and econometric models, which also apply mathematical models (in this instance based upon past market/price trends) to suggest future supplies of oil and gas attendant to exploratory and development efforts prompted by changes in price. The problem with each approach in making informed judgments about oil and gas activities is that it bases future projections partly on past experience. "Insofar as the past does not adequately represent the future, their estimates are likely to be in error" (Schanz, 1978, p. 18).

Each of these estimating methods is based upon different assumptions and methodologies, and the oil and gas resource estimates they produce are often in disagreement. This can lead to political conflicts about various development strategies (Wildavsky and Tenenbaum, 1981, p. 9-16). For example, in Lease Sale No. 42 (Georges Bank), opponents of leasing used lower estimates (and the impression of estimates) to argue that the value of fishery resources outweighed the potential of oil and gas resources (Colgan, 1982).

Resource Estimates

With the above general comments and caveats about estimating oil and gas reserves, we now turn to specific estimates for oil and gas in the northwest Atlantic. There are four lease areas in the ACSAR region which are thought to have oil and gas potential: the Southeast Georgia Embayment, the Blake Plateau, Baltimore Canyon, and Georges Bank, but only the latter three lease areas have large portions in waters deeper than 200 meters. Of these three, the Baltimore Canyon area is considered to have the greatest potential (Edgar and Bayer, 1979).

Blake Plateau

Under the five-year leasing program for 1982-1986 (BLM, 1982), the Blake Plateau has been combined with other areas of the South Atlantic planning area (i.e., South East Georgia Embayment). Based on U.S. Geological Survey data, resource estimates for that part of this broader area covered by Lease Sale No. 56 are:

	5%	Mean	95%
Oil (billions of barrels)	.8	1.4	2.1
Gas (trillions of cubic feet)	1.4	2.5	3.5

That is, if hydrocarbons are found, there is a 5% probability that there will be less than 0.8 bbl of oil and/or less than 1.4 tcf of gas; and a 95% probability that less than 2.1 bbl of oil and/or 3.5 tcf of gas will be found (BLM, 1981). The estimates as to the portion that may occur in the Blake Plateau are 300 million barrels of oil and 700 billion cubic feet of natural gas (Deis et al., 1982).

Baltimore Canyon

Currently there is no production or development in this area. As exploration continues, the efforts will move into deeper waters. Beginning with Lease Sale No. 59 and continuing with future offerings, the exploration effort (and development and production) will occur in waters of the continental slope. For Lease Sale No. 59, the U.S. Geological Survey provided estimates of oil and gas for this region. Using 5% and 95% confidence intervals, they estimated that the lease sale area may contain from 0.36 to 7.3 bbl of oil and from 1.9 to 28.5 tcf of oil (BLM, 1980). For Lease Sale No. 76, the mean estimates of recoverable resources, pending discovery, are 0.879 bbl of oil and 3.693 tcf of natural gas (Minerals Management Service (MMS), 1982, p. 7).

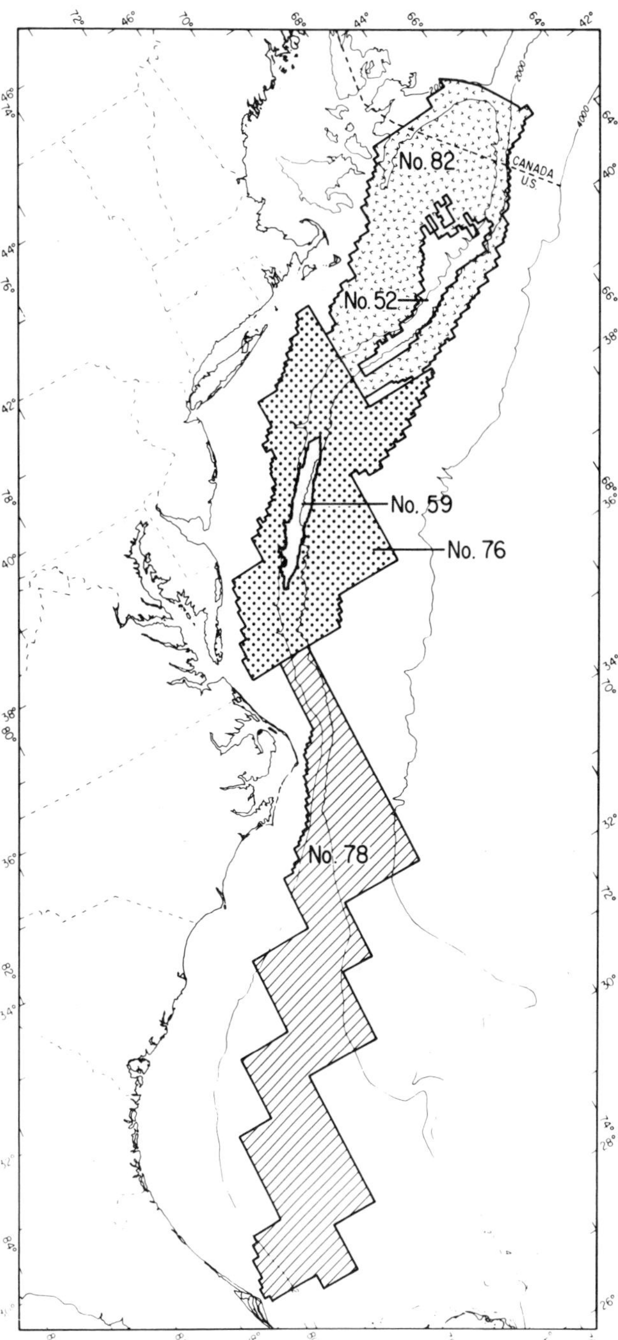

Figure 7.6 Oil and gas lease areas for the U.S. east coast and for the ACSAR area (modified from NOAA/CEQ, 1980).

Georges Bank

At this time, no development or production activity is occuring in this area; the trend is into deeper waters. Some of the tracts in Lease Offering No. 52 were in waters as deep as 2,800 meters. Estimates for the Lease Offering No. 52 area provided by the U.S. Geological Survey ranged from lows of 0.017 bbl for oil and 0.196 tcf for gas to highs of

6.35 bbl of oil and 13.49 tcf for gas. The conditional mean estimates were 1.73 bbl of oil and 5.25 tcf of gas (BLM, 1981). For the entire North Atlantic region, the risked resource estimates for water depths greater than 200 meters are 1 bbl of oil and 3.2 tcf of gas (USGS, 1981).

With the cancellation of Lease Offering No. 52 in 1983, the next lease offering scheduled for the Georges Bank area was the 1984 North Atlantic lease offering (also referred to as Lease Offering No. 82) involving more than 20 million acres. Much of this area is in waters shallower than 200 meters; the mean estimates for recoverable oil and gas are 210 million barrels and 4.9 trillion cubic feet, respectively. The Secretary of Interior announced in December of 1983 that Lease Offering No. 82, originally scheduled for April 1984, would be postponed until differences with the states could be worked out. Secretary Clark eventually set the lease offering for September 26, 1984. However, when the only bid received came from Greenpeace, an environmental activist group, the sale was cancelled. Various reasons were given for the lack of interest exhibited by the oil companies, including threats of future lawsuits, the lack of oil and gas discoveries under Lease Sale 42, and the continued prospect of low petroleum prices.

Finally, it should be noted that the October 12, 1984 decision by the International Court of Justice on the disputed boundary between the United States and Canada may affect oil and gas exploration in the Georges Bank region. With the new boundary, what previously would have been the northeast portion of Lease Offering No. 82 now falls under the jurisdiction and management of the Canadian Government (see Fig. 7.6).

Limiting Factors

Beyond the unknown resource quantities, there are two factors which may slow or prevent the development of oil and gas resources in the ACSAR area. One is technological, the other environmental. The technological limitation stems from the capability, or lack of capability, of the oil and gas industry to exploit resources in deep water. Currently, the industry can drill exploratory wells in waters up to 2400 meters. They expect to be able to extend this to 3000 meters within 5 years. But the limiting factor is the technology required for actual production in deep water. The deepest production facility in the world is only 900 meters; the industry hopes to be able to produce at 1800 meters within 10 years. The technological production problems are compounded by the transportation problems associated with deep-water oil and gas. If a major find were to occur, it is usually proposed that a pipeline be used to move the resource ashore. Yet current pipeline technology has only been applied in water depths to 600 meters. (BLM, 1981, p. 559). Thus, it will be some time before the industry has the facilities for developing oil and gas resources in much of the ACSAR area.

The environmental consideration relates largely to commercial fisheries. Specifically, a great number of political and juridical conflicts have developed over the issue of whether or not the production of OCS hydrocarbons will harm commercial fisheries. A major component of this controversy is the impact oil spills can have on fisheries stocks. Only recently has research begun into the probable impacts of oil spills on specific stocks within the ACSAR area (University of Rhode Island and Applied Science Associates, 1982, p. 8-9). The most recent work indicates that the impact of spills on specific stocks is related to the portion of the spawning cycle during which the spill occurs (University of Rhode Island and Applied Science Associates, 1982, p. 210-214).

OCS Mineral Development

Aside from oil and gas, the most publicized hard mineral resources with economic potential on the Atlantic continental margin are sand and gravel, phosphorite, and Blake Plateau manganese nodules. Other minerals are known to exist, such as placer gold and platinum, but their distribution, depth, and grade are for the most part poorly defined (Manheim and Hess, 1981). There is also some speculation that large deposits of metals such as copper and nickel may underlie the continental shelf. Depending on economic, political, and technological factors, these resources may be mined in the future (National Academy of Science, 1975), particularly as land-based supplies are exhausted or become non-exploitable because of environmental considerations (Manheim, 1979; Trondsen and Mead, 1977). A further factor which may induce mining for minerals in deeper waters of the continental margin (i.e., 200+ meters) is that many people would like to see decreased U.S. dependence on foreign mineral sources (Manheim and Hess, 1981).

Legal Structure Surrounding OCS Mining

Mining of Atlantic continental margin hard mineral resources would be covered by the same regulatory scheme, mentioned earlier, as applies to OCS oil and gas. The primary legislation is the Outer Continental Shelf Lands Act of 1953 as amended. Although national legislation has been adopted for the mining of manganese nodules (Deep Sea Bed Hard Minerals Resources Act), this would not pertain to any nodules which might be mined from the Blake Plateau; they would still be regulated by the OCSLA (Manheim and Hess, 1981; Anonymous, 1983).

As with oil and gas, hard minerals of the Atlantic coastal margin must be exploited according to the provisions of many other pieces of domestic legislation. The National Environmental Policy Act and the Coastal Zone Management Act of 1972 (as amended) are particularly relevant. The former because of its environmental impact statement requirement; the latter because of its requirement (as interpreted in recent court cases — see above) that any activity conducted under a federal license or permit must be consistent with the coastal zone management programs of affected states. Additionally, the techniques which would be used for ocean mining (National Academy of Sciences (NAS), 1975; Cruikshank, 1975), particularly dredging, would require permitting by the Army Corps of Engineers under provisions of the Rivers and Harbors Act of 1899, the Federal Water Pollution Control Act (amendments of 1972) and the Marine Protection, Research and Sanctuaries Act of 1972 (Pearce, 1979; Mazmanian and Nienaber, 1979).

While ocean minerals are to be mined under current provisions of the OCSLA, there is growing opinion that this must be changed. Either the OCSLA must be amended or new legislation must be passed in order to address the special problems of marine hard minerals as compared with OCS oil and gas (Anonymous, 1983).

Manganese Nodules and Pavements

In the ACSAR study area, the only manganese deposits with the potential for economic development are found on the Blake Plateau (Fig. 7.7). This resource appears in two forms: (1) pavement (layers of manganese-bearing deposits from 2 to 6+ cm thick) predominantly in water depths between 500-600 m); and (2) nodules found primarily in waters from about 675 to 1050 meters deep.

In the 14,000 km² area where manganese pavements and nodules have been identified, preliminary estimates indicate between 176 million and 528 million tons of pavement (based on high and low estimates of pavement thickness) and between 10 million and 100 million tons of dry nodules (based on high and low estimates of the fractional area of the bottom covered by nodules; Manheim et al., 1982). Although not as high in concentrations of economically valuable minerals as the prime nodules of the Clarion-Clipperton fracture zone in the Pacific, the Blake Plateau nodules contain higher levels of platinum than nodules from other parts of the world's oceans. Nonetheless, the Blake Plateau nodules were considered submarginal until 1978. Since that time, the ocean mining industry has indicated some interest in obtaining leases in this area (Manheim et al., 1982). One reason is that the manganese nodules of the Blake Plateau occur within 200 miles of the U.S., and thus are not shrouded in the jurisdictional uncertainty associated with international waters. While it has been uncertain what the conclusion of the Law of the Sea negotiation might mean for exploitation of deep-water nodules, there has been global unanimity on permitting coastal nations to control and regulate exploitation of resources within 200 miles of their shores, e.g., Blake Plateau nodules by the U.S. (Charles River Associates, Inc., 1979, p. 1-2).

An assessment of an hypothetical manganese nodule mining operation on the Blake Plateau was conducted by Charles River Associates, Inc. (1979). Based on their analysis, it would seem that the interest in Blake Plateau nodules is unwarranted in purely economic terms. Using different sets of assumptions for various development scenarios, they came to the conclusion that, "...a Blake Plateau manganese nodule project is likely to be a marginal investment, under the best of circumstances. In all other circumstances it is likely to be a submarginal investment (Charles River Associates, Inc., 1979, p. 5-1).

Phosphorites and Calcium Carbonate

Phosphorite

Phosphorite has important agricultural applications as a source of phosphate for fertilizers. Although land-based sources of phosphorite have not been exhausted, environmental-related conflicts have increased interest in marine deposits (Trondsen and Mead, 1977; National Academy of Sciences, 1975; Manheim and Hess, 1981). The only identified deposit of phosphorite with economic potential in the ACSAR area is found on the Blake Plateau (see Fig. 7.7). Present estimates indicate 2 billion metric tons of phosphorite nodules in the area (Manheim et al., 1980). The dollar value of these resources is not now estimated.

No current mining operations for phosphorite are reported in the ACSAR area, but such activities or proposals have been reported in southern California, Mexico, and New Zealand (Trondsen and Mead, 1977; Manheim and Hess, 1981). General studies on marine phosphorites or studies of problems associated with the resource in other geographical areas can be found in Mero, (1965), Manderson (1972), Elkins and Spangler (1968), Bowen (1972), and Sorenson and Mead (1969).

Calcium Carbonate

Pure deposits of calcium carbonate (95% $CaCO_3$) occur in shallow waters off southern Florida and on the Blake Plateau (see Fig. 7.7). At present, this resource is not mined in the ACSAR waters; however the Marcona Corporation is mining this resource in very shallow waters (approximately 3 meters) off the Bahama Islands (Manheim, 1979; Anonymous, 1978).

Sand and Gravel

Sand and gravel are a marine hard mineral resource which is now being exploited on the continental shelf. Although national land-based sand and gravel resources are considerable, there tend to be shortages in the major eastern metropolitan areas. As a result, there will be a continued interest in marine sand and gravel resources of the Atlantic margin (National Academy of Sciences, 1975). However, no exploitation of this resource is now occurring in waters deeper than 200 meters on the Atlantic continental margin (Pearce, 1979; Manheim, 1979; Manheim and Hess, 1981; Cruikshank and Hess, 1975; Schlee and Sanko, 1975) and very little future activity is likely to occur, given the distance from land and the vast sand and gravel deposits accessible on the adjacent shelf. Manheim (1979) identifies only one deposit in waters of the ACSAR area, located offshore from the South Carolina-Georgia border.

Placer Deposits

Placer deposits are minerals concentrated as a result of river transport processes. Therefore placers on the ocean bed largely are limited to portions of the continental margin exposed during former low stands of sea level, or a maximum

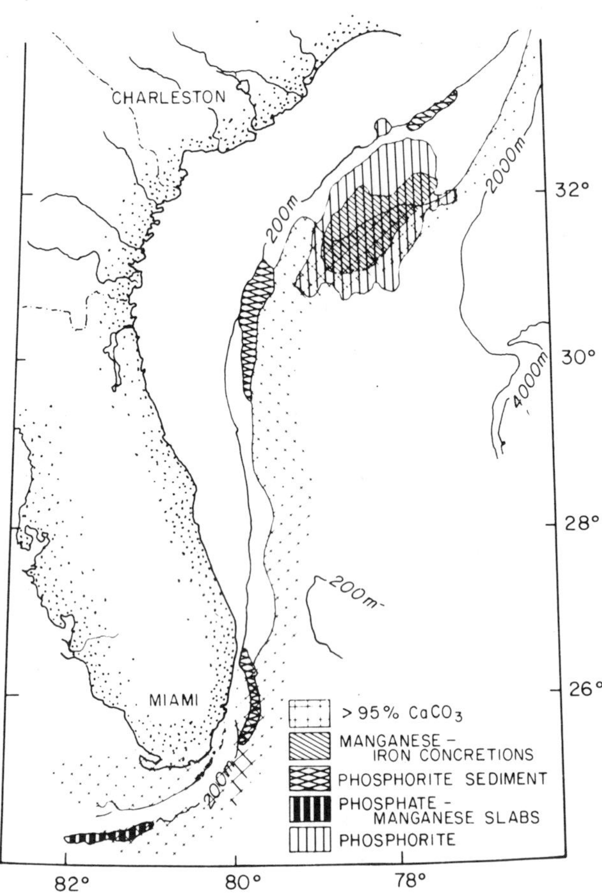

Figure 7.7 Phosphorite and manganese (a) and carbonate (b) deposits off the southeastern U.S. After Manheim et al. (1980) and Manheim (1979).

of 130 m below its present level (Emery and Uchupi, 1972). This suggests placers should not be found beyond 200 m depth and in fact, identified deposits on the Atlantic continental shelf are from depths of 20 m to 80-140 m (Milliman, 1972). Very little is known about the economic potential of the nearshore placer resources identified to date (Manheim and Hess, 1981), but preliminary findings indicate that economic deposits of placer minerals may occur off Virginia (Grosz and Ecowitz, 1983).

Ocean Dumping

The literature on ocean dumping has grown rapidly in the past few years and will continue to do so at least for the next few years. Of special relevance here are the proceedings of the International Ocean Dumping Symposiums (IODS) dealing with industrial wastes (Ketchum et al., 1981), industrial and sewage wastes (Duedall et al., 1983), dredged materials (Kester et al., 1983), radioactive wastes (Park et al., 1983) energy wastes (Duedall et al., in press), deep-sea waste disposal (Kester et al., in press) and nearshore ocean dumping (Ketchum et al., in press). A recent special IODS symposium at the University of Rhode Island was held to develop further comprehensive and coordinated strategies for ocean disposal, and to address the issue of safe disposal of wastes in the ocean. Studies on dumping in the New York Bight area, which borders the ACSAR, are collected in a recent volume edited by Mayer (1982). Champ and Park (1982) prepared a comprehensive bibliography on worldwide literature pertaining to ocean dumping. Implications of the concept of assimilative capacity to ocean dumping are discussed in the proceedings of a workshop held at Crystal Mountain, Washington (NOAA, 1979).

Waters of the U.S. east coast are dotted with marked disposal sites (active, inactive, and proposed) for industrial chemicals and other wastes, explosives, dredged materials, sewage sludge, construction debris and radioactive wastes (Fig. 7.8; NOAA/CEQ, 1980; OCS oil and gas lease sale EISs prepared by U.S. Department of the Interior). Many of these sites are located in the ACSAR area (Fig. 7.8).

Before 1972 there was no uniform national regulation of ocean disposal, and information on volumes and types of materials dumped before that time are sparse (U.S. Department of Commerce, 1979). Park and O'Connor (1981) indicate ocean dumping activity in the U.S. increased rapidly following World War II. Since regulation of ocean dumping under the Federal Water Pollution Control Act Amendment of 1972 ("Clean Water Act") and the Marine Protection, Research, and Sanctuaries Act (PL 92-532; "Ocean Dumping Act"), many valuable data have accumulated on the volume, characteristics and impacts of waste dumping. These acts established a national policy of strictly regulating ocean dumping by banning the dumping of chemical, biological, or radiological warfare agents and high-level radioactive wastes, and by authorizing a permit system for all other ocean dumping. Since 1972, EPA has approved ocean dump sites, and publishes announcements in the Federal Register (U.S. Environmental Protection Agency, 1977). In 1977 about 87% by volume of all ocean dumping of waste materials, other than dredge spoils, took place at dump sites located in the New York Bight off the coasts of New York and New Jersey, inshore of the ACSAR area (Anderson and Dewling, in Ketchum et al., 1981; Fig. 7.9).

Deep-water Dumpsite 106 (DWD 106)

Perhaps the best studied deep-water disposal site is located 106 nautical miles from the entrance to New York Harbor, in the ACSAR area. Water depth ranges from 1,800 m to 2,700 m over the approximately 1,700 km² area of this site. DWD 106 has received industrial and municipal wastes since 1972 and is under investigation to determine the likely upper limit of its capacity to receive wastes without undesirable impacts. Some of these studies address generic issues regarding the suitability of deep ocean sites to receive the residues of human activity and industry. The following synopsis is largely from Csanady et al. (NOAA, 1979).

Deep-water dumpsite 106 is located in the slope water gyre between the Gulf Stream and continental shelf systems, and thus is exposed to episodic passage of warm-core

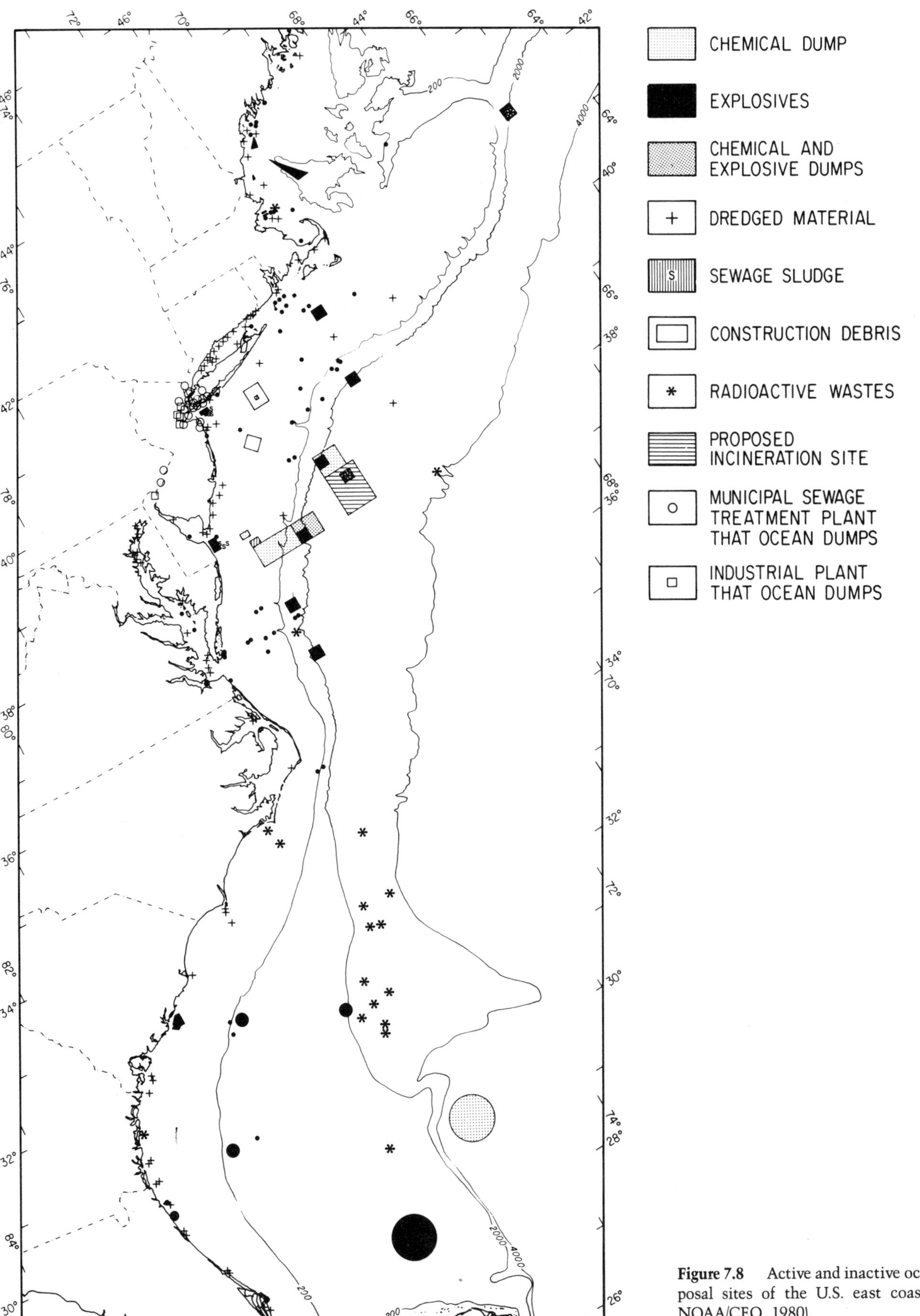

Figure 7.8 Active and inactive ocean disposal sites of the U.S. east coast (from NOAA/CEQ, 1980).

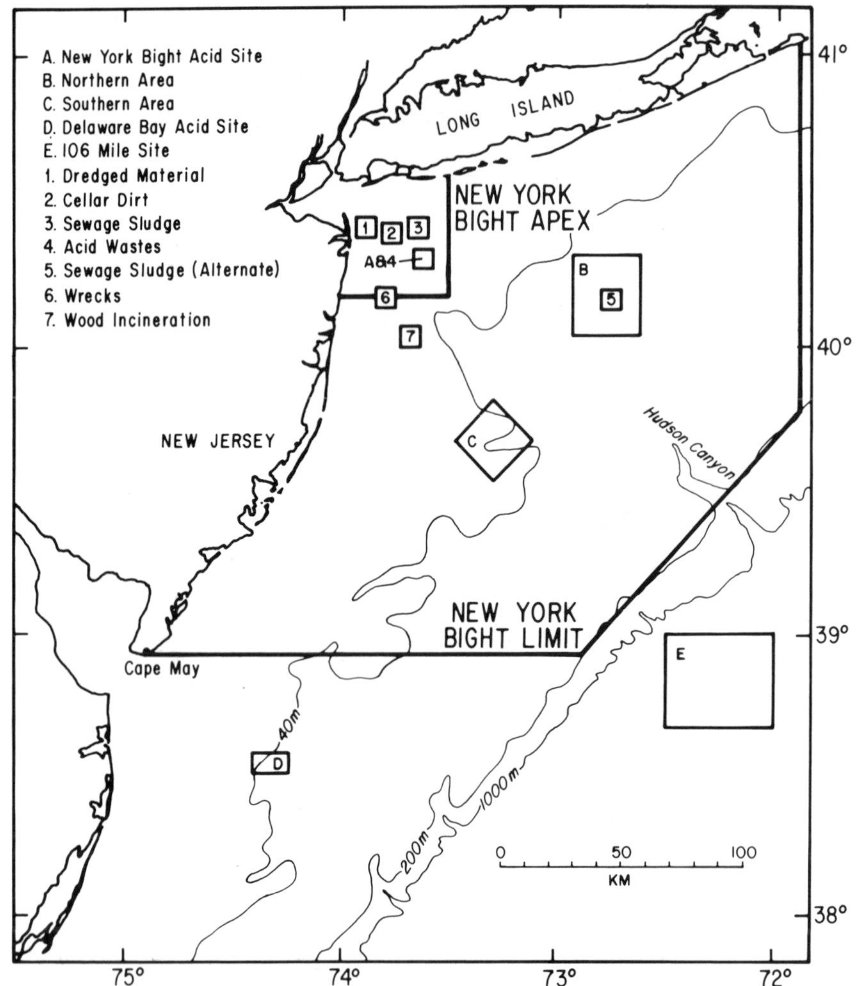

Figure 7.9 Active and inactive ocean disposal sites of the New York Bight area (modified from NOAA, 1979).

rings spun off from meanders of the Gulf Stream. These transient features contain areas of high current speed and are partly responsible for the highly dispersive nature of this site. Current velocities are typically 10 cm s^{-1} in the top 200 m of the water column and less at greater depths. The current associated with the slope gyre is about 100 km wide. Under the influence of the North Atlantic Deep Water (NADW), westward flow predominates on the average; it has been estimated that water from DWD 106 takes about 2 months to reach the Cape Hatteras area.

In 1978, 800,000 m^3 of industrial wastes were dumped at DWD 106. Most of this came from three plants. (1) The Dupont waste derived from titanium dioxide production is a strongly acid solution of iron chloride, containing chromium, vanadium, copper, zinc, nickel, lead, and traces of cadmium. At the dump site the acid is quickly neutralized by seawater and iron precipitates to form a flocculent particulate phase. The formation of these particles has important consequences regarding scavenging of heavy metals, availability of the waste materials to zooplankton and, potentially, the transport of wastes through the thermocline to the deep ocean. (2) The DuPont-Grasselli waste is an highly alkaline solution of sodium sulfate, containing some trace metals and organic compounds such as methyl sulfate and phenol. Upon contact with seawater, this alkaline material precipitates magnesium as magnesium hydroxide. Though expected to redissolve in normal seawater, this precipitate has been observed to persist. Strong mixing enhances dissolving of the particles. (3) American Cyanamid in 1978 dumped about 120,000 m^3 of a complex acidic solution containing 4% organic matter, derived from production of organophosphorus pesticides and chemicals associated with rubber production and the paper industry. No significant particulate phase is associated with dumping this material.

Wastes are carried to the dumpsite aboard approximately 4,000 m^3 barges or tankers and discharged from the moving vessel at a rate prescribed by EPA. This occurs over an approximately 45-km, U-shaped course, which causes an initial dilution by a factor of about 5,000. During the first hour or two the waste is distributed in the top 10 to 30 m during summer or in the top 150 to 200 m during winter. Very little, if any, of the waste penetrates into deep water, although particulates are known to sink slowly and accumulate on density surfaces (pycnoclines). Presumably this material is dispersed by horizontal mixing and advection processes.

The possible impact of dumping practices at DWD 106 is a particularly complex topic: (1) the water column at this site is inhabited by constantly changing assemblages of organisms, as waters of shelf, slope and Sargasso Sea origin alternately occupy the location; (2) the toxic fraction of the wastes involved is not precisely defined, and the composition of wastes can vary (this, in turn, could cause variations in its physical state after dumping); (3) as in other instances where biological systems are involved, potential impact is not limited to acute toxicity or abrupt death of adults of single species, but may involve long-term impact to certain life cycle stages or subtle changes in metabolism or the ability of a species or assemblage to persist over the long term. Methods used to assay marine organisms or biosystems for these kinds of effects are still primitive, and extrapolation of laboratory results to the field remains an area where major advances are needed. These are only a few of the complexities associated with this issue; in view of the overwhelming difficulty associated with defining the biological impacts of deep-water dumping, one of the most compelling arguments for continuing to use deep sites is that other environments are likely to be biologically richer and socially and economically more valuable.

Bioassays of the American Cyanamid waste indicated a detrimental effect at concentrations known to exist at DWD 106; however, the phytoplankter used in these tests was not a dominant species in the area. Other assays involving DuPont-Grasselli wastes and using two species of zooplankton showed a slight lethal effect and decreased feeding rate.

Areas of needed research include better definition and understanding of large-scale circulation affecting deep-water dumpsites; mixing processes distributing the wastes; cross-frontal and cross-pycnocline transport mechanisms; the fate of dumped chemicals; many aspects of the biological response and impact of wastes; and physical, chemical, and biological mechanisms for concentration of waste components.

Radioactive Waste Dumping

Between 1946 and 1962 (when land disposal was introduced) the U.S. Atomic Energy Commission permitted ocean disposal of about 120,000 curies of low-level radioactive wastes, largely contained in 55-gallon steel drums, at about 30 locations. This activity was phased out before 1970 when ocean dumping of radioactive materials was banned following recommendations of the newly created Council on Environmental Quality (Hurd, 1982).

EPA studies on radioactive dumping focus on three sites which contain 95% of radioactive materials dumped by the U.S.: the Farallon Island site 50 miles off the coast of California, and two sites about 120 miles and 200 miles off the Maryland-Delaware coast, respectively (Fig. 7.8; Chapter 5). EPA concludes from these studies that "there is no evidence of harm to humans or the marine environment from past U.S. ocean dumping of radioactive wastes" and cites a General Accounting Office report reaching the same conclusion (U.S. General Accounting Office, 1982). A more complete discussion of this is found in Chapter 4 of this book).

Other OCS Uses, Activities and Impacts

Other diverse uses and activities of the OCS have not attracted study by a large audience of academic researchers and as a result they are only briefly treated in the literature. In some cases, such as shipping, this may be because there is little associated controversy amenable to academic research. In other cases, such as military operations, what literature exists may not be open to the public. Nevertheless, some of these OCS uses probably involve economic consequences well in excess of those for fishing, mining, ocean dumping or other OCS activities considered at length in this report (see Pontecorvo et al., 1980).

Shipping

Shipping lanes for international transportation from ports along the U.S. east coast cross the ACSAR area at many places (Fig. 7.10). These shipping lanes connect 15 major U.S. ports with three principal passes leading to or through the Caribbean — the Straits of Florida, Mona Pass, and Anegada Pass — and four major world commodity (mainly oil) exchange regions — Northern Europe, the Mediterranean/North Africa, West Africa, and the Persian Gulf (via Cape Horn). Domestic shipping lanes most strongly affect the ACSAR area south of Cape Hatteras, where traffic lanes converge toward the Straits of Florida.

The smaller tankers along the east coast (i.e., 6,000 to 50,000 deadweight tons) are most involved in domestic traffic. Tankers of this size range also participate in international petroleum shipping from the Caribbean. On the other hand, large tankers (greater than 50,000 deadweight tons) serve only Portland (up to 100,000 tons), New York, and Philadelphia and involve cargos mainly from the Persian Gulf, North Africa, West Africa and (to a limited extent) the Caribbean.

NOAA/CEQ (1980) charts the flow of total commodities into east coast ports. As for petroleum, the ports of New York and Philadelphia lead the east coast in commodity shipping. The ranks of other ports depart widely; for example New Haven ranked third in petroleum imports but 13th for total commodities. The major ports for fish landings (see Fig. 7.2) differ even more widely. The major east coast fishing ports ranked by dollar landings or by quantity landed — New Bedford, Mass.; Gloucester, Mass.; Hampton Roads area, Va.; Rockland, Me.; Cape May-Wildwood, N.J.; etc. — are not included in the top 15 ports for oil or commodities.

Conflicts of shipping with other OCS uses could include collision danger for surface objects, whether moored, fixed to the bottom or dynamically positioned. Pollution resulting from vessel discharges at sea and from shipwrecks also poses potential problems. The U.S. Coast Guard, Office of Marine Environment and Systems, Pollution Response Branch operates a computerized Pollution Incident Reporting Service (PIRS) which logs and classifies oil spills. This involves rigorous documentation of spills within about 10 miles of land, with less complete coverage offshore where reporting may be spotty. NOAA is analysing operational discharges of ships (Basta, personal communication),

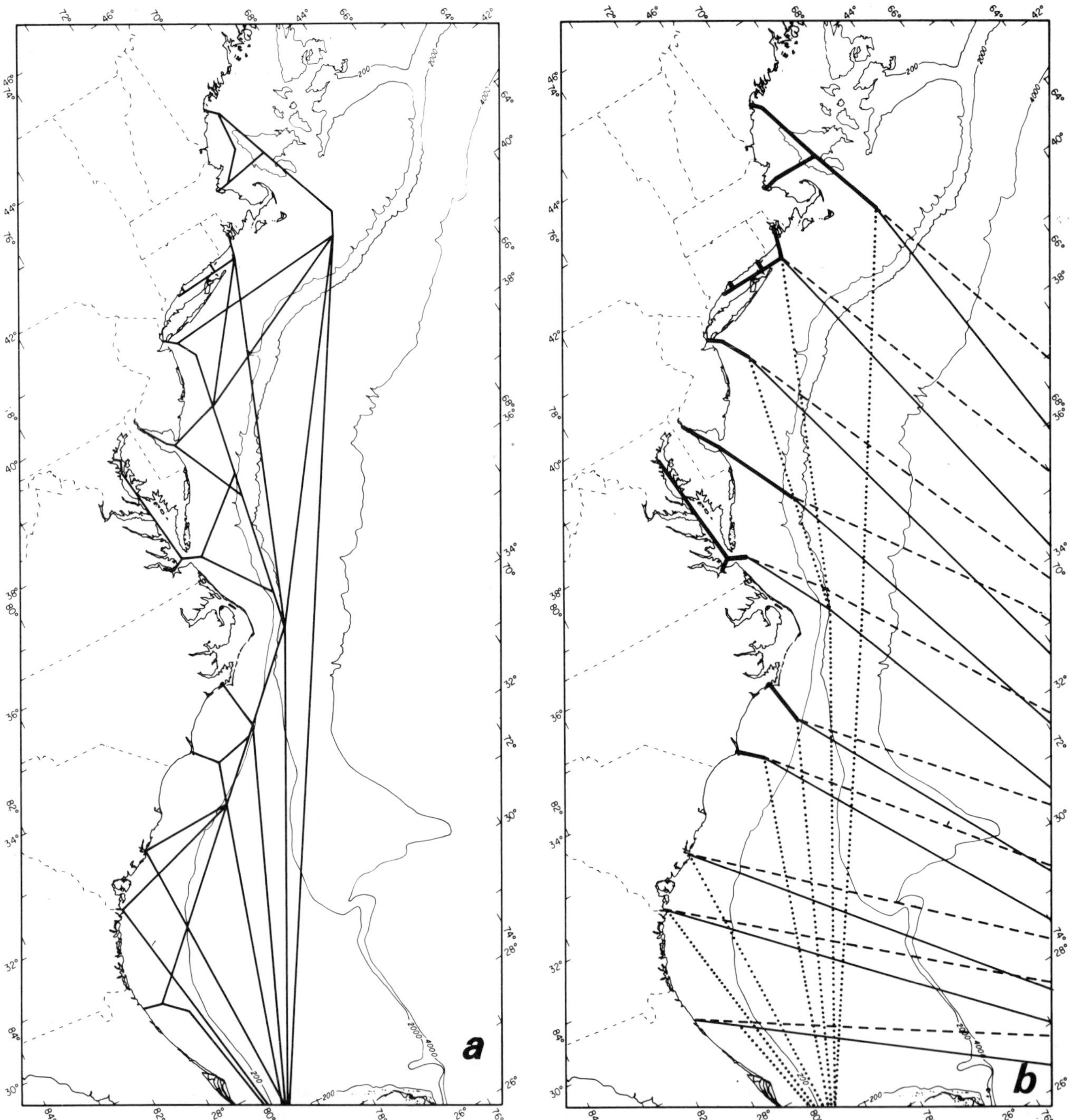

Figure 7.10 Major domestic (a), U.S.-Caribbean (b) and transatlantic (c) shipping lanes for the U.S. east coast. From D. Basta, personal communication.

which in combination with traffic patterns should provide spatial information on oil inputs to the ACSAR area originating from this source.

The best known tanker spill associated with the ACSAR area, the *Argo Merchant* spill, did not actually occur within ACSAR. Nevertheless, the slick was driven by prevailing winds and currents on a trajectory which probably took it into this area. During World War II, approximately 485,000 metric tons of oil was spilled within 50 miles of the U.S. Atlantic coast as a result of submarine attacks on shipping . Much of this oil, equivalent to 20 times the *Argo Merchant* spill, directly or indirectly became entrained into the Gulf Stream flow in the ACSAR area (Campbell et al., 1977). Most of these sinkings occurred during the first 6 months of 1942, with oil entering the marine environ-

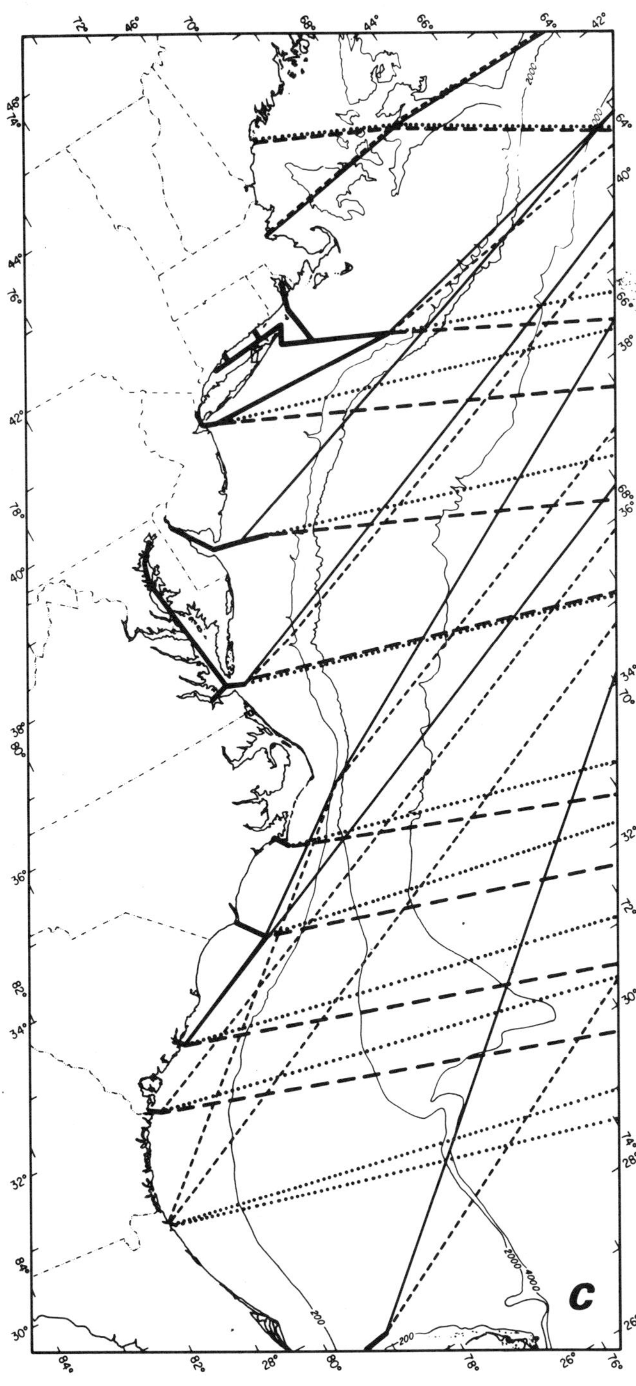

ment at a rate of almost one *Argo Merchant* cargo per week. Nothing is known regarding the impact of this oil on the environment.

A recent, unusual source of conflict between shipping and offshore oil exploitation arose when the State of New York Coastal Zone Management found Lease Sale No. 52 inconsistent with their coastal policies because of conflict with tanker traffic lanes.

Deep-water Terminals

Deep-water ports and offshore shipping terminals have been proposed mainly in connection with handling supertankers (Bragaw *et al.*, 1975). According to Ross (1978, p. 90) there are four reasons that deep-water terminals will be built: (1) the U.S. importation of petroleum will remain high or increase in future years; (2) supertankers have a clear economic advantage for moving petroleum; (3) U.S. ports are not currently suitable for offloading supertankers at berth and to dredge them would incur vast spoil disposal problems, not to mention the expense itself; and (4) the principal alternative to constructing offshore terminals would be to offload crude oil in neighboring countries, such as the Bahamas or Canada, and transship it to the U.S. in smaller tankers. However, following some well publicized accidents involving sinkings of supertankers, as well as the reopening of the Suez canal, there is some indication that supertankers may no longer be regarded as preferable to smaller tankers.

The most common types of offshore terminals proposed include platforms or structures rigidly fixed to the bottom (conventional piers, sea islands, sea island piers) and anchored mooring systems (multiple buoy berths "MBB" and single-point mooring systems "SPM"). Over 100 SPMs have been installed around the world and they are now accepted within the oil industry. The American Bureau of Shipping (1975) issued rules for building and classing SPMs.

Bragaw *et al.* (1975) indicate a likely location for a deep-water terminal would be near the Delaware estuary, where about 90% of U.S. east coast petroleum refinery capacity is sited. However, it seems unlikely that such a terminal in this area would be located in greater than 200 m depth in the OCS. It is also unlikely that present large or super-tanker traffic patterns would be strongly changed in the ACSAR area. The port of Philadelphia already receives about 75% of large tanker traffic.

Cables and Pipelines

Nine telecommunications cables, all constructed by Bell System, cross the ACSAR (Fig 7.11). The site and U.S. landfall for a tenth cable have been proposed. An additional trans-Atlantic telecommunications cable, TAT-8, has been proposed but neither the U.S. nor European terminals have been selected (AT&T Longlines, personal communication). Warnings to fishermen on NOS chart 13003 suggest the cables lie on the bottom or are not deeply buried.

There are no existing pipelines in this ACSAR area. Their most likely application would be in association with offshore petroleum production, for which pipelines are often assumed to be preferable to tankers. The technology of offshore hydrocarbon pipelines (excluding cryogenic substances) appears to be fairly well advanced, at least for continental shelf waters. The American Petroleum Institute (1976) and the Institute of Petroleum (1972) issued guidelines for their design, construction, operation and maintenance. One indication that pipelines in deep water are not yet feasible or not yet economical is that Norwegian oil produced in the North Sea is piped to Britain rather than across the Norwegian Trench (300 to 400 m deep) to Norway. A

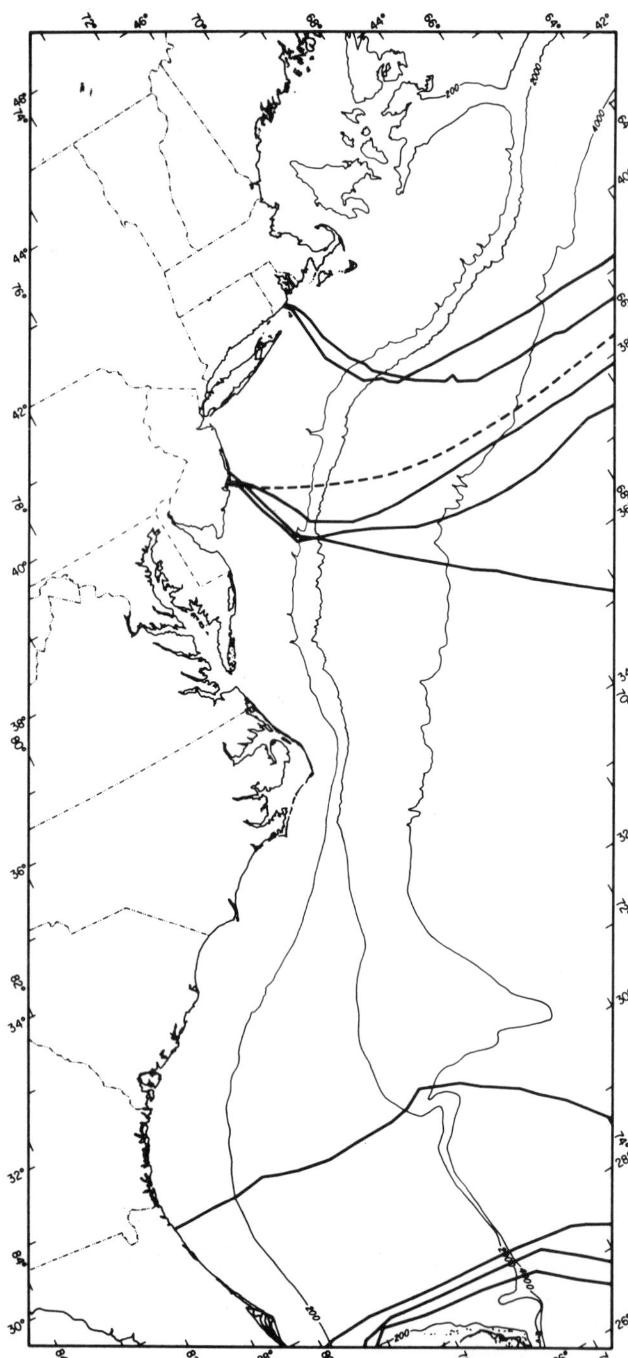

Figure 7.11 The location of telecommunication cables off the U.S. east coast. Dashed line indicates proposed cable (from AT&T Longlines, personal communication).

bibliography by Bowie and Wiegel (1977) compiles literature describing the design, construction, operation, and maintenance of pipelines up to 4 m in diameter in the ocean and rivers. A geographical index to marine pipeline locations is also included. A bibliography on offshore petroleum engineering by Chryssostomidis (1978) lists 51 general and specialized references on the topic of pipelines.

Restricted Zones and Obstructions

Restricted zones in the ACSAR include a rocket impact area near Cape Kennedy and bombing areas off the coast of Georgia (Fig 7.12). Surface uses are not obstructed by reefs or shoals in the ACSAR area but seafloor uses may find conflict with submerged objects of human origin. Ocean dump sites have already been discussed (see Fig. 7.8 and 7.9) and need to be taken into account in other future uses of the bottom. Shipwrecks are another potential obstruction to bottom activities in the OCS. Although their positions are less well known than for shallow waters, shipwrecks in the OCS have been plotted on the detailed maps accompanying the oil and gas lease sale environment impact statements. The Automated Wreck and Obstruction Information System (AWOIS), being developed by NOAA, is a computer-based file on shipwreck descriptions. The file can be searched by geographic coordinates. At present the coverage is best in shallow waters, but eventually AWOIS should be useful for the ACSAR as well.

Non-point Source Environmental Impacts and Consequences

A discussion of human impacts on the environment is not complete without mentioning non-point source pollutants. These are largely substances that are dispersed by atmospheric circulation, such as radioactivity from nuclear weapons testing and the nuclear industry, or chlorinated hydrocarbons and other organochlorine compounds, such as DDT and PCBs, manufactured for use as pesticides and in electrical components, respectively. In terms of radioactivity, surface waters of the Atlantic are known to contain measurable quantities of anthropogenic radionuclides such as tritium, cesium-137, strontium-90 and carbon-14. Except for carbon-14, these isotopes were dispersed in the atmosphere and deposited on the world's ocean surface largely by rain between 1954 and 1964; since then they have been mixed downward by natural processes and now occur to depths of at least 700 m in temperate areas of the North Atlantic (Broecker, 1974). Carbon-14 is taken up in the gaseous phase ($^{14}CO_2$) at the ocean surface in north temperate latitudes (Broecker, 1974).

Relatively few measurements are available of DDT, PCBs and other organochlorine chemicals for the ACSAR area, but available data indicate at least trace quantities of these substances even in deep water. A similar generalization can be made regarding some of the combustion products of petroleum hydrocarbons. A discussion of the few data can be found in Chapter 5.

Marine Mammals, Endangered Species and Special Habitats

Marine Mammals and Endangered Species

As a result of federal legislation, the effects of human activity on endangered species and marine mammals must be considered before development can go forward. The Endan-

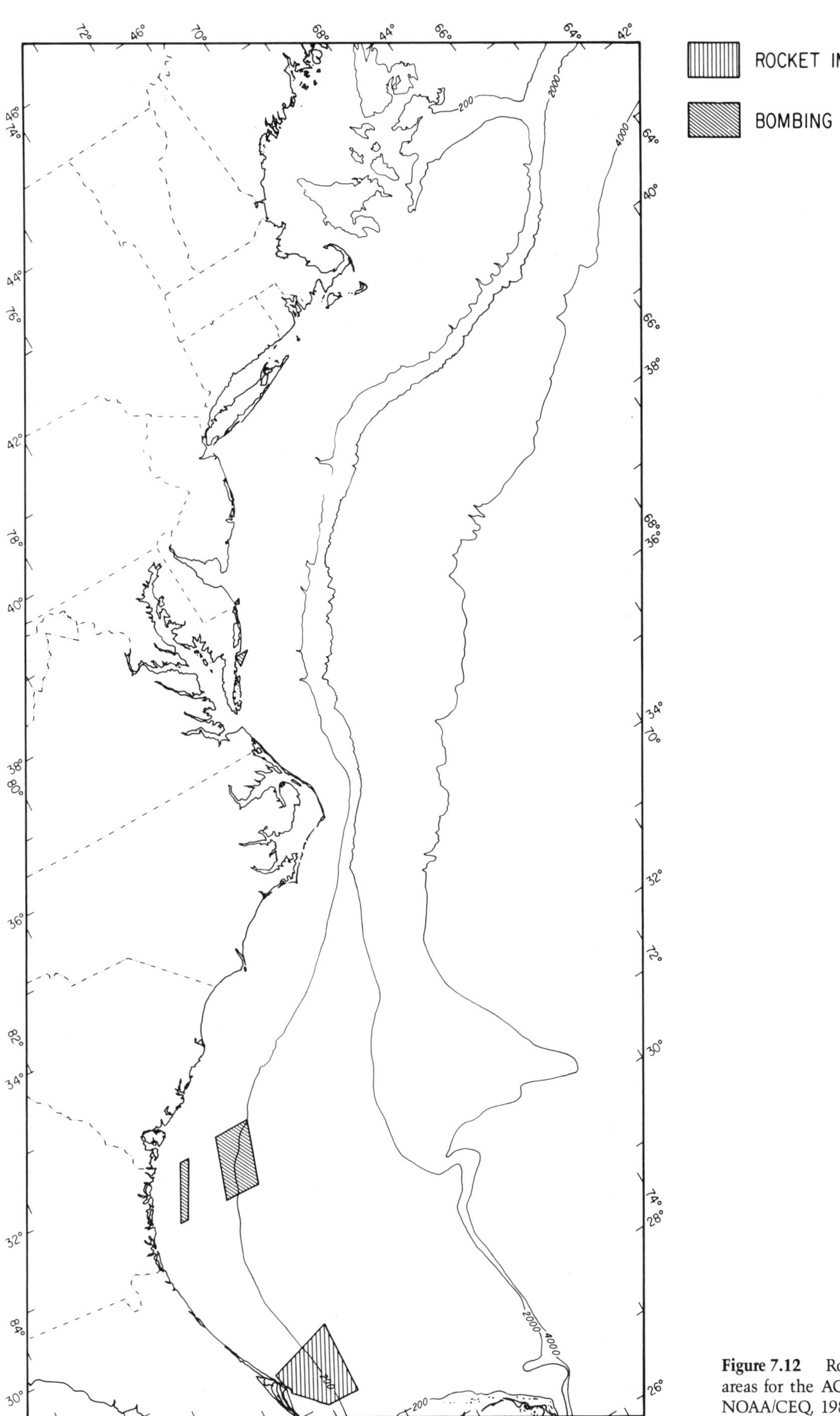

Figure 7.12 Rocket impact and bombing areas for the ACSAR area (modified from NOAA/CEQ, 1980).

gered Species Act of 1973 (16 U.S.C. 1531-1543) has as its stated purpose "to provide a means whereby the ecosystems upon which endangered species and threatened species depend may be conserved" (Sec.2.b.). It further notes all federal departments and agencies shall seek to conserve threatened and endangered species (Sec.2.c.). The Marine Mammals Protection Act of 1972 (16 U.S.C. 1361-1407) has as its specific purpose to protect and encourage the development of marine mammal stocks, within the parameters of sound resource management, so as to maintain the health and stability of the marine ecosystem (Sec.2.(6)).

For the study area, endangered species encompass marine mammals. Further, the category of marine mammals can be reduced to cetacean species, i.e., whales and dolphins, as no other mammals have been identified seaward of the 200-m isobath (NOAA/CEQ, 1980). The occurrences and distribution of mammals have been discussed in Chapter 6, based largely on the studies of CETAP (1981, 1982). Unfortunately, the CETAP work only covers the Mid-Atlantic and North Atlantic portions of the study area (34° to 45° North).

The EIS for Lease Sale No. 76 summarizes the cetaceans found in the western North Atlantic. Endangered species which have been sighted in 200+ meters of water include the sei, fin, humpback, right and sperm whales. It should be noted that though a specific cetacean may not have previously been sighted in the study area, this can not be taken to mean that the animal could not be present in the area (Moore, 1983).

While the baleens and odontocetes are the only endangered species which are present in significant numbers in the ACSAR area, this does not mean that they are the only ones which could be affected by human activity in the region. If a major oil spill were to occur from exploration or production activities, it is possible that that crude oil could foul the nearshore, shore, and estuarine ecosystems of the Atlantic coast. If this were to occur, marine mammals, turtles, birds, and shore plants might be affected. Many of these are on the threatened or endangered species lists (CETAP, 1981; CETAP, 1982; Bureau of Land Management, 1982; Bureau of Land Management, 1982; NOAA/CEQ, 1980).

Special Habitats

There are certain areas of the ocean bottom which, because of their unique characteristics as habitats for marine fauna, have been or may need to be considered in planning efforts for the management of the Atlantic continental margin. For the purposes here, two such habitats are considered — submarine canyons and colonies of deep-water corals.

Submarine Canyons

Recent actions by the states of New Jersey and Massachusetts in opposition to oil and gas leasing indicate the perceived importance of submarine canyons in the overall planning for the study area. In both cases, the states, using the consistency provisions of the Coastal Zone Management Act, blocked lease sales for oil and gas by the Federal government because of the inclusion of canyon tracts in the lease sales. In each instance, the states wanted the tracts deleted because of the canyons' importance for fisheries (lobster for Massachusetts; tilefish for New Jersey).

Ongoing research by Lamont-Doherty Geological Observatory indicates that the concerns of these two states may have some merit. Hecker et al. (1983) report the sightings of several species of either present or potential commercial value. Observations in Baltimore and Lydonia Canyons included sightings of lobster, shrimp (in association with coral colonies, e.g., *Paragorgia arborea*), several species of hake, flounder, tilefish and eels.

While presence of these and other commercial species has been established in the canyons and canyon heads, it would be premature to estimate the effect oil and gas development might have on these fish stocks. This area clearly needs additional research.

Deep-water Corals

Some deep-water coral species of the order *Scleractinia* (at least 14) have been identified in the study area (Cairns, 1981). Many occur as deep as 3,200 meters. Of the fourteen, thirteen are "ahermatypic", a term normally associated with non-reef-building species. Cairns (1981) reports that this is not always the case. There are some colonial deep-water ahermatypes which do create structures that are reeflike.

In addition to the Scleractinia, approximately 75 other deep-water corals have been identified in or near the ACSAR area. The highest concentrations of these corals occur between 600 and 800 m (Gulf of Mexico and South Atlantic Fishery Management Councils, 1982). Not much is known about the majority of deep-water corals; they are considered to be the most poorly understood corals in waters adjacent to the continental U.S. (Gulf of Mexico and South Atlantic Fishery Management Councils, 1982, p. 5-12).

In addition to their habitat value, some few deep-water corals have a direct economic value. The skeletons of these species can be cut and polished for use in jewelry manufacture. Within or bordering the ACSAR study area the following deep-water corals with economic potential have been identified: *Chrysogorgia desbonni*, *Candidella imbricata*, *Keratoisis flexibilis*, *Keratoisis ornata*, and *Distichpora foliacea*. As a result of distance from shore, deepwater corals are rather inconvenient to harvest. The use of deep-water submersibles for coral collection, which is current practice in Hawaii, could change this although submersibles have not been used for this purpose in the ACSAR area (Gulf of Mexico and South Atlantic Fishery Management Councils, 1982, p. 5-13).

Summary and Suggestions for Future Studies

8

Summary

Meteorology and Air-Sea Interactions

The central fact about the atmospheric circulation over the waters off the U.S. east coast is the importance of storms. Such short-term disturbances, with time scales of two to five days, tend to be hidden in standard climatological averages over seasons or years. However, their significance should not be overlooked when considering marine operations in the ACSAR region.

The general circulation pattern is dominated by the "Bermuda High", a high pressure system located in the subtropical gyre of the North Atlantic Ocean. Throughout most of the year this high produces southwesterly winds (that is, winds blowing towards the northeast) over the continental slope and rise. In winter the high weakens and is displaced southward by a low pressure system off Greenland, so that winds tend to be westerly or northwesterly. The storms which are superimposed on this mean picture can be considered in two major categories: severe tropical cyclones, and extratropical storms and cyclones.

1. A tropical cyclone is a warm-core storm with maximum sustained winds between 34 and 63 knots; in a hurricane the sustained winds exceed 64 knots. Although hurricanes have high wind speeds, their effects upon the offshore waters are mitigated by their small size (300-km radius) and rapid movement. About 85 percent of all tropical cyclones occur during August, September and October. Their number may range from two or three to as many as 22 per year, but many enter the Gulf of Mexico and do not affect the waters off the east coast. In one well-documented case, Hurricane Eloise mixed the upper water layers vertically to a depth of 70 m, bringing colder water into the mixed layer and causing the surface temperature to drop 2°C; the storm generated near-surface currents of the order of 2 knots, dominated by near-inertial frequencies. The effects of such storms decay rapidly after a few days, but currents can remain greater than normal for up to two weeks.
2. Extra-tropical storms occur mostly during the fall and winter but are not unusual as late as April. They have been subdivided into seven categories: large slow-moving storms which form in the Atlantic well east of ACSAR; wave developments along cold or stationary fronts over the southeast coastal states; wave developments in the Gulf of Mexico which can bring cold, dry continental air over warm ocean waters; depressions moving across the southern U.S. which intensify offshore; similar storms in which a strong secondary circulation develops, usually off Cape Hatteras and moving rapidly northeast; intense cyclones which develop over the Great Lakes and move northeast over land (these are less important offshore than along the coast); and finally, strong cold fronts accompanied by line squalls and severe local weather.

Such extra-tropical storms are much more common than tropical cyclones in the ACSAR region, about five per month in season. One 110-year study identified more than 70 such storms off southern New England, with a secondary maximum off Cape Hatteras. These storms generally increase in frequency to the north and display a local maximum near the coast at any latitude. The latter effect is apparently related to the abrupt change in surface roughness between land and water.

Precipitation over ACSAR reaches a maximum of 1.6 m yr^{-1} over the axis of the Gulf Stream east of Cape Hatteras, dropping to less than 1 m yr^{-1} near the coast. There is a slight bias toward more rainfall in winter. The area of maximum precipitation is also where oceanic heat loss to the atmosphere reaches a maximum of more than 200 watts m^{-2} yr^{-1}; that in turn is related to high evaporation which reaches 3.7 m yr^{-1} (more than twice the precipitation maximum) in the same area.

Seasonal cycles of heating/cooling and evaporation/precipitation affect the temperature, salinity and related hydrography of the upper layers of the ocean. Atmospheric forcing also generates currents in the upper ocean: there is an Ekman transport to the right of average wind direction but there are also storm-related events of short duration and large amplitude; such storms are capable of generating high-frequency currents with periods of one day and large surface gravity waves with periods around 10 seconds.

The upper ocean response to a low pressure system depends a great deal on the season. The first few storms of winter occur when sea surface temperature is still high and the mixed layer fairly shallow. The response includes mixed-layer deepening, rapid heat loss, and currents usually less than 1 knot. By the end of the winter the mixed layer has reached maximum depth and minimum temperature and the oceanic response is reduced accordingly.

Physical Oceanography

Water Masses

The character of the water overlying the Atlantic continental slope and rise changes dramatically at Cape Hatteras, where the Gulf Stream diverges from the coast. Southwest of Hatteras the Stream impinges directly upon the slope, while the Sargasso Sea overlies the Blake-Bahama Outer Ridge and associated mid-depth features. Northeast of Hatteras the Stream meanders offshore and the space between it and the shelf is occupied by Slope Water. There is continuity between the two regions only below about 2000 m, where the Western Boundary Undercurrent (WBUC) flows to the southwest along the upper rise.

The water masses of ACSAR can be subdivided vertically by temperature into three main classes: (1) the deep water, colder than 4°θ, which accounts for about two-thirds of all the water; (2) the thermocline, 4°θ to 17°C, which includes about one-quarter of the total; and (3) the warm water (warmer than 17°C) which accounts for about 8 percent. Deep and thermocline waters are found throughout the ASCAR region. The warm water layer is always present south of Cape Hatteras and in the Gulf Stream, but in the Slope Water region there is no water warmer than 17°C except in summer. Another principal difference is that because the Gulf Stream is marked by strongly sloping gradients, water of a given temperature and salinity occurs several hundred meters shallower in the Slope Water than in the Sargasso Sea.

The deep water in the ACSAR region consists almost entirely of North Atlantic Deep Water (NADW), formed by mixing of unknown proportions of five water masses which are the only sources of deep water in the North Atlantic; nevertheless it has a distinctive and tight potential temperature-salinity relationship. Traces of all five parent water masses can also be found in ACSAR: Antarctic Bottom Water (AABW), colder than 1.8°θ and fresher than 34.89‰, forms the very deepest water in the WBUC; Labrador Sea Water forms a salinity minimum in the Slope Water and even southwest of Cape Hatteras; the Mediterranean Water spreads in a broad high-salinity tongue across the ocean south of Bermuda and overlies the offshore Blake Plateau; and the two Norwegian Sea overflows combine to provide the core of the WBUC.

The thermocline region in ACSAR consists almost entirely of Western North Atlantic Water (WNAW), which also has a tight temperature-salinity correlation. However, because of the influx of Labrador Sea water, the Slope Water is slightly fresher than that in the Gulf Stream and Sargasso Sea. The warm water southwest of Cape Hatteras and in the Stream is also WNAW. Northeast of Hatteras, warm water is a seasonal phenomenon except in Gulf Stream rings.

Seasonal variability penetrates only the upper 200 m of the water column. Development of the seasonal thermocline is related to other near-surface features such as the "cold pool" which overlies the edge of the shelf and extends into deeper water, and the "warm band" which creates a special bottom habitat at the shelf edge. The shelf-slope front is a near-surface feature which fluctuates broadly in space and time, often extending many tens of kilometers beyond the 200-m curve. In deeper water the major fluctuations are related to the movement of Gulf Stream meanders and rings.

Currents

A unique feature of ACSAR is that it contains the strongest low frequency currents in the world ocean. The strongest of these is the Gulf Stream; much deeper is the Western Boundary Undercurrent. These flows and their associated eddy fields will be major sources of drag on any man-made structures in the region.

The Gulf Stream flows to the northeast along the continental slope south of Cape Hatteras. It is about 100 km wide, with the strongest currents (up to 5 knots) on the inshore side, and it extends to the bottom. North of Cape Hatteras the Stream separates from the slope and crosses over the WBUC in a manner not yet completely understood. Further downstream it again appears to extend to the bottom. The increase in depth is matched by an increase in volume transport so that the Gulf Stream off Cape Hatteras carries about twice as much as when it leaves the Florida Strait; it doubles again by 65°W, the eastern extremity of ACSAR. The WBUC flows to the southwest along the lower slope and rise in a stream about 50 km wide. It is the boundary current associated with the spread of the NADW, and it forms part of the generally westward flow found in the Slope Water.

Other strong currents in the region are associated with Gulf Stream rings which are formed by the breaking off of large meanders of the Stream in the region east of Cape Hatteras. Rings are formed both north and south of the Stream. Those north of the Stream usually contain a core of Sargasso Sea water and are known as warm-core rings (WCR). They rotate in a clockwise direction and migrate slowly westward through the Slope Water region until they either dissipate or are recaptured by the Gulf Stream. Cold-core rings (CCR) form south of the Stream, contain cold Slope Water and rotate in a counterclockwise direction; they are found in the ACSAR region only south of Cape Hatteras. Cold-core rings also generally move to the southwest and can coalesce with the Gulf Stream. Rings of both kinds may be 100-200 km in diameter with maximum surface currents up to 200 cm sec^{-1}. They extend to depths of 1000 m or more, but with diminished intensity.

Although the mean flow in the region is westward except in the Gulf Stream, fluctuations can be much stronger than the mean. The field of fluctuating flow is inhomogeneous both in depth and horizontally. This inhomogeneity has been tied to topographic Rossby wave dynamics through processes which may be related to the meandering of the Gulf Stream. Topographic waves appear to dominate motions in the deep water but become less energetic away from the bottom. In the thermocline the most energetic motions are related to WCRs.

Higher frequency currents (those with time scales shorter than about two days) in the North Atlantic seem to fit the "universal" Garrett-Munk spectrum which is related to the natural buoyancy frequency of a stratified fluid medium. It has also been established that internal wave energies are amplified on a sloping bottom when the bottom slope equals that characteristic slope of the internal wave. Such amplification has been found at the shelf break and deeper along the U.S. continental slope. Tidally-dominated currents occur in canyons; direction of net flow is not yet documented.

Geology

Bathymetry

The continental slope extends from the shelf-slope break (60 to 200 m, depending upon locality) to 2000 m; width varies from 10 to 50 km. Average gradient is 3-6°, but locally gradients can be nearly vertical. Between Georges Bank and Cape Hatteras, the slope is cut by at least 70 large canyons and numerous smaller canyons and gullies, many of which may be feeders to the large canyon systems. In contrast, there are few canyons on the Florida-Hatteras Slope, probably because it is affected by the Gulf Stream.

The continental rise, which extends from 2000 to 5000 m and deeper, is marked by a generally decreasing gradient. It is divided into the upper rise (2000-3000 m), central rise (3000-4000 m) and lower rise (greater than 4000 m). In all but a few localities, the lower rise is absent.

The Blake Plateau lies between 400 and 1200 m in water depth with fairly slight gradients (less than ½°). The seaward Blake Escarpment has an average slope of 15°, but is vertical in places. The Blake Spur (Nose) is a ENE trending protuberance with a gently-dipping top and steep sides. The Blake Outer Ridge (28-35°N) lies between 2000 and 4000 m, with gradients of 0.2° along the ridge axis and 2.5° on the sides. Finally, there are two groups of seamounts off the northeastern study area, the New England Seamount Chain, (e.g. Mytilus, Physalia, Balanus, etc.) and a smaller chain to the south (e.g. Caryn, Knauss, etc.).

Sediments

The "mud line" occurs at 250-300 m, below which silt and clay-size sediment predominate. On the Blake Plateau, however, biogenic sands (Foraminifera, pteropods) and gravels (corals, limestone) are the dominant sediments. Sand pockets also occur on the slope, the result of downslope mass movement. Local coarse sediments and rock outcrops occur in and near canyon walls.

Upper slope sediments are dominated by detrital sediments, with increasing carbonate content as water depth increases. Highest carbonate concentrations (greater than 95 percent) occur on the Blake Plateau. Detrital components tend to be less chemically and mineralogically mature in the north (feldspar, illite and chlorite) than in the south (quartz, montmorillonite, kaolinite), except for Tertiary outcrops on the Florida-Hatteras Slope (arkose/glauconite).

Accumulation rates vary from 2 to 40 cm 10^3 yr^{-1} on the slope, with slightly lower average values off the rise. Many areas of the slope have negative accumulation rates (i.e., erosion) since modern rivers contribute little sediment to the area, most terrigenous sediment comes from winnowing of outer-shelf and upper-slope sediments and from slope erosion.

Dynamic Sedimentary Processes

Undoubtedly the dominant sedimentary process in the ACSAR area is gravity controlled downslope movement. Slumps, slides, debris flows, and turbidity currents represent a continuum from thick cohesive movement to relatively thin non-viscous flow. Slumps occur as blocks of sediment, but may involve little downslope movement, while turbidity currents can be transported thousands of km. The slope between Cape Hatteras and 38°N appears to exhibit the most intense downslope movement of slumps and slides, although this is because it is the most intensively surveyed area. Debris flows may be more widely distributed, but they are not well documented.

Little is known about the erosive force or sediment transport where the Gulf Stream impinges upon the Blake Plateau and Florida-Hatteras Slope. However, there is much geologic evidence for erosion and sediment transport. Tidal currents in canyons have been shown to transport sediments; whether net transport is up- or down-canyon may depend upon the specific canyon in question.

Bottom boundary currents (involving the WBUC) may be the major bottom-forming mechanism on the continental rise, taking turbidity current deposits and transporting them to the south. The WBUC can be looked upon as a smoothing agent.

Sediment Mass Properties

Because of variable texture, composition and age, geotechnical properties of slope sediments have a predictably great variability. Rise sediments, on the other hand, are more homogeneous. Areas of underconsolidation (conducive to slumping) are of particular interest.

Structure

The basement lies 8 to 10 km beneath sea level on the continental rise. The continent-ocean transition in crust occurs at the inferred "basement ridge" beneath the continental slope and coincident with the East Coast Magnetic Anomaly; the ridge may be overlain by reefal limestones. The age of the crust at the Jurassic Magnetic Quiet Zone is 145-161 my, but initial separation of North America from Europe/North Africa may have begun 175 mya.

Sediment on the slope has a maximum thickness of 9 km, about 40 percent of which is Cenozoic in age. Primary structures are dominated by unconformities, which often are difficult to distinguish based on seismic profile records. Bedding is mostly conformable, progradational beneath the shelf but often erosional on the slope. Deeply buried reefs underlie the slope, although they outcrop at and near the Blake Escarpment.

There are local occurrences of faulting, but only one major surface fault that has subsurface expression — the outer edge of the northernmost Blake Plateau. Numerous erosional unconformities (via mass wasting and currents) occur, causing many of the seismic reflectors seen on records. Diapirs occur beneath the slope in both the north and south.

Based on many seismic lines plus a few available deep seismic multichannel lines and available borehole legs, a general stratigraphic synthesis can be offered. During the Jurassic, shallow water sediments were deposited at initial rifting. Shelf-edge reefs and carbonates tended to trap terrigenous sediments landward, although protruding fans suggest some sediments broke through to the deep sea. Black shales in the mid-Cretaceous were due to a shallower carbonate compensation depth and low-0_2 conditions. Shelf edge reefs off Georges Bank were overrun by terrigenous sediments due to climatic cooling and increased influx of land-derived sediment; this explains the development of offshore deep-sea fans. Chert characterizes the early to mid Eocene, while the advent of strong abyssal currents resulted in intense erosion of the slope and rise beginning in the Oligocene. Quaternary sedimentation has been characterized by numerous climate-controlled changes in sea level, the most recent of which lowered sea level by up to 120 m off New England. Presumably such oscillating sea levels have resulted in episodic influxes of sediment followed by periods with little or no influx of land-derived sediment.

Seismicity in the ACSAR area is generally poorly known. There is possible correlation with landward extensions of fracture zones and with landward zones of structural weakness, but further data are needed to prove this connection.

Chemistry

The chemical regime in the water column and bottom sediments is, in general, only partly understood in the ACSAR area. While regional concentrations are adequately documented for most parameters, fluxes and processes are not.

Water column distributions of dissolved nutrients depend upon both water mass and water depth. Surface concentrations are relatively depleted during warm weather months, and elevated in colder periods. Maxima for PO_4^{-3} and NO_3^- occur at about 300 m in slope waters and 1000 m in deeper waters; below these depths, concentrations remain more or less constant. In sediments in the ACSAR area, NO_3^- values indicate increased reducing conditions with decreasing water column depth. Estimated fluxes, however, have not been made.

Trace metal concentrations in seawater from the ACSAR area are given in various tables in this report, but many published figures are subject to analytical error, particularly through contamination, especially Fe, Cd, Cu, Pb, Hg, and Ni. Sediment trap data indicate that most particulate matter flux (and therefore most included trace metals) occurs in the finer-than-36 mm size fraction. Dissolved Fe and Mn also show increased reducing environments in shallower waters, whereas trace metals in particulate form show a relation with size.

Artificial radionuclides have been introduced to the area by fallout from nuclear weapons tests (largest), deep-water dumping, and accidents (smallest). Input from atmospheric testing, however, has been very small since 1962; the greatest impact was from New York to Washington, with smaller amounts to the north and south. Of the soluble nuclides, ^{137}Cs, ^{90}Sr, ^{3}H, and ^{14}C are considered water tracers, and as such have been studied extensively. Reactive nuclides (fission products and transuranic products) have settled rapidly to the bottom, where they have been mixed with marine sediments by physical and biological mixing. Studies at the *Thresher* crash site and dump sites off Delaware-Maryland show little apparent contamination, but this may be because of restricted data or the lack of adequate sampling. In contrast, the concentrations and distributions of naturally occurring radionuclides appear to be controlled by steady state processes which are at or trending towards equilibrium.

Research concerned with hydrocarbons has been concentrated on the shelf and nearshore areas, with the exception of pelagic tar surveys. The source of the hydrocarbons (biological, geochemical, anthropogenic) varies with the compound under consideration. The gases (C_1-C_4), for example, have significant inputs from biological and geochemical sources — methane being the best example. There are no published data for volatile compounds (C_5-C_{10}) nor heavier molecular weight compounds in the ACSAR area. Pelagic tar comes almost exclusively from oil introduced by man's activities; tars include both dispersed and dissolved particles.

Biology

The distribution of plants and animals on the slope and rise off the eastern United States is as complicated as the hydrography of the region. The Gulf Stream, which separates the Sargasso Sea and Slope Water, forms a sharp biogeographic boundary between the North Atlantic Temperate and North Atlantic Subtropical regions. The northern edge of the Gulf Stream is the southern limit for certain cold water organisms and the northern limit for other warm water ones. The current also serves to transport tropical Caribbean biota northward. These tropical biota are rapidly diluted and dispersed as the Stream entrains additional water with its burden of subtropical species; nevertheless, individual organisms of Caribbean origin occur well north of Cape Hatteras.

The picture is complicated further by the existence of Gulf Stream rings. Cold-core rings transport temperate pelagic organisms from the Slope Water into a subtropical environment, while warm-core rings have the opposite effect; both kinds of rings remove tropical plants and animals from the Gulf Stream. Subtropical biota introduced into the Slope Water by warm-core rings apparently survive better than temperate biota carried by cold-core rings, possibly because the former are sufficiently pervasive in the Slope Water.

Microbiology

Until recently it was believed that bacteria were not present in the oceanic water column in sufficient numbers to constitute a significant fraction of the biomass or to contribute significantly to respiratory processes. Newer work suggests that bacteria are both abundant and important. For example, one study indicates that they regulate the dissolved carbohydrate concentration of the Sargasso Sea. "Marine snow" in the ocean may or may not be an important site for bacterial growth.

New analytical techniques confirm that metabolism is extremely slow at high pressures and low temperatures. However some barophilic bacteria grow better at high pressures, and some cannot survive decompression. Also important is the availability of nutrients: bacteria grow best in deep water, in the guts of living organisms, or in carcasses.

Marine protozoa exhibit extreme variability in both abundance and size, and can have important roles in plankton communities with respect to respiration, filtration, and ingestion. Flagellates, by far the most numerous protozoa, form highly dynamic populations, with some exhibiting significant diel migrations and some attaching to particles and forming rich microenvironments. They may be the major consumers of bacteria in the plankton, and thus serve as a mechanism for making bacterial biomass available further up the food chain.

Foraminiferan species closely follow the hydrography of the ACSAR region, with abundances generally declining from Slope Water to the Sargasso Sea. Species diversity, however, increases towards the Sargasso Sea. Radiolaria tend to have similar distribution patterns although they are outnumbered in the Slope Water and dominant in warm water. Most pelagic marine ciliates are from the order Oligotrichida, including tintinnids. However, non-tintinnids constitute a larger fraction of the planktonic biomass.

Phytoplankton

Although poorly documented, the patchy distribution of phytoplankton in the continental slope and rise waters off the eastern U.S. is largely dictated by water mass properties and season. Seasonal development of a deep chlorophyll maximum, for example, is a direct result of a stable deep thermocline from late spring to early autumn. Similarly, phytoplankton concentrations (productivity?) can change markedly across boundaries separating oceanic fronts, as mixing/stratification change. The formation of Gulf Stream rings clearly affects phytoplankton distribution and productivity, although few quantitative data are available.

Species composition changes in an offshore direction, with a decreasing importance of diatoms coincident with increasing dominance of coccolithophoids. Again, however, species composition also varies with water mass and season. Highest productivity rates are measured during late winter and early spring, lowest rates in summer and early autumn.

Zooplankton

Zooplankton distributions also are strongly affected by rings, which create a mosaic of expatriated communities interspersed with endemic communities. The pattern constantly changes due to movement of the rings and their gradual hydrographic and biotic assimilation by surrounding waters. Cold-core rings originate in the biologically richer Slope Water and carry relatively high concentrations of zooplankton into the poorer Sargasso Sea. Warm-core rings begin with lower biomass concentrations but, once surrounded by Slope Water, their biomass increases more rapidly than the corresponding decline of biomass in cold-core rings.

Delineating the distribution, abundance, and seasonal cycles of zooplankton (particularly in the upper 1000 m) is affected by problems such as patchiness, hydrographic variability and net avoidance by large organisms. Nevertheless, some species favor the Slope Water and others are more abundant in the Sargasso Sea. Gulf Stream communities tend to resemble those in the Sargasso Sea more than the Slope Water, but that partly reflects sampling location. In the Slope Water numbers of individuals are substantially higher than in the Sargasso Sea, but also more variable, obscuring what appears to be a spring maximum and fall minimum in abundance. Some of the variability in the Slope Water undoubtedly results from intrusions of richer coastal waters.

Penetration of light appears to be the most important abiotic factor influencing the vertical distribution of zooplankton. However, temperature structure also can be important; for example, a sharp shallow thermocline can inhibit the daytime downward movement of some animals. Despite strong contrasts in the different hydrographic regimes in ACSAR, salinity does not appear important in controlling zooplankton distribution.

Neuston

Neuston, which occupies the upper 10 to 20 cm of the water column, is poorly defined in the ocean in general, and the ACSAR region in particular. Most neuston undergo diel migrations, with higher concentrations in more productive coastal waters. Larvae, zooplankton and mesopelagic fishes are common neuston components, as are plastics and tar balls (particularly in and near shipping lanes). The composition of the neuston community changes markedly with season, but data are still scarce.

Mesopelagic Fish

Mesopelagic fish are generally small, large-eyed, large-mouthed carnivores which inhabit the water column from the sea surface to about 1000 m. Many exhibit pronounced diel vertical migration, from the upper 100 m at night to deeper than 500 m at day. Many have gas-filled swimbladders, and are effective sound scatterers responsible for the "deep scattering layer".

Because the ACSAR region is environmentally so variable, it has a rich mesopelagic fish fauna, consisting of hundreds of species. Probably about 80 percent belong to the Myctophidae (lantern fish) or Gonostomatidae (pearlsides). Standing crops are several times greater in the Slope Water than in the Sargasso Sea, and there is a general inverse relation between mesopelagic fish and water depth.

Cetaceans

Although there is very little published work about cetaceans in the ACSAR region, it is possible to estimate abundances of different species on the basis of observations along the continental shelf and in deeper oceanic regions. Of the large whales, the most abundant are probably sperm and fin; saddleback dolphins and bottlenose dolphins are the most common smaller cetaceans.

Benthic Fauna

Both the abundance and size of bottom dwelling organisms decrease with increasing water depth, but the average weight of individual fish increases below 500 m. There are insufficient data to describe distributions in detail, but shifts in species composition occur both at depth and with latitude. Cape Hatteras appears to be a geographic boundary for upper slope fauna. Species diversity in most groups increases down to intermediate depth but declines in deeper water; the changes may be related to levels of environmental disturbance. Whole community respiration of benthic fauna appears to decrease with depth.

Birds

Seasonal distributions of seabirds are relatively low in the ACSAR compared to the adjacent continental shelf. Summertime concentrations are particularly low, presumably because much of their prey is confined to waters beneath the deep thermocline. Highest seabird concentrations generally occur in spring. Data are particularly sparse south of 35°N, thus limiting our conclusions about this area.

Human Activities and Impacts

The coastal waters of the eastern United States are among the most intensely used and managed in the world. Resources and environmental features of the ACSAR provide both opportunities and constraints with regard to human activities. Human activities are also governed by state, national and international laws. For example, a plethora of environmental protection legislation was passed by Congress during the late 1960's and the 1970's to prevent environmental damage in the marine environment. More recent emphasis focuses on resource management, with the Magnuson Fisheries Conservation Management Act of 1976 and the 1983 Presidential Proclamation establishing a 200-nautical mile Exclusive Economic Zone.

Fisheries

Active commercial fisheries in deep water on and over the edge of the continental shelf by U.S. fishermen include tilefish, billfish, sharks, lobster, red crab. Commercial fisheries for cod, haddock, silver hake, red hake, scup, sea bass, summer flounder and redfish also exist, although none of these are pursued predominantly by U.S. fishermen. Contribution to total U.S. fisheries from the ACSAR area is less than 0.1 percent of total catch.

Total catch from the deep-water areas averaged 105 MT per year from 1971 to 1981, down from 380 MT for U.S. commercial vessels between 1965 and 1969. Increases in operating costs — especially fuel after 1973 — probably explain the decline since the abundance of most species sought by U.S. fishermen has remained the same or increased since 19711. The Japanese fish in the slope area for squid, butterfish, mackerel and tuna; Canada fishes for silver hake, red hake, herring, mackerel, butterfish, and squid (*Illex* and *Loligo*). Europeans fish for silver hake, butterfish, and squid.

There is very little recreational fishing in the ACSAR region although many species of interest can be found there. High fuel costs in the last decade have discouraged party and charter boat fishing; private fishing boats occasionally fish for cod, pollock and tilefish. Other species available but not now fished in offshore waters include sea bass, bluefish, Atlantic Bonito, catfish, Atlantic croaker, cunner, American eel, summer flounder, winter flounder, flounders, hakes, herrings, Atlantic mackerel, mackerel and tuna, white perch, yellow perch, porgies, scup, sea robin, spotted sea trout, shark, dogfish, skates and rays, spot, striped bass, tautog, toadfish, weakfish, and windowpane.

Mineral Resources

Exploration wells in the northwest Atlantic have produced no oil and very little gas, and enthusiasm appears to have waned. Current appraisals of oil and gas potential suggest that any promise for the area lies in deeper waters, particularly in association with an ancient reef feature located beneath the continental slope. Recent and scheduled lease sales in this vicinity may determine the future of oil and gas development in the ACSAR area.

The major mineral resources in the ACSAR area include manganese nodules (on the Blake Plateau) and phosphorite. Although these resources are of commercial grade, their distance from shore and the depths of water in which they are found make it unlikely they will be exploited in the near future.

Waste Disposal

Areas in and adjacent to the ACSAR area receive large quantities of wastes as a result of ocean dumping. Most waste materials dumped in American waters, excluding dredge spoils, have been disposed of in the New York Bight, which lies inshore of the ACSAR area. The major deep water disposal site in ACSAR, Deepwater Dumpsite 106, has received industrial chemical wastes and explosives; an adjacent deepwater site is proposed for incineration of hazardous chemicals aboard special ships. Other human activities, largely during World War II, have resulted in spills of large quantities of oil, but the exact quantity and impact are poorly documented. Non-point source inputs of petroleum, other chemicals, and radioactivity have also affected the ACSAR area, but impact has not been documented.

Endangered Species

Certain species of whales found in the study area are listed as endangered species. Little is known about the potential effects which economic development, particularly oil and gas development, might have on these whales.

There are two types of special habitats which might influence development decisions in the ACSAR area. First are canyons and canyon heads. These areas are considered to be important habitats for some commercial species of fish and shellfish (e.g., tilefish and lobster). Studies are now underway to determine what effects oil and gas development might have on fish and shellfish populations using these areas. The second type of special habitats within ACSAR are deep-water coral (e.g., Scleractinia) areas. The Gulf of Mexico and South Atlantic Fishery Management Councils are now preparing a management plan which would include deep-water corals.

Human Activity

A number of other activities in the ACSAR are poorly documented in the scientific literature. Some of these, such as military activities and transportation, have very large economic impacts compared with uses and activities referred to above. Shipping traffic crosses the ACSAR area from 15 major commercial ports along the U.S. east coast, with the largest tankers running from the New York and Philadelphia area to north and west Africa and to routes to the Middle East around Cape Horn. Smaller tankers ply routes between several U.S. ports and the Caribbean and between domestic ports. Fishing vessels landing on the U.S. east coast mainly use different ports.

Ten telecommunications cables cross the ACSAR and two more are planned. The landfall of another has not yet been determined. Ship wrecks, rocket impact areas, bombing sites and other obstacles to ACSAR activities have been charted.

Future Studies

Despite the impressive number of research institutes and marine scientists that have studied and are studying the Atlantic continental slope and rise, and the very large number of scientific publications which have resulted from these studies, there are areas which are poorly known and merit further study. Most subject areas are known in general terms, but require more detailed study particularly in terms of scale and rates. In some areas, our knowledge is particularly sparse; for example, phytoplankton, neuston and radionuclides. Some of the suggestions for future study relate to all six general topics covered in this report, others are specific to a single topic or subtopic.

General Needs

With few exceptions the oceanographic and environmental phenomena in the ACSAR area have been defined; we generally know what and where processes occur. Three general areas, however, need to be addressed: meso- and micro-scale processes, fluxes, and more complete data utilization and interpretation.

Meso- and Micro-Scales

Two examples portray our state of knowledge in this realm: we know with relatively good accuracy the general bathymetry of the ACSAR area, but relatively little on horizontal scales finer than hundreds to thousands of meters. Similarly, finer-scale distributions of plankton and neuston are not known. Many other examples could be given in which far more detailed study is required to define distributions and processes on finer scales.

Clearly it is impossible to examine the entire region in fine scale; what is critical is to define carefully those experiments which will help clarify general principles or important processes. Furthermore, such studies require more sophisticated instrumentation than used in the past. Using again the example of bathymetry, meso- and micro-scales can be documented only by using accurate navigation, multibeam echosounders and deep-towed vehicles. Deep-towed vehicles only now are becoming available to the scientific community and a multi-beamed echosounder is now being used by the consortium of Lamont-Doherty Geological Observatory, University of Rhode Island and the Woods Hole Oceanographic Institution. Documentation of other oceanographic phenomena would require a significant increase in stations, current meters, etc. and/or greater sophistication in the measuring techniques.

Fluxes

The term "flux" refers to the rates of transport/transfer/reproduction of water, particles, chemical species or organisms. Where and when are various oceanographic components transported, and how? Specific examples will be given n a following section.

Data Utilization and Interpretation

In some instances, notably geology and meteorology, there exist large bodies of data which, if interpreted correctly, could aid immensely in our understanding of these systems. In some cases the data are in the hands of private companies (e.g., oil companies), but in others, the data are more accessible. For example, the data from the NOAA ocean buoys are available, but as yet not worked up. Similarly, the very large number of presently-available seafloor sediment samples (see Chapter 4) suggests that the need for (at least) more regional sampling may be small, but that complete analysis and interpretation of existing samples/data are also required. A final example is the apparent existence but present unavailability of fish landing data from areas other than the northeast; integrating these data with the northeast figures would give us a more complete picture of the fishing industry off the entire eastern U.S.

Many other examples could be cited, but the main point to be remembered is that interpreting existing data sets can be much less expensive in terms of both time and money than acquiring new data sets through new field studies.

Specific Studies

Unfortunately, in many areas, available data are not sufficient to document thoroughly the required parameters. This is particularly true of documenting smaller scales or fluxes, as mentioned in the preceding paragraphs. Many of these are necessarily multidisciplinary in approach. HEBBLE, PARFLUX and Warm Core Rings are three recent projects in which marine scientists of various disciplines have worked cooperatively to investigate complex oceanographic problems.

Meteorology

Perhaps the greatest need in understanding offshore meteorology is obtaining a better picture of storms and their effects on air-sea interactions. As pointed out in Chapter 2, time-averaged data are not particularly valuable because of the large short-term fluctuations following passage of a storm. In this instance, better utilization of existing long-term weather buoy data would be valuable.

Physical Oceanography

Recent studies are pointing the way to better documentation of the physical environment in the ACSAR area. This includes increased current-meter moorings in the Gulf Stream, moorings on the slope and in canyons, and detailed studies in such diverse water masses as the Slope Water and the Western Boundary Undercurrent. What is needed is better documentation of processes at boundaries between water masses, reactions of oceanographic parameters to the passage of storms, and the relation between physical processes and the biological environment.

Geology

The most obvious need is the understanding of meso- and micro-scale morphology, which requires bathymetric studies using far more sophisticated equipment and navigation than used previously. To understand downslope gravity movement, for instance, we need to delineate morphologic features on the scale of meters. Proposing sediment cores in relation to bottom morphology is particularly important, since zones of erosion and accumulation may lie directly adjacent to one another; a poor understanding of the morphology/shallow structure or poor navigation may result in sampling the wrong topographic features, with a corresponding erroneous interpretation of the remote sensing records. Utilizing present seismic records with deep drilling logs, and integrating these data with finer-scale studies, could result in a better understanding of both the structural history and the neotectonics of the area.

Chemistry

In most instances, chemical studies in the ACSAR area have been purely descriptive. Few have focused upon determining the processes governing distributions of various components, although proper documentation requires biological and sedimentological studies as well. Vertical flux of particulate matter, for example, has biological and geological implications as well as chemical. Similarly, remineralization rates are needed. These studies should be accompanied by a broad spectrum of ancillary measurements to characterize the sedimentary and biologic regimes. In sediment studies, for example, this would require defining redox conditions, major phase components (biogenic $Si0_2$, aluminosilicate, $CaC0_3$, organic carbon, leachable oxides of Fe and Mn), and mixing parameters using isotopes to elucidate benthic activity. Radionuclides have been studied less than other non-organic components, particularly with respect to influxes and removal rates.

In terms of hydrocarbons, there is a decided paucity of data for all types and molecular weight ranges in ACSAR. This includes dissolved, colloidal and particulate species, and is particularly true in terms of fluxes. Differentiating aeolian inputs of petroleum and pyrogenic hydrocarbons from exploration and production inputs may be particularly difficult, but needs to be undertaken.

Biology

In terms of phytoplankton, neuston and cetaceans, we know very little about the ACSAR area. The types of studies described for zooplankton in this report should be undertaken for other biologic components and in other oceanographic subenvironments. Problems such as patchiness, hydrographic variability, and net avoidance by large organisms are particularly important in interpreting data, and therefore should be studied in greater detail. Biological parameters should not be studied in isolation, but rather in concert with chemical, geological and physical oceanographic studies.

Biological sampling in the northwest Atlantic at depths greater than 1000 m is practically non-existent, either for plankton or fish. Clearly such studies are needed to understand the complete biological system in the ACSAR. Finally, there is a marked absence of coherent long-term time-series data with respect to population trends; such data should be integrated with flux/productivity studies in order to delineate the system.

Human Activities and Impact

Perhaps the most important aspect that needs further study is the impact of man and man's activities upon the ACSAR area. This includes sufficient understanding of the environment to determine the impact of previous activities (such as old dump sites) and predict the effect of new activities (such as drilling, mineral exploration, waste disposal). Predicting where we can expect maximum human activity and identifying major hazards (e.g., storms, slumps, slides, faulting, etc.) and where they are likely to occur should facilitate this process. Many of these studies are presently in progress, and the next few years should see synthesis and publication of their results.

References Cited

Aaron, J.M., B. Butman, M.H. Bothner, and R.E. Sylwester. 1980. Maps showing environmental conditions relating to potential geologic hazards on the United States northeastern Atlantic continental margin. U.S. Geol. Surv. Misc. Field Studies Map MF-1193. 3 sheets.

Allen, J.A. and H.L. Sanders. 1966. Adaptation to the abyssal life as shown by the bivalve *Abra profundorum* (Smith). Deep-Sea Res. 13:1174-1184.

Allen, J.A., and H.L. Sanders. 1969. Nucinella Serrei Lamy (Bivalvia: Protobranchia), A monomyarian Solemyid and possible living actinodont. Malacologia 7:381-396.

Allen, J.A. and H.L. Sanders. 1973. Studies on deep sea Protobranchia. The families Siliculidae and Lametilidae. Bull. Mus. Comp. Zool. 145:263-309.

Allen, J.A. and J.F. Turner. 1974. On the functional morphology of the family Verticordiidae (Bivalvia) with descriptions of new species from the abyssal Atlantic. Phil. Trans. Roy. Soc. Lond. B 268:401-536.

Allen, J.A. 1979. The adaptations and radiation of deep-sea bivalves. Sarsia 64:19-27.

Allen, J.A., and R. E. Morgan. 1981. The functional morphology of Atlantic deep water species of the families Cuspidariidae and Poromyidae (Bivalvia): An analysis of the evolution of the septibranch condition. Phil. Trans. Roy. Soc. Lond. B 294:413-546.

Aller, R.C. and J. K. Cochran. 1976. 234 Th/238 U disequilibrium in near-shore sediment: particle reworking and diagenetic time scales. Earth Planet. Sci. Lett. 29:37-50.

Altabet, M.A. and J.J. McCarthy. 1985a. Vertical patterns in ^{15}N natural abundance in PON from the surface waters of warm-core rings. J. Mar. Res., in press.

Altabet, M.A. and J.J. McCarthy. 1985b. Temporal and spatial variations in the natural abundance of ^{15}N in PON from a warm-core ring. Deep-Sea Res. in press.

Altabet, M.A. and W.G. Deuser. 1985. Seasonal variations in ^{15}N matural bundance in the flux of particulate matter to the deep ocean. Nature in press.

American Petroleum Institute. 1976. Recommended practice for design, construction, operation and maintenance of offshore hydrocarbon pipeline. First edition. API, RP1111, stock No. 831-11110.

Amos, A.F., A.L. Gordon, and E.D. Schneider. 1971. Water masses and circulation patterns in the region of the Blake-Bahama Outer Ridge: Deep Sea Res. 18:145-165.

Anderson, D.M. and F.M.M. Morel. 1978. Copper sensitivity of *Gonyaulax tamarensis*. Limnol. Oceanogr. 23:283-295.

Anderson, E.D., and F.P. Almeida. 1979. Assessment of Bluefish (*Pomatomus saltatrix*) of the Atlantic Coast of the United States. National Marine Fisheries Service, Laboratory Reference No. 79-19.

Anderson, O.R. and A.W.H. Bé. 1976. A cytochemical fine structure study of phagotrophy in a planktonic foraminifer, *Hastigerina pelagica* (d'Orbigny). Biol. Bull. 151: 437-449.

Anderson, O.R., M. Spindler, A.W.H. Bé, and C. Hemleben. 1979. Trophic activity of planktonic foraminifera. J. Mar. Biol. Ass. U.K. 59: 791-799.

Anderson, O.R. 1980 Radiolaria. In: Levandowsky, M. & S.H. Hutner (eds.). Biochemistry and physiology of protozoa. 2nd Ed. 3: 1-42.

Angill, J., J. Korshouer and G. Cottom. 1969. Quasibiennial variation in the 'centers of action". Month. Weath. Rev. 97:867-872.

Anonymous. 1978. Aragonite: White gold in the Bahamas. ESSO Carib. 1:4.

Anonymous. 1983. OCS Lands Act is unsuitable for Marine Hard Minerals Development. Strategic Materials Management 3(3):5-6.

Armi, L., and N.A. Bray. 1982. A standard analytic curve of potential temperature versus salinity for the western North Atlantic. Jour. Phys. Oceanogr. 12:384-387.

Aruga, Y., S. Ichimura, Y. Fujita, S. Shimura and Y. Yamaguchii. 1975. Characteristics of photosynthesis of planktonic marine blue-green algae, Trichodesmium. in Studies on the Community of Marine Pelagic Blue-green Algae. R. Marumo, (ed.), Ocean Research Institute, Univ. of Tokyo, Tokyo. pp. 48-55.

Atkinson, L.P. 1983. Distribution of Antarctic intermediate water over the Blake Plateau. J. Geophys. Res. 88:4699-4704.

Austin, J.A., Jr., E. Uchupi, D.R. Shaughnessy, III and R.D. Ballard. 1980. Geology of New England passive margin. AAPG Bull. 64:501-526.

Ayers, M.W. and O.H. Pilkey. 1981. Piston core and surficial sediment investigations of the Florida-Hatteras slope and inner Blake Plateau: appendix. in P. Popenoe (ed.), Appendices of environmental geologic studies on the southeastern United States Atlantic outer continental shelf 1977-1978, U.S. Geol. Surv. Open-File Rep. 81-0582-B, 5(1-5):14.

Ayala, F.J. and J.W. Valentine. 1978. Genetic variation and resource stability in marine invertebrates. In Marine Organisms, B. pp.23-51.

Azam, F., T. Fenchel, J.G. Field, J.S. Gray, L.A. Meyer-Reil and F. Thingstad. 1983. The ecological role of water-column microbes in the sea. Mar. Ecol Prog. Ser. 10: 257-263.

Backus, R.H. 1968. A Whole Ocean Polluted? Oceanus 15(1):1.

Backus, R.H., J.E. Craddock, R.L. Haedrich, and D.L. Shores. 1969. Mesopelagic fishes and thermal fronts in the western Sargasso Sea. Mar. Biol. 3: 87-106.

Backus, R.H., J.E. Craddock, R.L. Haedrich and B.H. Robison. 1977. Atlantic mesopelagic zoogeography. in Fishes of the Western North Atlantic. Mem. Sears. Found. Mar. Res. 1:266-287.

Backus, R.H. and J.E. Craddock. 1982. Mesopelagic fishes in Gulf Stream cold-core rings. J. Mar. Res. 40 (Suppl.):1-20.

Bacon, M.P., D.W. Spencer and P.G. Brewer. 1976. 210 Pb/226 Ra and 210 Po/210 Pb disequilibria in seawater and suspended particulate matter. Earth Planet. Sci. Lett. 32(2):277-296.

Badcock, J. 1970. The vertical distribution of mesopelagic fishes collected on the SOND cruise. J. Mar. Biol. U.K. 50:1001-1044.

Bailey, N.G. and J.M. Aaron. 1982a. High-resolution seismic-reflection profiles from the R/V Columbus Iselin, Cruise CI7-78-2, over the continental shelf and slope in the Georges Bank area:2.

Bailey, N.G. and J.M. Aaron. 1982b. High-resolution seismic-reflection profiles collected aboard R/V JAMES M. GILLISS, Cruise GS 7903-3, over the Atlantic continental slope and rise off New England. U.S. Geol. Surv. Open-File Report 82-0718:2.

Ball, M.M. 1978. Cruise Report R/V State Arrow, submersible 'Diaphus', South Atlantic Environmental Program, August 1978. U.S. Geol. Surv. (Woods Hole), unpubl. ms.

Ball, M.S. 1982. Good Old American Permits: Madisonian Federalism on the Territorial Sea and Continental Shelf. Environmental Law, 12:623-678.

Ballard, R. 1980. Mapping the Mid-Ocean Ridge: Proc. Offshore Tech. Conf. OTC-3682, Houston, TX, p. 55-65.

Bane, J.M., Jr. and D.A. Brooks. 1979. Gulf Stream meanders along the continental margin from the Florida Straits to Cape Hatteras. Geophys. Res. Lett. 6:280-282.

Bane, J.M., Jr., D.A. Brooks and K.R. Lorenson. 1981. Synoptic observations of three-dimensional structure, propagation and evolution of Gulf Stream meanders along the Carolina continental margin. J. Geophys. Res. 86, 6411- 6425.

Bane, J.M., Jr. 1983. Initial observations of the subsurface structure and short-term variability of the seaward deflection of the Gulf Stream off Charleston, SC. J. Geophys. Res. 88:4673-4684.

Barrett, J.R., Jr. 1965. Subsurface currents off Cape matteras. Deep-Sea Res. 12:173-184.

Barrett, J.R. 1971. Available potential energy of Gulf Stream rings. Deep-Sea Res. 18:1221-1231.

Barrett, J.R. and W. Schmitz. 1971. Transport float measurements and hydrographic station data from three sections across the Gulf Stream near 67ºW, R.V. Crawford Cruise 168, June-July 1968. Woods Hole Oceanogr. Inst. Ref. No. 71-66.

Bartsch, I. 1980. Fnf neue Arten der Gattung Halacarus (Acari, Halacaridae) aus dem Atlantik. [Five new species of the genus Halacarus (Acari, Halacaridae) from the Atlantic Ocean]. Zool. Scr. 10:203-215.

Bartsch, I. 1982. Drei Arten der Gattung Copidognathus (Acari, Halacaridae) aus dem Argentinischen Becken. Entom. Mitt. Zool. Mus. Hamburg Bd. 7:114.

Battaglia and J.A. Beardmore (eds.), Plenum Press, New York and London.

Baumgartner, A. and E. Reichel. 1975. The World Water Balance. Elsevier, 179 pp.

Bé, A.W.H. 1960. Ecology of recent planktonic foraminifera: Part 2. Bathymetric and seasonal distributions in the Sargasso Sea off Bermuda. Micropaleontol. 6: 373-392.

Bé, A.W.H., J.M. Forns, O.A. Roels. 1971. Plankton abundance in the North Atlantic Ocean. In J.D. Costlow, Jr. (ed.), Fertility of the Sea, 1 & 2. Symp. XI:308.Gordon and Breach Sci. Publ., NY.

Bé, A.W.H. and D.S. Tolderlund. 1971. Distribution and ecology of living planktonic foraminifera in surface waters of the Atlantic and Indian Oceans. in Funnell, B.M. & W.R. Fiedel (ed.). Micropaleontology of Oceans. pp. 105-149.

Bé, A.W.H., C. Hemleben, O.R. Anderson, M. Spindler, J. Hacunda and S. Tuntivate-Choy. 1977. Laboratory and field observations of living planktonic foraminifera. Micropaleontol. 23: 155-179.

Bé, A.W.H., H.J. Spero and O.R. Anderson. 1982. Effects of symbiont elimination and reinfection on the life processes of the planktonic foraminifera *Globigerinoides sacculifer*. Mar. Biol. 70:73-86.

Bé, A.W.H., O.R. Anderson, W.W. Faber, Jr. and D.A. Caron. 1983. Sequence of morphological and cytoplasmic changes during gametogenesis in the planktonic foraminifera *Globigerinoides sacculifer* (Brady). Micropaleontol. 29:310-325.

Beardsley, R.C. and W.C. Boicourt. 1981. On estuarine and continental shelf circulation in the Middle Atlantic Bight. In Evolution of Physical Oceanography, MIT Press:198-234.

Beers, J.R. and G.L. Stewart. 1969. Microzooplankton and its abundance relative to the larger zooplankton and other seston components. Mar. Biol. 4: 182-189.

Behr, L. 1979. Implementing Federal Consistency under the Coastal Zone Management Act of 1972. New York Sea Grant Law and Policy J. 3:1-76.

Belding, H.F. and W.C. Holland. 1970. Bathymetric maps: eastern continental margin, U.S.A. Scale 1:1,000,000. Am. Assoc. Petrol. Geol.

Bender, M.L. and C. Gagner. 1976. Dissolved copper, nickle and cadmium in the Sargasso Sea. J. Mar. Res. 34:327-339.

Bender, M.L., G.P. Klinkhammer and D.W. Spencer. 1977. Manganese in seawater and the marine manganese balance. Deep-Sea Res. 24(9):799-812.

Bennett, R.H., G.L. Freeland, D.N. Lambert, W.B. Sawyer and G.H. Keller. 1980. Geotechnical properties of surficial sediments in a mega-corridor: U.S. Atlantic continental slope, rise, and deep-sea basin. Mar. Geol. 38:123-140.

Benson, R. 1975. The origin of the psychrosphere as recorded in changes of deep-sea ostracod assemblages. Lethaea 8:69-83.

Benson, W.E., R.E. Sheridan, P. Enos, T. Freeman, F. Gradstein, I.O. Murdmaa, L. Pastouret, R.R. Schmidt, D.H. Stuermer, F.M. Weaver and P. Worstell. 1978a. Sites 389 and 390, north rim of Blake Nose. in P. Worstell (ed.), Deep Sea Drilling Project, Initial Rep. (44):337-393.

Benson, W.E., R.E. Sheridan, P. Enos, T. Freeman, F. Gradstein, I.D. Murdmaa, L. Pastouret, R.R. Schmidt, D.H. Stuermer, F.M. Weaver and P. Worstell. 1978b. Site 392: south rim of Blake Nose. in P. Worstell (ed.), Deep Sea Drilling Project, Initial Rep. (44):337-393.

Berger, W.H. and U. von Rad. 1972. Cretaceous and Cenozoic sediments from the Atlantic Ocean: Initial Reports of the Deep Sea Drilling Project 14:787-954.

Berggren, W.A., D.V. Kent, and J.A. Van Couvering. 1984a. Neogene geochronology and chronostratigraphy. in N.J. Snelling (ed.), Geochronology and the Geological Record: Geological Society of London, Special Paper, in press.

Berggren, W.A., D.V. Kent and J.J. Flynn. 1984b. Paleogene geochronology and chronostratigraphy: In N.J. Snelling (ed.), Geochronology and the Geological Record: Geological Society of London, Special Paper, in press.

Berk, S.G., D.C. Brownlee, D.R. Heinle, H.J. Kling and R.R. Colwell. 1977. Ciliates as a food source for marine planktonic copepods. Microb. Ecol. 4:27-40.

Betzer. 1978. In Chemical and biological benchmark studies. BLM/St-(78/28).

Bewers, J.M., B. Sundby and P.A. Yeats. 1976. The distribution of trace metals in the western North Atlantic off Nova Scotia. Geochim. Cosmochim. Acta 40(6):687-696.

Bidleman, T.F. and C.E. Olney. 1975. Long range transport of toxaphene insecticide in the atmosphere of the western North Atlantic. Nature 257:475-477.

Bidleman, T.F., C.P. Rice, C.E. Olney. 1976. High molecular weight chlorinated hydrocarbons in the air and sea: Rates and mechanisms of air- sea transfer. NSF/Intl. Decade of Ocean Exploration Pollution Transfer Workshop, 323-329.

Bidleman, T.F., E.J. Christensen, W.N. Billings and R. Leonard. 1981. Atmospheric transport of organochiorines in the North Atlantic gyre. J. Mar. Res. 39:443-464.

Bisagni, J.J. 1976. Passage of anticyclonic Gulf Stream eddies through deep water dumpsite 106 during 1974 and 1975. NOAA Tech. Rept. NMFS Circular 416:293-298.

Biscaye, P.E. 1965. Mineralogy and sedimentation of recent deep-sea clay in the Atlantic Ocean and adjacent seas and oceans: Geol. Soc. Am. Bull. 76:803-832.

Biscaye, P.E. and S.L. Eittreim. 1977. Suspended particulate loads and transports in the nepheloid layer of the abyssal Atlantic Ocean. Mar. Geol. 23:155-172.

Biscaye, P.E. S.R. Carson and G. Mathieu. 1980. Excess Radon in Shelf and Slope Waters of the New York Bight. United States Department of Energy. Conf-790382.

Blumberg, A.F. and G.L. Mellor. 1983. Diagnostic and prognostic numerical circulation studies of the South Atlantic Bight. J. Geophys. Res. 88:4579-4592.

Boehm, P.D. and R. Hirtzer. 1982. Gulf and Atlantic Survey for Selected Organic Pollutants in Finfish. NOAA Tech. Memorandum NMFS-F/NEC-13, U.S. Dept. of Commerce, Washington, DC.

Boltovskoy, E. 1973. Daily vertical migration and absolute abundance of living planktonic foraminifera. J. Foram. Res. 3: 89-94.

Bonin, D.J. and S.Y. Maestrini. 1981. Importance of organic nutrients for phytoplankton growth in natural environments: Implications for algal species competition. in T. Platt (ed.), Physiological bases of phytoplankton ecology. Can. Bull. Fish. Aquat. Sci. 210:279-291.

Booth, J.S., R.A. Farrow and T.L. Rice. 1981a. Geotechnical properties and slope stability analysis of surficial sediments on the Baltimore Canyon continental slope. U.S. Geol. Surv. Open-File Rep.81-733, 40 p.

Booth, J.S., R.C. Circé and A.G. Dahl. 1983a. Geotechnical characterizationand mass-movement potential of the United States Mid-Atlantic continental slope and rise. in B.A. McGregor (ed.), Environmental geologic studies on the United States mid and north Atlantic outer continental shelf area. U.S. Geol. Surv. Open-File Rep. 83-824, 2:4-1 to 4-124.

Booth, J.S., R.C. Circé and A.G. Dahl. 1983b. Geotechnical characterizationand mass-movement potential of the United States north Atlantic continental slope and rise. in B.A. McGregor (ed.), Environmental geologic studies on the United States mid and north Atlantic outer continental shelf area. U.S. Geol. Surv. Open-File Rep. 83-2824, 3:4-1 to 4-69.

Borror, A.C. 1980. Spatial distribution of marine ciliates; Micro-ecologic and biogeographic aspects of protozoan ecology. J. Protozool. 27: 10-13.

Bosart, L.F. 1981. The presidents day snowstorm of 18-19 February 1979: a subsynoptic-scale event. Month. Weath. Rev. 109:1542-1566.

Bothner, M.H., C.M. Parmenter and J.D. Milliman. 1981. Temporal and spatial variations in suspended matter in shelf and slope waters off northeastern United States: Estuar. Coastal and Shelf Sci. 13:213-234.

Bothner, M.H. R.R. Rendigs, E. Campbell et al. 1982a. The Georges Bank monitoring program: analysis of trace metals in bottom sediments. U.S. Geol. Surv. Final Rep. submitted to U.S. Bureau of Land Management. 62 p.

Bothner, M.H., C.M. Parmenter, R.R. Rendigs, B. Butman, L.J. Poppe and J.D. Milliman. 1982b. Studies of suspended matter along the north and middle Atlantic outer continental shelf. U.S. Geol. Surv. Open-File Rep. 82-938.

Bothner, M.H. and M.P. Bacon. 1983. Natural Radionuclides: book and atlas. Woods Hole Oceanographic Institution Coastal Research Center.

Bottazzi, E.M., B. Schreiber and V.T. Bowen. 1971. Acantharia in the Atlantic Ocean, their abundance and preservation. Limnol. Oceanogr. 16:677-684.

Bowen, R. 1972. Continental Shelf Phosphates — One Answer to Future Needs. Offshore Tech. Conf. Paper No. 1659.

Bowen, V.T. and W. Roether. 1973. Vertical distributions of Strontium 90, Cesium 137, and tritium near 45° North in the Atlantic. J. Geophys. Res. 78(27):6277-6285.

Bowen, V.T., V.E. Noshkin, H.L. Volchok, H.D. Livingston and K.M. Wong. 1974. Cesium-137 to Strontium-90 ratios in the Atlantic. 1966 through 1972. Limnol. Oceanogr. 19:670-681.

Bowen, V.T., H.D. Livingston and J.C. Burke. 1976.

Distribution of transuranium nuclides in sediment and biota on the North Atlantic Ocean. in Symposium on transuranium nuclides in the environment, proceedings. 17-21 November, 1975, San Francisco, CA. International Atomic Energy Agency p. 107-120.

Bowen, V.T. and H.D. Livingston. 1981. Radionuclide distribution in sediment cores retrieved from marine radioactive waste dumpsites. in International symposium on the impacts of radionuclide releases into the marine environment, Vienna, Austria, 6 October 1980. International Atomic Energy Agency:33-63.

Bowie, G.L. and R.L. Wiegel. 1977. Marine Pipelines: An Annotated Bibliography. CERC Misc. Report 77-2, 58 p. (NTIS: AD-A038, 747).

Bowin, C., W. Warsi and J. Milligan. 1982. Free-Air Gravity Anomaly Atlas ofthe World. Geol. Soc. Am. Map and Chart Series, No. MC-46:91

Boyd, S.H., P.H. Wiebe and J.L. Cox. 1978. Limits of *Nematoscelis megalops* in the Northwestern Atlantic in relation to Gulf Stream cold-core rings. Part II. Physiological and biochemical effects of expatriation. J. Mar. Res. 36:143-159.

Boyle, E.A. 1981. Cadmium, zinc, copper, and barium in foraminifera tests. Earth Planet. Sci. Lett. 53(1):11-35.

Boyle, E.A., S.S. Huested and S.P. Jones. 1981. On the distribution of copper, nickel, and cadmium in the surface waters of the North Atlantic and North Pacific Ocean. J. Geophys. Res. 86:8048-8066.

Bragaw, L.K., H.S. Marcus, G.C. Raffaele and J.R. Townley. 1975. The Challenge of Deepwater Terminals. Lexington Books, D.C. Heath and Co., Lexington, Mass. 162 pp.

Breeden, R. 1976. Federation and the Development of Outer Continental Shelf Mineral Resources. Stanford Law Review 28:1107-1159.

Brewer, W.C., Jr. 1976. Federal Consistency and State Expectations. Coastal Zone Management Journal, 2:315-324.

Brezenski, F.T. 1975. Analytical results for water-column samples collected at Deepwater Dumpsite 106. pp. 203-215 In NOAA, May 1974 Baseline Investigation of Deepwater Dumpsite 106. NOAA Dumpsite Evaluation Report 75-1. NOAA, Rockville, MD.

Brezenski, F.T. 1977. In NOAA/ODP Dumpsite Evaluation Rept. 77-1.Baltimore Canyon fish. Environ. Sci. and Technol. 13:878-879.

Broecker, W.S., A. Kaufman and R.M. Trier. 1973. The residence time of thorium in surface sea water and its implications regarding the rate of reactive pollutants. Earth Planet. Sci. Lett. 20:35-44.

Broecker, W.S. 1974. Chemical Oceanography. Harcourt Brace Janovich, Inc., New York, 214 pp.

Broecker, W.S., J. Goddard and J.L. Sarmiento. 1976. The distribution of .226.Ra in the Atlantic Ocean. Earth Planet. Sci. Lett. 32:220-235.

Brooks, D.A. and J.M. Bane, Jr. 1979. Gulf Stream deflection by a bottom feature off Charleston, South Carolina. Science 201:1225-1226.

Brooks, D.A. and J.M. Bane, Jr. 1981. Gulf Stream fluctuations and meanders over the Onslow Bay upper continental slope. J. Phys. Oceanogr. 11:247-256.

Brooks, D.A. 1983. The wake of hurricane Allen in the Gulf of Mexico. J. Phys. Oceanogr.

Brooks, D.A. and J.M. Bane, Jr. 1983. Gulf Stream meanders off North Carolina during winter and summer 1979. J. Geophys. Res. 88:4633-4650.

Brown, R.A., T.D. Searl, J.J. Elliott, P.H. Monaghan, D.E. Brandon. 1974. Measurement and interpretation of non-volatile hydrocarbons in the ocean. Part I. Measurements in Atlantic, Mediterranean, Gulf of Mexico, and Persian Gulf. Rept. for Jun 7 2-Dec 73, Exxon Prod. Res. Co.:221.

Brown, R.G.B., D.N. Nettleship, P. Germain, C.E. Tull and T. Davis. 1975. Atlas of eastern Canadian seabirds. Can. Wildl. Serv., Ottawa.

Brown, R.A. and H.L. Huffman, Jr. 1976. Hydrocarbons in open ocean waters. Science 191:847-849.

Brown, R.A. and R.J. Pancirov. 1979. Polynuclear aromatic hydrocarbons in Baltimore Canyon fish. Environ. Sci. Technol. 13:878-879.

Brown, R.G.B. 1977. Atlas of eastern Canadian seabirds. Supplement I (Bermuda-Halifax transects). Can. Wildl. Serv., Minister of Supply and Services, Canada.

Bryan, G.M. 1970. Hydrodynamic model of the Blake Outer Ridge: J. Geophys. Res. 75:4530-4537.

Bryan, G.M., R.G. Markl and R.E. Sheridan. 1980. IPOD site surveys in the Blake-Bahama Basin. in B.T.R. Lewis, P.D. Rabinowitz (eds.), Regional geophysical studies associated with IPOD site surveys. Mar. Geol. 35:43-63.

Budyko, M.I. 1963. Atlas of heat balance of the world (in Russian). Glav. Geofiz. Obo., Moscow, 69 pp. Translation by I.A. Donehoo, Weath. Bureau, WB/T-106, Washington, D.C., 25 pp.

Bumpus, D.F. 1965. Residual drift along the bottom on the continental shelf in the Middle Atlantic bight area: Limnol. Oceanogr. Suppl. 10:50-53.

Bumpus, D.F. and L.M. Lauzier. 1965. Surface circulation on the continental shelf off eastern North America between Newfoundland and Florida: Am. Geog. Soc., Serial Atlas of the Marine Environment, Folio 7:8.

Bumpus, D.F. 1973. A description of the circulation on the continental shelf of the east coast of the United States. Prog. Oceanogr. 6:111-157.

Bunce, E.T., K.O. Emery, R.D. Gerard, S.T. Knott, L. Lidz, T. Saito and J. Schlee. 1965. Ocean drilling on the continental margin. Science 150:709-716.

Bunker, A.F. 1975. Energy exchange at the surface of the western North Atlantic Ocean. Woods Hole Oceanogr. Tech. Rep. 75-3:111.

Bunker, A.F. 1976. Computations of surface energy flux and annual air-sea interaction cycles of the North Atlantic Ocean. Monthly Weather Review 104:1122-1140.

Bunker, A.F. and L.V. Worthington. 1976. Energy Exchange Charts of the North Atlantic Ocean. Bull. Am. Meteorol. Soc. 57:670-678.

Bunn, A.R. and B.A. McGregor. 1980. Morphology of the North Carolina continental slope, western North Atlantic, shaped by deltaic sedimentation and slumping. Mar. Geol. 37:253-266.

Bureau of Land Management (BLM). 1980. Draft EIS. Proposed 1981 OCS Oil and Gas Lease Sale Offshore the

Mid-Atlantic States. OCS Sale No. 59. U.S. Department of the Interior.

Bureau of Land Management (BLM). 1981. Department of the Interior Final Environmental Impact Statement, Proposed 1981 Outer Continental Shelf Oil and Gas Lease Sale 56. 2 volumes. U.S. Department of Interior, New Orleans, LA.

Bureau of Land Management (BLM). 1982. Final Supplement to the Final Environmental Statement: Proposed Five-Year OCS Oil and Gas Lease Sale Schedule January 1982-December 1986. 2 volumes. Washington, DC.

Burnett, B.R. 1977. Quantitative sampling of microbiota of the deep-sea benthos. I. Sampling techniques and some data from the abyssal central North Pacific. Deep-Sea Res. 24: 781-789.

Burnett, B.R. 1979. Quantitative sampling of microbiota of the deep-sea benthos. II. Evaluation of technique and introduction to the biota of the San Diego Trough. Trans. Amer. Micros. Soc. 98: 233-242.

Burnett, B.R. 1981. Quantitative sampling of nanobiota (microbiota) of the deep-sea benthos. III. The bathyal San Diego Trough. Deep-Sea Res. 28: 649-663.

Burney, C.M., K.M. Johnson, D.M. Lavoie and J.M. Sieburth. 1979. Dissolved carbohydrate and microbial ATP in the North Atlantic concentrations and interactions. Deep-Sea Res. 26:1267-1290.

Burney, C.M., P.G. Davis, K.M. Johnson and J.McN. Sieburth. 1981. Dependence of dissolved carbohydrate concentrations upon small scale nanoplankton and bacterioplankton distributions in the Western Sargasso Sea. Mar. Biol. 65:289-296.

Burns, K.A. and J.M. Teal. 1973. Hydrocarbons in the pelagic Sargassum community. Deep-Sea Res. 30:207-211.

Butcher, W.S. R.P. Anthony and J.B. Butcher. 1968. Distribution charts of oceanic birds in the North Atlantic. Woods Hole Oceanogr. Tech. Rep. 68-69.

Butler, J. N., B. F. Morris and J. Sass. 1973. Pelagic tar from Bermuda and the Sargasso Sea. Bermuda Biological Station for Research, St. George's West. Spec. Publ., 10, 346 pp.

Butler, J.N., J.C. Harris and K. Fine. 1973b. Preliminary gas chromatographicanalysis of pelagic tar samples from MARMAP survey:31.

Butler, J. N. 1975. Evaporative weathering of petroleum residues: the age of pelagic tar. Mar. Chem. 3:9-21.

Butler, J. N. 1976. Transfer of petroleum residues from sea to air: evaporative weathering. In Marine Pollutant Transfer, Chapter 9, H. L. Windom and R. A. Duce (eds.), Lexington Books, C. D. Heath and Co., Lexington, MA.

Butman, B., M. Noble and J. Moody. 1982. Observations of near-bottom currents at the shelf break near Wilmington Canyon. in J.M. Robb (ed.), Environmental Geologic Studies in the Mid-Atlantic Outer Continental Shelf Area: Results of 1978-1979 Field Seasons, U.S. Geol. Survey Final Report to U.S. Bureau of Land Management:3-1-3-58.

Buzas, M.A. and T.G. Gibson. 1969. Species diversity: Benthonic foram-inifera in western North Atlantic. Science 163:72-75.

Buzas, M.A. and S.J. Culver. 1980. Foraminifera: distribution of provinces in the western North Atlantic. Science 209:687-689.

Cairns, S.D. 1981. Marine Flora and Fauna of the Northeastern United States. Scleractinia, NOAA Tech. Report NMFS Circ. 438.

California v. Watt. 1982. No. 81-5B22 DC # CV 81-2080-MRP.

Campbell, B., E. Kern and D. Horn. 1977. Impact of oil spillage from World War II tanker sinking. NOAA Report (MITSG-77-4):96.

Campbell, R.A. 1983. Parasitism in the deep sea. Chapter 12. Deep-Sea Biology. In The Sea: Ideas and Observations on Progress in the Study of the Seas, Vol. 8, G.T. Rowe (ed.), John Wiley and Sons, pp. 473-552.

Capriulo, G.M. 1982. Feeding of field collected tintinnid microzooplankton on natural food. Mar. Biol. 71: 73-86.

Caron, D.A., A.W.H. Be and O.R. Anderson. 1982a. Effects of variations in light intensity on life processes of the planktonic foraminifer *Globigerinoides sacculifer* in laboratory culture. J. Mar. Biol. Ass. U.K. 62: 435-451.

Caron, D.A. and A.W.H. B. 1983. Predicted and observed feeding rates of the spinose planktonic foraminifer *Globigerinoides sacculifer*. Bull Mar. Sci. in press.

Caron, D.A. 1984. The role of heterotrophic microflagellates in plankton communities. Ph.D. thesis, Woods Hole Oceanographic Institution and Massachusetts Instiute of Technology, 268 pp.

Caron, D.A. and A.W.H. Bé. 1984. Predicted and observed feeding rates of the spinose planktonic foraminifera *Globigerinoides sacculifer*. Bull. Mar. Sci. 35:1-10.

Carpenter, E. J. 1972. Nitrogen fixation by a blue-green epiphyte on pelagic Sargassum. Science 178: 1207-1208.

Carpenter, E.J. and J.J. McCarthy. 1975. Nitrogen fixation and uptake of combined nitrogenous nutrients by *Oscillatoria thiebautii* in the western Sargasso Sea. Limnol. Oceanogr.20:389-401.

Carpenter, E. J. 1976. Plastics, pelagic tar and other litter. In Strategies for Marine Pollution Monitoring, Chapter 5, E. D. Goldberg, (ed.), Wiley-Interscience Publication. John Wiley and Sons, New York.

Carpenter, E. J. and C. C. Price IV. 1977. Nitrogen fixation, distribution, and production of *Oscillatoria* (Trichodesmium) in the northwestern Atlantic Ocean and Caribbean Sea. Limnol. Oceanogr. 22: 60-72.

Carpenter, E.J., G.R. Harbison, L.P. Madin, N.R. Swanberg, D.C. Biggs, E.M. Hulburt, V.L. McAlister and J.J. McCarthy. 1977. Rhizosolenia Mats. Limnol. Oceanogr. 22:739-741.

Carpenter, E. J. 1983. Physiology and ecology of marine planktonic Oscillatoria (Trichodesmium). Marine Biology Letters 4: 69-85.

Carpenter, G.B. 1981a. Coincident sediment slump/clathrate complexes on the U.S. Atlantic continental slope. Geo-Marine Letters 1:29-32.

Carpenter, G.B. 1981b. Potential geologic hazards and constraints for blocks in proposed south Atlantic OCS oil and gas lease Sale 56. U.S. Geol. Surv. Open-File Rep. 81-019:36 and 286.

Carpenter, G.B., A.P. Cardinell, D.K. Francois, L.K. Good, R.L. Lewis and N.T. Stiles. 1982. Potential geologic hazards and constraints for blocks in proposed North Atlantic OCS oil and gas lease sale 52. U.S. Geol. Surv. Open-File Rep. 82-0036:54.

Carson, S.R., P.E. Biscaye, G. Mathieu, N. Milford. 1979.

Excess radon in shelf and slope waters of the New York Bight. EOS 60:279-280.

Cartwright, D.E., B.D. Zetler and B.V. Hamon. 1979. Pelagic Tidal Constants,IAPSO Publication Scientifique No. 30.

Cerame-Vivas, M.J. and I.E. Gray. 1966. The distributional pattern of benthic invertebrates of the continental shelf off North Carolina: Ecology 47:260-270.

CETAP (Cetacean and Turtle Assessment Program). 1981. A characterization of marine mammals and turtles in the Mid- and North-Atlantic Area of the U.S. Outer Continental Shelf. Annual Report for 1979. Proposed for U.S. Department of the Interior, Bureau of Land Management under Contract #AA551-T8-48.

CETAP (Cetacean and Turtle Assessment Program). 1982. A characterization of marine mammals and turtles in the Mid- and North Atlantic Areas of the U.S. Outer Continental Shelf. Final Report. Porposed for the U.S. Department of the Interior, Bureau of Land Management, under Contract #AA551-CT8-48.

Champ, M.A. and P.K. Park. 1982. Global Marine Pollution Bibliography, Ocean Dumping of Municipal and Industrial Wastes. IFI/Plenum, New York, 399 p.

Chardy, P. 1974. Two new abyssal isopoda from the North Atlantic belonging to the genus *Janirella*. Crustaceana 26:172-178.

Chardy, P. 1976. *Storthyngura magnifica* n. sp. abyssal isopod from the North Atlantic. Crustaceana 30:287-291.

Charles River Associates. 1979. Economic feasibility of mining Blake Plateau manganese nodules, CRA Report 461.

Charm, W.B., W.D. Nesteroff and S. Valdes. 1969. Drilling on the continental margin off Florida: Detailed stratigraphic description of the JOIDES cores on the continental margin off Florida. U.S. Geol. Surv. Prof. Paper 581-D, 13 p.

Cheney, J. 1982. The spatial and temporal abundance patterns of chaaetognaths in the western north Atlantic Ocean. Ph.D. Thesis. Mass. Inst. Tech./Woods Hole Oceanogr. Inst.:337.

Chenoweth, S. 1978. F.Z. Zooplankton (Chapter 7.0 Biological Oceanography) In 'Summary of Environmental Information on the Continental Slope - Canadian - United States Border to Cape Hatteras, North Carolina. Submitted to the BLM by TRIGOM, Sout Portland, ME, 7-39 - 7-96.

Cherry, R.D. and L.V. Shannon. 1974. The alpha radioactivity of marine organisms. Atom. En. Rev. 12:3-45.

Cherry, R.D. and M. Heyraud. 1982. Evidence of high radiation in certain mid-water organisms, Science 218:54-56.

Child, C.A. 1982. Deep-sea Pycnogonida from the North and South Atlantic Basins. Smithsonian Contrib. to Zool., 349.

Chryssostomidis, M. 1978. Offshore Petroleum Engineering. A Bibliographic Guide to Publications and Information Sources. MIT Sea Grant Report No. MIT-SG-78-5. Nichols Publishing Co., NY.

Chung, Y., H. Craig, T.L. Ku, J. Goddard and W.S. Broecker. 1974. Radium - 226 measurements from three Geosecs intercalibration stations. Earth Planet. Sci. Lett. 23(1):116-124.

Cifelli, R. 1962. Some dynamic aspects of the distribution of planktonic foraminifera in the western North Atlantic. J. Mar. Res. 20:201-212.

Cifelli, R. 1965. Planktonic foraminifera from the western North Atlantic. Smithson. Misc. Collect. 148(4):1-36.

Cifelli, R. and R.N. Sachs Jr. 1966. Abundance relationship of planktonic foraminifera and radiolaria. Deep-Sea Res. 13: 751-753.

Clapp, R.B., R.C. Banks, D. Morgan-Jacobs and W.A. Hoffman. 1982. Marine birds of the southeastern United States and Gulf of Mexico. Part I, Gaviiformes through Pelecaniformes. U.S. Fish and Wildlife Service, Office of Biol. Serv., Wash., D.C., FWS/OBS-82/01, 637 pp.

Clark, A.M. 1977. Notes on deep-water Atlantic Crinoidea. Bull. British Mus. (Nat. Hist.) Zool. 31(4):159-186.

Clark, R.C., Jr. and D.W. Brown. 1977. Petroleum: properties and analyses in biotic and abiotic systems. In Effects of Petroleum on Arctic and Subarctic Marine Environments and Organisms, Chapter 1, D. C. Malins, (ed.), Academic Press, Inc., New York.

Clark, S.H. and B.E. Brown. 1976. Changes in biomass of finfishes and squids from the Gulf of Maine to Cape Hatteras, 1963-74, as determined from research vessel survey. Fish. Bull. 75(1):1-21.

Clarke, G.L. 1940. Comparative richness of zooplankton in coastal and offshore areas of the Atlantic. Biol. Bull. 78:226-255.

Claypool, G.E. and I.R. Kaplan. 1974. The origin and distribution f methane in marine sediments. in I.R. Kaplan, (ed.), Natural gases in marine sediments: New York, Plenum Press:99-139.

Cleary, W.J. and J.R. Conolly. 1974. Hatteras deep-sea fan. J. Sediment. Petrol. 44::1140-1154.

Cleary, W.J., O.H. Pilkey and M.W. Ayers. 1977. Morphology and sediments of three ocean basin entry points Hatteras Abyssal Plain. J. Sediment. Petrol. 47:1157-1170.

Climatic Study of the Near Coastal Zone East Coast of United States. 1976. Prepared by the Naval Weather Service Detachment, Asherville, NC, published by the Director, Naval Oceanogr. Meteor.

Climatic Summaries of NOAA Data Buoys. 1983. U.S. Dept. Commerce, NOAA National Weather Service, NOAA DATA Buoy Center, NSTL Station, MS.

Cline, J.D. and F.A. Richards. 1972. Oxygen deficient conditions and nitrate reduction in the eastern tropical North Pacific Ocean. Limnol. Oceanogr. 17:885-900.

Colgan, C. 1982. 'The Search for an Ocean Management Policy: The Georges Bank Case." p. 29-70. In J. Goldstein (ed.), The Politics of Offshore Oil, New York, Praeger.

Colton, J.B., Jr., F.D. Knapp and B.R. Burns. 1972. Plastic particles in surface waters of the northwestern Atlantic. Science 185:491-498.

Cochran, J. K. 1980. The flux of 226 Ra from deep-sea sediments. Earth Planet. Sci. Lett. 49:381-392.

Cochran, J.K. 1982. The oceanic chemistry of the U- and Th- series nuclides. in Ivanovich, M. and R.S. Harmon, (ed.) Uranium series disequilibrium: applications to environmental problems. Clarendon Press, Oxford.:384-430.

Cochran, J.K. and H.D. Livingston. 1983. Artificial radionuclide distributions in the Georges Bank region. in

Georges Bank and its surroundings: book and atlas. Woods Hole Oceanogr. Inst. Coastal Res. Center.

Colton, J.B., Jr. 1968. A comparison of current and long-term temperatures of continental shelf waters Nova Scotia to Long Island. ICNAF Res. Bull. 5:110-129.

Colton, J.B., et al. 1974. Plastic particles in surface waters of the Northwestern Atlantic. Science 185:491-497.

Colucci, S.J. 1976. Winter cyclone frequencies over the eastern United States and adjacent western Atlantic, 1964-1973. Am. Meteor. Soc. Bull. 57:548-553.

Columbo, P.E., S.R. Carson, G. Mathieu. 1980. Excess radon in shelf and slope waters of the New York Bight. U.S. Dept. of Energy, Conf. 790382.

Combs, E.R. Inc. 1977. Venture Analysis and Feasibility Study Relating England Fisheries Development Program and the National Marine Fisheries Service; NMFS contract no. 03-7-073-35121.

Connell, J.H. 1978. Diversity in tropical rain forests and coral reefs. Science 199:1302-1310.

Conover, R.J. 1982. Interrelations between microzooplankton and other plankton organisms. Ann. Inst. Oceanogr. Paris. 58(S):31-45.

Cooper, C. and B. Pearce. 1982. Numerical simulations of hurricane-generated currents. J. Phys. Oceanogr. 12, 1071-1091.

Cooper, R.A., and J.R. Uzmann. 1977. Ecology of juvenile and adult clawed lobsters, *Homarus americanus, Homarus gammarus*, and *Nephrops norregicus* - A review In B.F. Phillips and J.S. Cobb (eds.), Workshop on lobster and rack lobster ecology and physiology. Commonw. Sci. Ind. Pres. Org., Div. Fish. Oceanogr. Circ. 7:187-208.

Coull, B.C., R.L. Ellison, J.W. Fleeger, R.P. Higgins, W.D. Hope, W.D. Hummon, R.M. Rieger, W.E. Sterer, H. Thiel and J.H. Tietjen. 1977. Quantitative estimates of the meiofauna from the deep sea off North Carolina, USA. Mar. Biol. 39:233-240.

Cowles, T.J. 1982. Vertical distribution patterns of copepods in Gulf Stream cold-core rings (abs). EOS 63:60.

Cox, J.L. and P.H. Wiebe. 1979. Origins of plankton in the Middle Atlantic Bight. Est. Coast. Mar. Sci. 9:509-527.

Cox, J.L., P.H. Wiebe, P.B. Ortner and S.H. Boyd. 1982. Seasonal development of subsurface chlorophyll maxima in Slope Water and Northern Sargasso Sea in the northwestern Atlantic Ocean. J. Biol. Oceanogr. 1:271-285.

Craddock, J.E. 1968. Neuston Fishing. Oceanus 15(1):14-15.

Cramp, S. (chief ed.). 1977. Handbook of the birds of Europe, the Middle East and North Africa. v. 1, Ostrich-Ducks. Oxford Univ. Press, Oxford.

Cruickshank, M.J. 1975. Technological and enviornmental considerations in the exploration and exploitation of marine minerals, Ph.D. Thesis. Univ. Wisconsin at Madison.

Cruikshank, M.J. and H.D. Hess. 1975. Marine sand and gravel mining, Oceanus 19:32-44.

Csanady, G.T., G. Flierl, D. Karl, D. Kester, T. O'Connor, P. Ortner and W. Philpot. 1979. Deepwater Dumpsite 106. in U.S. Department of Commerce, 1979. Assimilative Capacity of U.S. Coastal Waters for Pollutants. Working Paper No. 1: Federal Plan for Ocean Pollution Research Development and Monitoring, FY 1981-1985. Proceedings of a Workshop, Crystal Mountain, Washington, July 29-August 4, 1979.

Cuhel, R. L., H. W. Jannasch, C. D. Taylor and R. S. Lean. 1983. Microbial growth and macromolecular synthesis in the northwestern Atlantic Ocean. Limnol. Oceanogr. 18: 1-18.

Culver, S.J. and M.A. Buzas. 1982. Recent benthic foraminiferal provinces between Newfoundland and Yucatan. Geol. Soc. Am. Bull. 93:269-277.

Cushing, D.H. 1975. Marine Ecology and Fisheries, Cambridge Univ. Press.

Cutler, E.B. and N.A. Duffy. 1972. A new species of Phascolion Sipuncula from the western North Atlantic. Proc. Biol. Soc. Wash. 85(6):71-75.

Cutler, E.B. 1973. Sipuncula of the western North Atlantic. Bull. Am. Mus. Nat. Hist. 152(3):107-204.

Cutler, E.B. and K. Doble. 1979. North Carolina continental slope zoogeographical barrier. Deep-Sea Res. 26:851-853.

Dalrymple, G.B., C.S. Gromme and R.W. White. 1975. K-Ar age and paleomagnetism of diabase dikes and sills in Liberia: Initiation of central Atlantic rifting. Geol. Soc. Am. Bull. 86:399-411.

Datta-Gupta, A.K. 1981. Atlantic Echiurans 1. Report on 22 species of deep sea Echiurans of the North and South Atlantic Ocean. Bull. Mus. Natl. Hist. Nat. Sect. a Zool. Biol. Ecol. Anim. 3:353-378.

David, P.M. 1965. The surface fauna of the ocean. Endeavour 24:95-100.

Davis, P.G., D.A. Caron and J.M. Sieburth. 1978. Oceanic Amoebae from the North Atlantic culture distribution and taxonomy. Trans. Am. Microsc. Soc. 97:73-88.

Davis, P.G. 1982. Bacterivorous flagellates in marine waters. Ph.D. Thesis, Univesity of Rhode Island, 218p.

Davis, P.G., D.A. Caron, P.W. Johnson and J.McN. Sieburth. 1985. Phototrophic and apochlorotic components of picoplankton and nanoplankton in the North Atlantic: geographic, vertical, seasonal and diel distributions. Mar. Ecol. Prog. Ser. 21:15-26.

Dayal, R., A. Okubo, I.W. Duedall and A. Ramamoorthy. 1979. Radionuclide redistribution mechanisms at the 2800-m Atlantic nuclear waste disposal site. Deep-sea Res. 26A:1329-1345.

Deevey, G.B. 1971. The annual cycle in quantitiy and composition of the zooplankton of the Sargasso Sea off Bermuda. I. The upper 500 m. Limnol. Oceanogr. 16: 219-240.

Deevey, G.B. and A.L. Brooks. 1971. The seasonal cycle in quantity and composition of the zooplankton of the Sargasso Sea off Bermuda. II. The surface to 2000 m. Limnol. Oceanogr. 16:927-943.

Deevey, G.B. and A.L. Brooks. 1977. Copapods of the Sargasso Sea off Bermuda: Species composition, and vertical and seasonal distribution between the surface and 2000 m. Bull. Mar. Sci. 27(2):256-291.

Deis, J.L., F.N. Kurz and E.O. Porter. 1982. A revision of outer Continental Shelf Oil and Gas Activities in the South Atlantic (U.S.) and their onshore impacts; a summary report July 1980. South Atlantic Summary report 2, Outer Continental Shelf Oil and Gas Information Program,

U.S.G.S. Open-file Rept. 82-15, MMS U.S. Department of Interior, 50 pp.

Deller, W.R. 1980. Federation and offshore oil and gas leasing: Must federal tract selections and lease stipulations be consistent with state coastal zone management programs? Univ. Calif., Davis Law Review, 14:105-123.

Desbruyeres, D., J.Y. Bervas and A. Khripounoff. 1980. Un cas de colonisation rapide d'un sediment profond. Oceanologica Acta 3(3).

Deuser, W. and E. H. Ross. 1980. Seasonal change in the flux of organic carbon to the deep Sargasso Sea. Nature 283:364-365.

Devito, C.M. 1981. Radionuclides as tracers of sediment transport processes in North Carolina continental slope sediments. Master's thesis, Univ. of North Carolina, Chapel Hill.

Dickson, R.R. and J. Namias. 1976. North American influences on the circulation and climate of the North Atlantic sector. Monthly Weather Rev. 104:1255-1265.

Dill, R.F. 1966. Sand flows and sand falls. in R.W. Fairbridge (ed.), Encyclopedia of Oceanography, Reinhold Publ. Co., New York, p. 763-765.

Dillon, W.P. and H.B. Zimmerman. 1970. Erosion by biological activity in two New England submarine canyons. J. Sedim. Petrol. 40:542-547.

Dillon, W.P., P. Popenoe, J.A. Grow, K.D. Klitgord, B.A. Swift, C.K. Paull and K.V. Cashman. 1982. Growth faulting and salt diapirism: their relationship and control in the Carolina Trough, eastern North America. Am. Assoc. Petrol. Geol. Mem. 34, p. 21-46.

Dillon, W.P., P. Popenoe, J.A. Grow, K.D. Klitgord, B.A. Swift, C.K. Paull and K.V. Cashman. 1983. Growth faulting and salt diapirism: their relationship and control in the Carolina Trough, eastern North America: Am. Assoc. Pet. Geol. AAPG Hedberg Symposium Volume.

Dorman, C.E. and R.H. Bourke. 1981. Precipitation over the Atlantic Ocean, 30°S to 70°N. Monthly Weather Rev. 109:554-563.

Doyle, L.J., O.H. Pilkey and C.C. Woo. 1979. Sedimentation on the eastern United States Continental Slope, In Doyle, L.J., and Pilkey, O.H. (eds.), Geology of Continental Slopes. Soc. Econ. Paleontol. Mineral. Spec. Publ. 27:119-129.

Doyle, R.W. 1972. Genetic variation in *Ophiomssium lymuni* (Echinodermala) population in the deep sea. Deep-Sea Res. 19:661-664.

Dreisigacker, E. and W. Roether. 1978. Tritium and strontium-90 in North Atlantic surface water. Earth Planet. Sci. Lett. 38(2):301-312.

Duce, R.A., G.L. Hoffman and W.H. Zoller. 1974. Atmospheric trace metals at remote northern and southern hemisphere sites: Pollution or Natural? Science 187:59-61.

Duce, R. A. and R. B. Gagosian. 1982. The input of atmospheric n-ClO to n-C30 alkanes to the ocean. J. Geophys. Res. 87:7192-7200.

Duce, A., G. Quinn, L. Wade. 1974. Residence time of non-methane hydrocarbons in the atmosphere. Mar. Pollution Bull. 5(4):59-61.

Duedall, I.W. et al. (eds.). 1983. Industrial and Sewage Wastes in the Ocean, Volume I. Proceedings of the 2nd International Ocean Dumping Symposium, April 1980, Woods Hole, Mass.

Duedall, I.W., B.H. Ketchum, P.K. Park and D. Kester, (eds.). 1983. Wastes in the Ocean, vol 1, Industrial and Sewage Wastes in the Ocean. Wiley, New York, 300 pp.

Duedall, I.W., B.H. Ketchum, P.K. Park and D. Kester (1984). Wastes in the Ocean vol. IV. Energy Wastes in the Ocean. Wiley, New York, 864 pp.

Dugdale, R. C., D. W. Menzel and J. H. Ryther. 1961. Nitrogen fixation in the Sargasso Sea. Deep-Sea Res. 7: 298-300.

Dunstan, W. M. and J. Hosford. 1977. The distribution of planktonic blue-green algae related to the hydrography of the Georgia Bight. Bull. Mar. Sci. 27: 824-829.

Dyer, R.S. 1976. Environmental surveys of two deep-sea radioactive waste disposal sites using submersibles. Page 317-338, In Symposium on the management of radioactive wastes from the nuclear fuel cycle, proceedings, vol. II. 22-26 March 1976, Vienna, Austria. International Atomic Energy Agency, Vienna, Austria.

EG&G Environmental Consultants. 1978. Analysis report: Interaction of a Gulf Stream eddy with Georges Bank. Appendix D of Eight Quarterly Progress Report, New England Outer Continental Shelf Physical Oceanography Program, submitted to Bureau of Land Management by EG&G Environmental Consultants, Waltham, MA.

Edgar, N.T. and K.C. Bayer. 1979. Assessing oil and gas resources on the U.S. continental margin. Oceanus 22(3):12-22.

Edsall, D.W. 1978. Southeast Georgia embayment high-resolution seismic reflection survey. U.S. Geol. Surv. Open-File Rep. 78-800, 92 p.

Edsall, D.W. 1980. Southeast Georgia Embayment high-resolution seismic-reflection survey. in P. Popenoe (ed.), Final report: environmental studies, southeastern United States Atlantic outer continental shelf 1977: geology. U.S. Geol. Surv. Open-File Rep. 80-146, 9(1-9):28.

Egelson, D.C. 1981. Acoustic interval velocities of the continental rise sediments of the western North Atlantic from Cape Hatteras to Cape Cod. Unpublished. Masters Thesis. Univ. of Rhode Island, Kingston:79 p.

Eittreim, S., P.E. Biscaye and A.F. Amos. 1975. Benthic nepheloid layers and the Ekman thermal pump, Geol. Soc. Am. Abstr. Programs 7:1066-1067.

Eittreim, S.L., E.M. Thorndike and L. Sullivan. 1976. Turbidity distribution in the Atlantic Ocean. Deep Sea Res. 23:1115-1127.

Elbrachter, M. and R. Boje. 1978. On the ecological significance of *Thalassiosira partheneia* in the Northwest African upwelling area. In: Boje, R. & M. Tonczak (eds.). Upwelling ecosystems. Springer-Verlag, Berlin:24-31.

Elkins, C. and M.B. Sprangler. 1968. The potential for marine mining of phosphates and some implications for Federal Policies and Programs, National Planning Assoc.

Embley, R.W. and R.D. Jacobi. 1977. Distribution and morphology of large submarine sediment slides and slumps on Atlantic continental margins. Mar. Geotech. 2:205-228.

Embley, R.W. 1980. The role of mass transport in the distribution and character of deep-ocean sediments with special reference to the North Atlantic. Mar. Geol. 38:23-50.

Embley, R.W. 1982. Anatomy of some Atlantic margin sediment slides and some comments on ages and mechanisms. in S. Saxov and J.K. Nieuwenhuis (eds.), Marine Slides and Other Mass Movements, Plenum Publ. Corp., NY:189-213.

Emerson, S., R. Jahnke, M. Bender, P. Froelich, G. Klinkhammer, C. Bowser and G. Setlock. 1980. Early diagenesis in sediments from the eastern equatorial Pacific: I. Pore water nutrient and carbonate results. Earth Planet. Sci. Lett. 49:57-80.

Emerson, S., V. Grundmanis and D. Graham. 1982. Carbonate chemistry in marine pore waters: MANOP (Manganese Nodule Project) sites C and S. Earth Planet. Sci. Lett. 61:220-232.

Emery, K.O. and D.A. Ross. 1968. Topography and sediments of a small area of the continental slope south of Martha's Vineyard. Deep-Sea Res. 15:415-422.

Emery, K.O., Elazar Uchupi, J.D. Phillips, C.O. Bowin, E.T. Bunce and S.T. Knott. 1970. Continental rise of eastern North America. Am. Assoc. Petrol. Geol. Bull. 54:44-108.

Emery, K.O. and E. Uchupi. 1972. Western North Atlantic Ocean: Topography, rocks, structure, water, life and sediments. Am. Assoc. Petrol. Geol. Mem. 17, 532 p.

Eppley, R.W. 1972. Temperature and phytoplankton growth in the sea. Fish. Bull. 70:1063-1085.

Eppley, R.W. 1981. Relations between nutrient assimilation and growth in phytoplankton with a brief review of estimates of growth rate in the ocean. in T. Platt (ed.), Physiological Bases of Phytoplankton Ecology:251-263.

Ericson, D.B., M. Ewing and B.C. Heezen. 1952. Turbidity currents and sediments in North Atlantic. Am. Assoc. Petrol. Geol. Bull. 36:489-511.

Erséus, C.. 1979a. Taxonomic revision of the marine genus Phallodrilus Pierantoni (Oligochaeta, Tubificidae), with descriptions of thirteen new species. Zool. Scripta. 8:187-208.

Erséus, C.. 1979b. Taxonomic revision of the marine genera Bathydrilus Cook and Macroseta Ersus (Oligochaeta, Tubificidae), with descriptions of six new species and subspecies. Zool. Scripta 8:139-151.

Erséus. C. 1982. *Atlantidrilus*, A new genus of deep-sea tubificidae (Oligochaeta). Sarsia 67:43-46.

Ewing, J.I., M. Ewing and R. Leyden. 1966. Seismic-profiler survey of Blake Plateau: Am. Assoc. Petrol. Geol. Bull. 50:1948-1971.

Ewing, J.I. and C.D. Hollister. 1972. Regional aspects of deep sea drilling in the western North Atlantic. in C.D. Hollister and others, Initial Reports of the Deep Sea Drilling Project, v. XI. Washington, D.C., U.S. Govt. Printing Office:951-973.

Ewing, V.M.. 1984a. Gravity Anomalies. In E. Uchupi and A.N. Shor (eds.), Eastern North American Continental Margin and Adjacent Ocean Floor, 39° to 46° and 64° to 74°W: Atlas 3, Ocean Margin Drilling Program, Regional Atlas Series. Marine Science International, Woods Hole, MA, p. 2.

Ewing, V.M.. 1984b. Gravity Anomalies. In: J.I. Ewing and P.D. Rabinowitz (eds.), Eastern North American Continental Margin and Adjacent Ocean Floor, 34° to 41°N and 68° to 78°W: Atlas 4, Ocean Margin Drilling Program, Regional Atlas Series. Marine Science International, Woods Hole, MA, p. 2.

Ewing, V.M. 1984c. Gravity Anomalies. In: G.M. Bryan and J.R. Heirtaler (eds.), Eastern North American Continental Margin and Adjacent Ocean Floor, 28° to 36°N and 70° to 82°W: Atlas 5, Ocean Margin Drilling Program, Regional Atlas Series. Marine Science International, Woods Hole, MA, p. 2.

Fairbanks, R.G. and P.H. Wiebe. 1980. Foraminifera and chlorophyll maximum: vertical distribution, seasonal succession and paleoceanographic significance. Science 209:1524-1526.

Fairbanks, R.G., P.H. Wiebe and A.W.H. Be. 1980. Vertical distribution and isotopic composition of living planktonic foraminifera in the Western North Atlantic. Science 207:61-63.

Falkowski, P.G. 1980. Light-shade adaptation in marine phytoplankton. in P.G. Falkowski (ed.), Primary productivity in the sea. Plenum:99-119.

Farrington, J.W. and J.M. Teal. 1972. Summary of intercalibration measurements and analysis of open ocean organisms for recently biosynthesized and petroleum hydrocarbons. In Baseline Studies of Pollutants in the Marine Environment, Background Papers. IDOE-NSF, Washington, D.C., p. 583-631.

Farrington, J.W. and P.A. Meyers. 1975. Hydrocarbons in the marine environment. in G. Eglinton (ed.), Environmental Chemistry, Chemical Society, London, Chapter 5, Volume 1.

Farrington, J.W. and B.W. Tripp. 1977. Hydrocarbons in western North Atlantic surface sediments. Geochim. Cosmochim. Acta 41:1627-1641.

Farrington, J. W., N. M. Frew, P. M. Gschwend and B. W. Tripp. 1977. Hydrocarbons in cores of northwestern Atlantic coastal and continental margin sediments. Est. Coast. Mar. Sci. 5:793-808.

Farrington, J. W.. 1980. An overview of petroleum hydrocarbons in the marine environment. In Petroleum: in the Marine Environment, Chapter 1, Advances in Chemistry Series No. 185, L. Petrakis and F. Weiss (eds.), Amer. Chem, Soc. Washington, D. C.

Farrington, J. W. 1982. Sources and distribution processes of chemical contaminants in the coastal zone. Paper presented in the Session 'Pollution of the Economic Zone" Law of the Sea Institute Conference, Halifax, Nova Scotia, Canada.

Farrington, J.W., J.M. Capuzzon, T.M. Leschine and M.A. Champ. 1982. Ocean dumping. Oceanus 25(4):39-50.

Fawley, R.B., S. Cibik, C.K. Rutledge and H.G. Marshall. 1980. Observations of phytoplankton composition on the southeastern continental shelf (Abst). Va. J. Sci. 31:4.

Fedor, L.S., T.W. Godbey, J.F.R. Gower, R. Guptill, G.S. Hayne, C.L. Ruffenach and E.J. Walsh. 1979. Satellite altimeter measurements of sea state: An algorithim comparison. J. Geophys. Res. 84:3991-4001.

Feeley, H.W., G.W. Kipphut, R.M. Trier and C. Kent. 1980.

^{228}Ra and ^{228}Th in coastal waters. Estuarine Coastal Mar. Sci. 11:179-205.

Fenchel, T. 1977. The significance of bactivorous protozoa in the microbial community of detrital particles. in Cairns, J.Jr. (ed.). Aquatic microbial communities. Garland Publishing, New York:529-544.

Fenchel, T. 1982a. Ecology of heterotrophic microflagellates. I. Some important forms and their functional morphology. Mar. Ecol. Prog. Ser. 8: 211-223.

Fenchel, T. 1982b. Ecology of heterotrophic microflagellates. II. Bioenergetics and growth. Mar. Ecol. Prog. Ser. 8:225-231.

Fenchel, T. 1982c. Ecology of heterotrophic microflagellates. IV. Quantitative occurrence and importance as bacterial consumers. Mar. Ecol. Prog. Ser. 9: 35-42.

Fenner, P., G. Kelling and D.J. Stanley. 1971. Bottom currents in Wilmington submarine canyon. Nature Phys. Sci. 229:52-54.

Ferguson, R. L. and A. V. Palumbo. 1979. Distribution of suspended bacteria in neritic waters south of Long Island during stratified conditions. Limnol. Oceanogr. 24:697-705.

Field, M.E. and O.H. Pilkey. 1969. Feldspar in Atlantic continental margin sands off the southeastern United States. Geol. Soc. Am. Bull. 80:2097-2102.

Field, M.E. and O.H. Pilkey. 1971. Deposition of deep-sea sands: comparison of two areas of the Carolina continental rise. J. Sedim. Petrol. 41:526-536.

Fisher, A. 1972. Entrainment of shelf water by the Gulf Stream northeast of Cape Hatteras. J. Geophys. Res. 77:3248-3255.

Fitzgerald, R.A., D.C. Gordon, Jr. and R.E. Cranston. 1974. Total mercury in seawater in the northwest Atlantic Ocean. Deep-Sea Res. 21-139-144.

Fitzgerald, J.L. and J.L. Chamberlin. 1983. Anticyclonic warm-core Gulf Stream rings off the northeastern United States in 1980. Ann. Biol. 37:41-47.

Fletcher, J.B., M.L. Sbar and L.R. Sykes. 1978. Seismic trends and travel-time residuals in eastern North America and their tectonic implications. Geol. Soc. Am. Bull. 89:1656-1676.

Flood, R.D. and A.N. Shor. 1984. Bathymetry. In G.M. Bryan and J.R. Heirtzler (eds.), Eastern North American Continental Margin and Adjacent Ocean Floor, 28° to 36°N and 70° to 82°W: Atlas 5, Ocean Margin Drilling Program, Regional Atlas Series. Marine Science International, Woods Hole, MA, p. 1.

Fofonoff, N.P. and F. Webster. 1971. Current measurements in the western Atlantic. Phil Trans. R. Soc. Lond. A. 270:423-436.

Fofonoff, N.P. 1980. The Gulf Stream System. in Evolution of Physical Oceanography, Scientific Surveys in Honor of Henry Stommel, M.I.T. Press, Cambridge, Mass.:112-139.

Fogg, G. E., W. D. P. Stewart, P. Fay and A. E. Walsby. 1973. The blue-green algae. Academic Press 1-495 p.

Ford, W.L., J.R. Longard and R.E. Banks. 1952. On the nature, occurrence, andorigin of cold low salinity water along the edge of the Gulf Stream. J. Mar. Res. 11:281-293.

Forristall, G.Z., E.G. Ward, L.E. Borgman and V.J. Cardone. 1978. Storm wave kinematics, presented at 1978 Offshore Technology Conference, Houston, 8-11 May.

Fournier, R.O., J. Marra, R. Bohrer and M. van Det. 1977. Plankton dynamics and nutrient enrichment of the Scotian Shelf. J. Fish. Res. Board Can. 34:1004-1018.

Fournier, R.O., M. Van Det, J.S. Wilson and N.B. Hargreaves. 1979. Influence of the shelf-break front off Nova Scotia on phytoplankton standing stock in winter. J. Fish. Res. Bd. Can. 36:1228-1237.

Freeman, Bruce L. and Lionel A. Walford. 1974. The Angler's Guide to the United States Atlantic Coast; Section III. U.S. Department of Commerce, National Oceanic and Atmospheric Administration, NMFS, Washington, D.C., U.S. Gov. Printing office.

Fritz, S.J. and O.H. Pilkey. 1975. Distinguishing bottom and turbidity current coarse layers on the continental rise. J. Sedim. Petrol. 45:57-62.

Fuglister, F.C. and L.V. Worthington. 1947. Hydrography of the Western Atlantic; meanders and velocities of the Gulf Stream. Woods Hole Oceanogr. Inst. Ref. No. 47-9.

Fuglister, F.C. and L.V. Worthington. 1951. Some results of a multiple ship survey of the Gulf Stream. Tellus 3:1-14.

Fuglister, F.C. 1963. Gulf Stream '60. in Prog. Oceanogr. 1:265-373.

Fuglister, F.C. 1972. Cyclonic rings formed by the Gulf Stream 1965-66. in Studies in physical oceanography. A tribute to Georg Wüst on his 80th birthday, A.L. Gordon (ed.), Gordon and Breach, New York, I:137-168.

Fuglister, F.C. 1977. A cyclonic ring formed by the Gulf Stream 1967. A Voyage of Discovery. Pergamon Press, 177-198.

Gagosian, R.B., O.C. Zafiriou, E.T. Peltzer and J.B. Alford. 1982. Lipids in aerosols from the tropical North Pacific: Temporal variability. J. Geophys. Res. 87:133-144.

Gardiner, L.F. 1975. The systematics, postmarsupial development, and ecology of the deep-sea family Neotanaidae (Crustacea: Tanaidacea). Smithsonian Contrib. Zool. 170:1-265.

Gardner, W. 1977. Fluxes, dynamics and chemistry of particulates in the ocean. Ph.D. thesis, MIT/WHOI, 394 pp.

Garrett, C. and W. Munk. 1972. Space-time scales of internal waves. Geophys. Fluid Dyn. 3:225-264.

Garrett, C. and W. Munk. 1975. Space-time scales of internal waves: A progress report. J. Geophys. Res. 80:291-297.

Gatien, M. G. 1976. A study in the slope water region south of Halifax. J. Fish. Res. Board Can. 33:2213-2217.

GEBCO. 1982. General bathymetric chart of the oceans (sheet 5.08, North Atlantic): Canadian Hydrographic Service, Ottawa; Scale 1:10 million, 1 sheet.

General Accounting Office (GAO). 1981. Issues in leasing offshore lands for oil and gas development, report to the Congress of the U.S. by the Comptroller General, EMD-81-59.

Gerrior, P. 1981. The Distribution and Effects of Fishing on the Deep Sea Red Crab, *Geryon quinquedens* (Smith) off Southern New England. Unpublished Ms. thesis: Southeastern Massachusetts University, Dartmouth, MA.

1979. Observation of a subsurface oil-rich layer in the open ocean. Science 205:999-1001.

Harvey, G. R. and W. G. Steinhauer, 1976. Transport pathways of polychlorinated biphenyls in Atlantic water. J. Mar. Res. 34:561-575.

Harvey, G. R. and W. G. Steinhauer, 1976. Biogeochemistry of PCB and DDT in the North Atlantic. In Environmental Biogeochemistry, Carbon, Nitrogen Phosphorus, Sulfur and Selenium Cycles, Chapter 15, Volume 1, J. 0. Nriagu (ed.), Ann Arbor Science Publishers, Ann Arbor, Michigan.

Harvey, G. R., A. G. Requejo, P. A. McGillivary and J. Tokar, 1979. Observation of a subsurface oil-rich layer in the open ocean. Science 205:999-1001.

Hathaway, J.C. 1971. Data file, continental margin program, Atlantic coast of the United States, 2, Sample collection and analytical data. Woods Hole Oceanogr. Inst. Ref. No. 71-15:489.

Hathaway, J.C. 1972. Regional clay-mineral facies in estuaries and continental margin of the United States east coast. Geol. Soc. Am. Mem. 133:293-316.

Hathaway, J.C., J.S. Schlee, C.W. Poag, P.C. Valentine, E.G.A. Weed, M.H. Bothner, F.A. Kohout, F.T. Manheim, R.Schoen, R.E. Miller and D.M. Schultz (eds.) 1976. Preliminary summary of the 1976 Atlantic Margin Coring Project of the U.S. Geological Survey. U.S. Geol. Surv. Open-File Rep. 76-844, 217 p.

Hathaway, J.C., C.W. Poag, P.C. Valentine, R.E. Miller, D.M. Schultz, F.T. Manheim, F.A. Kohout, M.H. Bothner and D.A. Sangrey. 1979. U.S. Geological Survey core drilling on the Atlantic Shelf. Science 206:515-527.

Haury, L.R., J.A. McGowan and P.H. Wiebe. 1978. Patterns and processes in the time-space scales of plankton distributions. in J.H. Steele (ed.), Spatial pattern in plankton communities. Plenum Press, New York. pp. 277-327.

Hausknecht, K.A., 1977. Results of studies on the distribution of some transition and heavy metals at Deepwater Dumpsite 106. pp. 499-542 in NOAA, Baseline Report of Environmental Conditions in Deepwater Dumpsite 106. v. III. Contaminant inputs and chemical characteristics. NOAA Dumpsite Evaluation Report 77-1. NOAA, Rockville, MD.

Hausknecht, K.A. and D.R. Kester, 1981. 106 mile ocean waste disposal site chemical data report, In NOAA/ODP Dumpsite Evaluation Rept. Ɛ1-1, pp. 70-79.

Hayden, B.P. 1981. Secular variation in Atlantic coast extratropical cyclones. Monthly Weather Review 109:159-167.

Hebert, P.J. 1974. North Atlantic tropical cyclones. 1973. Mariner's Weather Log, 18:8-16.

Hebert, P.J. 1976. North Atlantic tropical cyclones. 1975. Mariner's Weather Log, 20:63-73.

Hebert, P. and G. Taylor. 1979. Everything you always wanted to know about hurricanes - Part II. Weatherwise 32:100-107.

Hebert, P.J. 1980. Annual Data and Verification Tabulation Atlantic Tropical Cyclones 1979. National Hurricane Center, Coral Gables, FL, Tech. Memo:84.

Hecker, B.M., G. Blechschmidt and P. Gibson. 1980. Final Report for the Canyon Assessment Study in the mid- and north Atlantic Areas of the U.S. outer continental shelf. Submitted to Bureau of Land Management under Contract No. BLM-AA551-CT8-49, 388 p.

Hecker, B. and D.T. Logan. 1981. First interim report for the Canyon and Slope Processes Study. Prepared for the Bureau of Land Management Under Contract No. BLM AA851-CTO-59(March, 1971):65 p.

Hecker, B., D.T. Logan, J. Stepien, W.D. Gardner and K. Hunkins. 1981. Second interim report for the Canyon and Slope Processes Study: Prepared for the Bureau of Land Management Under Contract No. BLM AA851-CTO-59(Nov., 1981)135 and 3 maps.

Hecker, B. 1982. Possible benthic fauna and slope instability relationships. in S. Saxov, and J. Nieuwenhuis (eds.), Marine Slides and other Mass Movements. Plenum Publ. Corp.:335-347.

Hecker, B., D.T. Logan and F.E. Gandorillos, 1983. Epifauna, In Fourth Interim Report for the Canyon and Slope Processes Study. Prepared for the Bureau of Land Management under MMS Contract No. AA851-CT0-59. Lamont-Doherty Geological Observatory, NY.

Hecker, B., D.T. Logan, F.E. Gandarillas and P.R. Gibson, 1983. Megafaunal assemblages in Lydonia Canyon, Baltimore Canyon, and selected slope areas. Chapter 1, Biological Processes. In Volume III, Canyon and Slope Processes Study, pp. 1-140.

Heezen, B.C. and M. Ewing. 1952. Turbidity currents and submarine slumps and the Grand Banks earthquake. Am. J. Sci. 250:849-873.

Heezen, B.C., Marie Tharp, and Maurice Ewing. 1959. The floors of the oceans, 1. The North Atlantic. Geol. Soc. Am. Spec. Paper 65:122.

Heezen, B.C., C.D. Hollister and W.F. Ruddiman. 1966. Shaping of the continental rise by deep geostrophic contour currents. Science 152:502-508.

Heezen, B.C. and C.D. Hollister. 1971. The face of the deep. New York, Oxford Univ. Press:759 p.

Heinbokel, J.F. 1978. Studies on the functional role of tintinnids in the Southern California Bight. I. Grazing and growth rates in laboratory cultures. Mar. Biol. 47: 177-189.

Heinbokel, J.F. and J.R. Beers. 1979. Studies on the functional role of tintinnids in the Southern California Bight. III. Grazing impact of natural assemblages. Mar. Biol. 52: 23-32.

Heirtzler, J.R., P.T. Taylor, R.D. Ballard and R.L. Houghton. 1977a. The 1974 ALVIN Dives on Corner Rise and New England Seamounts. Woods Hole Oceanogr. Inst. Tech. Rep. 77-8, 59 p.

Hempel, G. and H. Weikert. 1972. The neuston of the subtropical and boreal Northeastern Atlantic Ocean. A review. Mar. Biol. 13:70-88.

Henry, V.J., C.J. McCreery, F.D. Foley and D.R. Kendall, 1981. Ocean bottom survey of the Georgia Bight, In P. Popenoe, (ed.), Environmental geologic studies on the southeastern Atlantic Outer Continental Shelf, 1977-1978. U.S. Geol. Surv. Open-File Rept. 81-582A:6-1 - 6-85.

Herman, A.W. and K.L. Denman. 1979. Intrusions and vertical mixing at the shelf/slope water front south of Nova Scotia. J. Fish. Res. Bd. Can. 36:1445-1453.

Grice, G.D. and A.D. Hart. 1962. The abundance, seasonal occurrence, and distribution of the eipzooplankton between New York and Bermuda. Ecol. Monogr. 32:287-309

Grow, J.A., R.E. Mattick and J. Schlee. 1977. Depth conversion of multi-channel seismic-reflection profiles over Atlantic outer continental shelf and upper continental slope between Cape Hatteras and Georges Bank. Am. Assoc. Pet. Geol. Bull. 61:790-791.

Grow, J.A., C.O. Bowin and D.R. Hutchinson. 1979. The gravity field of the U.S. Atlantic continental margin. in C.E. Keen (ed.), Crustal properties across passive margins, Tectonophys. 59:27-52.

Grow, J.A., D.R. Hutchinson, K.D. Klitgord, W.P. Dillon and J.S. Schlee. 1983. Representative multichannel seismic reflection profiles over the U.S. Atlantic continental margin. In A.W. Bally (ed.), Am. Assoc. Petr. Geol. Seismic Structural Atlas 2.2.3 (Passive margins).

Grosz A.E. and E.C. Escowitz. 1983. Economic heavy minerals of the U.S. Atlantic Continental Shelf, in W.F. Tanner, ed., Sixth Symposium on Coastal Sedimentology. Southeastern Geologic Studies Association, Florida State University, Tallahassee, FL.

Gschwend, P.M., O.C. Zafiriou, R.F.C. Mantoura, R.P. Schwarzenbach and R.B. Gagosian. 1982. Volatile organic compounds at a coastal site. I. Seasonal variations. Environ. Sci. Technol. 16:31-38.

Gulf of Mexico and South Atlantic Fishery Management Councils. 1982. Fishery Management Plan Final Environmental Impact Statement for Coral and Coral Reefs of the Gulf of Mexico and South Atlantic.

Haas, L.W. and K.L. Webb. 1979. Nutritional mode of several non-pigmented microflagellates from the York River Estuary, Virginia. J. Exp. Mar. Biol. Ecol. 39: 125-134.

Haas, L.W. 1982. Improved epifluorescent microscopic technique for observing planktonic micro-organisms. Ann. Inst oceanogr. Paris. 58(S):261-266.

Hachey, H.B. 1955. Water replacements and their significance to a fishery. Deep-Sea Res. suppl. 3:68-73.

Haedrich, R.L., G.T. Rowe and P.T. Polloni. 1975. Zonation and faunal composition of epibenthic populations on the continental slope south of New England. J. Mar. Res. 33:191-212.

Haedrich, R.L. and P.T. Polloni. 1976. A contribution to the life history of a small Rattail Fish Coryphaenoides-Carapinus. Bull. South Calif. Acad. Sci. 75:203-211.

Haedrich, R.L. and G.T. Rowe, 1977. Megafaunal biomass in the deep sea. Nature 269:141-142.

Haedrich, R.L., G.T. Rowe and P.T. Polloni. 1980. The megabenthic fauna in the deep sea south of New England, USA. Mar. Biol. 57:165-179.

Haidvogel, D.B. 1983. The dynamics of the Gulf Stream System: observations and models. EOS 64:1027.

Hall, R.W. and H.R. Ensminger (eds). 1979. Potential geologic hazards and constraints for blocks in proposed Mid-Atlantic OCS oil and gas lease sale 49: U.S. Geol. Surv. Open-File Rep. 79-264:176.

Halliwell, G.R., Jr. and C.N.K. Mooers. 1979. Space-time structure and variability of the shelf water-slope water and Gulf Stream surface temperature fronts and associated warm-core eddies. J. Geophys. Res. 84:7707-7725.

Hamilton, P. 1982. Analysis of current meter records at the northwest Atlantic 2800 meter radioactive waste dumpsite. EPA 520/1-82-002, 98 p.

Hamilton, R.D. and J.E. Preslan. 1969. Cultural characteristics of a pelagic marine Hymenostome ciliate, Uronema sp.. J. Exp. Mar. Biol. Ecol. 4:90-99.

Hammond, E.E., H.J. Simpson and G.G. Mathieu. 1977. J. Geophys. Res., 82:3913.

Harbison, G.R., L.P. Madin and N.R. Swanberg. 1978. On the natural history and distribution of oceanic Ctenophores. Deep-Sea Res. 25:233-256.

Harding, G.C.H. 1974. The food of deep-sea copepods. J. Mar. Biol. Ass. U.K. 54:141-155.

Hardy, E.B., P.W. Krey and H.L. Volchok. 1973. Global inventory and distribution of fallout plutonium. Nature 241, 444.

Hardy, E.P., P.M. Krey and H.L. Volchok. 1972. Global inventory and distribution of Pu-238 from SNAP 9A, U.S.A.E.C. Health Safety Lab. Rep. HASL-250.

Harland, W.B., A.V. Cox, D.G. Llewellyn, C.A.G. Pickton, A.G. Smith and R. Walters. 1982. A geologic time scale: Cambridge Univ. Press, New York.

Harris, G.P. 1978. Photosynthesis, productivity and growth. The physiological ecology of phytoplankton. Ergeb. Limnol. 10:1-171.

Harris, G.P. 1980. The measurement of photocynthesis in natural populations of phytoplankton. in I. Morris (ed.), The physiological ecology of phytoplankton. Univ. of Calif. Press, pp. 129-187.

Harris, R., C. Smith, R. Bieri, C. Ruddell and H. Kator, 1978. Middle Atlantic outer continental shelf environmental studies. Vol. II-B. Chemical and biological benchmark studies. BLM/St-(78/28):688.

Hart, A.D. and J.A. Wormuth. 1982. Pelagic amphipods of Gulf Stream cyclonic rings. EOS 63:60.

Hartman, O. 1965. Deep water benthic polychaetous annelids off New England to Bermuda and other North Atlantic areas. Allan Hancock Found. Publ. Occas. Pap. 28:1-378.

Hartman, O. and K. Fauchald. 1971. Deep water benthic Polychaetous Annelids off New-England to Bermuda and other North Atlantic areas Part 2. Allan Hancock Monogr. Mar. Biol. 6:1-327.

Harvey, G. R., W. G. Steinhauer and J. M. Teal, 1972. Polychlorobiphenyls in North Atlantic Ocean water. Science 180:179-180.

Harvey, G. R., H. P. Miklas, V. T. Bowen and W. G. Steinhauer, 1974. Observations on the distribution of chlorinated hydrocarbons in Atlantic Ocean organisms. J. Mar. Res. 32:103-118.

Harvey, G. R., W. G. Steinhaver and H. P. Miklas, 1974. Decline of PCB concentrations in North Atlantic surface water. Nature 252:387-388.

Harvey, G.R. and W.G. Steinhauer. 1976. Biogeochemistry of PCB and DDT in the North Atlantic. Environ. Biogeochem. 1:203-221.

Harvey, G.R., A.G. Requejo, P.A. McGillivary and J. Tokar.

Geyer, R.A. (ed.). 1980. Marine Environmental Pollution, 1. Hydrocarbons. Elsevier Oceanography Series 27A. Elsevier Scientific Publishing Co., New York.

Geyer, R.A. and C.P. Giammona. 1980. Naturally occurring hydrocarbons in the Gulf of Mexico and Caribbean Sea, pp. 37-196. In Marine Environmental Pollution. I. Hydrocarbons, R. Geyer (ed.), Elsevier, New York.

Gibbs, R.J., J.E. Lathrop, O. Farah and G.B. Robinson. 1979. Suspended material and composition at the shelf break in the Wilmington Canyon area. EOS 60:281.

Gilbert, L.E. and W.P. Dillon. 1981. Bathymetric map of the Blake Escarpment. U.S. Geol. Surv. Misc. Field Studies Map MF-1362, 1 sheet.

Given, M.M. 1977. Mesozoic and Early Cenozoic geology of offshore Nova Scotia. Bull. Can. Petrol. Geol. 25:63-91.

Glibert, P.M. 1982. Regional studies of daily, seasonal and size fraction variability in ammonium remeneralization. Mar. Biol. 70:209-222.

Glibert, P.M. and J.J. McCarthy. 1984. Uptake and assimilation of ammonium and nitrate by phytoplankton: indices of nutritional status for natural assemblages. J. Plank. Res. 6:677-697.

Glibert, P.M., M.R. Dennett and J.C. Goldman. 1985. Inorganic carbon uptake by phytoplankton in Vineyard Sound, Massachusetts. II. Comparative primary productivity and nutritional status of winter and summer assemblages. J. Exp. Mar. Biol. Ecol. 86:101-118.

Godshall, F.A., R.G. Williams, J.M. Bishop, F. Everdale and S.W. Fehler. 1980. A climatologic and oceanographic analysis of the Georges Bank Region of the Outer Continental Shelf. Final report to the Bureau of Land Management, U.S. Department of the Interior. Nat. Ocean. Atmos. Admin. Environ. Data Information Sevice 290 p.

Gold, K. 1970. Cultivation of marine ciliates (Tintinnida) and heterotrophic flagellates. Helgolander wiss. Meers. 20:264-271.

Gold, K. and E.A. Morales. 1975. Seasonal changes in lorica sizes and the species of Tintinnida in the New York Bight. J. Protozool. 22:520-528.

Goldman, J.C. and J.H. Ryther. 1976. Temperature-influenced species competition in mass cultures of marine phytoplankton. Biotech. Bioeng. 18:1125-1144.

Goldman, J.C., J.J. McCarthy and D.G. Peavey. 1979. Growth rate influence on the chemnical composition of phytoplankton in oceanic waters. Nature 279:210-215.

Goldman, J.C. and P.M. Glibert. 1983. Kinetics of inorganic nitrogen uptake by phytoplankton. In E.G. Carpenter and D.G. Capne (eds.), Nitrogen in the Marine Environment, Academic Press, New York.

Goldman, J.C. 1984. Conceptual role for microaggregates. Bull. Mar. Sci. 35:in press.

Goldman, J.C. and D.A. Caron. 1985. Experimental studies on an omnivorous microflagellate: Implications for grazing and nutrient regeneration in the marine microbial food chain. Deep-Sea Res.

Gordon, A.L. and F. Aikman. 1981. Mid-Atlantic Bight pycnocline salinity maximum. Limnol. Oceanogr. 26:123-130.

Gordon, D.C., P.D. Keiser and J. Dale. 1974. Estimates using fluorescence spectroscopy of the present state of petroleum hydrocarbon contamination in the water column of the Northwest Atlantic Ocean . Mar. Chem. 2:251-262.

Gorsline, D.S. 1963. Bottom sediments of the Atlantic shelf and slope off the southern United States. J. Geol. 71:422-440.

Gowing, M.M. and M.W. Silver. 1982. Fecal pellet breakdown: An inside or an outside job? EOS 63:54.

Grant, G.C. 1977. Middle Atlantic Bight Zooplankton: Seasonal bongo and neuston collections along a transect off southern New Jersey. Special Report in Applied Marine Science and Ocean Engineering No. 173 VA Inst. Marine Sci.

Grant, G.C., et al. 1979. Middle Atlantic Zooplankton: 2nd Year Results and a Discussion of the Two-Year BLM-VIMS Survey. Special Report in Applied Marine Science and Ocean Engineering, VA Inst. Marine Sci. No. 192.

Grassle, J.F. 1967. Influence of environmental variations on species diversity in benthic communities of the continental shelf and slope. Ph.D. Thesis. Duke University, 195 pp.

Grassle, J.F. 1972. Species diversity, genetic variability and environmental uncertainty. in B. Battaglia (ed.), Fifth European Marine Biological Symposium:19-26.

Grassle, J.F. and H.L. Sanders. 1973. Life histories and the role of disturbance. Deep-Sea Res. 20:643-659.

Grassle, J.F., H.L. Sanders, R.R. Hessler, G.T. Rowe and T. McLellan. 1975. Pattern and zonation: a study of the bathyal megafauna using the research submersible ALVIN. Deep-Sea Res. 22:457-481.

Grassle, J.F. 1977. Slow recolonisation of deep-sea sediment. Nature 265:618-619.

Grassle, J.F. and J.P. Grassle. 1978. Life histories and genetic variation in marine invertebrates. pp. 347-364. In Marine Organisms, B. Battaglia and J.A. Beardmore (eds.), Plenum Press, New York and London.

Grassle, J.F., H.L. Sanders and W.K. Smith. 1979. Faunal changes with depth in the deep-sea benthos. Ambio Spec. Rep. (6):47-50.

Gray, J.S. 1974. Animal-sediment relationships. Oceanogr. Mar. Biol. Ann. Rev. 12:223-261.

Gray, W. 1979. Hurricanes: their formation, structure, and likely role in the tropical circulation, in Meteorology Over the Tropical Oceans. Roy. Meteor. Soc.:115-218.

Greig, R.A., D.R. Wenzloff, A. Adams, B. Nelson and C. Shelpuk. 1977. Trace metals in organisms from ocean disposal sites of the Middle Atlantic Eastern United States. Arch. Environm. Contam. Toxicol. 6:395-409.

Greig, R.A. and D. Wenzloff. 1977. Final report on heavy metals in small pelagic finfish, euphausiid crustaceans and apex predators, including sharks as well as on heavy metals and hydrocarbons (C_{15+}) in sediments collected at stations in and near Deepwater Dumpsite 1067. pp. 547-564 in NOAA, Baseline Report of Environmental Conditions in Deepwater Dumpsite 106. v. III. Contaminant inputs and chemical characteristics. NOAA Dumpsite Evaluation Report 77-1. NOAA, Rockville, MD.

Herman, S.S. and J.A. Mihursky. 1964. Infestation of the copepod *Acartia tonsa* with the stalked ciliate *Zoothamnium*. Science 146: 543-544.

Hessler, R.R. 1967. A record of Serolidae (Isopoda) from the North Atlantic Ocean. Crustaceana 12:159-162.

Hessler, R.R. 1968. The systematic position of Dactylosylis Richardson (Isopoda, Asellota). Crustaceana 14:143-146.

Hessler, R.R. 1970a. The Desmosomatidae (Isopoda, Asellota) of the Gay Head-Bermuda transect. Bull. Scripps Inst. Oceanogr., 15:1-185.

Hessler, R.R. 1970b. A new species of Serolidae (Isopoda) from bathyal depths of the equatorial Atlantic Ocean. Crustaceana 18:227-232.

Hessler, R.R. and H.L. Sanders. 1967. Faunal diversity in the deep sea. Deep-Sea Res. 14:65-78.

Hessler, R.R., G.D. Wilson and D. Thistle. 1979. The deep-sea isopods: A biogeographic and phylogenetic overview. Sarsia 64:67-75.

Hilland, J.E. 1983. Variation in the shelf-water front position from Georges Bank to Cape Romain in 1980. Annales Biologiques 37:38-40.

Hinga, K.R., P.G. Davis and J.McN. Sieburth. 1979. Enclosed chambers for the reverse flow concentration and selective filtration of particles. Limnol. Oceanogr. 24: 536-540.

Hinga, K. R., J. M. Sieburth and G. R. Heath, 1979. The supply and use of organic material at the deep-sea floor. J. Mar. Res. 37:557-579.

Hirche, H.J. von. 1974. Die Copepoden Eurytemora affina Poppe und Acartia tonsa Dana und ihre besiedlung durch Myoschiston centropagidarum Precht (Peritricha) in der Schlei. Kieler Meeresforsch. 30:43-64.

Ho, F.P., R.W. Schwerdt and H.V. Goodyear. 1975. Some climatological characteristics of hurricanes and tropical stormes, Gulf and East coasts of the United States, 1975. NOAA, National Weather Service Technical Report NWS 15, June 1975, 87 pp.

Hobbie, J. E., O. Holm-Hansen, T. T. Packard, L. R. Pomeroy, R. W. Sheldon, J. P. Thomas and W. J. Wiebe. 1972. A study of the distribution and activity of microorganisms in ocean water. Limnol. Oceanogr. 17: 544-555.

Hodson, R. E. 1981. Dissolved ATP utilization by free living and attached bacterioplankton. Mar. Biol. 64: 43-52.

Hogg, N.G. 1981. Topographic waves along 70°W on the Continental Rise. J. Mar. Res. 39:627-649.

Hogg, N.G. 1983. A note on the deep circulation of the western North Atlantic: Its nature and causes. Deep-Sea Res. 30:945-961.

Holland, W.R. and L.B. Lin. 1975a. On the generation of mesoscale eddies and their contribution to the oceanic general circulation, I, A preliminary numerical experiment. J. Phys. Oceanogr. 5:642-657.

Holland, W.R. and L.B. Lin. 1975b. On the generation of mesoscale eddies and their contribution to the oceanic general circulation, II, A parameter study. J. Phys. Oceanogr. 5:658-669.

Holland, W.R. 1978. The role of mesoscale eddies in the general circulation of the ocean - numerical experiments using a wind-driven quasi-geostrophic model. J. Phys. Oceanogr. 8:363-392.

Holland, W.R., D.E. Harrison and A.D. Semtner, Jr. 1983. Eddy-resolving models of large-scale ocean circulation. Chapter 17, in Eddies in Marine Science, A.R. Robinson, (ed.), Springer-Verlag, Berlin, Heidelberg.

Hollister, C.D. and B.C. Heezen. 1972. Geologic effects of ocean bottom currents: western North Atlantic, in A.L. Gordon (ed.), Studies in physical oceanography — a tribute to George Wüst on his 80th birthday. New York, Gordon and Breach:659.

Hollister, C.D., J.I. Ewing and others. 1972a. Shipboard report, site 107-Upper Continental Rise. Initial Reports of the Deep Sea Drilling Project, v. XI, p. 351-364.

Hollister, C.D. 1973. Atlantic Continental Shelf and Slope of the United States - Texture of Surface Sediments from New Jersey to Southern Florida. U.S. Geol. Surv. Prof. Paper 529-M.

Hollister, C.D., R.D. Flood, D.A. Johnson, P.F. Lonsdale and J.B. Southard. 1974. Abyssal furrows and hyperbolic echo traces on the Bahama outer ridge. Geology:6.

Hollister, C.D., J.B. Southard, R.D. Flood and P.E. Lonsdale. 1976. Flow phenomena in the benthic boundary layer and bed forms beneath deep-current systems. in I.N. McCave (ed.), The Benthic Boundary Layer. Plenum Publ. Corp., NY:183-204.

Hollister, C.D. and I.N. McCave, 1984. Sedimentation under deep-sea storms. Nature 309:2200-225.

Honjo, S. and M.R. Roman. 1978. Marine copepod fecal pellets: Production, preservation and sedimentation. J. Mar. Res. 36:45-57.

Honjo, S. 1980. Material fluxes and modes of sedimentation in the mesopelagic and bathypelagic zones. J. Mar. Res. 38:53-97.

Hope, W.D. and D.G. Murphy. 1969a. *Rhaptothyreys typicus* n.g., n.sp. an abyssal marine nematode representing a new family of uncertain taxonomic position. Proc. Biol. Soc. Wash., 82:81-92.

Hope, W.D. and D.G. Murphy. 1969b. *Syringonomus typicus* n.g., n.sp. (Enoplida: Leptosomatidae) a marine nematode inhabiting arenaceous tubes. Proc. Biol. Soc. Wash., 82:511-518.

Hope, W.D. and D.G. Murphy. 1970. A redescription of *Enoplus groenlandicus* Ditlevsen, 1926 (Nematoda: Enoplidae). Proc. Biol. Soc. Wash., 83:227-240.

Hope, W.D., 1977. *Deontostorma coptochilus* n. sp., a marine nematode (Leptosomatidae) from the foot cavity of the deep-sea anemone *Actinauge longicornis* (Verrill, 1882). Proc. Biol. Soc. Wash. 90:946-962.

Hopkins, T.S. and N. Garfield, III. 1981. Physical origins of Georges Bank water. J. Mar. Res. 39:465-500.

Horn, D.R., M.N. Delach and B.M. Horn. 1974. Physical properties of sedimentary provinces, North Pacific and North Atlantic Oceans. in Deep-Sea Sediments: Physical and Mechanical Properties: Relationships Between Physical, Mechanical and Geologic Properties, Mar. Sci. 2:417-441.

Hotchkiss, F.S. and C. Wunsch. 1982. Internal waves in Hudson Canyon with possible geological implications. Deep Sea Res. 29:415-442.

Houghton, R.L., J.R. Heirtzler, R.D. Ballard and P.T. Taylor.

1977. Submersible observations of the New England Seamounts. Naturwiss. 64:348-355.

Houghton, R.W., R. Schlitz, R. Beardsley, B. Butman and J.L. Chamberlin. 1982. The Middle Atlantic Bight cold pool: evolution of the temperature structure during summer 1979. J. Phys. Oceanogr.

Hulburt, E.M., J.H. Ryther and R.R.L. Guillard. 1960. The phytoplankton of the Sargasso Sea off Bermuda. J. Du Conseil 25:115-128.

Hulburt, E.M. 1962. A note on the horizontal distribution of phytoplankton in the open ocean. Deep-Sea Res. 9:72-74.

Hulburt, E.M. and R.S. MacKenzie, 1971. Distribution of Phytoplankton species at the western margin of the North Atlantic Ocean. Bull. Mar. Sci. 21:603-612.

Hulburt, E.M. 1983. The capacity for change and the unpredictability of the phytoplankton of the east coast of the United States. J. Plank. Res. 5:35-42.

Hulsemann, J. 1967. The continental margin off the Atlantic coast of the United States: carbonate in sediments, Nova Scotia to Hudson Canyon: Sedimentol. 8:121-145.

Human Sciences Research Inc. 1979. Marine Recreational Fishery Statistics Survey, Atlantic and Gulf Coasts. Prepared for the U.S. Department of Commerce, NMFS, Resource Statistics Division, F/SR1.

Humes, A. 1974. New cyclopoid copepods associates with an abyssal holothurian in the eastern North Atlantic. J. Nat. Hist. 8:101-117.

Hunkins, K.L. 1983. Circulation and upwelling in Baltimore Canyon. EOS 64:1059.

Hunkins, K., W.d. Gardner, J.C. Stepien, B. Hecker, D.T. Logan and F.E. Gandanillas. 1983. Fourth interim report for the canyon and slope processes study. Prepared for the Bureau of Land Management under MMS Contract No. AA851-CTO-59(Jan., 1983)405.

Hunt, J.M. 1979. Petroleum Geochemistry and Geology. W.H. Freeman and Company, San Francisco.

Hunt, J.M. and J.K. Whelan. 1979. Volatile organic compounds in Quaternary sediments. Organnic Geochem. 1:219:224.

Huntsman, S.A. and W.G. Sunda. 1980. The role of trace metals in regulating phytoplankton growth with emphasis on Fe, Mn and Cu. in I. Morris (ed.), The Physiological ecology of phytoplankton, Univ. Calif. Pres,:285.

Hurd, M. 1982. Testimony. In Ocean Dumping: Hearings before Committee on Merchant Marine and Fisheries. House of Representatives, Ninety-Seventh Congress, Second session, Serial No. 97-40(No. 141):45-50.

Hurley, R.J. 1964. Bathymetric data from the search for USS "Thresher": Internat. Hydrog. Rev. 41:43-52.

I.O.C. 1982. Report of Fourth Session of the Working Committee for the Global Investigation of Pollution in the Marine Environment (GIPME). IOC-UNESCO, Paris.

Institute of Petroleum. 1972. IP Model code of safe practices in the petroleum industry. 34th revised edition. Applied Science Publishers. Part 6: Petroleum Pipelines Safety Code, 3rd edition (1967). Part 8: Drilling Production and Pipeline Operations in Marine Areas Safety Code.

Iselin, C.O'D. 1936. A study of the circulation of the western North Atlantic. Papers in Phys. Oceanogr. Meteorol. 4(4):101.

Iselin, C.O'D. and F.C. Fuglister. 1948. Some recent developments in the study of the Gulf Stream. J. Mar. Res. 7:317-329.

Istoshin, Y.V. 1961. Formative area of "eighteen-degree" water in the Sargasso Sea. Okeanologiia 1:600-607. (Translation in Deep-Sea Research 9:384-390.)

Jackson, D.W. 1982. Atlantic Ocean Disposal Sites Literature Review. Sandia National Laboratories, Albuquerque, N.M. Report SAND82-7025.

Jahn, A.E. 1976. On the midwater fish fauna of Gulf Stream rings with respect to habitat differences between Slope Water and Northern Sargasso Sea. Ph.D. Thesis, Woods Hole Oceanographic Institution, 173 pp.

Jahn, A.E. and R.H. Backus. 1976. On the mesopelagic fish faunas of slope water Gulf Stream and northern Sargasso Sea. Deep-Sea Res. 23(3):223-234.

Jamart, B.M., D.F. Winter, K. Banse, G.C. anderson and R.K. Lane. 1977. A theoretical study of phytoplankton growth and nutrient distribution in the Pacific Ocean off the northwestern U.S. Coast. Deep-Sea Res. 24:753-773.

Jannasch, H.W., K. Eimhjellen, C.O. Wirsen and A. Farmanfarmaian. 1971. Microbial degradation of organic matter in the deep sea. Science 171:672-675.

Jannasch, H. W. 1973. Bacterial content of particulate matter in offshore surface waters. Limnol. Oceanogr. 18: 340-342.

Jannasch, H. W. and C. O. Wirsen. 1973. Deep-sea microorganisms: in situ response to nutrient enrichment. Science 180:641-643.

Jannasch, H. W. and C. O. Wirsen. 1977. Microbial life in the deep sea. Sci. Amer. 236: 42-52.

Jannasch, H. W. 1979. Microbial turnover of organic matter in the deep sea. BioScience 29: 228-232.

Jannasch, H. W. and C. O. Wirsen. 1980. Studies on the microbial turnover of organic substrates in deep sea sediments. in R. Dumas (ed.), Biogeochimie de la matiere organique a l'interface eau-sediment marin. Actes des Colloques de C.N.R.S. 293: 285-290.

Jannasch, H. W. and C. O. Wirsen. 1982. Microbial activities in undecompressed and decompressed deep-sea water samples. Appl. Environ. Microbiol. 43: 1116-1124.

Jannasch, H. W., C. O. Wirsen and C. D. Taylor. 1982. Deep-sea bacteria: isolation in the absence of decompression. Science 216:1315-1317.

Jansa, L.F. and J.A. Wade. 1975. Geology of the continental margin off Nova Scotia and Newfoundland, In Offshore geology of eastern Canada regional geology. Can. Geol. Surv. Paper 74-30:51-105.

Jansa, L.F., P. Enos, B.E. Tucholke, F.M. Gradstein and R.E. Sheridan. 1979. Mesozoic-Cenozoic sedimentary formations of the North American Basin: western north Atlantic. in M. Talwani, W. Hay, W.B.F. Ryan (eds.), Deep drilling results in the Atlantic Ocean: continental margins and paleoenvironment. Am. Geophys. Union Maurice Ewing Ser., 3:1-57.

Jansa, L.F. 1981. Mesozoic carbonate platforms and banks

of the eastern North American margin. Mar. Geol. 44:97-117.

Jenkins, W.J. and P.B. Rhines. 1980. Tritium in the deep North Atlantic ocean. Nature 286:877-880.

Jenkins, W.T. 1982. Oxygen utilization rates in the North Atlantic subtropical gyre: Implications for primary production in "oligotrophic" systems. Nature 300:246-248.

Jessop, A.M., M.A. Hobart and J.G. Sclater. 1976. The world heat flow data collection-1975: Geothermal Series No. 5; Energy, Mines, and Resources, Canada, Ottawa, 125 pp.

Johnson, D.A. and P.F. Lonsdale. 1976. Erosion and sedimentation around Mytilus Seamount, New England continental rise. Deep-sea Res. 23:429-440.

Johnson, L.R. 1979. Mineralogical dispersal patterns of North Atlantic deep-sea sediments with particular reference to eolian dusts. Mar. Geol. 29:335-345.

Johnson, P. W. and J. McN. Sieburth. 1979. Chroococcoid cyanobacteria in the sea: A ubiquitous and diverse phototrophic biomass. Limnol. Oceanogr. 24:928-935.

Johnson, P.W., H. Xu and J.McN. Sieburth. 1982. The utilization of chroococcoid cyanobacteria by marine protozooplankters but not by calanoid copepods. Ann. Inst. Oceanogr., Paris 58(S):297-308.

Jonas, R.B. and F.K. Pfaender. 1976. Chlorinated hydrocarbon pesticides in western North Atlantic Ocean. Environ. Sci. Technol. 10:770-773.

Jones, N.S. and H.L. Sanders. 1972. Distribution of Cumacea in the deep Atlantic. Deep-Sea Res. 19:737-745.

Jones, N.S. and H.L. Sanders. 1972. Distribution of Cumacea in the deep Atlantic. Deep-Sea Res. 19:737-745.

Jones, N.S. 1973. Some new Cumacea from deep water in the Atlantic. Crustaceana 25:297-319.

Jones, N.S. 1974. *Campylaspis* species (Crustacea: Cumacea) from the deep Atlantic. Bull. Brit. Mus. (Nat. Hist.). Zool. 27:249-300.

Jones, R.O., W.J. Mead and P.E. Sorensen. 1979. The Outer Continental Shelf Lands Act Amendments of 1978. Nat. Res. J. 19:885-908.

Joseph, A.B., P.F. Gustafson, I.R. Russell, E.A. Schuert, H.L. Volchok and A. Tamplin. 1971. Sources of radioactivity and their characteristics. Chapter 2, in Radioactivity in the Marine Environment, National Academy of Science Report.

Joyce, T.M. and P.H. Wiebe. 1983. Warm Core Rings of the Gulf Stream. Oceanus 26(2):34-44.

Joyce, T.M., R.W. Schmitt and M.C. Stalcup. 1984. Influence of the Gulf Stream upon the short term evolution of a Warm Core Ring. Austr. J. Mar. Freshwater Res. 34:515-524.

Joyce, T.M., R.H. Backus, K. Baker, P. Blackwelder, O. Brown, T. Cowles, R. Evans, G. Fryxell, D. Mountain, D. Olson, R. Schlitz, R. Schmitt, P. Smith, R. Smith and P.H. Wiebe. 1984. Rapid evolution of a Gulf Stream warm-core ring. Nature 308:837-840.

Judd, J.B., W.C. Smith and O.H. Pilkey. 1970. The environmental significance of iron-stained quartz grains on the southeastern United States Atlantic shelf. Mar. Geol. 8:355-362.

Jumars, P.A. 1976. Deep-sea species diversity: does it have a characteristic scale? J. Mar. Res. 34:217-246.

Kadko, D. 1980. 230 Th, 226 Ra and 222 Rn in abyssal sediments. Earth Planet. Sci. Lett. 49:360-380.

Kalenak, L.A. and H.G. Marshall. 1981. Fall phytoplankton distribution in Northeastern waters of the continental shelf. Va. J. Sci. 32:103.

Keer, F.R. and A.P. Cardinell. 1981. Potential geologic hazards and constraints for blocks in proposed Mid-Atlantic OCS Oil and Gas Lease Sale 59. U.S. Geol. Surv. Open-File Rep. 81-725:109.

Keller, G.H. and R.H. Bennett. 1970, Variations in the mass physical properties of selected submarine sediments. Mar. Geol. 9:215-223.

Keller, G.H., D. Lambert, G. Rowe and N. Staresinic. 1973. Bottom Currents in the Hudson Canyon. Science 180:181-183.

Keller, G.H. and F.P. Shepard. 1978. Currents and sedimentary processes in submarine canyons off the Northeast United States. in D.J. Stanley and G. Kelling (eds.), Sedimentation in submarine canyons, fans, and trenches. Dowden, Hutchinson & Ross, Inc.:15-31.

Keller, G.H., D.N. Lambert and R.H. Bennett. 1979. Geotechnical properties of continental slope deposits Cape Hatteras to Hydrographer Canyon North Carolina, U.S.A. in L.J. Doyle, O.H. Pilkey (eds.), Soc. Econ. Paleontol. Mineral. Sp. Publ. 27:131-152.

Kelling, G.H. and D.J. Stanley. 1970. Morphology and structure of Wilmington and Baltimore submarine canyons, eastern United States. J. Geol. 78:637-660.

Kelling, G. and D.J. Stanley. 1976. Sedimentation in canyon, slope and Base-of Slope Environments. in D.J. Stanley and D.J.P. Swift (eds.), Marine Sediment Transport and Environmental Management, John Wiley and Sons, Inc., p. 379-439.

Kelly, P., S.D. Sulkin and W.F. Van Heukelem. 1982. A dispersal model for larvae of the deep sea red crab Geryon quinquedens based upon behavioral regulation of vertical migration in the hatching stage. Mar. Biol. 72:35-43.

Kensley, Brian. 1982. Deep-water Atlantic Anthuridea (Crustacea: Isopoda). Smithsonian Contrib. Zool.:346.

Kester, D.R. and R.A. Courant. 1973. A summary of chemical oceanographic conditions: Cape Hatteras to Nantucket shoals. in Coastal and Offshore Inventory: Cape Hatteras to Nantucket shoals. Mar. Exp. Sta., Grad. Sch. Oceanogr. Univ. of R.I., Mar. Publ. Ser. 2-1 2-36.

Kester, D.R., K.A. Hausknecht and R.C. Hittinger. 1977. Recent analyses of copper, cadmium, and lead at Deepwater Dumpsite 106. pp. 543-546 in NOAA, Baseline Report of Environmental Conditions in Deepwater Dumpsite 106. NOAA Dumpsite Evaluation Report 77-1. NOAA, Rockville, MD.

Kester, D.R., R.C. Hittinger and P. Mukherji. 1981. Transition and heavy metals associated with acid-iron waste disposal at Deep Water Dumpsite 106. In B.H. Ketchum, D.R. Kester, and P.K. Park (eds.), Ocean Dumped Industrial Wastes. Plenum Press. New York, NY:215-232.

Kester, D.R., B.H. Ketchum and I.W. Duedall (eds.). 1983. Wastes in the Ocean, vol 2, Dredged Materials Disposal in the Ocean, Wiley, New York, 299 pp.

Kester, D.R., B.H. Ketchum, I.W. Duedall and P.K. Par. 1985.

Wastes in the Ocean, vol. V. Deep Sea Waste Disposal. Wiley, New York, 432 pp.

Ketchum, B.H. and N. Corwin. 1964. The persistence of "winger" water on the continental shelf south of Long Island, New York. Limnol. Oceanogr. 9(9):467-475.

Ketchum, B.H. and J.H. Ryther. 1965. Biological, chemical and radiochemical studies, 1958-1960. Woods Hole Oceanogr. Inst. Tech. Rep. 65-47 .

Ketchum, B.H., D.R. Kester and P. K. Park (eds.). 1981. Ocean Dumping of Industrial Wastes. Proceedings of the 1st International Ocean Dumping Symposium, October 10-13, 1978, University of Rhode Island, West Greenwich, R.I. Plenum Press, New York. 525 p.

Ketchum, B.H., D.R. Kester, I.W. Duedall and P.K. Park (in press). Wastes in the Ocean, vol. VI. Near Shore Waste Disposal, Wiley, New York, 640 pp.

Khripounoff, A., D. Desbruyeres and P. Chardy. 1980. Les peuplements benthiques de la faille Vema: donnees quantitatives et bilan d'energie en milieu abyssal. Oceanologica Acta 3:187-198.

Kimor, B. 1981. The role of phagotrophic dinoflagellates in marine ecosystems. Kieler Meeresforsch. 5: 164-173.

Kirby, J.R., J.M. Robb and J.C. Hampson, Jr. 1982. Detailed bathymetry of the U.S. Continental Slope between Lindenkohl Canyon and South Toms Canyon offshore New Jersey. U.S. Geol. Surv. Misc. Field Studies Map MF-1443.

Klasik, J.A. and O.H. Pilkey. 1975. Processes of sedimentation of the Atlantic continental rise off the southeastern U.S. Mar. Geol. 19:69-89.

Klinck, J.M., L.J. Pietrafsa and G.S. Jarowitz. 1981. Continental shelf circulation induced by a mooring localized wind stress. J. Phys. Oceanogr. 11:836-848.

Klitgord, K.D. and J.C. Behrendt. 1977. Aeromagnetic anomaly map of the United States Atlantic continental margin, U.S. Geol. Surv. Misc. Field Studies Map MF-913.

Klitgord, K. and J.C. Behrendt. 1979. Basin structure of the U.S. Atlantic continental margin. in J.S. Watkins, L. Montadert, and P.W. Dickerson (eds.), Geological and Geophysical Investigations of Continental Margins. Am. Assoc. Petrol. Geol. Mem. 29:85-112.

Klitgord, K.D. and J.A. Grow. 1980. Jurassic seismic stratigraphy and basement structure of the western Atlantic magnetic quiet zone. Am. Assoc. Petrol. Geol. Bull. 64:1658-1680.

Knauss, J.A. 1969. A note on the transport of the Gulf Stream. Deep-Sea Res. Suppl. 16:117-123.

Knebel, H.J. and D.W. Folger. 1976. Large sand waves on the Atlantic outer continental shelf around Wilmington Canyon, off eastern United States. Mar. Geol. 22:M7-M15.

Knebel, H.J. 1979. Anomalous topography on the continental shelf around Hudson Canyon. Mar. Geol. 33:M67-M75.

Knebel, H.J. and B. Carson. 1979. Small-scale slump deposits, middle Atlantic continental slope, off eastern United States. Mar. Geol. 29:221-236.

Kofoid, C.A. and O. Swezy. 1921. The free-living unamoured Dinoflagellata. University of California Press, Berkeley, 562p.

Kopylov, A.I., A.F. Pasternak and Y.V. Moiseyev. 1981. Consumption of zooflagellates by planktonic organisms. Oceanol. 21: 269-271.

Korzum, V.I., (ed). 1974. Atlas of World Water Balance. Hydrometeor Publ. House, Moscow, 65 pp.

Krishnaswami, S., L.K. Benninger, R.C. Aller and K.L. Von Damm. 1980. Atmospherically-derived radionuclides as tracers of sediment mixing and accumulation in near-shore marine and lake sediments: evidence from ^{7}Be, ^{210}Pb, and $^{239/240}$Pu. Earth Planet. Sci. Lett. 47(3):307-318.

Krueger, R.B. and L.H. Singer. 1979. An analysis of the Outer Continental Shelf Lands Act Amendments of 1978. Nat. Resour. J. 19:909:927.

Kupferman, S.L., H.D. Livingston and V.T. Bowen. 1979. A mass balance for 137 ^{Cs} and 90^{Sr} in the North Atlantic Ocean. J. Mar. Res. 37(1):157-199.

LaFlamme, R. E. and R. A. Hites. 1978. The global distribution of polycyclic aromatic hydrocarbons in recent sediments. Geochim. Cosmochim. Acta 43:1687-1691.

Lahey, W. L. 1982. Ocean Dumping of Sewage Sludge: The Tide Turns from Protection to Management. The Harvard Environmental Law Review, 6(2):295-431.

Lai, D. 1983. The Blake Plateau Florida Current, Ph.D. thesis, University of Rhode Island, 200 pp.

Lai, D.Y. and P.L. Richardson. 1977. Distribution and movement of Gulf Stream rings. J. Phys. Oceanogr. 7:670-683.

Laine, E.P. 1980. New evidence from beneath the western North Atlantic for the depth of glacial erosion in Greenland and North America. Quat. Res. 14:188-198.

Lambert, D.N., R.H. Bennett, W.B. Sawyer and G.H. Keller. 1981. Geotechnical properties of continental upper rise sediments-Veatch Canyon to Cape Hatteras. Mar. Geotechnol. 4:281-306.

Lamontagne, R.A., J.W. Swinnerton, V.J. Linnebom and W.D. Smith. 1973. Methane concentrations in various marine environments. J. Geophys. Res.:5317-5323.

Lancelot, Y. and J.I. Ewing. 1972. Correlation of natural gas zonation and carbonate diagenesis in Tertiary sediments from the north-west Atlantic. Deep Sea Drill. Proj., Initial Rep. 11:791-799.

Lange, A.M., S.A. Murawski, M.P. Sissenwine, R.L. Mayo and B.E. Brown. 1981. Fishery trends off the northeaster coast of the United States, 1964-80. NMFS/NOAA; Northeast Fisheries Center Laboratory Ref. Doc. No. 81-17.

Laval, M. 1968. Zoothamnium pelagicum du plessis, cilie peritriche planctonique: Morphologie, croissance et comportement. Protistologia 4:333-363.

Laval-Peuto, M. 1982. Methods of taxonomy and selection of criteria for determination of marine planktonic protozoa. Ann. Inst. Oceanogr. Paris. 58(S):151-168.

Lawrence, M.B. 1977. Atlantic hurricane season of 1976. Monthly Weather Review, 105:497-507.

Lee, D.S. and J. Booth, Jr. 1979. Seasonal distribution of offshore and pelagic birds in North Carolina waters. Am. Birds 33:715-721.

Lee, D.S. and R.A. Rowlett. 1979. Additions to the seabird fauna of North Carolina. Chat 43:1-9.

Lee, J.J. 1980. Nutrition and physiology of the foraminifera. Biochem. Phys. Protozoa 3: 43-66.

Lee, T.N. 1975. Florida Current spin-off eddies. Deep-Sea Res. 22:753-765.

Lee, T.N. and P.A. Mayer. 1977. Low-frequency current variability and spin-of eddies off southeast Florida. J. Mar. Res. 35:193-220.

Lee, T.N., L.P. Atkinson and R. Legeckis. 1981. Observations of a Gulf Stream frontal eddy on the Georgia continental shelf, April 1977. Deep-Sea Res. 28:347-378.

Lee, T.N. and L.P. Atkinson. 1983. Low-frequency current and temperature variability from Gulf Stream frontal eddies and atmospheric forcing along the southeast U.S. Outer Continental Shelf. J. Geophys. Res. 88:4541-4567.

Lee, T.N. and E. Waddell. 1983. On Gulf Stream variability and meanders over the Blake Plateau at 30°N. J. Geophys. Res. 88:4617-4631.

Leetmaa, A. and A.F. Bunker. 1978. Updated charts of the mean annual wind stress, convergences in the Ekman Layers, and Sverdrup transports in the North Atlantic. J. Mar. Res. 36:311-322.

Legeckis, R. 1979. Satellite observations of Gulf Stream interaction with changes in bottom topography. EOS 60:295.

Levine, N.D., J.O. Corliss, F.E.G. Cox, G. Deroux, J. Grain, B.M. Honigberg, G.F. Leedale, A.R. Loeblich III, J. Lom, D. Lynn, E.G. Merinfeld, F.C. Page, G. Poljansky, V. Sprague, J. Vavra and F.G. Wallace. 1980. A newly revised classification of the protozoa. J. Protozool. 27: 37-58.

Levy, E.M. and A. Walton. 1976. High seas oil pollution particulate petroleum residues in the North Atlantic. J. Fish. Res. Bd. Can. 33:2781-2791.

Levy, E.M., M. Ehrhardt, D. Kohnke, E. Sobtchenko, T. Suzuoki and A. Tokuhiro. 1981. Global oil pollution. Results of MAPMOPP, the IGOSS pilot project on marine pollution (petroleum) monitoring. Intergovernmental Oceanographic Commission, Paris. 35 p.

Lewis, J.B. 1954. The occurrence and vertical distribution of the Euphausiacea of the Florida Current. Bull. Mar. Sci. 4:265-301.

Li, W. K. W., D. V. Subba-Rao, W. G. Harrison, J. C. Smith, J. J. Cullen, B. Irwin and T. Platt. 1983. Autotrophic picoplankton in the tropical ocean. Science 219:292-295.

Li, Y.-H. G. Mathieu, P. Biscaye and H. James Simpson. 1977. The flux of 226 Ra from estuarine and continental shelf sediments. Earth Planet. Sci. Lett. 37:237-241.

Lisitzin, A.P. 1972. Sedimentation in the World Ocean. Soc. Econ. Paleont. Mineralog. Special Publ. 17:218.

Livingston, H.D. and V.T. Bowen. 1979. Pu and 137 Cs in coastal sediments. Earth Planet. Sci. Lett. 43:29-45.

Livingston, H.D. and Jenkins, W.D. 1982. Radioactive tracers in the Ocean. in The Future of Oceanography, Proceedings of a Symposium, Woods Hole, MA, 1980. Springer-Verlag, N.Y.

Louis, J.P. and P.C. Smith. 1982. The development of the barotropic radiation field of an eddy over a slope. J. Phys. Oceanogr. 12, 56-73.

Louis, J.P., B.P. Petrie and P.C. Smith. 1982. Observations of topographic Rossby waves on the Continental Margin off Nova Scotia. J. Phys. Oceanogr. 12:47-55.

Lowrie, Allen, Jr. and B.C. Heezen. 1967. Knoll and sediment drift near Hudson Canyon. Science 157:1552-1553.

Luyten, J.R. 1977. Scales of motion in the deep Gulf Stream and across the continental rise. J. Mar. Res. 35:49-74.

Luyten, J.R. and H.M. Stommel. 1983. The density jump across the Little Bahama Bank. J. Geophys. Res. 89:2097-2100.

Luyten, J.R. and H.M. Stommel. 1985. Upstream effects of the Gulf Stream on the structure of the mid-ocean thermocline. Prog. in Oceanogr. 14:387-399.

Lyall, A.K., D.J. Stanley, H.N. Giles and A. Fisher, Jr. 1971. Suspended sediment and transport at the shelf-break and on the slope, Wilmington Canyon area, eastern U.S.A. Mar. Tech. Soc. J. 5:15-27.

Mackensie, A. S., S. C. Brassell, G. Eglinton and J. R. Maxwell. 1982. Chemical fossils: The geological fate of steroids. Science 217:491-504.

MacIlvaine, J.C. 1973. Sedimentary processes on the Continental slope off New England. Unpubl. Ph.D. Dissertation, Woods Hole Oceanogr. Inst. and Mass. Inst. Tech.:211 p.

Macintyre, I.G. and J.D. Milliman. 1970. Physiographic features on the outershelf and upper slope, Atlantic continental margin, southeastern United States: Geol. Soc. Am. Bull. 81:2577-2598.

Macintyre, R.M. 1979. Timing of igneous events near the margins of continents bordering the North Atlantic. in Mesozoic and Tertiary volcanism in the North Atlantic and neighbouring regions: proceedings of the Flett symposium, G.B. Geol. Surv. Bull. 70:6-7.

Maciolek, N.J. 1981. A new genera and species of Spienidae (Annelida: Polychaeta) from the North and South Atlantic. Proc. Biol. Soc. Wash. 94:228-239.

Maestrini, S.Y. and D.J. Bonin. 1981. Competition among phytoplankton based on inorganic macronutrients. in T. Platt (ed.), Physiological bases of phytoplankton ecology. Can. Bull. Fish. Aquat. Sci. 210:264-278.

Mague, T. H., M. M. Weare and O. Holm-Hansen. 1974. Nitrogen fixation in the north Pacific Ocean. Mar. Biol. 24: 109-119.

Malahoff, A., R.W. Embley, R.B. Perry and C. Fefe. 1980. Submarine mass-wasting of sediments on the continental slope and upper rise south of Baltimore Canyon. Earth Planet. Sci. Lett. 49:1-7.

Malahoff, A., R.W. Embley, and D.J. Fornari. 1982. Geomorphology of Norfolk and Washington Canyons and the surrounding continental slope and upper rise as observed from DSRV ALVIN: in R.W. Scrutton and M. Talwani (eds.), The Ocean Floor (John Wiley and Sons, Ltd.) New York:97-111.

Malone, T.C. 1976. Phytoplankton productivity in the apex of the New York Bight: Environmental regulation of productivity/chlorophyll *a*. Limnol. Oceanogr. Spec. Symp. 2:260-272.

Malone, T.C., T.S. Hopkins, P.G. Falkowski and T.E. Whitledge. 1983. Production and transport of phytoplankton biomass over the continental shelf of the New York Bight. Cont. Shelf Res. 1:305-337.

Manderson, M.C. 1972. Commercial development of offshore marine phosphates. Offshore Tech. Conf. Paper 1658.

Manheim, F.T. 1967. Evidence for submarine discharge of water on the Atlantic continental slope of the southern United States, and suggestions for further search. New York Acad. Sci. Trans. Ser. 2, 29:839-853.

Manheim, F.T., R.H. Meade and G.C. Bond. 1970. Suspended matter in surface waters of the Atlantic continental margin from Cape Cod to the Florida Keys. Science 167:371-376.

Manheim, F.T. 1979. Potential hard mineral and associated resources on the Atlantic and Gulf Continental margins. U.S. Dept. of Commerce, NTIS (PB81-192643).

Manheim, F.T., R.M. Pratt and P.F. McFarlin. 1980. Composition and origin of phosphorite deposits of the Blake Plateau. in Y.K. Bentor (ed.), Marine phosphorites; geochemistry, occurrence genesis, Spec. Publ., Soc. Econ. Paleontol. Mineral. 29:117-137.

Manheim, F.T. and H.D. Hess. 1981. Hard Mineral Resources Around the U.S. Continental Margin. 13th Ann. Offshore Techn. Conf. 4:129-138.

Manheim, F.T., P. Popenoe, W. Siapno and C. Lane. 1982. Manganese-phosphorite deposits of the Blake Plateau. in P. Halbach, and P. Winter (eds.), Marine mineral deposits: new research results and economic prospects, Proceedings of the Clausthaler Workshop held in Sept. 1982, Marine Rohstoffe und Meevestechnik, Verlag Glckauf Gmbh, Essen, 6:9-44.

Mantoura, R.F.C., P. M. Gschwend, 0. C. Zafiriou and K. R. Clarke. 1982. Volatile organic compounds at a coastal site. II. Short-term variations. Environ. Sci. Technol. 16:38-42.

Markle, D.F. and J.A. Musick. 1974. Benthic slope fishes found at 900 meter depth along a transect in the western North Atlantic Ocean. Mar. Biol. 26:225-233.

Marshall, H. G. 1971. Composition of phytoplankton off the southeastern coast of the United States. Bull. Mar. Sci. 21: 806-825.

Marshall, H.G. 1976. Phytoplankton distribution along the eastern coast of the U.S.A. I. Phytoplankton distribution. Mar. Biol. 38:81-89.

Marshall, H. G. 1981. Occurrence of blue-green algae (Cyanophyta) in the phytoplankton off the southeastern coast of the United States. J. Plankton Res. 3: 163-166.

Marshall, H.G. 1982. Phytoplankton distribution along the eastern coast of the USA. IV. Shelf waters between Cape Lookout, North Carolina, and Cape Canaveral, Florida. Proc. Biol. Soc. Wash. 95:99-113.

Marshall, H.G.. 1984. Phytoplankton distribution along the eastern coast of the U.S.A. Part V. Seasonal density and cell volume patterns for the northeastern continental shelf. J. Plank. Res. 6:169-193.

Martens, C.S., G.W. Kipphut and J.V. Klump. 1980. Sediment-water chemical exchange in the coastal zone traced by *in situ* radon-222 flux measurements. Science 208:285-288.

Martin, P.J. 1982. Mixed-layer simulation of buoy observations taken during hurricane Eloise. J. Geophys. Res. 87, 409-427.

Marumo, R. and S. Nagasawa. 1976. Seasonal variation of the standing crop of a pelagic blue-green alga, *Trichodesmium*, in the Kuroshio water. Bull. Plankt. Soc. Japan 23: 19-25.

Mather, J.R., H. Adams III and G. A. Yoshioka. 1964. Coastal storms of the Eastern United States. J. Appl. Meteor. 3, 693-706.

Mathews, T.D. and O. Pashuk. 1977. A description of oceanographic conditions off the southeastern United States during 1973. South Carolina Marine Resources Center Tech. Rep. 19.

Mayer, D.A., H.O. Mofjeld and K.D. Leaman. 1981. Near-inertial internal waves observed on the outer shelf in the middle Atlantic Bight in the wake of Hurricane Belle. J. Phys. Oceanogr. 11:87-106.

Mayer, G.F. (ed.). 1982. Ecological Stress and the New York Bight: Science and Management. Estuarine Research Federation, Belle W. Baruch Institute for Marine Biology and Coastal Research, University of South Carolina, Columbia, S.C.:715.

Mazmanian, D.A. and J. Nienaber. 1979. Can Organizations Change?: Environmental Protection, Citizen Participation and the Corps of Engineers, Brookings Institution, Washington, DC.

McCarthy, J.J. 1980. Nitrogen and phytoplankton ecology. in I. Morris (ed.), The physiological ecology of phytoplankton. Blackwell, Oxford, England:292-233.

McCarthy, J.J. 1981. The kinetics of nitrogen utilization. in T. Platt (ed.), Physiological bases of phytoplankton ecology. Can. Bull. Fish. Aquat. Sci. 210:211-233.

McCarthy, J. J. and E. J. Carpenter. 1979. Oscillatoria (Trichodesmium) thiebautii (Cyanophyta) in the central North Atlantic Ocean. J. Phycol. 15:75-82.

McClellan, H.J. 1957. On the distinctness and origin of the slope water off the Scotian Shelf and its easterly flow south of the Grand Banks. J. Fish. Res. Bd. Canada 10:155-176.

McGowan, J.A. and D.M. Brown. 1966. A new opening/closing paired zooplankton net. Univ. Calif. Scripps Inst. Oceaanogr. Ref. 66-23.

McGowan, W.E., W.A. Saner and G.L. Hufford. 1974. Tar ball distribution in the western North Atlantic. NTIS Report: 30.

McGregor, B.A. and R.H. Bennett. 1977. Continental slope sediment instability northeast of Wilmington Canyon. Bull. Am. Assoc. Petrol. Geol. 61:918-928.

McGregor, B.A. 1979a. Current meter observations on the U.S. Atlantic continental slope: variation in time and space. Mar. Geol 29:209-219.

McGregor, B.A. 1979b. Variations in bottom processes along the U.S. Atlantic continental margin. in J.S. Watkins, L. Montadert, P.W. Dickerson (eds.), Geological and geophysical investigations of continental margins, Am. Assoc. Petrol. Geol. Mem. 29:139-149.

McGregor, B.A. and R.H. Bennett. 1979. Mass movement of sediment on the continental slope and reise seaward of the Baltimore Canyon Trough. Mar. Geol. 33:163-174.

McGregor, B.A., R.H. Bennett and D.N. Lambert. 1979. Bottom processes, morphology, and geotechnical properties of the continental slope south of Baltimore Canyon. Appl. Ocean Res. 1:177-187.

McGregor, B.A. 1982a. 3.5 kHz data collected in the

Wilmington Canyon area during 1980, ENDEAVOR Cruise 80-EN-056. U.S. Geol. Surv. Open-File Rep. 82-0498:4.

McGregor, B.A. 1982b. Morphologic setting for core sites on the slope in the Baltimore Canyon Trough. U.S. Geol. Surv. Open-File Rep. 82-0710:31.

McGregor, B.A., J.C. Hampson, Jr. and W.B.F. Ryan. 1982. Sidescan data collected in the Wilmington Canyon area durnig 1980, Gyre cruise 80-G-8B. U.S. Geol. Surv. Open-File Rep. 82-499:4.

McGregor, B.A., W.L. Stubblefield, W.B.F. Ryan and D.C. Twichell. 1982b. Wilmington submarine canyon: a marine fluvial-like system. Geology 10:27-30.

McIver, R.D. 1982. Role of naturally occurring gas hydrates in sediment transport, Am. Assoc. Petrol. Geol. Bull. 66:789-792.

McLellan, T. 1977. Feeding strategies of macrourids. Deep-Sea Res. 24:1019-1036.

Menzel, D.W. and J.H. Ryther. 1960. The annual cycle of primary production in the Sargasso Sea off Bermuda. Deep-Sea Res. 6:351-367.

Menzel, D.W. and J.H. Ryther. 1961. Annual variations in primary production of the Sargasso Sea off Bermuda. Deep-Sea Res. 7:282-288.

Mero, J.O. 1965. The mineral resources of the sea. Elsevier, Amsterdam, NY:312.

Meyers, E.P. and C.G. Gunnerson. 1976. Hydrocarbons in the Ocean. MESA Special Report, U.S. Dept. of Commerce, NOAA Environmental Research Laboratories, Boulder, CO.

Mid-Atlantic Fishery Management Council. 1978. Final Environmental Impact Statement/Fishery Management Plan for the Squid Fishery of the Northwest Atlantic Ocean Supplement 1.

Mid-Atlantic Fishery Management Council. 1980a. Pelagic Sharks Fishery Management Plan, Appendix II. Fisheries.

Mid-Atlantic Fishery Management Council. 1980b. Bluefish management plan, Appendix I: Biological Information on the Bluefish on the Western North Atlantic. Appendix II: Fisheries.

Miller, K.G. and B.E. Tucholke. 1983. Development of Cenozoic abyssal circulation south of the Greenland-Scotland Ridge. In M.H. Bott, S. Saxov, M. Talwani, and J. Thiede (eds.), Structure and Development of the Greenland-Scotland Ridge. Plenum Publ. Corp.:549-589.

Miller, S.L. 1974. The nature and occurrence of clathrate hydrates. in I.R. Kaplan (ed.), Natural gases in marine sediments. New York, Plenum Press:151-1777.

Milliman, J.D., F.T. Manheim, R.M. Pratt and E.F.K. Zarudzki. 1967. Alvin dives on the continental margin off the southeastern United States July 2-13, 1967. Woods Hole Oceanogr. Inst., Ref. No. 67-80:64 pp.

Milliman, J.D. and K.O. Emery. 1968. Sea levels during the past 35,000 years. Science 162:1121-1123.

Milliman, J.D. 1972. Atlantic continental shelf and slope of the United States - Petrology of the sand fraction of sediments, northern New Jersey to southern Florida. U.S. Geol. Surv. Prof. Pap. 529-J:40.

Milliman, J.D., O.H. Pilkey and D.A. Ross. 1972. Sediments of the continental margin of the eastern United States. Geol. Soc. Am. Bull. 83:1315-1334.

Milliman, J.D. and R.H. Meade. 1983. World-wide delivery of river sedimentto the oceans. J. Geol. 91:1-21.

Mills, C.A. and P. Rhines. 1979. Deep western boundary current at the Blake-Bahama outer ridge: current meter and temperature observations. Woods Hole Oceanogr. Inst. Tech. Rep. 79-85:77.

Mills, E.L. 1967. Deep-sea amphipods from the western North Atlantic Ocean. l. Ingolfiellidae and an unusual new species in the gammaridean family Pardaliscidae. Can. J. Zool. 45:347-355.

Mills, E.L. 1971. Deep-sea Amphipoda from the western North Atlantic Ocean. The Family Ampeliscidae. Limnol. Oceanogr. 16:357-386.

Mills, E.L. 1972a. Deep-sea Amphipoda from the western North Atlantic Ocean. Caprellidea. Can. J. Zool. 50(4):371-383.

Mills, E.L. 1972b. T.R.R. Stebbing, the Challenger and knowledge of deep-sea Amphipoda. Proc. Roy. Soc. Edinburgh (B), 72:69-87.

Mills, E.L., K. Pittman and B. Munroe. 1982. Effect of preservation on the weight of marine benthic invertebrates. Canadian J. Fish. Aquat. Sci. 39:221-224.

Minerals Management Service (MMS). 1982. Department of the Interior Final Environmental Impact Statement, Proposed 1983 Outer Continental Shelf Oil and Gas Lease Sale Offshore the Mid-Atlantic States OCS Sale No. 76. U.S. Department of the Interior, New York.

Mizenko, D. and J.L. Chamberlin. 1979. Anticyclonic Gulf Stream eddies off the northeastern United States during 1976. NOAA Tech. Rep. NMFS Circ. 427:259-280.

Mognard, N.M., W.J. Campbell, R.E. Cheney and J.G. Marsh. 1983. Southern Ocean mean monthly waves and surface winds for winter 1978 by Seasat radar altimeter. J. Geophys. Res. 88:1736-1744.

Mommessin, P.R. and J.C. Raia. 1974. Chemical and physical characterization of tar samples from the marine environment. Final Rep.:210.

Monniot, C. and F. Monniot. 1968. Les ascides de grandes profoundeurs recoltees par les navire oceanographique american ATLANTIS II (Premier note). Bull. Inst. Oceanogr. Monoco 67:1-48.

Monniot, C., and F. Monniot, 1970, Les ascides de grandes profoundeurs recoltees par les navires ATLANTIS, ATLANTIS II, et CHAIN (2 eme note). Deep-Sea Res. 17:317-336.

Monniot, C. and F. Monniot. 1975. Abyssal tunicates: an ecological paradox. Ann. Inst. Oceanogr. Paris 51:99-129.

Monniot, C. and F. Monniot. 1976. Tuniciers abyssaux du bassin Argentin rcolts par l'"Atlantis II". Bull. Mus. Natn. Hist. Nat. Paris, Ser. 3, No. 387, Zoologie 269:629-662.

Monniot, C. and F. Monniot. 1978. Recent work on the deep-sea Tunicates. Oceanogr. Mar. Biol. Ann. Rev. 16:181-228.

Monniot, F. 1971. Les ascides de grandes profondeurs recoltees par les navaires Atlantis II et Chain. 3e note. Cahiers de Biologie Marine 12:457-468.

Monniot, F. 1979. Faunal affinities among abyssal Atlantic basins. Sarsia 64:93-95.

Mooers, C.N.K., J. Fernandez-Partagas and J.F. Price. 1976. Meteorological forcing fields of the New York Bight (First Year's Progress Report) Tech. Rept., RSMAS, U. Miami TR76-8, Miami, FL 1151 pp.

Mooers, C.N.K., R.W. Garvine and W.W. Martin. 1979. Summertime synoptic variability of the middle Atlantic shelf water/slope water front. J. Geophys. Res. 84:4837-4855.

Moore, H.B. 1941. Notes on the distribution of oceanic birds in the North Atlantic 1937-1941. Proc. Linn. Soc. N.Y. 52-53:53-62.

Moore, H.B. 1951. The seasonal distribution of oceanic birds in the western North Atlantic. Bull. Mar. Sci. 1:1-14.

Moore, H.B. 1953. Plankton of the Florida Current. II. Siphonophora. Bull. Mar. Sci. 2: 559-573.

Moore, H.B., H. Owre, E.C. Jones and T. Dow. 1953. Plankton of the Florida Current. III. The control of the vertical distribution of zooplankton in the daytime by light and temperature. Bull. Mar. Sci. 3: 83-95.

Moore, H.B. and E.G. Corwin. 1956. The effects of temperature, illumination, and pressure on the vertical distribution of zooplankton. Bull. Mar. Sci. 6: 273-287.

Moore, H.B., and D.L. O'Berry. 1957. Plankton of the Florida Current. IV. Factors influencing the vertical distribution of some common copepods. Bull. Mar. Sci. 7: 297-287.

Morey-Gaines, G. In press. Heterotrophic nutrition. in F.J.R. Taylor, (ed.). The biology of dinoflagellates.

Morita, R. Y. 1976. Survival of bacteria in cold and moderate hydro-static pressure environments with special reference to psychrophilic and barophilic bacteria. 26th Symp. Soc. Gen. Microbiol. pp. 279-298. in The Survival of Vegetative Microbes. T. Gray and J. R. Postgate (eds.), Cambridge Univ. Press, New York.

Morris, B.F. 1971. Petroleum: Tar Quantities Floating in the Northwestern Atlantic Taken with a New Quantitative Neuston Net. Science 173:430-432.

Morris, B.F. 1975. The Neuston of the Northwest Atlantic. Unpubl. Ph.D. Thesis, Dalhousie Univ., Halifax. 285 p.

Morris, B.F., J. Cadwallader, J. Geiselman and J.N. Butler. 1976. Transfer of petroleum and biogenic hydrocarbons in the Sargassum community. In Windom, H.L., R.A. Duce (ed.), Marine pollutant transfer. Chapter 10, D. C. Heath and Co., Lexington Books, Lexington, MA.:235-259.

Morzer Bruyns, W.F.J. 1967. Black-capped Petrels *(Pterodroma hasitata)* in the Atlantic Ocean. Ardea 55:270.

Mountain, G.S. 1981. Stratigraphy of the Western North Atlantic based on the study of reflection profiles and DSDP results. Unpubl. Ph.D. Thesis, Columbia Univ. 316 p.

Mountain, G.S. 1983. Post-Eocene stratigraphy of USGS line 25 and adjacent region: U.S. Geol. Surv. Misc. Field Studies, in press.

Mountain, G.S. and B.E. Tucholke. 1984. Mesozoic and Cenozoic geology of the U.S. continental slope and rise. in C.W. Poag (ed.), Stratigraphy and Depositional History of the U.S. Atlantic Margin Van Nostrand-Reinhold, in press.

Mountain, G.S. and B.E. Tucholke. 1985. Mesozoic and Cenozoic geology of the U.S. continental slope and rise. in C.W. Poag (ed.), Stratigraphy and Depositional History of the U.S. Atlantic Margin Van Nostrand-Reinhold, in press.

Mukherji, P. and D.R. Kester. 1978. Mercury in Gulf Stream ring, Franklin. EOS 59:1099.

Mukheriji, P. and D.R. Kester. 1979. Mercury distribution in the Gulf Stream. Science 204:64-66.

Mukherji, P., D.R. Kester, L.M. Petrie and R.C. Hittinger. 1979. Cadmium and mercury in Gulf Stream rings. EOS 60(18):296.

Mullins, H.T., A.C. Neumann, R. Jude Wilber, A.C. Hine and S.J. Chinburg. 1980. Carbonate sediment drifts in northern Straits of Florida. Am. Assoc. Petrol. Geol. Bull. 64:1701-1717.

Murphy, L.S., G.T. Rowe and R.L. Haedrich. 1976. Genetic variability in deep sea echinoderms. Deep-Sea Res. 23:339-348.

Murphy, R.C. and L.S. Mowbray. 1951. New light on the cahow, Pterodroma chaow. Auk 68:266-280.

Murray, J.W. and V. Grundmanis. 1980. Oxygen consumption in pelagic marine sediments. Science 209:1527-1530.

Musick, J.A. 1976. Community structure of fishes on the continental slope and rise off the Middle Atlantic Coast of the U.S. Manuscript presented at the Jiont Oceanographic Assembly, Edinburgh, Sept. (copies available from J.A. Musick, Virginia Inst. of Marine Sci., Gloucester Point, VA, 23062, U.S.A.).

NMFS. 1960-1980. Fish Distribution Charts; Groundfish Surveys; National Marine Fisheries Service, Woods Hole, MA 02543

NOAA. 1977. Summarization and interpretation of historical physical oceanographic and meteorologic information for the mid-Atlantic region. Final Rept. to BLM. AA550-1A6-12. 295 p.

NOAA. 1978. Proceedings of a Workshop on Scientific Problems Relating to Ocean Pollution. NOAA, U.S. Dept. of Commerce, Washington, DC.

NOAA. 1978. Atlantic billfishes and sharks: Preliminary Fishery Management Plan, U.S. Dept. Commerce, Federal Register, Jan 27 1978, Part IV.

NOAA. 1979. Proceedings of a Workshop on Assimilative Capacity of U.S. Coastal Waters for Pollutants. U.S. Dept. of Commerce, Washington, D.C.

NOAA. 1980. Marine Environmental conditions off the coasts of the U.S. - Jan. 1978-March 1979. NOAA Technical Memo. NMFS-OF-5.

Nalewajko, C. and D.R.S. Lean. 1980. Phosphorus. in I. Morris (ed.), The physiological ecology of phytoplankton. Univ. Calif. Press, pp. 235-258.

Nastav, F.L., G.F. Merrill, T.A. Nelson and R.H. Bennett. 1980. Research Elements, Study Areas, Sample Locations, Tracklines, and Publications of the Marine Geotechnical Seafloor Stability Program 1967-1980. NOAA Rep.:107.

National Academy of Sciences. 1975. Mining in the Outer Continental Shelf and in the Deep Ocean. Wash., DC:119.

National Academy of Science. 1975. Petroleum in the Marine Environment. National Academy of Sciences, Washington, D. C.

National Academy of Science. 1979. Polychlorinated

Biphenyls. National Academy of Sciences, Washington, D. C.

National Advisory Committee on the Oceans and Atmosphere. 1981. The role of the ocean in a waste management strategy. GPO, Washington, D.C. 103 pp.

National Marine Fisheries Service (NMFS). 1977. Final Environmental Impact Statement/Preliminary Fishery Management Plan for the Hake Fisheries of the Northwestern Atlantic. Northeast Fisheries Center, NOAA, Woods Hole, MA.

National Marine Fisheries Service (NMFS). 1982. Fisherise of the United States. 1981. Current Fishery Statistics No. 8200. U.S. Dept of Commerce, NOAA, National Marine Fisheries Service, Washington, DC, 131 pp.

National Ocean Survey. 1975 to 1981. Bathymetric maps of the U.S. Atlantic margin (Numbers NH16-12, NH17-3, NH17-6, NH18-7, NI17-9, NI17-12, NI18-7, NI18-10, NJ18-1, NJ18-2, NJ18-3, NJ18-4, NJ18-5, NJ18-6, NJ18-8, NJ18-9, NJ18-11, NJ18-12, NJ19-1, NJ19-2, NK19-4, NK19-9, NK19-10, NK19-11, and NK19-12) U.S. Dept. of Commerce, NOAA, Washington, D.C., scale 1:250,000.

National Oceanic and Atmospheric Administration and Council on Environmental Quality (NOAA/CEQ). 1980. Eastern United States Coastal and Ocean Zones Data Atlas. Prepared for Council on Environmental Quality and Office of Coastal Zone Management. NOAA Washington, DC.

Needham, H.D., D. Habib and B.C. Heezen. 1969. Upper Carboniferous palynomorphs as a tracer of red sediment dispersal patterns in the northwest Atlantic. J. Geol. 77:113-120.

Neff, J.M. 1979. Polycyclic Aromatic Hydrocarbons in the Aquatic Environment. Applied Science Publishers, Ltd., London.

Neiheisel, J. 1979, Sediment characteristics of the 2800 meter Atlantic Nuclear Waste Disposal Site; radionuclide retention potential. EPA, Office of Radiation Programs, Washington, D.C. Tech. Note ORP/TAD-79-10.

Neumann, A.C., J.W. Kofoed and G.H. Keller. 1977. Lithoherms in the Straits of Florida. Geology 5:4-10.

Neumann, C.J. and G.W. Cry. 1978. A revised Atlantic tropical cyclone climatology. Mariner's Weather Log, 22:231-236.

Neumann, C.J., G.W. Cry, E.L. Caso and B.R. Jarvinen. 1978. Tropical cyclones of the North Atlantic Ocean, 1871-1977. National Climatic Center:176.

Neumann, C.J., et al. 1981. Tropical cyclones of the North Atlantic Ocean, 1871-1980. National Climatic Center:174.

New England Fishery Management Council. 1978. Briefing Document: Silver Hake. NEFMC Res. Doc. 78 SH 8.1.

New England Fishery Management Council. 1982. Draft American Lobster Fishery Managment Plan.

Nicholson, L.E. and R.P. Ruais. 1979. Description of the Recreational Fisheries for Cod, Haddock, Pollock and Silver Hake of the Northeast Coast of the U.S. Report to the New England Fishery Management Council.

Normark, W.R. 1970. Growth patterns of deep sea fans. Am. Assoc. Petrol. Geol. Bull. 54:2170-2195.

Ohwada, K., P. S. Tabor and R. R. Colwell. 1980. Species composition and barotolerance of gut microflora of deep sea benthic macrofauna collected at various depths in the Atlantic Ocean. Appl. Environ. Microbiol. 40:746-755.

O'Leary, D.W. 1982a. Midrange sidescan-sonar data from the continental slope off Georges Bank, between Lydonia and Oceanographer Canyons. U.S. Geol. Surv. Open-File Rep. 82-600:4.

Olson, D.G., R.W. Schmitt, M. Kennelly and T.M. Joyce. 1983. Long term physical evolution of Warm Core Ring 82B. EOS 64:1073.

Orr, C.D., R.M.P. Ward, N.A. Williams and R.G.B. Brown. 1982. Migration patterns of Red and Northern Phalaropes in southwest Davis Strait and in the northern Labrador Sea. Wilson Bull. 94:303-312.

Ortner, P.B. 1977. Investigations into the seasonal deep chlorophyll maximum in the Western North Atlantic, and its possible significance to regional food chain relationships. Ph.D. thesis, Woods Hole Oceanogr. Inst.

Ortner, P.B. 1978. Investigations into the seasonal deep chlorophyll maximum in the western North Atlantic, and its possible significance to regional food chain relationships. Woods Hole Oceanogr. Inst. Tech. Rep. 78-59:327.

Ortner, P.B., P.H. Wiebe and J.L. Cox. 1978. Relationships between oceanic epizooplankton distributions and the seasonal deep chlorophyll maximum in the Northwestern Atlantic Ocean. J. Mar. Res. 38:507-531.

Ortner, P.B., E.M. Hulburt and P.H. Wiebe. 1979. Gulf Stream rings, phytohydrography, and herbivore habitat contrasts. J. Exp. Mar. Biol. Ecol. 39:101-124.

Ortner, P.B., P.H. Wiebe, L. Haury and S. Boyd. 1979. Variability in zooplankton biomass distribution in the northern Sargasso Sea: The contribution of Gulf Stream cold core rings. Woods Hole Oceanogr. Inst. Tech. Rep. 79-14:18.

Ortner, P.B., P.H. Wiebe and J.L. Cox. 1980. Relationships between oceanic epizooplankton distributions and the seasonal deep chlorophyll maximum in the Northwestern Atlantic Ocean. J. Mar. Res. 38:507-531.

Ostlund, H.G., H. G. Dorsey and C.G. Rooth. 1974. Geosecs North Atlantic radiocarbon and tritium results. Earth Planet. Sci. Lett. 23:69-86.

Ostlund, H.G., H.G. Dorsey, C.G. Rooth. 1976. Geosecs Atlantic radiocarbon and tritium results (Miami), NSF/IDOE-78-13, 96 p.

Ou, H.W. and R.C. Beardsley. 1980. On the propagation of free topographic Rossby waves near continental margins. Part 2. Numerical model. J. Phys. Oceanogr. 10:1323-1339.

Ou, H.W., J.A. Vermersch, W.S. Brown and R.C. Beardsley. 1980. New England shelf/slope experiment, February to August, 1976, Data Report: The Moored Array. Woods Hole Oceanogr. Inst. Tech. Rep. 80-3:59.

Overall, M.P. 1968. Mining phosphorite from the Sea. Ocean Industry.

Packard, T.T. and P.J. LeB. Williams. 1981. Rates of respiratory oxygen consumption and electron transport in surface seawater from the Northwest Atlantic. Oceanol. Acta 4:351-358.

PACODF. 1981. Transient tracers in the ocean: Preliminary Hydrographic Report. Data report of the Physical and Chemical Oceanographic Data Facility, U. Calif. San Diego.

Palmer, R.S., (ed). 1962. Handbook of North American Birds. v. 1. Yale Univ. Press, New Haven, CT.

Pamatmat, M.M. 1973. Benthic community metabolism on the continental terrace and in the deep-sea in the North Pacific. Int. Revue ges. Hydrobiol. 58:345-368.

Park, P.K., and T.P. O'Connor. 1981. Ocean dumping research: historical and international development. p. 3-23 in Ketchum, B.H., D.R. Kseter and P.K. Park (eds.), 1981. Ocean Dumping of Industrial Wastes. Proceedings of the 1st International Ocean Dumping Symposium, October 10-13, 1978, University of Rhode Island, West Greenwich, RI. Plenum Press, NY. 525 pp.

Park, P.K. et al. (eds.). 1983. Radioactive Waste and the Oceans, Volume III. Proceedings of the 2nd International Ocean Dumping Symposium, Woods Hole, Mass.

Park, P.K., D.R. Kester, I.W. Duedall and B.H. Ketchum, (eds.). 1983. Wastes in the Ocean, vol. 3, Radioactive Waste in the Ocean, Wiley, New York 522 pp.

Parker, C.E. 1971. Gulf Stream rings in the Sargasso Sea. Deep-Sea Res. 18:981-993.

Paras-Carayannis, G. 1973. Ocean dumping in the New York Bight: An assessment of environmental studies. Tech. Memorandum No. 39, U.S. Coastal Engineering Research Center, 159 pp.

Parsons, C.L. 1979. GEOS-3 wave height measurements: an assessment during high sea state conditions in the North Atlantic. J. Geophys. Res. 84:4011-4020.

Paull, C.K. and W.P. Dillon. 1980. Erosional origin of the Blake Escarpment: an alternative hypothesis. Geology 8:538-542.

Paull, C.K. and W.P. Dillon. 1981. Erosional origin of the Blake Escarpment: an alternative hypothesis: Reply. Geology 9:338-339.

Pearce, J.B., J. Thomas and R. Greig. 1975. Preliminary investigation of benthic resources at Deepwater Dumpsite 106. pp. 217-228 in NOAA, May 1974 Baseline Investigation of Deepwater Dumpsite 106. NOAA Dumpsite Evaluation Report 75-1. NOAA, Rockville, MD.

Pearce, J.B. 1979. Trace metals in living marine resources taken from North Atlantic waters. in N.-P. Luepke (ed.), Monit. Environ. Mater. Specimen Banking, Proc. Int. Workshop:505-515.

Perkins, H. and M. Wimbush. 1976. A cyclonic mini-eddy near the Blake escarpment. Geophys. Res. Lett. 3:625-628.

Premuzic, E.T., C.M. Benkovitz, J.S. Gaffney and J.J. Walsh. 1982. The nature and distribution of organic matter in the surface sediments of world oceans and seas. Organic Geochem. 4:63-77.

Permuzic, E. T. 1980. Organic carbon and nitrogen in the surface sediments of world oceans and seas: distribution and relationship to bottom topography. Report BNL 51084. Environmental Chemistry Division and Oceanographic Sciences Division, Department of Energy and Environment, Brookhaven National Laboratories, Upton, New York 11973.

Perry, R.K., H.S. Fleming, P.R. Vogt, N.Z. Cherkis, R.H. Feden, J. Thiede and J.E. Strand. 1980. North Atlantic Ocean: Bathymetric and Plate Tectonic Evolution: Naval Research Laboratory, Acoustics Division, Environmental Sciences Branch, Scale 1:87 million, 1 sheet.

Petersen, H. 1975. Micronutrient analysis of seawater samples taken at Deepwater Dumpsite 106 — May 1974. pp. 189-202 in NOAA, 1974 Investigation of Deepwater Dumpsite 106. NOAA Dumpsite Evaluation Report 75-1. NOAA, Rockville, MD.

Peterson, M.N.A. 1976. Soils study continental margin sites. Bearing capacity study of seafloor soils, middle Atlantic Ridge, Atlantic Ocean. Scripps Inst. Oceanogr. Rep. DSDP-TR-9. 110 pp. NTIS: PB257 027 76-24, 8J.

Petrie, B. 1975. M2 surface and internal tides on the Scotian Shelf and Slope. J. Mar. Res. 33:303-323.

Phillips, J.D., A.H. Driscoll, K.R. Peal, W.M. Marquet and D.M. Owen. 1978. A new undersea geological survey tool: ANGUS Deep-Sea Res. 26:211-225.

Phillips, J.H. 1963. The pelagic distribution of the Sooty Shearwater, Procellaria grisea. Ibis 105:340-353.

Pierce, E.L. 1951. The chaetognatha of the west coast of Florida. Biol. Bull. 100:206-228.

Pierce, E.L. and M.L. Wass. 1962. Chaetognatha from the Florida Current and coastal water of the southeastern Atlantic states. Bull. Mar. Sci. 12:403-431.

Pilkey, O.H., D. Schnitker and D.R. Devear. 1966. Olites on the Georgia continental shelf edge. J. Sed. Petrol. 36:462-467.

Pilkey, O.H., B.W. Blackwelder, L.J. Doyle, Ernest Estes and P.M. Terlecky. 1969. Aspects of carbonate sedimentation on the Atlantic continental shelf off the southern United States: J. Sed. Petrol. 39:744-768.

Pinet, P.R. and Popenoe, P. 1981. Shallow seismic stratigraphy and post-Albian geologic history of the northern and central Blake Plateau (abs.). Geol. Soc. Am. Abstract with Programs 13:529.

Pinet, P.L., P. Popenoe, M.L. Otter and S.M. McCarthy. 1981a. An assessment of potential geologic hazards of the northern and central Blake Plateau. in P. Popenoe, (ed.), Environmental geologic studies on the southeastern Atlantic outer continental shelf, 1977-1978. U.S. Geol. Surv. Open-File Rep.p. 8-1 to 8-48.

Pinet, P.R., P. Popenoe, S.M. McCarthy and M.L. Otter. 1981b. Seismic stratigraphy of the northern and central Blake Plateau. in P. Popenoe, (ed.), Environmental geologic studies on the southeastern Atlantic outer continental shelf, 1977-1978. U.S. Geol. Surv. Open-File Rep. 81-0582-A, p. 7-1 to 7-91.

Pinet, P.R., P. Popenoe and D.F. Neilligan. 1981c. Gulf Stream: Reconstruction of Cenozoic flow patterns over the Blake Plateau. Geology 9:266-270.

Pinet, P.R. and P. Popenoe. 1982. Blake Plateau; control of Miocene sedimentation patterns by large-scale shifts of the Gulf Stream axis. Geology 10:257-259.

Pingree, R.D., P.R. Pugh, P.M. Holligan and G.R. Forster. 1975. Summer phytoplankton blooms and red tides along tidal fronts in the approaches to the English Channel. Nature 258:672-677.

Platt, T. and K. Denman. 1980. Patchiness in phytoplankton distribution. In I. Morris (ed.), The Physiological ecology of phytoplankton. Univ. Calif. Press.:413-432.

Poag, C.W. and R.E. Hall. 1979. Foraminiferal biostratigraphy paleoecology, and sediment accumulation rates. in P.A. Scholle, (ed.), Geological studies of the cost GE-1 well, U.S. South Atlantic outer continental shelf area. U.S. Geol. Surv. Circ. 800:49-63.

Poag, C.W. 1980a. Foraminiferal stratigraphy and paleoecology. in R.E. Mattick, J.L. Hennessy, (eds.), Structural framework, stratigraphy, and petroleum geology of the area of oil and gas lease sale No. 49 on the U.S. Atlantic continental shelf and slope, U.S. Geol. Survey. Circ. 812:35-48.

Poag, C.W. 1980b. Foraminiferal stratigraphy, paleoenviornments, and depositional cycles in the outer Baltimore Canyon Trough. In P.A. Scholle (ed.), Geological studies of the COST No. B-3 well, United State Mid-Atlantic Continental Slope area. U.S. Geol. Surv. Circ. 833:44-65.

Poag, C.W. 1982. Stratigraphic reference section for Georges Bank Basin - depositional model for New England passive margin. Am. Assoc. Petrol. Geol. Bull. 66:1021-1041.

Poag, C.W. 1982b. Foraminiferal and seismic stratigraphy, paleoenviornments, and depositional cycles in the Georges Bank basin. In P.A. Scholle and C.R. Wenkam (eds.), Geologic studies of the COST Nos. G-1 and G-2 wells, United States North Atlantic Outer Continental Shelf. U.S. Geol. Surv. Circ. 861:43-91.

Pollard, R.T. and R.C. Millard, Jr. 1970. Comparison between observed and simulated wind-generated inertial oscillations. Deep-Sea Res. 17:813-821.

Pomeroy, L.R. and R.E. Johannes. 1968. Occurrence and respiration of ultraplankton in the upper 500 meters of the ocean. Deep-Sea Res. 15:381-391.

Pomeroy, L.R. 1974. The ocean's food web, a changing paradigm. Bioscience 24:499-504.

Pomeroy, L.R. and D. Deibel. 1980. Aggregation of organic matter by pelagic tunicates. Limnol. Oceanogr. 25:643-652.

Pontecorvo, G., M. Wilkinson, R. Anderson and M. Holdowsky. 1980. Contribution of the Ocean Sector to the U.S. Economy. Science 208:1000-1006.

Popenoe, P. 1980. Single-channel seismic-reflection profiles collected on the northern Blake Plateau, 29 September to 19 October 1978. U.S. Geol. Surv. Open-File Rep. 80-1265:4.

Popenoe, P., B. Butman, C.K. Paull, M.M. Ball and S.L. Pfirman. 1981a. Interpretation of graphic data on potential geologic hazards on the southeastern United States Atlantic continental shelf. U.S. Geol. Surv. Misc. Field Studies Map MF-1276:3 sheets.

Popenoe, P., K.V. Cashman, D. Chayes and W.B.F. Ryan. 1981b. Mid-range sidescan-sonar profiles covering parts of proposed tracts and contiguous areas, OCS Sale #56, Manteo, Cape Fear and adjacent quadrangles, North Carolina. U.S. Geol. Surv. Open-File Rep. 81-554:4.

Popenoe, P., E.L. Coward and K.V. Cashman. 1982c. A regional assessment of potential environmental hazards to and limitations on petroleum development of the southeastern United States Atlantic continental shelf, slope, and rise offshore North Carolina. U.S. Geol. Surv. Open-File Rep. 82-136:67.

Poppe, L.S. 1981. Data file Atlantic Margin Coring Project (AMCOR) of the U.S. Geological Survey. U.S. Geol. Surv. Open-File Rep. 81-239:96.

Posmentier, E.S. and R.W. Houghton. 1981. Springtime evolution of the New England shelf break front. J. Geophys. Res. 86:4253-4259.

Powers, K.D. and E.H. Backus. 1981. The relationship of marine birds to oceanic fronts off the northeastern United States. U.S. Dept. of Energy, Office of Energy Res., Washington, D.C., Unpubl. Rep.

Powers, K.D. 1982. A comparison of two methods of counting birds at sea. J. Field Ornith. 53:209-222.

Powers, K.D. 1983. Pelagic distributions of marine birds off the northeastern United States. U.S. Dept. of Energy, Office of Energy Res., Washington, D.C.

Powers, K.D. and R.G.B. Brown. In press. Seabirds. Chapter 7.5 in Georges Bank R.H. Backus (ed.), MIT Press, Cambridge, MA.

Powles, H. and B.W. Stender. 1976. Observations on composition, seasonality and distribution of ichthyoplankton from MARMAP cruises in the South Atlantic Bight in 1973. South Carolina Mar. Res. Center Tech. Rep. 11.

Prahl, F.G. 1982. The geochemistry of polycyclic aromatic hydrocarbons in Columbia River and Washington coastal sediments. Ph.D. Dissertation, University of Washington.

Pratt, R.M. and B.C. Heezen. 1964. Topography of Blake Plateau. Deep-Sea Res. 11:721-728.

Pratt, R.M. 1966. The Gulf Stream as a graded river. Limnol. Oceanogr. 11:60-67.

Pratt, R.M. 1967. The seaward extension of submarine canyons off the northeast coast of the United States. Deep-Sea Res. 14:409-420.

Pratt, R.M. 1968. Atlantic continental shelf and slope: physiography and sediments of the deep-sea basin. U.S. Geol. Surv. Prof. Paper 529-B:44.

Price, J.F. 1981. Upper ocean response to a hurricane. J. Phys. Oceanogr. 11:153-175.

Price, J.F., C.N.K. Mooers and J.C. Van Leer. 1978. Observation and simulation of mixed-layer deepening. J. Phys. Oceanogr. 8:582-599.

Queffeulou, P. 1983. SEASAT Wave height measurements: A comparison with sea-truth data and a wave forecasting model - Application to the geographic distribution of strong sea states in storms. J. Geophys. Res. 88:1779-1788.

Rankin, M.N. and E.A.G. Duffey. 1948. A study of the bird life of the North Atlantic. Brit. Birds 41 (suppl.):1-42.

Rawson, M.D. and W.B.F. Ryan. 1978. Ocean floor sediment

and polymetallic nodules (Map): Office of the Geographer, U.S. Dept. of State, Scale 1:23,230,300.

Reed, R.K. and W.P. Elliot. 1977. A comparison of oceanic precipitation as measured by gauge and assessed from weather reports. J. Appl. Meteor. 16:983-986.

Reed, R.K. and W.P. Elliott. 1979. New precipitation maps for the North Atlantic and North Pacific oceans. J. Geophys. Res. 84:7839-7846.

Reitan, C.H. 1974. Frequencies of cyclones and cyclogenesis for North America, 1951-1970. Month. Weath. Rev. 102:861-868.

Resio, D.T. and B.P. Hayden. 1975. Recent secular variations in mid-Atlantic winter extratropical storm climate. J. Appl. Meteorol. 14:1223-1234.

Rex, M.A. 1972. Species diversity and character variation in some western North Atlantic deep sea gastropods. Harvard Univ. Ph.D. Thesis, Zoology, X.

Rex, M.A. 1973. Deep-sea species diversity: decreased gastropod diversity at abyssal depths. Science 181:1051-1053.

Rex, M.A. 1976. Biological accommodation in the deep-sea benthos: comparative evidence on the importance of predation and productivity. Deep-Sea Res. 23:975-987.

Rex, M.A. 1977. Zonation in deep-sea gastropods: The importance of biological interactions to rates of zonation. In B.F. Keegan, P.O. Geidigh and P.J.S. Boaden (eds.), Biology of Benthic Organisms. Pergamon Press, NY. pp. 521-530.

Rex, M.A., C.A. Van Ummersen and R.D. Turner. 1979. Reproductive pattern in the abyssal snail *Benthonella tenella,* in S.E. Stancyk, (ed.), Reproductive Ecology of Marine Invertebrates, Univ. South Carolina Press, The Belle W. Baruch Library in Mar. Sci. 9:173-188.

Rex, M.A. 1981. Community structure in the deep-sea benthos. Ann. Rev. Ecol. Syst. 12:331-353.

Rex, M.A. and A. Waren. 1982. Planktotrophic development in deep-sea prosobranch snails from the western North Atlantic. Deep-Sea Res. 29:171-184.

Rex, M.A. 1983. Geographic patterns of species diversity in the deep-sea benthos. In The Sea, v. 8, G.T. Rowe (ed.), John Wiley and Sons, PP. 453-472.

Reyss, D. 1974. A contribution to the study of deep sea Cumacea from the North Atlantic Genus *Makrokylindrus*. Crustaceana 26:5-28.

Reyss, D. 1978. Cumacea of the depths of the North Atlantic Family of Lampropidae. Crustaceana 35:1-21.

Rhines, P. 1971. A note on long-period motions at Site D. Deep-Sea Res. 18:21-26.

Rhodehamel, E.C. 1977. Lithologic descriptions. In P.A. Scholle (ed.). Geological studies on the COST No. B-2 Well, U.S. Mid-Atlantic outer continental shelf area. U.S. Geol. Surv. Circ. 750:15-22.

Rhodehamel, E.C. 1979. Lithologic descriptions. In P.A. Scholle (ed.), Geological Studies of the COST GE-1 well. U.S. Geol. Surv. Circ. 800:24-36.

Richardson, M.J., M. Wimbush and L. Mayer. 1981. Exceptionally strong near-bottom flows on the continental rise of Nova Scotia. Science 213:887-888.

Richardson, P.L. and J.A. Knauss. 1971. Gulf Stream and Western Boundary Undercurrent observations at Cape Hatteras. Deep-Sea Res. 18:1089-1109.

Richardson, P.L., A.E. Strong and J.A. Knauss. 1973. Gulf Stream eddies: recentobservations in the western Sargasso Sea. J. Phys. Oceanogr. 3:297-301.

Richardson, P.L. 1977. On the crossover between the Gulf Stream and the Western Boundary Undercurrent. Deep-Sea Res. 24:139-159.

Richardson, P.L., R.E. Cheney and L.V. Worthington. 1978. A census of Gulf Stream rings, spring 1975. J. Geophys. Res. 83:6136-6144.

Richardson, P.L. 1980. Gulf Stream ring trajectories. J. Phys. Oceanogr. 10:90-104.

Richardson, P.L. 1981. Gulf Stream trajectories measured with free-drifting buoys. J. Phys. Oceanogr. 11:999-1010.

Richardson, P.L., J.F. Price, W.B. Owens, W.J. Schmitz, H.T. Rossby, A.M. Bradley, J.R. Valdes and D.C. Webb. 1981. North Atlantic Subtropical Gyre: SOFAR floats tracked by moored listening stations. Science 213:435-437.

Richardson, P.L. 1983a. Eddy kinetic energy in the North Atlantic from surface drifters. J. Geophys. Res. 88:4355-4367.

Richardson, W.S., W.J. Schmitz, Jr. and P.P. Niiler. 1969. The velocity structure of the Florida Current from the Straits of Florida to Cape Fear. Deep-Sea Res. 16:225-231.

Riley, G.A. 1946. Factors affecting phytoplankton populations on Georges Bank. J. Mar. Res. 6:54-73.

Ring Group. 1981. Gulf Stream cold-core rings: their physics, chemistry, and biology. Science 212:1091-1100.

Robb, J.M. 1980a. High-resolution seismic-reflection profiles collected by the R/V James M. Gilliss, cruise GS 7903-4, in the Baltimore Canyon outer continental shelf area, offshore New Jersey. U.S. Geol. Surv. Open-File Rep. 80-934:3.

Robb, J.M. 1980b. High-resolution seismic-reflection profiles collected by the R/V Columbus Iselin, cruise C1 7807-1, in the Baltimore Canyon Outer Continental Shelf area, offshore New Jersey. U.S. Geol. Surv. Open-File Rep. 80-935:3.

Robb, J.M. and J.R. Kirby. 1980. Maps showing kinds and sources of environmental geologic and geophysical data collected by the U.S. Geological Survey in the Baltimore Canyon Trough Area. U.S. Geol. Surv. Misc. Field Studies Map MF-1210, 4 sheets.

Robb, J.M., J.R. Kirby and J.C. Hampson. 1981a. Bathymetric map of the continental slope and uppermost continental rise between Lindenkohl Canyon and South Toms Canyon, offshore eastern United States. U.S. Geol. Surv. Misc. Field Studies Map MF-1270, 1 sheet.

Robb, J.M., J.C. Hampson, Jr. and D.C. Twichell. 1981b. Geomorphology and sediment stability of a segment of the U.S. continental slope off New Jersey. Science 211:935-937.

Robb, J.M., W.B.F. Ryan and J.C. Hampson, Jr. 1981c. Description of mid-range sidescan-sonar data from the continental slope offshore New Jersey, collected by R/V GYRE, Cruise 80-G-8A. U.S. Geol. Surv. Open-File Rep. 81-1328:7.

Robb, J.M., J.C. Hampson, Jr., J.R. Kirby and D.C. Twichell.

1981d. Geology and potential hazards of the continental slope between Lindenkohl and South Toms canyons, offshore mid-Atlantic United States. U.S. Geol. Surv. Open-File Rep. 81-0600:36.

Robb, J.M. and J.C. Hampson, Jr. 1983. Processes creating canyons and the complex submarine landscape of the continental slope off New Jersey. EOS 64:1051.

Robb, J.M., J.R. Kirby, J.C. Hampson, Jr., P.R. Gibson and B. Hecker. 1983. Furrowed outcrops of Eocene chalk on the lower continental slope offshore New Jersey. Geology 11:182-186.

Roberts, B.B. 1940. The life cycle of Wilson's Petrel *Oceanites oceanicus* (Kuhl). Brit. Graham Land Exped. 1934-1937, Sci. Rep. 1:141-194.

Robertson, J.R. 1983. Predation by estuarine zooplankton on tintinnid ciliates. Est. Coast. Shelf Sci. 16:27-36.

Robinson, M.K., R.A. Bauer and E.H. Schroeder. 1979. Atlas of North Atlantic-Indian Ocean Monthly Mean Temperatures and Mean Salinities of the Surface Layer. Rep. NOO-RP-18:1339.

Roether, W. and K.O. Munnich. 1972. Tritium profile at the Atlantic 1970 Geosecs test cruise station. Earth Planet. Sci. Lett. 16:127-130.

Rooth, Claes G. and H. Göte Ostlund. 1972. Penetration of tritium into the Atlantic thermocline. Deep-Sea Res. 19:481-492.

Ross, D.A. 1970. Atlantic continental shelf and slope of the United States, heavy minerals of the continental margin from southern Nova Scotia to northern New Jersey. U.S. Geol. Surv. Prof. Pap. 529-G:4.

Ross, D.A. 1978 Opportunities and uses of the ocean. Springer-Verlag, New York, 320 pp.

Ross, D.A. and J.C. MacIlvaine. 1978. Sedimentary processes on continental slope of New England. Am. Assoc. Petrol. Geol. Bull. 62:559.

Rowe, G.T. and R.J. Menzies. 1969. Zonation of large benthic invertebrates in the deep-sea off the Carolinas. Deep-Sea Res. 16:531-537.

Rowe, G.T. 1971. Observations on bottom currents and epibenthic populations in Hatteras Submarine Canyon. Deep-Sea Res. 18:569-581.

Rowe, G.T. 1972. The exploration of submarine canyons and their genthic faunal assemblages. Proc. Roy. Soc. Edinburgh, Ser. A, 73:159-169.

Rowe, G.T. and C.H. Clifford. 1978. Sediment data from short cores taken in the northwest Atlantic Ocean. Woods Hole Oceanogr. Inst. Tech. Rep. 78-46:58.

Rowe, G.T. and W.D. Gardner. 1979. Sedimentation rates in the slope water of the northwest Atlantic Ocean measured directly with sediment traps. J. Mar. Res. 37:581-600.

Rowe, G.T., P.T. Polloni and R.L. Haedrich. 1982. The deep-sea macrobenthos on the continental margin of the northwest Atlantic Ocean. Deep-Sea Res. 29:257-278.

Rowe, G.T. 1983. Biomass and production of the deep-sea macrobenthos. Chapter 3. Deep-Sea Biology. In The Sea: Ideas and Observations on Progress in the Study of the Seas, vol. 8, G.T. Rowe (ed.), John Wiley and Sons, pp. 97-122.

Rowlett, R.A. 1973. Sea Birds wintering off Maryland shores, 1972-73. Maryland Birdlife 29:88-102.

Rowlett, R.A. 1980. Observations of marine birds and mammals in the northern Chesapeake Bight. U.S. Fish and Wildl. Serv., Biol. Serv. Progr. FWS/OBS-80/04.

Ruddiman, W.F. 1977. North Atlantic ice-rafting: a major change at 75,000 years before the present. Science 196:12008-1211.

Ryan, W.B.F., M.B. Cita, E.L. Miller, D. Hanselman, W.D. Nesteroff, B. Hacker and M. Nibbelonk. 1978. Bedrock geology in New England submarine canyons. Oceanologica Acta 1:233-254.

Ryan, W.B.F., G. Blechschmidt and P.R. Thompson. 1980. Technical aspects. in B. Hecker, G. Blechschmidt and P. Gibson, Final report for the canyon assessment study in the Mid- and North-Atlantic areas of the U.S. Outer Continental Shelf. Prepared for the Bureau of Land Management – 1-22.

Ryther, J.H. 1959. Potential productivity of the Sea. Science 130:602-608.

Ryther, J.H. and D.W. Menzel. 1960. The seasonal and geographical range of primary production in the western Sargasso Sea. Deep-Sea Res. 6:235-238.

Ryther, J.A. 1963. Geographic variations in production. In M.N. Hill (ed.), The Seas, vol. 2. Interscience Publ., New York.

Saint. John, P.A. 1958. A volumetric study of zooplankton distribution in the Cape Hatteras area: Limnol. Oceanogr. 3:387-397.

Sanders, H.L., R.R. Hessler and G.R. Hampson. 1965. An introduction to the study of deep-sea benthic faunal assemblages along the Gay Head-Bermuda transect: Deep-Sea Res. 12:845-867.

Sanders, H.L. 1968. Marine benthic diversity: A comparative study. Am. Naturalist 102:243-282.

Sanders, H.L. and R.R. Hessler. 1969a. Ecology of the deep-sea benthos. Science 163:1419-1424.

Sanders, H.L. and R.R. Hessler. 1969b. Diversity and composition of the abyssal benthos. Science 166:1034.

Sanders, H.L. and J.A. Allen. 1973. Studies on deep sea Protobranchia. Prologue and the Pristiglomidae. Bull. Mus. Comp. Zool. 145:237-261.

Sanders, H.L. 1977. Evolutionary ecology and the deep-sea benthos. Acad. Nat. Sci. Special Publ. 12:223-243.

Sanders, H.L. and J.A. Allen. 1977. Studies on the deep sea Protobranchia (Bivalvia). The Family Tindariidae and genus Pseudotindaria. Bull. Mus. Comp. Zool. 148(2):23-59.

Sansone, F.J. and C.S. Martens. 1982. Volatile fatty acid cycling in organic-rich marine sediments. Geochim. Cosmochim. Acta 46:1575-1589.

Santschi, P.H., Y. Li, J.J. Bell, R.M. Trier and K. Kawtaluk. 1980. Pu in coastal marine environments. Earth Planet. Sci. Lett. 51:248-265.

Sauer, T.C., Jr. 1981. Volatile organic compounds in open ocean and coastal surface waters. Organic Geochem. 3:91-101.

Saunders, P.M. 1971. Anticyclonic eddies formed from shoreward meanders of the Gulf Stream. Deep-Sea Res. 18:1207-1219.

Sayles, F.L. 1981. The composition and diagenesis of interstitial solutions-II. Fluxes and diagenesis at the water-sediment interface in the high latitude North and South Atlantic. Geochim. Cosmochim. Acta 45:1061-1084.

Sayles, F.L. and H.D. Livingston. 1983. The distribution of $^{239,\ 240}Pu$, ^{137}Cs and ^{55}Fe in continental margin sediments: relation to sedimentary redox environment, In Proceedings of the 4th International Ocean Disposal Symposium, Plymouth, England, I. Duedall (ed.), in press.

Scanlon, K.M. 1982a. Geomorphic features of the western North Atlantic continental slope between Northeast Channel and Alvin Canyon as interpreted from GLORIA II long-range sidescan sonar data. U.S. Geol. Surv. Open-File Rep. 82-728:9.

Scanlon, K.M. 1982b. GLORIA sidescan and seismic data collected by the DESV STARELLA along the continental slope and upper continental rise of the eastern United States in 1979. U.S. Geol. Surv. Open-File Rep. 82-1095:3.

Schanz, John J., Jr. 1978. Oil and Gas Resources - Welcome to Uncertainty. Resources, (58). Resources for the Future.

Schaule, B.K. and C.C. Patterson. 1981. Lead concentrations in the Northeast Pacific: evidence for global orthropogenic perturbations. Earth Planet. Sci. Lett. 54:97-116.

Schell, W.R. 1980. Radionuclides at the deep water disposal sites located near the Farallon Islands in the Pacific and at the Mouth of the Hudson Canyon in the Atlantic. 2nd Int. Ocean Dumping Symp. Woods Hole, MA.

Scheltema, R.S. 1968. Ocean Insects. Oceanus 14(3).

Scheltema, R.S. 1972. Reproduction and dispersal of bottom dwelling deep-sea invertebrates: a speculative summary. Barobiol. & Exper. Biol. Deep Sea:58-66.

Scheltema, A.H. 1981. Comparative morphology of the radulae and alimentary tracts in the Aplacophora. Malacologia 20:361-383.

Schlee, J.S. 1968. Sand and gravel on the continental shelf off the northeastern United States: U.S. Geol Surv. Circ. 602:9.

Schlee, J.S. and R.M. Pratt. 1970. Atlantic continental shelf and slope of the United States - Gravels of the northeastern part. U.S. Geol. Surv. Prof. Paper 529-H:39.

Schlee, J. 1973. Atlantic continental shelf and slope of the United States sediment texture of the northeastern part. U.S. Geol. Surv. Prof. Paper 529-L:64.

Schlee, J. and P. Sanko. 1975. Sand and gravel. MESA New York Bight Atlas Mono. 21, NY Sea Grant, Institute, Albany, NY:26.

Schlee, J.S., W.P. Dillon and J.A. Grow. 1979. Structure of the continental slope off the eastern United States. in L.J. Doyle, O.H. Pilkey (eds.), Geology of continental slopes. Soc. Econ. Paleontol. Mineralog. Special Publ. 27:95-117.

Schmitz, W.J., Jr., A.R. Robinson and F.C. Fuglister. 1970. Bottom velocity observations directly under the Gulf Stream. Science 170:1192-1194.

Schmitz, W.J., Jr. 1974. Observations of low-frequency current fluctuations on the continental slope and rise near site D. Woods Hole Oceanogr. Inst. Tech. Rep. 74-53:24.

Schmitz, W.J., Jr. 1977. On the deep general circulation in the western North Atlantic, J. Mar. Res. 35:21-28.

Schmitz, W.J., Jr. 1978. Observations of the vertical distribution of low frequency kinetic energy in the western North Atlantic. J. Mar. Res. 36:295-310.

Schmitz, W.J., Jr. 1980. Weakly depth-dependent segments of the North Atlantic circulation. J. Mar. Res.38:111-133.

Schmitz, W.J., Jr., W.R. Holland and J.F. Price. 1983. Mid-latitude meso-scale variability. Rev. Geophys. Space Physics 21:1109-1119.

Schmitz, W.J., Jr. 1984. Abyssal eddy kinetic energy in the North Atlantic. J. Mar. Res. 42:509-536.

Schneider, E.D., P.J. Fox, C.D. Hollister, H.D. Needham and B.C. Heezen. 1967. Further evidence of contour currents in the western North Atlantic. Earth Planet. Sci. Lett. 2:351-359.

Schoener, A. 1967. Post-larval development of five deep-sea ophiuroids. Deep-Sea Res. 14:654-660.

Schoener, A. 1968. Evidence for reproductive periodicity in the deep sea. Ecol. 49:81-87.

Schoener, A. 1969. Atlantic ophiuroids: some post-larval forms. Deep-Sea Res. 16:127-140.

Schoener, A. 1972. Fecundity and possible mode of development of some deep-sea ophiuroids. Limnol. Oceanogr. 17:193-199.

Schopf, T.J.M. and J.L. Gooch. 1971. A natural experiment using deep-sea invertebrates to test the hypothesis that genetic homozygosity is proportional to environmental stability. Biol. Bull. 141:401.

Schouten, H. and K. Klitgord. 1977. Mesozoic magnetic anomalies, western North Atlantic . U.S. Geol. Surv. Misc. Field Studies Map MF-915, scale: 1:2,000,000.

Schroeder, E.H., H. Stommel, D. Menzel and W. Sutcliffe, Jr. 1959. Climatic stability of eighteen degree water at Bermuda. J. Geophys. Res. 64:363-366.

Schroeder, E.H. 1963. Serial Atlas of the Marine Environment, Folio 2, North Atlantic Temperatures at a depth of 200 meters. Amer. Geograp. Soc. N.Y.

Schroeder, E.H. 1966. Average surface temperatures of the western North Atlantic. Bull. Mar. Sci. 16:302-323.

Scranton, M.I. and P.G. Brewer. 1977. Occurrence of methane in the near-surface waters of the western subtropical North Atlantic. Deep-Sea Res. 24:127-138.

Scranton, M.I. 1977. The marine geochemistry of methane. Ph.D. thesis, Massachusetts Institute of Technology/Woods Hole Oceanographic Institution Joint Program in Oceanography.

Sedberry, G.R. and J.A. Musick. 1978. Feeding strategies of some demersal fishes of the continental slope and rise off the Mid-Atlantic coast of the USA. Mar. Biol. 44:357-375.

Sen-Gupta, B.K. and D.P. Strickert. 1982. Living benthic foraminifera of the Florida-Hatteras slope: distribution trends and anomalies. Geol. Soc. Am. Bull. 93:218-224.

Shapiro, Michael E. 1981. The Federal Consistency Requirement. ALI-ABA Course of Study, Coastal Zone Management, March 26-27, Atlantic City, NJ:51-75.

Shepard, F.P. and R.F. Dill. 1966. Submarine canyons and other sea valleys. Chicago, Rand McNally & Co.:381.

Shepard, F.P. 1975. Progress of internal waves along submarine canyons. Mar. Geol. 19:131-138.

Shepard, F.P. 1976. Tidal components of currents in submarine canyons. J. Geol. 84:343-350.

Shepard, F.P., N.F. Marshall, P.A. McLonghlin and F.G.

Sullivan. 1979. Currents in submarine canyons and other sea valleys. Am. Assoc. Petrol. Geol. Studies in Geology No. 8.

Sheridan, R.E., C.L. Drake, J.E. Nafe and J. Hennion. 1965. Seismic refraction measurements of the continental margin east of Florida. Geol. Soc. Am. Spec. Paper 82:183.

Sheridan, R.E., et al. 1982. Early history of the Atlantic Ocean and gas hydrates on the Blake Outer Ridge: Results of the Deep Sea Drilling Project Leg 76. Geol. Soc. Am. Bull. 93:876-885.

Sherr, B.F., E.B. Sherr and T. Berman. 1982. Decomposition of organic detritus: A selective role for microflagellate protozoa. Limnol. Oceanogr. 27: 765-769.

Shipley, T.H., R.T. Buffler and J.S. Watkins. 1978. Seismic stratigraphy and geologic history of Blake Plateau and adjacent western Atlantic continental margin. Am. Assoc. Petrol. Geol. Bull. 62:792:812.

Shipley, T.H., M.H. Houston, R.T. Buffler, F.J. Shaub, K.J. McMillen, J.W. Ladd and J.L. Worzel. 1979. Seismic evidence for widespread possible gas hydrate horizons on continental slopes and rises. Am. Assoc. Petrol. Geol. Bull. 63:2204-2213.

Sholkovitz, E.R., J.K. Cochran and A.E. Carey. 1983. Laboratory studies of radionuclide diagenesis in nearshore sediments: I. The artificial radionuclides, ^{239,240}Pu, ^{37}Cs and ^{55}Fe. Geochim. Cosmochim. Acta in press.

Shor, A.N. 1984. Bathymetry. In E. Uchupi and A.N. Shor (eds.), Eastern North American Continental Margin and Adjacent Ocean Floor, 39° to 46° and 64° to 74°W: Atlas 3, Ocean Margin Drilling Program, Regional Atlas Series. Marine Science International, Woods Hole, MA, p. 1.

Shor, A.N. and R.D. Flood. 1984. Bathymetry. In J.I. Ewing and P.D. Rabinowitz (eds.), Eastern North American Continental Margin and Adjacent Ocean Floor, 34° to 41°N and 68° to 78°W: Atlas 4, Ocean Margin Drilling Program, Regional Atlas Series. Marine Science International, Woods Hole, MA, p. 1.

Siebenaller, J.F. and R.R. Hessler. 1977. The Nannoniscidae (Isopoda, Asselota): Hebefustis N. Gen and Nannoniscoides Hansen. Trans. of the San Diego Soc. Nat. History 19(2):17-44.

Siebenaller, J.F. and R.R. Hessler. 1981. The genera of the Nannoniscidae (Isopoda, Asellota). Trans. of the San Diego Soc. Nat. History 19(16):227-250.

Sieburth, J.McN., P.-J. Willis, K.M. Johnson, C.M. Burney, D.M. Lavoie, K.R. Hinga, D.A. Caron, F.W. French, III, P.W. Johnson, et al. 1976. Dissolved organic matter and heterotrophic micro neuston in the surface micro layers of the North Atlantic. Science 194:1415-1418.

Sieburth, J.McN., V. Smetacek and J. Lenz. 1978. Pelagic ecosystem structure: Heterotrophic compartments of the plankton and their relationship to plankton size fractions. Limnol. Oceanogr. 23: 1256-1263.

Sieburth, J.McN. 1979. Sea microbes. Oxford University Press, New York:491.

Sieburth, J.McN. 1984. Protozoan bacterivory in pelagic marine waters. In J.E. Hobbie and P.J. LeB. Williams (eds.). Heterotrophic activity in the sea. Plenum Press, New York, pp. 405-444.

Sieburth, J.McN. In press. Protozoan bacterivory in pelagic marine waters. in J.E. Hobbie, and P.J.leB. Williams (eds.). Heterotrophic activity in the sea, Plenum Press, New York.

Silver, M. W. and A. L. Alldredge. 1981. Bathypelagic marine snow: deep-sea algal and detrital community. J. Mar. Res. 39: 501-530.

Silver, M.W., M.M. Gowing, D.C. Brownlee and J.O. Corliss. 1982. Protozoan colonists of sinking detritus. EOS 63:959.

Simoneit, B.R.T. and M.A. Mazurek. 1981. Air Pollution: the organic components. In Critical Reviews in Environmental control. CRC Press, 2:219-276.

Sims, G.G., J.R. Campbell, F. Zemlyak and J.M. Graham. 1977. Organochlorine residues in fish and fishery products from the Northwest Atlantic. Bull. Environ. Contam. Toxicol. 18:697-705.

Sinderman, C.J. 1979. Status of northwest Atlantic herring stocks of concern to the United States. Tech Ser. Rep. NOAA-NMF Northeast Fish. Cent.

Singer, J.J. L.P. Atkinson, J.O. Blanton and J.A. Yoder. 1983. Cape Romain and the Charleston Bump: Historical and recent hydrographic observations. J. Geophys. Res. 88:4685-4697.

Slater, R.A. 1981. Submersible observations of the sea floor near the proposed Georges Bank lease sites along the North Atlantic outer continental shelf and upper slope. U.S. Geol. Surv. Open-File Rep. 81-0742:65.

Slater, R.A., D.C. Twichell and J.M. Robb. 1981. Submersible observations of potential geologic hazards along the mid-Atlantic outer continental shelf and uppermost slope. U.S. Geol. Surv. Open-File Rep. 81-0968:50.

Smith, K.L. and J.M. Teal. 1973. Deep-sea benthic community respiration: an *in situ* study at 1850 meters. Science 179:282-283.

Smith, K.L., Jr. 1978. Benthic community respiration in the Northwest Atlantic Ocean in-situ measurements from 40 to 5200 meters. Mar. Biol. 47:337-348.

Smith, K.L., Jr., G.A. White, M.B. Laver and J.A. Haugsness. 1978. Nutrient exchange and oxygen consumption by deep-sea benthic communities: preliminary in situ measurements. Limnol. Oceanogr. 23:997-1005.

Smith, K.L., Jr. and S.G. Horrigan. 1983. Benthic boundary layer in the central and eastern North Pacific. II. Sediment community oxygen consumption and nutrient exchange. Limnol. Oceanogr.

Smith, K.L., Jr. and K.R. Hinga. 1983. Sediment community respiration in the deep sea. Chapter 8. Deep-Sea Biology. In The Sea: Ideas and Observations on Progress in the Study of the Seas, vol. 8, G.T. Rowe (ed.), John Wiley and Sons, pp. 331-370.

Smith, P.C. and B.D. Petrie. 1982. Low-frequency circulation at the edge of the Scotian Shelf. J. Phys. Oceanogr. 12:28-46.

Smith, W.E.T. 1960. Earthquakes of eastern Canada and adjacent areas, 1928-1959. Publ. Dominion Obs., Ottawa 32 3:87-121.

Smyth, P.O. 1980. Callinectes (Decapoda: Portunidae) Larvae in the Middle Atlantic Bight, 1975-1977. Fish. Bull. 78(2).

Sorokin, Y.I. 1977. The heterotrophic phase of plankton succession in the Japan Sea. Mar. Biol. 41: 107-117.

Sorokin, Y.I. 1981. Microheterotrophic organisms in marine

ecosystems. in A.R. Longhurst (ed.). Analysis of marine ecosystems, Academic Press, London:293-342.

Sournia, A. 1970. La Cyanophyceae Oscillatoria (Trichodesmium) dans la plancton marin. Nova Hedwigia 15:1-12.

South Atlantic Fishery Management Council. 1979. Fishery Management Plan for the Atlantic Billfishes: White Marlin, Blue Marlin, Sailfish and Spearfish. November 1979.

Southward, E.C. 1968. On a new genus of pogonophore from the western Atlantic Ocean, with descriptions of two new species. Bull. Mar. Sci. 18:182-190.

Southward, E.C. and T. Brattegard. 1968. Pogonophora of the northwest Atlantic: North Carolina Region. Bull. Mar. Sci. 18:836-875.

Southward, E.C. 1971. Pogonophora of the northwest Atlantic: Nova Scotia to Florida: Smithsonian Contrib. Zool. 88:29.

Spencer, A. 1979. A Compilation of Moored Current Meter Data, Whitehorse Profiles and Associated Oceanographic Observations. Volume XX. Rise Array, 1974. Woods Hole Oceanogr. Inst. Tech. Rep. 79-56:72.

Spies, R.B., P.H. Davis and D.H. Stuermer. 1980. Ecology of a submarine petroleum seep off the California Coast. pp. 208-263. In Environmental Pollution. I. Hydrocarbons, R. Geyer (ed.), Elsevier, New York.

Spindler, M., O.R. Anderson, C. Hemleben and A.W.H. Be. 1978. Light and electron microscopic observations of gametogenesis in Hastigerina pelagica (Foraminifera). J. Protozool. 25:427-433.

Stalcup, M.C., T.M. Joyce, R.W. Schmitt and J.A. Dunworth. 1982. Warm Core Ring Cruise No. 1. R/V Endeavor Cruise no. 74. Woods Hole Oceanogr. Tech. Rep. 82-35:133.

Stanley, D.J., H. Sheng and C.P. Pedraza. 1971. Lower continental rise east of the middle Atlantic states: predominant sediment dispersal perpendicular to isobaths. Geol. Soc. Am. Bull. 82:1831-1840.

Stanley, D.J. 1974a. Pebbly mud transport in the head of Wilmington Canyon. Mar. Geol. 16:1-8.

Stanley, D.J. and G.L. Freeland. 1978. The erosion-deposition boundary in the head of Hudson submarine canyon defined on the basis of submarine observations. Mar. Geol. 26:37-46.

Stanley, D.J. and C.M. Wear. 1978. The "mud-line": an erosion depostion boundary on the upper continental slope. Mar. Geol. 28:19-29.

Stanley, D.J., H. Sheng, D.N. Lambert, P.A. Rona, D.W. McGrail and J.S. Jenkyns. 1981. Current-influenced depositional provinces, continental margin off Cape Hatteras, identified by petrologic method. Mar. Geol. 40:215-235.

Steele, J.H. 1976. Patchiness. In Cushing and Walsh. The ecology of the seas. W.B. Saunders Co.:98-115.

Stefansson, U. and L.P. Atkinson. 1968. Physical and chemical properties of the shelf and slope waters off North Carolina. Duke Univ. Marine Lab. Tech. Rep.:230.

Stefansson, U. and L.P. Atkinson. 1971. Nutrient density relationships in the western North Atlantic between Cape Lookout and Bermuda. Limnol. Oceanogr. 16:51-59.

Stefansson, U., L.P. Atkinson and D.F. Bumpus. 1971. Seasonal studies of hydrographic properties and circulation of the North Carolina Shelf and Slope waters. Deep-Sea Res. 18:383-420.

Stegeman, J.J. 1981. Polynuclear aromatic hydrocarbons and their metabolism in the marine environment. in Polycyclic Hydrocarbons and Cancer. Chapter 1, volume 3. Academic Press 3(1):60.

Stepien, J.C. 1980. The occurrence of chaetognaths, pteropods, and euphausiids in relation to deep flow reversals in the Straits of Florida. Deep-Sea Res. 27:987-1011.

Stepien, J. and B. Hecker. 1981. Interim Report II, BLM Contract BLM AA851-CTO-59.

Stepien, J. 1982. Interim Report III, BLM Contract BLM AA851-CTO-59.

Stepien, J. 1983. Interim Report IV, BLM Contract BLM AA851-CTO-59.

Stetson, T.R., D.F. Squires and R.M. Pratt. 1962. Coral banks occurring in deepwater on the Blake Plateau. Am. Mus. Novitates 2114:1-39.

Stetson, T.R., Elazar Uchupi and J.D. Milliman. 1969. Surface and subsurface morphology of two small areas of the Blake Plateau. Gulf Coast Assoc. Geol. Socs. Trans. 19:131-142.

Stick, L.V. and C.C. Johnson. 1978. Relation and contribution of slope water particulate trace metals to North Atlantic continental shelf. EOS 59:1098.

Stommel, H. 1965. The Gulf Stream: a Physical and Dynamical Description, University of California and Cambridge University Press, 2nd ed., 248 pp.

Stommel, H., P. Niiler and D. Anati. 1978. Dynamic topography and recirculation of the North Atlantic. J. Mar. Res. 36:449-468.

Stoner, A. W., M. W. Farmer and S. E. Humphris. 1985. Evidence for increases in the concentration of pelagic tar in the northwest Atlantic, Caribbean Sea, and Gulf of Mexico. Deep-Sea Res. submitted.

Stout, V. F. 1980. Organochlorine residues in fishes from the northwest Atlantic Ocean and Gulf of Mexico. Fish. Bull. 78:51-58.

Stout, J.D. 1980. The role of protozoa in nutrient cycling and energy flow. Adv. Microbiol. Ecol. 4:1-50.

Stow, D.A.V. 1976. Deep water sands and silts on the Nova Scotian continental margin. Marit. Sediments 12:81-90.

Stow, D.A.V. 1979. Distinguishing between fine grained turbidites and contourites on the Nova Scotian deep water margin. Sedimentology 26:371-387.

Stubblefield, W.L., B.A. McGregor, E.B. Forde, D.N. Lambert and G.F. Merrill. 1982. Reconnaissance in DSRV Alvin of a "fluvial-like" meander system in Wilmington Canyon and slump features in South Wilmington Canyon. Geology 10:31-36.

Sumpter, R. 1979. Baltimore Canyon future clouded by sparse yield. Oil Gas J. 77(18):119-123.

Sunda, W. and R.R.L. Guillard. 1976. The relationship between cupric ion activity and the toxicity of copper to phytoplankton. J. Mar. Res. 34:511-529.

Sutter, J.F. and T.E. Smith. 1979. $^{40}Ar/^{39}Ar$ ages of diabase intrusions from the Newark trend basins in Connecticut and Maryland. Initiation of central Atlantic rifting. Am. J. Sci. 279:808-831.

Swallow, J.C. 1955. A neutral-buoyancy float for measuring deep currents. Deep-Sea Res. 3:74-81.

Swallow, J.C. and L.V. Worthington. 1961. An observation of a deep countercurrent in the western North Atlantic. Deep-Sea Res. 8:1-19.

Swanberg, N.R. 1979. The ecology of colonial radiolarians: Their colony morphology, trophic interactions and associations, behavior, distribution and the photosynthesis of their symbionts. Ph.D. Thesis. Woods Hole Oceanogr. Inst. Mass. Inst. Tech. Joint Program, Woods Hole, MA:202 .

Sweet, W. 1981. Air-sea interaction effects in the lower troposphere across the north wall of the Gulf Stream. Monthly Weather Review 109:1042-1052.

Swift, D.G. 1980. Vitamins and phytoplankton growth. in I. Morris (ed.), The physiological ecology of phytoplankton. Univ. Calif Press:329-370.

Swinnerton, J.W. and R.A. Lamontagne. 1974. Oceanic distributions of low molecular-weight hydrocarbon. Baseline measurements. Environmental Science and Technology 8:657-663.

Sylwester, R.E., W.P. Dillon and J.A. Grow. 1979. Active growth fault on seaward edge of Blake Plateau. in D. Gill and D.F. Merriam (eds.), Geomathematical and Petrophysical Studies in Sedimentology, Pergamon Press, NY:197-209.

Tabor, P. S. and R. R. Colwell. 1978. Initial investigations with a deep ocean in situ sampler. Proc. MTS/IEEE Oceans '76.

Tabor, P. S., K. Ohwada and R. R. Colwell. 1981. Filterable marine bacteria found in the deep sea: distribution, taxonomy and response to starvation. Microb. Ecol. 7:67-83.

Takahashi, K., and S. Honjo. 1981. Vertical flux of Radiolaria: A taxon-quantitative sediment trap study from the western tropical Atlantic. Micropaleontol. 27:140-190.

Talley, L.D. and M.S. McCartney. 1982. Distribution and circulation of Labrador Sea Water. J. Phys. Oceanogr. 12:1189-1205.

Taylor, F.J.R., D.J. Blackbourn and J. Blackbourn. 1971. The red-water ciliate Mesodinium rubrum and its "incomplete symbionts". A review including new ultrastructural observations. J. Fish. Res. Bd. Can. 28:391-407.

Taylor, F.J.R. 1982. Symbioses in marine microplankton. Ann. Inst. Oceanogr. Paris 58(S): 61-90.

Teal, J.M. 1976. Hydrocarbon uptake by deep-sea benthos. In Sources, Effects and Sinks of Hydrocarbons, Proceedings of a Symposium. Am. Inst. Biol. Sci. Washington, D.C.:358:372.

Therriault, J-C. and T. Platt. 1981. Environmental control of phytoplankton patchiness. Can. J. Fish. Aquat. Sci. 38:638-641.

Thiel, J. 1979. Structural aspects of the deep-sea benthos. In The Deep-Sea - Ecology and Exploitation. Ambio Spec. Rept. 6:25-31.

Thistle, D. and R.R. Hessler. 1976. The origin of a deep-sea family, the Ilyarachnidae (Crustacea, Isopoda). Syst. Zool. 25:110-116.

Thistle, D. and R.R. Hessler. 1977. A revision of Betamorpha (Isopoda: Asellota) in the world ocean with three new species. Zool. J. Linn. Soc. 60:273-295.

Thompson, P.R., G. Blechschmidt and W.B.F. Ryan. 1980. East Coast submarine canyons. in B. Hecker, G. Blechschmidt and P. Gibson, Final report for the canyon assessment study in the Mid- and North-Atlantic areas of the U.S. Outer Continental Shelf. Prepared for the Bureau of Land Management:73.

Thompson, R. 1971a. Topographic Rossby waves at a site north of the Gulf Stream. Deep-Sea Res. 18:1-20.

Thompson, R. 1971b. Why there is an intense eastward current in the North Atlantic but not in the South Atlantic. J. Phys. Oceanogr. 1:235-237.

Thompson, R. and J.R. Luyten. 1976. Evidence for bottom-trapped topographic Rossby waves from sample moorings. Deep-Sea Res. 23:629-635.

Thompson, R. 1977. Observations of Rossby waves near Site D. Prog. Oceanog. 7:1-28.

Thunnell, R.C. and S. Honjo. 1981. Planktonic foraminiferal flux to the deep ocean: sediment trap results from the tropical Atlantic and the central Pacific. Mar. Geol. 40:237-253.

Tietjen, J.H. 1971. Ecology and distribution of deep-sea meiobenthos off North Carolina. Deep-Sea Res. 18:941-957.

Tissot, B.P. and D.H. Welte. 1978. Petroleum Formation and Occurrence. Springer-Verlag, New York.

Toonkel, L.E. 1980. Quarterly deposition at world land site. Environmental Quarterly Report EML-370 (Appendix). Environmental Measurements Laboratory, U.S. Dept. of Energy, New York.

Trier, R.M., W.S. Broecker and H.W. Feely. 1972. Radium - 228 profile at the second Geosecs intercalibration station, 1970, in the North Atlantic. Earth Planet. Sci. Lett. 16:141-145.

Tripp, B.W., J.W. Farrington and J.M. Teal. 1981. Unburned coal as a source of hydrocarbons in surface sediments. Mar. Poll. Bull. 12:122-126.

Trondsen, Eilif and W.J. Mead. 1977. California offshore phosphorite deposits — An economic evaluation. U. of Calif., Santa Barbara, Mar. Sci. Inst. Sea Grant Publ. 59:188. La Jolla, CA.

Tropical Cyclones of the North Atlantic Ocean, 1871-1980. 1978. U.S. Dept. Commerce, NOAA, National Weather Service, Environmental Data and Information Service, Asherville, NC.

Trumbull, J.V.A. and M.J. McCamis. 1967. Geological exploration in an east coast submarine canyon from a research submersible. Science 158:370-372.

Tucholke, B.E., W.R. Wright and C.D. Hollister. 1973. Abyssal circulation over the Greater Antilles Outer Ridge. Deep-Sea Res. 20:973-995.

Tucholke, B.E. 1975. Sediment distribution and deposition by the Western Boundary Undercurrent the Great Antilles Outer Ridge. J. Geol. 83:177-208.

Tucholke, B.E., G.M. Bryan and J.I. Ewing. 1977. Gas-hydrate horizons detected in seismic-profiler data from the western North Atlantic. Am. Assoc. Petrol. Geol. Bull 61:698-707.

Tucholke, B.E. 1979a. Furrows and focussed echoes on the Blake Outer Ridge. Mar. Geol. 31:13-20.

Tucholke, B.E. 1979b. Geologic significance of sedimentary reflectors in deep western North Atlantic. Am. Assoc. Petrol. Geol. Bull. 63:543.

Tucholke, B.E. and P.R. Vogt. 1979. Western North Atlantic: sedimentary evolution and aspects of tectonic history. Initial Reports of the Deep Sea Drilling Project 43:791-825.

Tucholke, B.E. and G.S. Mountain. 1979. Seismic stratigraphy, lithostratigraphy and paleosedimentation patterns in the North American basin. Deep Drilling Results in the Atlantic Ocean: Continental Margins and Paleo-environment. Am. Geophys. Union M. Ewing Series 3:58-86.

Tucholke, B.E., C.D. Hollister, P.E. Biscaye, W.D. Gardner and L.G. Sullivan. 1979. Zonation and effects of abyssal currents on the Nova Scotian continental rise. EOS 60:855.

Tucholke, B.E. 1980. Acoustic environment of the Hatteras and names abyssal plains, western North Atlantic Ocean, determined from velocities and physical properties of sediment cores. J. Acoust. Soc. Am. 68:1376-1390.

Tucholke, B.E., R.E. Houtz and W.J. Ludwig. 1982. Sediment thickness and depth to basement in the western North Atlantic Ocean basin. Am. Assoc. Petrol. Geol. Bull. 66:1384-1395.

Tucholke, B.E. and E.P. Laine. 1983. Neogene and Quaternary development of the lower continental rise off the central U.S. East Coast. Am. Assoc. Petr. Geol. Mem. (Hedberg Symp. Volume):295-305.

Turner, R.D. 1973. Wood-boring bivalves, opportunistic species in the deep sea. Science 180:1377-1379.

Turner, R.D. 1977. Wood, mollusks, and deep-sea food chains. Bull. Am. Malacol. Union:13-19.

Turner, S.C., E.D. Anderson and S.J. Wilk. 1981 A Preliminary Analysis of the Status of the Tilefish Population in the Southern New England-Middle Atlantic Region. NMFS Laboratory Reference Document 81-03.

Tuttle, J.H. and H.W. Jannasch. 1976. Microbial utilization of Thio sulfate in the Deep Sea. Limnol. Oceanogr. 21:697-701.

Twichell, D.C. and D.G. Roberts. 1982. Morphology, distribution, and development of submarine canyons on the United States Atlantic continental slope between Hudson and Baltimore Canyons. Geology 10:408-412.

Twichell, D.C. 1983. Geology of the head of Lydonia Canyon, U.S. Atlantic Outer Continental Shelf. Mar. Geol.

U.S. Bureau of Land Management (BLM). 1981. Compilation of Laws Related to Mineral Resources Activities on the Outer Continental Shelf, Volumes I and II. U.S. Geological Survey, U.S. Department of Interior. National Technical Information Service, Springfield, Virginia. 647 p.

U.S. Deptartment of Commerce. 1979. Assimilative Capacity of U.S. Coastal Waters for Pollutants. Working Paper No. 1: Federal Plan for Ocean Pollution Research Development and Monitoring, FY 1981-1985. Proceedings of a Workshop, Crystal Mountain, Washington, July 19-August 4, 1979.

U.S. Department of the Navy. 1982. Draft Environmental Impact Statement on the Disposal of Decommissioned, Defuelled Naval Submarine Reactor Plants, Washington, DC.

U.S. Environmental Protection Agency. 1977. Ocean Dumping, Final Revision of Regulations and Criteria. Federal Register 42(7):2485.

U.S. Naval Climatic Atlas of the World. 1974. North Atlantic Ocean, vol. 1, revised 1971, Naval weather service Command, Ref. No. NAVAR 50-1C-528.

U.S. Naval Oceanographic Office. 1968. Oceanographic Atlas of the North Atlantic Ocean, Section III. ICE Publ. 700.

U.S. Naval Oceanographic Office. 1984. Standard Navy ocean area world relief maps (North Atlantic Ocean - sheets NA6, NA9 and 9A), Mercator projection, scale 1″=1°: World Bathymetric Unit, U.S. Naval Oceanographic Office, Bay St. Louis, MS, (unpubl.).

U.S. Naval Weather Service Command. 1970. Summary of synoptic Meteorological observations. N. America coastal marine areas. 2-3-4.

Uchupi, E. 1965. Maps showing relation of land and submarine topography, Nova Scotia to Florida: U.S. Geol. Surv. Misc. Geol. Inv. Map I-451, scale, 1:1,000,000, 3 sheets.

Uchupi, E. 1967a. The continental margin south of Cape Hatteras, North Carolina: shallow structure: Southeastern Geol. 8:155-177.

Uchupi, E. 1967b. Slumping on the continental margin southeast of Long Island, New York: Deep-Sea Res. 14:635-638.

Uchupi, E. and K.O. Emery. 1967. Structures of continental margin off Atlantic coast of United States. Bull. Am. Assoc. Petrol. Geol. 51:223-234.

Uchupi, E. 1968a. Atlantic continental shelf and slope: Physiography: U.S. Geol. Survey Prof. Paper 529-C, 30.

Uchupi, E. 1970. Atlantic continental shelf and slope of the United States: shallow structure. U.S. Geol. Surv. Prof. Paper 529-I:44.

Uchupi, E. 1971. Bathymetric atlas of the Atlantic, Caribbean and Gulf of Mexico: Woods Hole Oceanogr. Inst. Ref. 71-72.

Uchupi, E. and J.A. Austin, Jr. 1979. The geologic history of the passive margin off New England and the Canadian Maritime Provinces. in C.E. Keen (ed.), Crustal properties across passive margins. Tectonophys. 59:53-69.

Uchupi, E., J.A. Austin, Jr. and D.H. Gever. 1982a. Salt diapirism and associated faulting beneath the eastern end of Georges Bank. Northeastern Geol. 4:20-22.

Uchupi, E., J.P. Ellis, J.A. Austin, Jr., G.H. Keller and R.D. Ballard. 1982b. Mesozoic-Cenozoic regressions and the development of the margin off northeastern North America. in R.A. Scrutton and M. Talwani (eds.), The Ocean Floor, John Wiley and Sons, Ltd. New York:81-95.

Uda, M. 1959. Water mass boundaries. "Siome." Frontal theory in oceanography. Fish. Res. Bd. Can. Rep. Ser. No. 51, (seminar 2):11p.

University of Rhode Island and Applied Science Associates, Inc. 1982. Assessing the Impact of Oil Spills on a Commercial Fishery, U.S. Department of the Interior,

New York Outer Continental Shelf Office, Contract No. AA851-CTO-75.

Uzmann, J.R., R.A. Cooper, R.B. Theroux and R.L. Wigley. 1977. Synoptic comparison of three sampling techniques for estimating abundance and distribution of selected megafauna: submersible vs. camera sled vs. otter trawl. Mar. Fish. Res. Paper 1273, Mar. Fish. Rev. 39(12):11-19.

Vail, P.R., R.M. Mitchum, Jr. and S. Thompson, III. 1977. Seismic stratigraphy and global changes of sea level, Part 4: Global cycles of relative changes of sea level, in C.E. Payton, (ed.), Seismic stratigraphy — applications to hydrocarbon exploration. Am Assoc. Petrol. Geol. Mem. 26:83-97.

Vail, P.R. and R.G. Todd. 1981. Northern North Sea Jurrassic unconformities, chronostratigraphy and sea level changes from seismic stratigraphy. in Petroleum geology of the continental shelf of northwest Europe: London, Heyden and Son Ltd.:216-235.

Valentine, P.C., J.R. Uzmann and R.A. Cooper. 1980. Geology and biology of Oceanographer submarine canyon. Mar. Geol. 38:283-312.

Valentine, P.A., R.A. Cooper and J.R. Uzmann. 1984. Submarine topography, surficial geology and fauna of the northern part of Oceanographer Canyon. U.S. Geol. Surv. Misc. Field Studies Map MF-1531, 5 sheets.

Van Hinte, J.E. 1976. A Jurassic time scale. American Assoc. Petrol. Geol. Bull. 60:489-497.

Vassallo, K., R.D. Jacobi and A.N. Shor. 1984a. Echo character, microphysiography, and geologic hazards. In J.I. Ewing and P.D. Rabinowitz (eds.), Eastern North American Continental Margin and Adjacent Ocean Floor, 34° to 41°N and 68° to 78°W: Ocean Margin Drilling Program, Regional Atlas Series, Atlas 4, Marine Science International, Woods Hole, MA, p. 31.

Vassallo, K., R.D. Jacobi and A.N. Shor. 1984b. Echo character, microphysiography, and geologic hazards. In G.M. Bryan and J.R. Heirtzler (eds.), Eastern North American Continental Margin and Adjacent Ocean Floor, 28° to 36°N and 70° to 82°W: Ocean Margin Drilling Program, Regional Atlas Series, Atlas 5, Marine Science International, Woods Hole, MA, p. 40.

Veatch, A.C. and P.A. Smith. 1939. Atlantic submarine valleys of the United States and the Congo Submarine Valley. Geol. Soc. Am. Spec. Paper 7:101.

Vild, B.F. 1979. State Government and OCS Policy: An Analysis of the Outer Continental Shelf Lands Act ant the 1978 Amendments. Marine Affairs J. 6:39-59.

Vishniac, W. 1971. Limits of microbial production in the oceans. in D.E. Hughes and A.H. Rose (eds.), Microbes and biological productivity. Cambridge Univ. Press:355-366.

Vogt, P.R. 1973. Early Events in the Opening of the North Atlantic. in Implications of Continental Drift to the Earth Sciences, Rift Margins and Continental Edge Structures 2:693-712.

Volkmann, G. 1962. Deep current observations in the western North Atlantic. Deep-Sea Res. 9:493-500.

Von Arx, W.S., D.F. Bumpus and W.S. Richardson. 1955. On the fine-structure of the Gulf Stream front. Deep-Sea Res. 3:46-65.

Wade, T. L. and J. G. Quinn. 1975. Hydrocarbons in the Sargasso Sea surface microlayer. Mar. Polution Bull. 6:54-57.

Wade, T. L., J. G. Quinn, W. T. Lee and C. W. Brown. 1976. Source and distribution of hydrocarbons in surface waters of the Sargasso Sea. In Sources, Effects and Sinks of Hydrocarbons in the Environment, Proceedings of a Symposium. American Institute of Biological Sciences, Washington, DC, pp. 271-286.

Wakeham, S.G. and J.W. Farrington. 1980. Hydrocarbons in contemporary aquatic sediments. In Contaminants and Sediments, Chapter 1, volume 1, R.A. Baker (ed.), Ann Arbor Publishers, Ann Arbor, MI.

Wakeham, S.G., A.C. Davis and J.T. Goodwin. 1982. Volatile organic compounds in marine experimental ecosystems and the estuarine environment - initial results. in Marine Mesocosms. Biological and Chemical Research in Experimental Ecosystems, G.D. Grice and M.R. Reeve, (eds.), Springer-Verlag, New York:137-151.

Wallace, G.T., Jr., G.L. Hoffman and R.A. Duce. 1977. The influence of organic matter and atmospheric deposition on the particulate trace metal concentration of northeast Atlantic surface water. Mar. Chem. 5:143-170.

Warm Core Rings Executive Committee. 1982. Multidisciplinary program to study warm core rings. EOS 63: 834-836.

Warme, J.E., T.B. Scanland and N.F. Marshall. 1971. Submarine Canyon erosion. Contribution of Marine Rock Burrowers. Science 173:1127-1129.

Warme, J., R. Slater and R.A. Cooper. 1978. Bioerosion in submarine canyons. in D.J. Stanley and G. Kelling (eds.), Sedimentation in Submarine Canyons, Fans and Trenches, Dowden, Hutchinson and Ross, Stroudsburg, PA:65-70.

Warren, B.A. and G.K. Volkmann. 1968. Measurement of the volume transport of the Gulf Stream south of New England. J. Mar. Res. 26:110-126.

Waterbury, J. B., S. W. Watson, R. R. L. Guillard and L. E. Brand. 1979. Widespread occurrence of a unicellular marine, planktonic, cyano bacterium. Nature 277:293-294.

Waterbury, J. B., S. W. Watson and F. Valois. 1980. Preliminary assessment of the importance of *Synechococcus* spp. as oceanic primary producers. in P. Falkowski (ed.), Primary Production in the Sea. Plenum Press, N.Y.:516-577.

Watts, D.R. and W.E. Johns. 1982. Gulf Stream meanders: observations on propagationand growth. J. Geophys. Res. 87:8456-9476.

Webster, F. 1961a. A description of Gulf Stream meanders off Onslow Bay. Deep-Sea Res. 8:130-143.

Webster, F.A. 1961b. The effect of meanders on the kinetic energy balance of the Gulf Stream. Tellus 13:391-401.

Webster, F. 1969. Vertical profiles of horizontal ocean currents. Deep-Sea Res. 16:85-98.

Webster, F. 1971. On the intensity of horizontal ocean currents. Deep-Sea Res. 18:885-893.

Wenner, C.A. and J.A. Musick. 1977. Biology of the morid fish *Antimora rostrata* in the western North Atlantic. J. Fish. Res. Bd. Can. 34:2362-2369.

Wenner, E.L. 1978. Comparative biology of four species of glyphocrangonid and crangonid shrimp from the continental slope of the Middle Atlantic Bight. Can. J. Zool. 56:1052-1065.

Wenner, E.L. 1979a. Biology of deep sea lobsters of the family Polychelidae Crustacea Decapoda from the western North Atlantic. U.S. Natl. Mar. Fish. Serv. Fish Bull. 77:435-444.

Wenner, E.L. 1979b. Distribution and reproduction of Nematocarcinid shrimp Decapoda Caridea from the Northwestern North Atlantic. Bull. Mar. Sci. 29:380-393.

Weston, F.M. 1953. Red Phalarope *(Phalaropus fulicarius)* wintering near Pensacola, Florida. Auk 70:491-492.

Whittle, K.J., R. Hardy, A.V. Holden, R. Johnston and R.J. Pentreath. 1977. Occurrence and fate, organic and inorganic contaminants in marine animals. Annals of the New York Academy of Sciences 298:47-79.

Whitton, B.A. 1973. Freshwater plankton. in The Biology of Blue-Green Algae. N. G. Carr and B. A. Whtton (eds.), University of California Press, Berkeley:353-367.

Wiebe, W. J. and L. R. Pomeroy. 1972. Microorganisms and their association with aggregates and detritus in the sea: a microscopic study. Mem. Ist. Ital. Idrobiol. 29:325-352.

Wiebe, P. 1976. The biology of cold-core rings. Oceanus 19(3):69-76.

Wiebe, P.H., E.M. Hulburt, E.J. Carpenter, A.E. Jahn, G.P. Knapp, III, S.H. Boyd, P.B. Ortner and J.L. Cox. 1976a. Gulf Stream cold-core rings: large-scale interaction sites for open ocean plankton communities. Deep-Sea Res. 23:695-710.

Wiebe, P.H., K.H. Burt, S.H. Boyd and A.W. Morton. 1976b. A multiple opening/closing net and environmental sensing system for sampling zooplankton. J. Mar. Res. 34:313-326.

Wiebe, P.H. and S.H. Boyd. 1978. Limits of *Nematoscelis megalops* in the Northwestern Atlantic in relation to Gulf Stream cold core rings. I. Horizontal and vertical distributions. J. Mar. Res. 36:119-142.

Wiebe, P.H., L.P. Madin, L.R. Haury, G.R. Harbison and L.M. Philbin. 1979. Diel vertical migration by *Salpa aspera* and its potential for large-scale particulate organic matter transport to the deep sea. Mar. Biol. 53:249-255.

Wiebe, P.H. 1981a. A conceptual model of structure in oceanic plankton communities. in G. Magazzu and L. Guglielmo, eds. and publ., The use of mathematical simulation models in the study of bilogy in marine ecosystems:23-40.

Wiebe, P.H. 1981b. Interactions between field data acquisition and modelling in the study of the biology of Gulf Stream rings. in G. Magazzu and L. Guglielmo, eds. and publ., The use of mathematical simulation models in the study of biology in marine exosystems:107-125.

Wiebe, P.H. 1982a. Gulf Stream Rings. Scientific American 246:60-70.

Wiebe, P.H., S.H. Boyd, B.M. Davis and J.L. Cox. 1982. Avoidance of towed nets by the euphausiid *Nematoscelis megalops*. Fish. Bull. 80:75-91.

Wiebe, P.H. and G.R. Flierl. 1983. Euphausiid invasion/dispersal in Gulf Stream cold core rings. Aust. J. Mar. Freshwat. Res. 34:625-652.

Wiebe, P.H., A.W. Morton, A.M. Bradley, R.H. Backus, J.E. Craddock, T.J. Cowles, V.A. Barber and G.R. Flierl. in press. New developments in the MOCNESS, an apparatus for sampling zooplankton and micronekton. Mar. Biol.

Wiebe, P.H., V.A. Barber, S.H. Boyd, C.S. Davis and G.R. Flierl. in press. Evolution of zooplankton biomass structure in warm-core rings. J. Geophys. Res.

Wigley, R.L., R.B. Theroux and H.E. Murray. 1975. Deep-sea Red Crab. 'Geryon quinquedens', Survey off Northeastern United States. Mar. Fish. Rev. 37(8):1-21.

Wildavsky, A. and E. Tenenbaum. 1981. The politics of Mistrust: Estimating American Oil and Gas Resources. Sage Publications, Beverly Hills, CA.

Willebrand, J. 1978. Temporal and spatial scales of the wind field over the North Pacific and North Atlantic. J. Phys. Oceanogr. 8:1080-1094.

Williams, P.J. leB. 1981a. Microbial contribution to overall marine plankton metabolism: Direct measurements of respiration. Oceanolog. Acta 4:359-364.

Williams, P.J. leB. 1981b. Incorporation of microheterotrophic processes into the classical paradigm of the planktonic food web. Kieler Meeresforsch. 5:1-28.

Wilson, G.D. and R.R. Hessler. 1974. Some unusual Paraselloidea (Isopoda, Asellota) from the deep benthos of the Atlantic. Crustaceana27:47-67.

Wilson, G.D. 1976. The systematics and evolution of Haplomunna and its relatives (Isopoda, Haplomunnidae, new family). J. Nat. Hist. 10:569-580.

Wilson, G.D. 1980a. Incipient speciation in a deep sea Eurycopid Isopod Crustacea. Am. Zool. 20:815.

Wilson, G.D. 1980b. New insights into the colonization of the deepsea: Systematics and zoogeography of the Munnidae and the Pleurogoniidae comb. nov. (Isopoda; Janiroidea). J. Nat. Hist. 14:215-236.

Wilson, G.D. and R.R. Hessler. 1980. Toxonomic characters in the morphology of the genus *Eurycope* (Crustacea, Isopoda) with a redescription of E. cornvta Sars 1864. Cah. Biol. Mar. 21:241-263.

Wilson, G.D. and R.R. Hessler. 1981. A revision of the genus *Eurycope* (Isopoda, Asellota) with descriptions of three new genera. J. Crust. Biol. 1:401-423.

Wilson, R.D., P.H. Monaghan, A. Osanik, L.C. Price and M.A. Rogers. 1974. Natural marine oil seepage. Science 184:857-865.

Windom, H., F. Taylor and R. Stickney. 1973. Mercury in North Atlantic Plankton. J. du Conseil International pour l'Exploration de la Mer, 35(1):18-21.

Windom, J. and R.G. Smith. 1979. Copper concentrations in surface waters off the southeastern Atlantic coast, U.S.A. Mar. Chem. 7:157-163.

Wingate, D.B. 1964. Discovery of breeding Black-capped Petrels on Hispaniola. Auk 81:147-159.

Wirsen, C. O. and H. W. Jannasch. 1976. Decomposition of solid organic materials in the deep sea. Environ. Sci. Technol. 10:880-886.

Withee, G.W. and A. Johnson. 1976. Data reports: Buoy observations during Hurricane Eloise (September 19 to October 11, 1975). U.S. Dept. Commerce, NOAA, NSTL Station, Mississippi:21.

Wormelle, R.L. 1962. A survey of the standing crop of plankton of the Florida Current. VI. A study of the distribution of pteropods of the Florida Current Bull. Mar. Sci. Gulf and Carribbean. 12:95-136.

Wormuth, J.H. 1981. Vertical distributions and diel migrations of *Euthecosomata* in the northwest Sargasso Sea. Deep-Sea Res. 28A:1493-1515.

Worthington, L.V. 1959. The 18-degree water in the Sargasso Sea. Deep-Sea Res. 5:297-305.

Worthington, L.V. and W.G. Metcalf. 1961. The relationship between potential temperature and salinity in deep Atlantic water. Rapports et proces-verbaux I.C.E.S. 149:122-128.

Worthington, L.V. 1964. Anomalous conditions in the Slope Water area in 1959. J. Fish. Res. Bd. Can. 21:327-333.

Worthington, L.V. and H. Kawai. 1972. Comparison between deep sections across the Kuroshio and the Florida current and Gulf Stream. in H. Stommel and K. Yoshida, (eds.), Kuroshio, its physical aspects. Univ. Tokyo Press:371-385.

Worthington, L.V. 1976. On the North Atlantic circulation. Johns Hopkins Oceanogr. Studies No. 6, 110 p. Publ. The Johns Hopkins Univ. Press, Baltimore.

Wright, W.R. and L.V. Worthington. 1970. The water masses of the North Atlantic Ocean: a volumetric census of temperature and salinity. Serial Atlas of the Marine Environment, Amer. Geogr. Soc. Folio No. 19.

Wright, W.R. 1976a. Physical Oceanography. Chapter 4 in: Summary of environmental information on the continental shelf - Canadian/U.S. border to Cape Hatteras, N.C. The Research Institute of the Gulf of Maine. 4 volumes. NTIS No. PB 284001/AS.

Wright, W.R. 1976b. The limits of shelf water south of Cape Cod, 1941 to 1972. J. Mar. Res. 34:1-14.

Wright, W.R. and C.E. Parker. 1976. A volumetric temperature-salinity census for the Middle Atlantic Bight. Limnol. Oceanogr. 21:563-571.

Wunsch, C. 1969. Progressive internal waves on slopes. J. Fluid Mech. 35:131-144.

Wunsch, C. and R. Hendry. 1972. Array measurements of the bottom boundary layer and the internal wave field on the continental slope. J. Geophys. Fluid Mechanics 4:101-145.

Wunsch, C. 1976. Geographical variability of the internal wave field: A Search for Sources and Sinks. in J. Phys. Oceanogr. 6:471-485.

Wynne-Edwards, V.C. 1935. On the habits and distribution of birds on the North Atlantic. Proc. Boston Soc. Nat. Hist. 40:233-346.

Wyrtki, K., L. Magaard and J. Hager. 1976. Eddy energy in the oceans. J. Geophys. Res. 81:2641-2646.

Yingst, J.Y. and R.C. Aller. 1982. Biological activity and associated sedimentary structures in hebble-area deposits, western North Atlantic. Mar. Geol. 48:7-15.

Zaitsev, Y.P. 1970. Marine Nuestonology. Naukova Dumka, Kiev. Israel Program for Scientific Translations. IPST Cat. No. 5976. U.S. Dept. of Commerce.

Zarudzki, E.F.K. and E. Uchupi. 1968. Organic reef alignments on the continental margin south of Cape Hatteras. Geol. Soc. Am. Bull. 79:1867-1870.

ZoBell, C. E. 1946. Marine Microbiology. Chronica Botanica Co., Waltham, Mass.

ZoBell, C. E. 1970. Pressure effects on morphology and life processes. in High Pressure Effects on Cellular Processes, A. Zimmerman (ed.), Academic Press, New York.

Zsolnay, A. 1977. Tar "specks" in the North Atlantic. Mar. Pollution. Bull. 8:116.

Subject and Geographic Index

Note that in most cases genera and species of organisms are not listed. Please refer to class or order.

A

B

C

D

E

F

G

H

I

J

K

L

M

N

O

P

Q

R

S

T

U

V

W

X

Z